AF657638

METAMORPHISM AND CRUSTAL EVOLUTION

METAMORPHISM AND CRUSTAL EVOLUTION

PAPERS IN HONOUR OF
PROF. R.S. SHARMA

EDITOR

H. THOMAS

Published by

ATLANTIC
PUBLISHERS & DISTRIBUTORS
B-2, Vishal Enclave, Opp. Rajouri Garden,
New Delhi-110027
Phones : 25413460, 25429987, 25466842

Sales Office
7/22, Ansari Road, Darya Ganj,
New Delhi-110002
Phones : 23273880, 23275880, 23280451
Fax : 91-11-23285873
web : www.atlanticbooks.com
e-mail : info@atlanticbooks.com

Copyright © Atlantic Publishers and Distributors, 2005

ISBN 81-269-0436-4

All rights reserved. No part of this publication may be reproduced, stored in a retrieval system, transmitted or utilized in any form or by any means, electronic, mechanical, photocopying, recording or otherwise, without the prior permission of the copyright owner. Application for such permission should be addressed to the publisher.

Printed in India
at Nice Printing Press, Delhi

Dedicated to my Guru
Professor R.S. Sharma, F.N.A., F.A.Sc.

ACADEMIC BIOGRAPHY OF PROFESSOR R.S. SHARMA, FNA, FASc

Professor Ram Swaroop Sharma was born on 10th of July 1937 in Jaipur, Rajasthan, where he had his early education up to B.Sc. Thereafter he went to Banaras Hindu University (BHU), Varanasi and did his M.Sc. in Geology in 1960. He obtained his Ph.D. from the University of Basel, Switzerland, under the able guidance of Professor Eduard Wenk with a specialization in mineralogy and metamorphic petrology. He joined BHU as a permanent faculty member in 1968 and served his *alma mater* until his superannuation as Professor of Geology in 1997. During his academic career in BHU, Dr. Sharma received several recognitions. He was honoured twice Alexander Von Humboldt Foundation Fellowship to work (1976 and 1982) in Germany. He also received Royal Society Research Scholarship in 1983 to work in Leicester, U.K. for one year. Professor Sharma was elected Fellow of the Indian Academy of Sciences in 1991 and Fellow of the Indian National Science Academy in the following year. His research contributions are mainly on the metamorphic and tectonic evolution of the high-grade gneisses and granulites from the Banded Gneissic Complex of Rajasthan, NW India. For the first time, he presented mineralogic criteria for distinguishing the polymetamorphosed Archaean gneisses from the Proterozoic cover rocks that show a single regional metamorphism and constitute the Aravalli mobile belt in Rajasthan. On the basis of mineralogical and geothermobarometric studies he deduced an anticlockwise P-T path for the Rajasthan granulites which places constraints on the tectono-metamorphic evolution of the Precambrian rocks from this part of the Indian shield. Based on the geological, structural, geochronological and metamorphic data, Professor Sharma has established a working model for the evolution of the Proterozoic Aravalli mobile belt. He has published more than 100 research papers in national and international journals. Professor Sharma has also been a great teacher and was invited twice as a visiting Faculty by Curtin University of Technology, Perth, Western Australia. He has also served DST, CSIR, INSA and some University Academic bodies in different capacities. Presently he is INSA Senior Scientist at the Department of Geology, University of Rajasthan, Jaipur, since 2003. His current research work focuses on the interrelationships between deformation and metamorphism of the Precambrian rocks of Rajasthan. Academic and professional acclaim always followed him.

Prof. Sharma's relationship with his students was challenging, demanding and yet forgiving, loving and supporting in a most intimate and indescribable way. He communicated an unspoken criterion of excellence, which drove one never to be satisfied with anything less than the very best effort—and even then to realize the weakness and inadequacy in what was being presented. But this high requirement for academic excellence was never directed toward any shallow or surface intellectual achievement, its goal was always the student's own spiritual awakening and self-understanding. The student who came into contact with Professor Sharma could have the rare experience of perceiving that such a forbidding challenge could also become exciting opportunity. From him one also learns to be open to all philosophical viewpoints, and to make every effort to understand each position in its own terms.

Professor Sharma is now sixty-seven, an age when activities of men are generally marked by ebbing. Professor Sharma's inquisitive academic spirit, however, remains unaffected. I wish him many more years of good health and good science! Here, I wish to end this note with the Vedic sayings:

Kurvanneveha Karmāṇi jijīviṣecchatam samāḥ/
Evam tvayi nānyatheto'sti na karma lipyate nare//

Be desirous of living a hundred years following the path of action like this and while doing so there is nothing else for you here in this world (and therefore), no action will be binding on being.

HAREL THOMAS

SELECTED RESEARCH PAPERS

Sharma, R.S., 1968. Normative mineral proportions of granite and migmatite rocks in phase diagrams. *Indian Mineralogist,* **vol. 9**, 14-19.

Sharma, R.S., 1969. On banded gneisses and migmatites from Lavertezzo and Rozzera, Valle Verzasca, Canton Ticino. *Schweiz. Mineral. Petrogr. Mitt.*, **vol. 49**, 199-270.

Sharma, R.S., 1970. Plagioclase composition and some aspects of gneissic rocks of Almora-Dudatoli region, Kumaun Himalaya. *J. Adv. Centre Geol., Chandigarh,* **vol. 7**, 1-13.

Sharma, R.S. & Upadhyaya, T.P., 1975. Multiple deformation in the Precambrian rocks to the southeast of Ajmer, Rajasthan, India. *J. Geol. Soc. India,* **vol. 16**, 428-40.

Sharma, R.S. & Narayan, V., 1975. Wollastonite paragenesis in the regional metamorphic terrain, southeast of Beawar, Rajasthan, India. *Neues. Jb. Mineral. Mh.*, **vol. 12**, 561-69.

Sharma, R.S. & Narayan, V., 1975. Distribution of elements between coexisting garnet-biotite and muscovite-biotite pairs from polymetamorphic schists of southeast Beawar, Rajasthan, India. *Schweiz. Petrogr. Mineral. Mitt.*, **vol. 55**, 61-67.

Sharma, R.S. & Narayan, V., 1975. Petrology of polymetamorphic schists from an Archaean complex terrain, southeast of Beawar, Rajasthan, India. *Neues. Jb. Mineral. Abh.*, **vol. 124**, 190-222.

Sharma, R.S., 1977. Deformational and crystallization history of the Precambrian rocks in north-central Aravalli mountain, Rajasthan, India. *Precamb. Res.*, **vol. 4**, 133-62.

Sharma, R.S. & Althaus, E., 1977. Stability relations of sapphirine in the Banded gneissic complex of Bhilwara district, India. *Fortsch. Mineralogie,* **vol. 55**, 128-29.

Sharma, R.S. & Roy, A.K., 1979. On the polymineralic paragneisses from Karera, District Bhilwara, Rajasthan: I-Mineralogy and Petrography. *Indian Jour. Earth Sci.*, **vol. 6**, 67-81.

Sharma, R.S., 1979. Calcite fabric in relation to the tectonic setting of the Precambrian rocks around Karera, District Bhilwara, Rajasthan, India. *J. Geol. Soc. India,* **vol. 20**, 339-48.

Sharma, R.S. & Roy, A.K., 1980. On the polymineralic paragneisses from Karera, District-Bhilwara, Rajasthan: II Phase relations and paragenesis. *Indian Jour. Earth Sci.*, **vol. 14**, 131-45.

Sharma, R.S., 1981. Mineralogy of scapolite-bearing rock from Rajasthan, northwest peninsular India. *Lithos.*, **vol. 14**, 165-72.

Sharma, R.S. & MacRae, N.D., 1981. Paragenetic relations in gedrite-cordierite – staurolite – biotite – sillimanite - kyanite gneiss at Ajitpura, Rajasthan. *Contrib. Mineral. Petrol.*, **vol. 78**, 48-60.

Sharma, R.S., 1982. Petrology of scapolite-bearing rock from Karera, District-Bhilwara, Rajasthan. *J. Geol. Soc. India,* **vol. 23**, 319-29.

Sharma, R.S., 1983. Basement-cover rocks relations in north-central Aravalli ranges: A tectonic and metamorphic synthesis. In Recent Researches in Geology 10 (Structure and Tectonics of the Precambrian rocks), ed. S. Sinha Roy. *Hindustan Publ. Corpn. Delhi,* 53-71.

Sharma, R.S. & Roy, A.K., 1983. On the polymineralic paragneisses from Karera, District-Bhilwara, Rajasthan: III Metamorphic conditions and Petrogenesis. *Indian Jour. Earth Sci.*, **vol. 10**, 152-69.

Sharma, R.S., 1983. Pressure-temperature-time indicator in granulites from Rajasthan, NW India. *Geophy. J. Royal. Soc.*, **vol. 73**, 291.

Sharma, R.S. & Windley, B.F., 1984. Mineral parageneses and metamorphic reactions in metasedimentary enclaves from the Archaean gneiss complex of north-west India. *Mineral. Mag.*, **vol. 48**, 195-209.

Sharma, R.S. & Joshi, M., 1984. Sand Mata granulite: a case of deep-level exposure of Precambrian crust in NW Indian shield. In: *Crustal evolution of the Indian shield and its bearing on Metallogeny,* Calcutta, 35-36.

Sharma, R.S., Althaus, E. & Wilson, R.N., 1985. Kornerupine in aluminous enclaves of the Archaean gneiss complex from Rajasthan, NW India: paragenesis and and stability. *Terra. Cognita.*, **vol. 5**, 328.

Sharma, R.S. & Roy, A.B., 1986. Evolution of Precambrian Lithoshere: western India. The Indian Lithoshere, *Indian National Science Academy,* New Delhi, 1-14.

Sharma, R.S., 1986. Mineralogy and metamorphic reactions in the basic intrusion from Sand Mata area, Rajasthan. *Indian Mineralogist,* **vol. 27**, 47-61.

Sharma, R.S., 1987. Geology of Sand Mata Complex. In Guide book on Field Excursion in south-central Rajasthan, IGCP-215: Proterozoic foldbelts. *Geol. Surv. India, Jaipur,* 19-23.

Sharma, R.S., Sills, J.D. & Joshi, M., 1987. Mineralogy and metamorphic history of Norite dykes within granulite facies gneisses from Sand Mata, Rajasthan. *Mineral. Mag.*, **vol. 51**, 207-15.

Sharma, R.S., 1988. Granulites from northwest Indian shield: Their differences and similarities with Southern Granulite Terrain. *J. Geol. Soc. India,* **vol. 31**, 136-38.

Sharma, R.S., 1988. Patterns of metamorphism in the Precambrian rocks of the Aravalli Mountain Belt. *Geol. Soc. India, Mem.*, **7**, 33-75.

Sharma, R.S. & Mall, A.P., 1988. Coronas in olivine metagabbros from the proterozoic Chhotanagpur terrain of Mathurapur, Bihar, India. *Lithos,* **vol. 21**, 291-300.

Sharma, R.S. & West, W.D., 1989. A note on the metamorphism of the carbonate rocks of the Sausar Group in the north-western part of the Nagpur District. *J. Geol. Soc. India,* **vol. 33**, 496-502.

Sharma, R.S., 1989. Geotectonic evolution of the Aravalli Mountain belt. In: Current Trends in Geology 12 (Metamorphism, Ophiolite and Orogenic belts) *Today & Tomorrow Publ. New Delhi,* 211-21.

Sharma, R.S., 1990. Metamorphic evolution of rocks from Rajasthan craton, NW Indian shield. In: Developments in Precambrian geology (Precambrian continental crust and its Economic Resources). *Elesevier Publ., Amsterdam,* 349-66.

Sharma, R.S., Joshi, M. & Thomas, H., 1992. Granulite facies metamorphism in the Archaean gneiss complex from north-central Rajasthan. *Proc. Nat. Acad. Sci., India,* **vol. 63(A)**, 167-87.

Sharma, R.S., 1993. Deep continental crust: an example from Rajasthan, NW India Newsletter, *DST, Govt. of India, New Delhi,* **vol. 3, No. 1**, 6-10.

Sharma, R.S., 1993. Fluid mobility in deep crustal rocks: a review based on equilibrium, experimental and field studies. In: Fluid Processes in the Earth Crust. *Nat. Geophy. Res. Inst., Hyderabad,* 48-53.

Sharma, R.S., 1995. An evolutionary model for the Precambrian crust of Rajasthan: some petrological and geochronological considerations. *Geol. Soc. India, Mem.*, **vol. 31**, 91-115.

Sharma, R.S., 1999. Crustal development in Rajasthan craton. *Indian J. Geology,* **vol. 71**, 65-80.

Sharma, R.S., 1999. The tell-tale of the granulites from southern India and Rajasthan. Platinum Jubilee Lecture, Earth Sciences Section, 86th Indian Sci. Congr. Chennai. *Indian Sci. Congr. Assoc. Calcutta,* part III, 55-76.

Sharma, R.S., 2000. P-T-t evolution of orogenic belts: a case study of the Aravalli mobile belt. Key-Note paper *12th Indian Geol. Congr. Udaipur*, 1-3.

Sharma, R.S., 2000. Symphony of the Himalayan metamorphic rocks. *J. Panjab Univ. Chandigarh,* **vol. 7**, 1-3.

Sharma, R.S., 2003. Evolution of Proterozoic fold belts of India: a case of the Aravalli Mountain belt of Rajasthan, NW India. *Geol. Soc. India, Mem.*, **vol. 52**, 145-62.

Sharma, R.S. & Pandit, M.K., 2003. Evolution of early continental crust: a review *Current Sci.*, **vol. 84, No. 8**, 995-1001.

Contents

Section – III
Himalayan Metamorphism

Section – IV
UHP/UHT Metamorphism

Section – V
Geochronology & Geochemistry

PREFACE

This Volume is a collection of papers contributed by the eminent earth scientists from India and abroad to felicitate Professor Ram Swaroop Sharma's, *FNA, FASc* on his superannuation from illustrious service in the field of earth science. Professor Sharma is one of the most famous teachers and earth scientist in the field of mineralogy and metamorphic petrology. A large number of students queued up to take M.Sc. dissertation and Ph.D. Degree under his able supervision and guidance because of his dynamic personality. I was one of the fortunate student having received my M.Sc. dissertation and Ph.D. degree under his skilful guidance.

He has also served the DST, CSIR, INSA and the academic bodies of several Universities in different capacities. Since 2003 he has been serving as the INSA Senior Scientist at the Department of Geology, University of Rajasthan, Jaipur. His current research work focuses on the interrelationships between deformation and metamorphism of the Precambrian rocks of Rajasthan.

The papers for this Volume were invited under the theme, "Metamorphism and Crustal Evolution" with a view to bring out the latest developments in the field of metamorphic petrology. About thirty-three world-renowned earth scientists have contributed their research findings for this volume. The Volume is classified into five sections, namely Metamorphism, Fluid Processes, Himalayan Metamorphism, UHP/UHT Metamorphism, and Geochronology & Geochemistry.

The Section on Metamorphism starts with the paper contributed by *Johnson and Michael Brown*. According to them many migmatites and granulites preserve evidence of post-thermal-peak retrograde reaction. They concluded that a_{H_2O} high anatexis gives rise to low volume, essentially immobile melts and there may be only limited macroscopic evidence for melting in the rocks. Volatile-fluxed melting is likely to be rare in nature given the low porosity of high-grade metamorphic rocks, although granite crystallizing within the system may locally supply for H_2O-rich volatile phase to drive retrogression.

Tatyana N. Polyakova, Konstantin A. Savko, and *Vyacheslav Yu. Skryabin* have inferred the phase relations and mineral compositions as well as the metamorphic condition of the different types of metapelites within Tim-Yastrebovsky Paleorift, Voronezh crystalline massif, and East-European Platform. From the investigation, they found that the presence of oversaturated silica metapelites and the rare occurrence of their high alumina varieties are the evidence of low maturity and small amount of weathering products of Archaean basement rocks because the Early Proterozoic metapelites were formed at their re-deposition. The vast occurrence of high potassium metapelites reveals that the Archaean rocks had undergone the processes of migmatization and granitization. They concluded that the area was affected by two metamorphic episodes: early regional metamorphism in the greenschist facies conditions at 2250-2086 Ma and the younger more high-temperature contact metamorphism at 2086±5 Ma.

Faryad suggested that carbonate-evaporitic rocks are the best reservoir of fluids (CO_2, F, Cl, S, and B) of various compositions that enter mineral structure during metamorphism. These minerals have retrograde origin. His work is mainly aimed to summarize mineralogy of both metamorphosed carbonates and evaporites that are part of volcano-sedimentary sequences of the Sare Sang series. This is to confirm or otherwise the postulated origin of volatiles from carbonate-evaporite source rocks and implications for their infiltration to the host silicate rocks during metamorphism. Variations in halogen contents and their preferential partitioning in coexisting silicate minerals are also discussed.

Pereira has discussed the origin of PEÑA Negra Migmatite and anatectic granitoids on the basis of petrogenesis. According to her these are formed *in situ* by anatexis and correspond to melt/restite segregations in two different conditions, namely high melt-fraction in the case of the granodiorite, and low melt fraction in the case of the leucogranites (The leucogranites display large enclaves of sillimanite-rich composition, representing the restite fraction). Granodiorite was generated as consequence of crustal thickening under condition of T = 750 to 800°C and P = 4kb, coinciding with the second phase of the Hercynian orogeny.

Thomas has discussed the polymetamorphism in the Archaean gneiss complex of Shivpura Gyangarh, District Bhilwara, Rajasthan. The pelitic granulite documents the chronological development of Al-silicate phase in order: $kyanite_1 \rightarrow sillimanite \rightarrow kyanite_2$. The basic granulite exhibits coronitic textures wherein an early assemblage of hypersthene and plagioclase become incompatible in the second metamorphic event to form garnet-diopside-quartz corona. The norite dykes show both igneous and metamorphic paragenesis, with garnet at the interface of hypersthene and plagioclase, suggesting its development during the waning stage of granulite facies metamorphism. On the basis of estimated P-T and textural evidences and the crystallization sequence of the Al-silicate minerals a movement path has been derived, which is reverse to the "normal" P-T path documented for collision belts placing considerable constraints on the geological evolution of the Precambrian rocks from this part of the Indian Shield.

Patel studied the granulite of Bolangir anorthositic complex in the Orissa sector of Eastern Ghats mobile belt on the basis of mineralogy and P-T condition. He found that the cooling occurred at pressure of 7.5 kbar from a minimum temperature of 740°C. His studies show the near-isobaric cooling path interpreted as part of clockwise P-T trajectory.

The second Section starts with the *William Lamb*'s paper. His paper explores the possibility of using equilibria involving carbonates and other minerals common to many quartz-feldspathic granulite facies gneisses to determine peak metamorphic CO_2 activities. Small amount of carbonate minerals are often present in granulite facies, feldspathic gneisses, and the presence of these minerals may allow the determination of CO_2 activities using a variety of CO_2-buffering equilibria. However, it is often not possible to use the presence of small amounts of carbonates to determine peak metamorphic fluid compositions, and it is also inadvisable to use fluid inclusions that are intimately associated with these carbonates to determine peak metamorphic fluid compositions. It is interesting to note that some of the CO_2-rich fluid inclusions that are associated with carbonates have other characteristics, such as textures and fluid densities, that have sometimes been taken as indicative of primary, or peak metamorphic, fluid inclusions. Thus, calculation of the stability of carbonates in quartzo-feldspathic granulite facies gneisses demonstrates that the identification of those fluid inclusions that actually contain peak metamorphic fluids is difficult.

The third Section starts with the paper of *Yoshino* and *Okudaira*. They discussed the polyphase deformation model on the basis of three tectonic stages: crustal thickening, crustal thinning, and collision. They suggested that the crustal thickening could trigger lithospheric delamination and gravity collapse of thickened crust. These relationships suggest gravitational collapse of the thickened crust as a possible cause for exhumation of deep-seated rocks of the Kohistan-Arc.

Joshi and *Tiwari* have discussed the metamorphic evolution of Almora gneisses and described detailed petrology of these metapelitic gneisses with particular emphasis on the relationship between deformation and metamorphism. The Almora gneisses were metamorphosed under high-T, intermediate-P conditions and represent the culmination of regional metamorphism in Almora Nappe. They have given a P-T array for the Almora gneisses which underwent isobaric cooling following the peak metamorphism.

Verma, Nag, and *Sengupta* have discussed the distinction between the Lower Paleozoic and Tertiary Granites of the Himalayas with the help of primary muscovite present in the Mandi granite. They concluded that Mandi granite has undergone a period of metamorphism synchronous with that of the country rocks. Petrographic studies on the Tertiary granites indicate the occurrence of muscovite in them as a primary igneous phase along with quartz, feldspar, biotite, garnet, etc. They extrapolate their result to the entire Himalayas, as the granites of each category are similar to each other. Muscovite in the Mandi granite is formed as a result of metamorphism and can be successfully used as a distinguishing factor for differentiating Tetiary and Paleozoic granites in the Himalayas.

Dubey, Catlos, and *Sharma,* discussed the Modeling of P-T-t paths constrained by mineral chemistry, P-T and monazite dating of metapelites in relationship to MCT activity in Sikkim, Eastern Himalayas. The different geothermometers revealed temperatures up to 465 to 527°C in garnet zone and 555 to 576°C in core in staurolite zone. The maximum pressure in core of garnets in staurolite zone was inferred upto 6.6kb. Monazites dated in the thin section indicate that the Sikkim Himalayan MCT shear was active during 22-20 Ma and 14-10Ma. The age indicates inverted metamorphism generated *via* accretion of successive tectonic slivers of the Lesser Himalayas to the hanging wall.

In the fourth Section *Liou, Tsujimori, Katayama,* and *Maruyama* suggested that the continental rocks are subjected to UHP metamorphism at T ranging from –700 to 950°C and P > 2.8 to 5.0Gpa, corresponding to depth of – 100 to 150 km. They assumed these UHP units are subsequently exhumed to crustal depth and sujected to intense hydration and the amphibolite facies overprint. Their finding of coesite inclusion in the rare Himalayan ecologite and country rock gneisses is a typical example. Himalayan UHP eclogite and adjacent gneisses were formed at mantle depth, depths > 100 km at 46 to 52 Ma.

Mohan, Sharma, and *Singh* also reviewedthe regional metamorphism of deep crustal rocks which have gained temperature high enough to stabilize critical associations of sapphirine-quartz, spinel-quartz, high aluminous orthopyroxene-sillimanite-quartz and osumalite at T>900°C at P>9 kbar led to the recognition of UHT metamorphism.

Fifth Section starts with paper contributed by *Roy, Alfred Kroner, Lavl,* and *Purohit*. They investigated single zircon dating (mean age of 757.8 ± 0.9 Ma) of a hypersthene-bearing granitoid of the Balaram-Abu Road area in order to resolve the temporal relationships with the Delhi Supergroup and the possible tectonic implications of the granulite wedge in southern part of the Aravalli Mountains. Based on the similarity of granite ages, they correlate the Balaram-Abu Road thermal event with the Malani magmatism, which has been interpreted as Neoproterozoic plume-related crustal heating process.

Prakash and *Thomas* studied the high-grade granulites from Kodaikanal in southern India, of which the mafic granulites are found to be metamorphosed of tholeiitic basalts that show REE fractionation and Fe-enrichment trend. The bulk composition of sapphirine-granulites from the area is unusual. These rocks are rich in Al_2O_3, (MgO + FeO), but poor in CaO. The SiO_2-content is rather high compared to most silica-deficient granulites and the (Na_2O + K_2O) content is also raised on account of the presence of phlogopite, K-feldspar and plagioclase in sapphirine granulites. Different scholars have suggested that these unusual sapphirine granulite formation is probably due to the dehydration melt reaction, i.e. bt + sil = spr +crd +kfs + L.

Mahadevan discussed the lower crust of the Indian Shield, and its evolution with the help of geophysical and petrological data. The Indian shield had a polychronous evolution with a core region of Archaean- Early Proterozoic rocks, encircled by high-grade domains (HGDs) evolved largely in the Proterozoic. The continuous epeirogenic processes in the Indian shield may also be still effecting uplifts as evidenced by the continuing uplift of the Western ghats, the Satpura range, and the Chhotanagpur Plateau. His paper mainly discusses the recent insights gained into the characteristic of the large exhumed crust of south India and compare them with the known feature of the lower crustal rocks in other parts of the shield to arrive at a comprehensive picture.

It is my pleasure to pay my respect and regards to all eminent Professors, who have reviewed papers that form parts of this volume. Among them: Prof. R.K. Lal (BHU, India), Prof. M. Dolores Pereira Gomez (Salamanca, Spain), Prof. A.B. Roy (Udaipur, India), Prof. P.K. Verma (Delhi, India), Prof. M. Joshi (BHU, India), Dr. A. Mohan (BHU, India), Dr. S.C. Patel (IIT Bombay, India), Dr. C.S. Dubey (Delhi, India) are to name a few to whom, I am grateful for their critical observations and meaningful suggestions. My deepest and most sustained obligations are due to my revered teacher, Professor *R.S. Sharma.* I am very deeply indebted to him for his constant and relentless guidance to me throughout my academic endeavours.

I am very grateful to my esteemed friend Dr. A.D. Sharma (Department of Philosophy, Sagar University), for his constant inspiring to bring out this Volume successfully.

I acknowledge Prof. P.O. Alexander, Prof. V.D. Choubey (Emeritus scientist), and Dr. K.R. Gupta (DST, New Delhi) for their continuous encouragement and inspiration during the preparation of this Volume.

I am greatly indebted to my wife Dr. Rajkumari Tamphasana and daughter Priyanka for their help and co-operation in completing the Volume, and finally credit also goes to all my family members for their patience and cooperation.

HAREL THOMAS
Editor

SECTION – I

METAMORPHISM

1

THE GRANULITE FACIES, CRUSTAL MELTING, AND PROGRADE AND RETROGRADE PHASE EQUILIBRIA IN METAPELITES AT THE AMPHIBOLITE TO GRANULITE FACIES TRANSITION

JOHNSON, TIM E. AND BROWN, M.*

ABSTRACT

Prograde metamorphic evolution involves the progressive liberation and loss of a volatile phase, and at upper amphibolite to granulite facies conditions the generation, segregation and potential loss of melt, taking with it H_2O formerly structurally bound in micas. Nonetheless, many migmatites and granulites preserve evidence of post-thermal-peak retrograde reaction, the extent of which is influenced by several factors, including: (1) the shape of the *P-T* path, particularly the ΔP between the prograde and retrograde segments, and its relation to low variance equilibria in the system; (2) the rate of cooling; and, (3) the availability of fluid (H_2O-rich volatile phase or melt) for fluid-consuming reactions along the retrograde segment of the *P-T* path. In general, reaction reversal or 'back reaction' requires the *P-T* path to approximate isobaric heating and cooling, without significant ΔP, and evolved fluid (H_2O-rich volatile phase or melt) to remain within the equilibration volume. The larger the ΔP in the *P-T* evolution, the more chance there is of crossing different reactions along the retrograde segment from those crossed along the prograde segment. Above the solidus, melt segregation and loss potentially changes both the bulk composition and the size of the equilibration volume. Since the proportion of melt to residue may constantly change, the composition of the equilibration volume may likewise change throughout the supra-solidus segment of the retrograde *P-T* evolution. If some melt has been lost from the system but the ΔP between the prograde and retrograde segments is small, retrograde reaction to a mineral assemblage similar to the subsolidus prograde assemblage may still occur, but reaction may be partial rather than complete. On the one hand, the common preservation of apparently close-to-thermal-peak mineral assemblages in migmatite and granulite may suggest that extensive reaction with melt is uncommon, implying melt isolation or loss prior to crossing potential melt-consuming reactions. On the other hand, the limited occurrence of UHT granulites within common granulites may record retrogression of formerly more extensive UHT mineral assemblages facilitated by retention of

* Laboratory for Crustal Petrology, Department of Geology, University of Maryland, College Park, MD 20742, USA.
Corresponding author: Telephone: +1-301-405-4080; Fax: 301-314-7970; Email: mbrown@geol.umd.edu

some melt within the system (although it may also reflect the scarcity of the high-Mg bulk compositions necessary to record a mineralogical imprint of UHT conditions). Water dissolved in melt is transported through the crust to be exsolved on crystallization at the solidus appropriate to the intrinsic aH_2O. This recycled water causes retrogression at subsolidus conditions. We illustrate these themes with modelled phase relations in the (Mn)NCKFMASH system, using examples of upper amphibolite to granulite facies anatectic migmatites from both contact and regional metamorphic environments (contact aureole beneath the Bushveld Complex, South Africa; Acadian metamorphism of Maine, USA; and, the Variscan metamorphic belt of Southern Brittany, France). These examples have been chosen to illustrate different *P-T* evolution and variable degrees of melt loss and volatile-phase interaction; consequently, they exhibit variable retrograde reaction.

KEY WORDS

Acadian; Brittany, France; Bushveld aureole; contact metamorphism; granulite facies; Maine, USA; melt; metapelite; phase equilibria; prograde reaction; pseudosection; *P-T* path; retrograde reaction; vapour phase; Variscan.

INTRODUCTION

The preservation of 'dry' granulite facies mineral assemblages that are characteristic of the lower continental crust requires the loss of substantial volumes of silicate melt if retrogression to 'hydrated' amphibolite facies mineral assemblages is to be avoided (e.g. White & Powell, 2002). In orogens, melt generation, segregation, ascent and emplacement leads to large-scale differentiation of the continental crust as a result of which a residual and 'dry' lower crust underlies a less dense and 'hydrated' upper crust. Heat advected with migrating granite magma enhances metamorphism of upper-crustal rocks, particularly in high-temperature – low-pressure (HT – LP) metamorphic belts. The resulting compositional, thermal and rheological heterogeneities within the continental crust have important consequences for tectonics.

The past decade has seen the range of pressure (*P*) — temperature (*T*) conditions within which granulite facies mineral assemblages are stable extended to incorporate the new sub-facies of high-pressure (HP) and ultrahigh-temperature (UHT) granulites (e.g. Harley, 1998, O'Brien & Rötzler, 2003). Consequently, we now refer to 'common' granulite facies metamorphism, characterized by the presence of coexisting orthopyroxene + clinopyroxene + plagioclase in metabasic bulk compositions, HP granulite facies metamorphism, characterized by orthopyroxene-free garnet + clinopyroxene + plagioclase + quartz assemblages in metabasic bulk compositions, and UHT granulite facies metamorphism, characterized by the assemblages orthopyroxene + sillimanite + quartz (at high *P* and high *T*), or garnet + cordierite + sillimanite or quartz (at intermediate *P* and high *T*), or sapphirine + quartz (at high *P* and extreme *T*) or spinel + quartz (at low *P* and extreme *T*) in metapelitic bulk compositions.

Estimating *T* and *P* of peak metamorphism in granulites is difficult, due to resetting of Fe-Mg exchange thermometers along the retrograde *P-T* evolution, which may lead to an underestimate of *T* and consequent underestimate of *P*. However, Pattison *et al.* (2003) recently have calculated the *T-P* conditions of common granulites using data from the literature and garnet-orthopyroxene thermobarometry corrected for retrograde exchange. Based on this new compilation, the *T-P* range of common granulites is *T* of 800-900°C and *P* of 0.4-1.0 GPa. The *T-P* conditions of the full granulite facies now include the HP granulite sub-facies, with T of 750°-1050°C and P > 1.0 GPa (O'Brien & Rötzler, 2003), and the UHT granulite sub-facies, with T > 900°C and maximum values of > 1150°C (Harley & Motoyoshi, 2000; Moraes *et al.*, 2002). In addition to regional granulite facies terrains, low-P granulite and UHT granulite mineral assemblages may be developed in contact aureoles surrounding intrusions of high-temperature magma (e.g. Pattison & Tracy, 1991; Westphal *et al.*, 2003).

At metamorphic conditions of the upper amphibolite facies and above, prograde metamorphic reaction commonly is complete and mineral assemblages generally reflect peak temperature conditions. This occurs because the rate-limiting step for reaction is volume diffusion, which increases exponentially with increasing temperature. Additionally, melting is ubiquitous under upper amphibolite and granulites facies conditions in fertile crustal protoliths. Diffusion is orders of magnitude faster in the presence of a fluid phase, which further ensures that the residual mineral assemblages left after melt loss from the system reflect peak temperature conditions.

Unfortunately, as a consequence of prograde metamorphic reactions going to completion and extensive melting at granulite facies conditions, most, if not all, of the prograde metamorphic history is obliterated, with the exception of mineral and fluid inclusions in porphyroblastic phases. In spite of this difficulty, it is clear that many granulites followed some form of gross clockwise P-T path (Brown, 2001a; 2003), although the prograde path is rarely well constrained and the retrograde path may vary from decompression with cooling (e.g. Galán & Marcos, 2000) to decompression with heating (e.g. O'Brien, 2000) or may involve a combination of near-isothermal decompression segments with near-isobaric cooling segments (e.g. Brown & Raith, 1996; Raith *et al.*, 1997). Examples of upper amphibolite to granulite facies terrains that record evidence of gross counterclockwise P-T paths are rarer, but include the Pikwitkonei granulite domain, Canada (Mezger *et al.*, 1990), the Acadian metamorphic belt of New England (Schumacher *et al.*, 1990; Kohn *et al.*, 1997; Johnson *et al.*, 2003a) and HT-LP metamorphic complexes in the Lachlan orogen (e.g. Morand, 1990; Johnson & Vernon, 1995).

In many metamorphic rocks, disequilibrium at the thin-section scale is implied by various microstructural and/or compositional features; the most obvious manifestation of which is compositional zoning of phases, particularly garnet, in which rates of volume diffusion are slow [~10^{-23} to $10^{-22} m^2 s^{-1}$ (Ayers & Vance, 1997)], and reaction microstructures between reactant and product phases (Brown, 2001a, 2002), features that are ubiquitously developed along the retrograde P-T evolution. However, in some migmatites and metapelitic granulites in which a melt phase is inferred to have been present around P-T conditions of the thermal peak, but from which melt is inferred to have been lost at or immediately post peak, solid-solution phases (particularly ferromagnesian peritectic phases such as orthopyroxene or garnet) commonly are unzoned (except for narrow retrograde rims; e.g. Moraes *et al.*, 2002). This implies that close to the thermal peak the effects of disequilibrium are minimized, and that a semi-quantitative examination of suprasolidus and near-solidus processes using equilibrium phase diagrams is justified.

UHT metamorphic mineral assemblages identified to date occur in small or limited areas within larger terrains of common granulite facies mineral assemblages. This may suggest that part, or all, of these larger terrains of common granulite have been retrogressed from UHT metamorphic mineral assemblages. However, retrograde processes at these extreme conditions tend to obliterate much of the evidence for peak metamorphism. For this reason, it is instructive to examine prograde and retrograde processes across the amphibolite to granulite facies transition. In this contribution we review crustal melting of metapelitic rocks within the sillimanite stability field around the amphibolite to granulite facies transition and discuss the prograde and retrograde evolution of such rocks.

We illustrate processes at amphibolite to granulite facies metamorphic conditions using equilibrium pseudosections in pressure (*P*) — temperature (*T*) — composition (*X*) space. The pseudosections were calculated with THERMOCALC 3.1 (Powell & Holland, 1988) using the Holland & Powell internally consistent thermodynamic dataset (1998, updated 19-09-99; see also White *et al.*, 2001). We illustrate processes using field examples of anatectic metapelitic migmatites from both contact and regional metamorphic terrains, with examples from the Bushveld Complex in South Africa, the Acadian metamorphic belt in Western Maine (U.S.A.) and the Southern Brittany metamorphic belt, part of the Armorican Massif in France.

We consider the amphibolite to granulite facies transition to occur at temperatures exceeding the breakdown of the phase association biotite + sillimanite (to form garnet or garnet + cordierite; e.g. Waters, 1988), or, at lower-*P*, at the appearance of orthopyroxene in rocks that correspond to the bulk composition of a common sub-aluminous metapelite. Migmatite terminology follows Mehnert (1968), Brown (1973) and Ashworth (1985), and mineral abbreviations are after Kretz (1983).

MELTING AND RETROGRADE PROCESSES IN METAPELITES

Before discussing specific examples, we give an overview of near-solidus and supra-solidus phase equilibria with reference to a *P-T* pseudosection (Fig. 1a, b) constructed in the NCKFMASH (Na_2O-CaO-K_2O-FeO-MgO-Al_2O_3-SiO_2-H_2O) model system (cf., White *et al.*, 2001). The bulk composition for which the diagram is drawn is a sub-aluminous metapelite, an average of 554 analyses taken from the literature [Shaw, 1956 (n = 155); Senior & Leake, 1978 (n = 120); Atherton & Brotherton, 1982 (n = 230); Dalrymple, 1995 (n = 15); Solar & Brown, 2001 (n = 13); Tinkham, 2002 (n = 21)].

Fig. 1a shows the calculated multivariant phase stability fields at high-*T* subsolidus and suprasolidus conditions at $P \leq 10$ kbar. Fig. 1b shows the types of equilibria and their associated peritectic phases, the amphibolite to granulite facies transition in sub-aluminous metapelitic rocks and exemplar *P-T* paths that are discussed subsequently. The dashed lines on Fig. 1a show melt proportion isopleths (in mol. %) that are calculated on a one oxygen basis; mol. % can be taken to approximate vol. %. The bulk composition contains just enough H_2O to saturate the solidus at 10 kbar; as H_2O solubility in melt decreases with decreasing *P* (e.g. Holtz & Johannes, 1994), the melt proportion isopleths within the low-*P* cordierite field in Fig. 1b are artificially high. Whilst Fig. 1a is instructive, it is important to emphasise that this *P-T* pseudosection shows phase relations and melt volumes for a single bulk composition, and the specific details of processes discussed in the following section will vary with bulk composition (see White & Powell, 2002; Johnson *et al.*, 2003a, b).

Melting equilibria

A fundamental control on the melting equilibria encountered by a metapelitic rock is the shape of the path that it follows in *P-T* space. With reference to Fig. 1a & b, the low-variance point labelled 'IP1' (at approximately 650°C, 3.5 kbar) provides an important control on prograde (and retrograde) evolution (e.g. Spear *et al.*, 1999; Brown, 2002). The *P-T* location of 'IP1' is insensitive to moderate changes in bulk composition. Rocks following

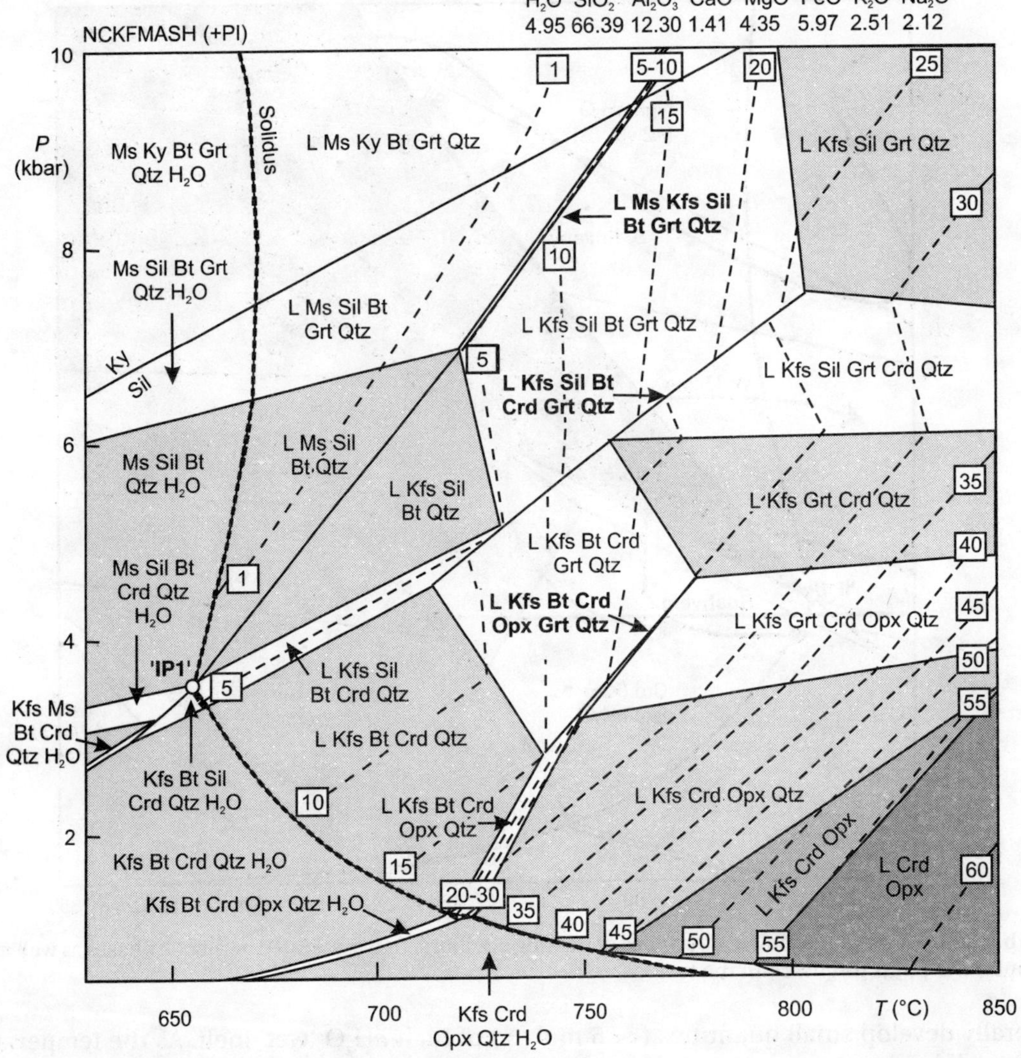

Fig. 1a. *P-T* pseudosection for an average metapelite composition (n = 554). (a) Calculated stable assemblages; the numbered dashed lines show molar melt proportion isopleths (in mol. %). The depth of shading reflects increasing variance; the darkest field is sexivariant.

prograde *P-T* paths at P > 'IP1' will contain muscovite at the *T* of the solidus and, if temperatures are sufficiently high, will undergo volatile phase-absent melting that first consumes muscovite, followed by reactions consuming biotite. At the *T* of the solidus at P < 'IP1', rocks lack muscovite and volatile phase-absent melting will be via biotite breakdown alone. Close to the solidus, partial melting will generally give rise to high aH_2O melts; at higher *T*, volatile-phase absent mica breakdown produces lower aH_2O melts.

High aH_2O melting

Rocks that attain a thermal peak not far in excess of the solidus, i.e. within the high aH_2O field (grey area) or within the cordierite (+ K-feldspar) field (pale blue area) in Fig. 1b, will

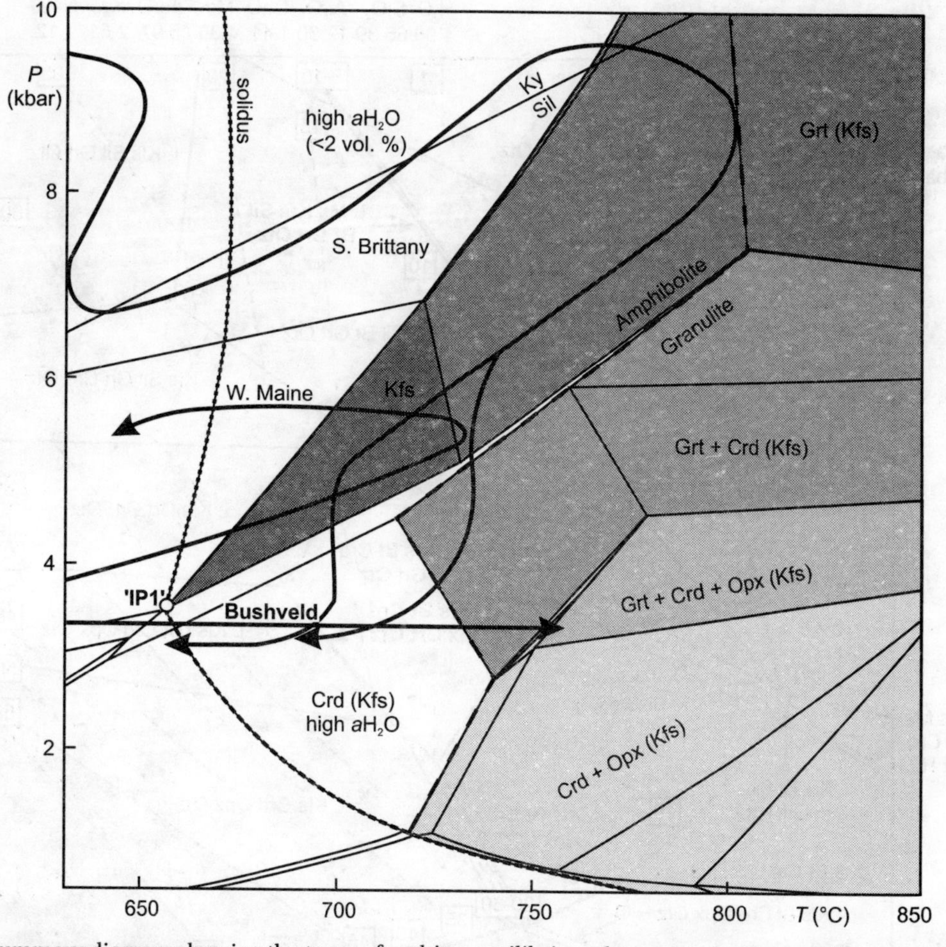

Fig. 1b. Summary diagram showing the types of melting equilibria and associated peritectic phases as well as the exemplar *P-T* paths discussed in the text.

generally develop small quantities (<< 5 mol. %) of high aH_2O 'wet' melt. As the temperature of the solidus is attained, congruent H_2O-present ($aH_2O = 1$) melting produces a volume of melt that is directly proportional to the quantity of free H_2O available — commonly < or << 1 vol. %. Once all H_2O has been consumed, a further rise in T will produce small quantities of melt by incongruent, volatile phase-absent reaction, and the H_2O activity of the system, although remaining high, falls below unity. The small melt volumes characteristic of high aH_2O anatexis may not form an interconnected network (e.g. Laporte *et al.*, 1997), which coupled with the negative volume change ($\Delta V_{(r)}$) of volatile phase-present melting equilibria (e.g. Brown, 1994a; Clemens & Droop, 1998), means that such small volumes of high aH_2O 'wet' melt are unlikely to segregate from the source. If an external supply of H_2O-rich volatile phase is available (in most natural cases we consider this to be unlikely), larger volumes of melt may be produced (see e.g. Johnson *et al.*, 2001). Larger melt fractions may exceed the permeability threshold, potentially enabling segregation and ascent, especially since most crustal melting occurs during orogenic deformation (e.g. Brown, 1994b). However, as the solubility of H_2O in granitic melt increases with P (e.g. Holtz & Johannes, 1994), H_2O saturated or near saturated melts will boil and crystallize as they begin to ascend (Cann, 1970), commonly forming small volume sheets and plutons of peraluminous leucogranite within the migmatite host.

Moderate to low aH_2O melting

The majority of intracrustal melt production from metapelitic rocks results from the incongruent breakdown of micas in the absence of a free volatile phase (dehydration melting). Most volatile phase-absent melting occurs within low-variance fields (di- or trivariant in NCKFMASH) that have a steep or moderate positive dP/dT slope and steep melt proportion isopleths (Fig. 1a).

At $P >$ 'IP1' incongruent reaction consuming muscovite occurs over a narrow T interval to produce moderate aH_2O melts (Fig. 1a). The amount of melt produced will be proportional to the muscovite content of the rock and, for any particular bulk composition, will increase with P. In the absence of an external source of H_2O, a maximum of 15 mol. % melt is predicted from muscovite-consuming volatile phase-absent reaction in common metapelitic rocks below 10 kbar (Fig. 1a). The positive $\Delta V_{(r)}$ of muscovite-consuming melting equilibria favour melt-induced cracking and segregation (Brown, 1994a; Rushmer, 2001).

The incongruent breakdown of biotite occurs over a much wider T interval and may occur at pressures both above and below 'IP1'. As Fe and particularly Mg are relatively insoluble in granitic melts, biotite breakdown is characterized by the production of ferromagnesian peritectic products (Fig. 1b) — garnet ± orthopyroxene at high-P (>5 kbar); garnet + cordierite ± orthopyroxene at moderate-P (3-7 kbar); and cordierite + orthopyroxene at low-P (<4 kbar). Volatile phase-absent reactions consuming biotite are not associated with a pronounced $\Delta V_{(r)}$ (Brown, 1994a; Rushmer, 2001) and melt may be unable to segregate unless the rocks are actively deforming. In nature, deformation generally accompanies melting so that the melt produced by volatile phase-absent reactions consuming biotite, which is predicted to be in excess of 50 mol. % at 850°C (Fig. 1a), within the middle of the range for granulite facies conditions, commonly can escape from the source to form upper-crustal peraluminous leucogranites. Retention of this melt volume in the source will favour the formation of diatexite (e.g. Sawyer, 1998), which may ascend by bulk flow, allowing differentiation into a mafic diatexite residue and a more mobile leucodiatexite magma that ultimately may fractionate to peraluminous leucogranite. Low aH_2O ('dry') melts have the potential to ascend large distances before they attain H_2O saturation and crystallize (e.g. Cann, 1970; Clemens & Droop, 1998).

The steep to moderate, generally positive slopes of the melt molar proportion isopleths within the field of low aH_2O melting (Fig. 1a) imply that additional melt commonly will be produced as rocks decompress if cooling rates are sufficiently slow. Consequently, maximum melt production may not correspond to the maximum T attained by the rock. This has important ramifications for high-T regional metamorphism within 'normal' collisional orogens that are characterized by clockwise P-T loops (e.g. Thompson & Connolly, 1995), in which melting after the thermal peak occurs along the steeply negative dP/dT (near isothermal decompression) segment of the exhumation path. In near-isobaric P-T paths characteristic of contact metamorphism, maximum melt production will correspond to peak T.

Retrograde processes

The processes that lead to retrograde reaction in melt-bearing metamorphic rocks have been extensively discussed in the recent literature (Kriegsman & Hensen, 1998; Spear *et al.*, 1999; Kriegsman, 2001; Thompson, 2001; White *et al.*, 2001; Brown, 2002; Cenki *et al.*, 2002; Moraes *et al.*, 2002; White & Powell, 2002). The nature and degree of retrograde reaction in rocks are influenced by a number of factors. Of particular importance are: the volume of the system under consideration and the (effective) bulk composition of that system, which may be strongly influenced by grain size (Stüwe, 1997) and, where appropriate, loss of melt (White & Powell, 2002); the retrograde segment of the P-T path followed by the rock that is a function of the rates of exhumation and cooling; strain-rates; and whether, when and where a fluid phase (volatile species or melt) was present.

In migmatites and granulites, the two most important driving forces for retrograde reaction are (i) the melt phase (at suprasolidus conditions), and (ii) the H_2O-rich volatile phase that is released as melts cool to the temperature of the contemporary solidus. Suprasolidus retrograde reaction is controlled by the nature of the melting equilibria (high, moderate or low aH_2O, congruent or incongruent, muscovite- and/or biotite-consuming), the abundance of melt and the degree, scale and rapidity over which melt migrates within the system. An important consequence of melt segregation, and ascent to shallower crustal levels is that it leads to scale-dependent changes in the bulk composition, thermal structure and rheology of the system. Retrograde reaction with the volatile phase that is released as melts crystallize will also be dependent on the nature of fluid transport, and may occur *in situ*, or may cause retrogression of proximal (commonly overlying) subsolidus rocks or may lead to anatexis of rocks into which it infiltrates (e.g. Thompson, 2001; Johnson *et al.*, 2003b).

Rocks that cool to the solidus at $P >$ 'IP1' will potentially develop retrograde muscovite; those that cool below 'IP1' will lack muscovite unless a volatile phase is available at lower subsolidus T (e.g. Spear *et al.*, 1999; Fig. 1a). The small volumes of melt that are expected from high aH_2O (low-T) melting in the absence of an external supply of H_2O are unlikely to segregate or be extracted from the source. Discrete macroscopic leucosomes are unlikely to form, although the former presence of melt may be inferred using microstructural criteria for the former presence of melt pockets, which are characterized by low dihedral angles with phases inferred to have been solid at the time of textural equilibration with melt in the pocket (e.g. Ashworth & McLennan, 1985; Sawyer, 2001; Marchildon & Brown, 2002; Johnson *et al.*, 2003b).

As volatile phase-absent melting is characterized by much larger volumes of melt, the effects of the change in bulk composition that are a consequence of melt loss are more pronounced. If the melt remains within the equilibration volume, the bulk composition of the system is unchanged and rocks that may have attained granulite facies conditions will be retrograded to amphibolite facies subsolidus assemblages (e.g. McGregor & Friend, 1997). However, if some melt is removed, the bulk composition will change and parageneses that developed close to the thermal peak will likely be preserved; the degree of preservation is largely a function of the amount of melt both generated and subsequently lost from the system (e.g. White & Powell, 2002). The effect of compositional changes in the protolith bulk composition on phase equilibria cannot be illustrated with reference to an iso-compositional pseudosection (Fig. 1a, b).

FIELD EXAMPLES OF MELTING AND RETROGRADE PROCESSES

We illustrate prograde and retrograde processes with reference to three field examples, migmatites formed by low volume H_2O-present melting from the aureole of the Bushveld Complex (Johnson *et al.*, 2003b), anatectic migmatites formed by mica breakdown within regional metamorphic terrains in Western Maine (e.g. Solar & Brown, 2001; Johnson *et al.*, 2003a) and Southern Brittany (e.g. Jones & Brown, 1990; Brown & Dallmeyer, 1996; Johnson & Brown, in review), and by mica breakdown within the aureole of the Bushveld Complex (Johnson *et al.*, 2003b). For the sake of brevity, we do not describe in detail the regional context of each of these terrains; more detailed information is available in the papers cited above and the references therein.

High aH_2O melting

We take migmatites in the aureole of the c. 2.06 Ga Bushveld Complex (Walraven *et al.*, 1990) as an example of high aH_2O melting (Johnson *et al.*, 2003b). An unusual but critical feature of the Bushveld Complex is that the aureole rocks underlie the ultramafic rocks of the sill-like Rustenburg Layered Suite, which is the putative heat source. Consequently, the thermal gradient was inverted relative to the normal situation during regional metamorphism; underlying subsolidus rocks that were dehydrating, provided a supply of H_2O-rich volatile phase that variably interacted with overlying, suprasolidus rocks. The aureole rocks record P of about 3 kbar (Waters & Lovegrove, 2002), consistent with a lack of primary muscovite at the anatectic front, enabling examination of phase relations with T-X pseudosections.

In outcrops at the outer edge of the migmatite front, shallowly-dipping biotite-rich (± andalusite) metapelitic rocks of the Silverton Formation are characterized by sinuous fibrolite seams, thin (~10 mm) quartz-sillimanite veins and discordant podiform leucosomes (Fig. 2a, b). A variably oriented foliation is present. Shallowly-dipping to subvertical fibrolite

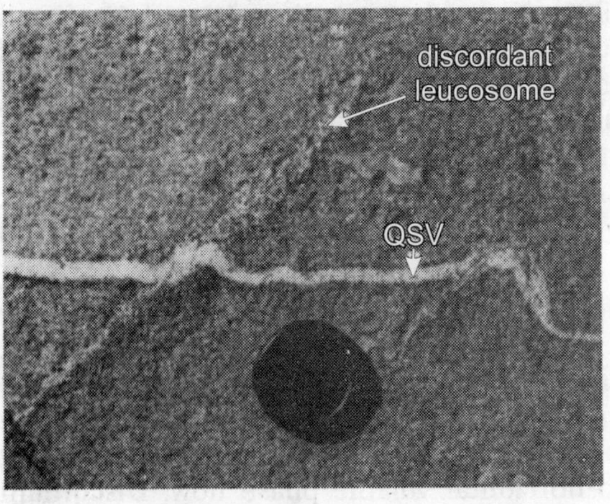

(a)

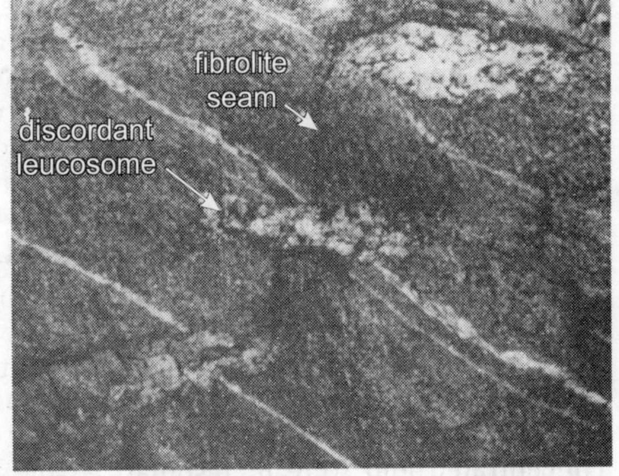

(b)

Fig. 2. Photographs to show petrographic aspects of lower-grade migmatites in the aureole of the Bushveld Complex. (a) Sub-horizontal quartz-sillimanite vein with well-defined melanosome exhibits gentle, upright folds. A discordant leucosome is aligned parallel to the weak foliation. Numerous millimetric leucosomes patches are just visible within the mesosome. (b) Cross-cutting pegmatitic discordant leucosomes within a tension gash array. The leucosome pods are connected by thin fibrolite seams. The seams are sub-parallel to the steep foliation, which becomes more moderately inclined away from these areas.

seams (Fig. 2b, c) are abundant and show a close relationship with the foliation. Leucosome, which comprises << 5 vol. % of the outcrop, commonly is concentrated parallel to the foliation and around small antiformal fold closures in quartz-sillimanite veins (Fig. 2a). The fibrolite seams are spatially related to the discordant leucosomes, into which they splay and terminate (Fig. 2b). The spatial proximity of fibrolite seams, quartz-sillimanite veins and discordant leucosomes suggest a genetic relationship.

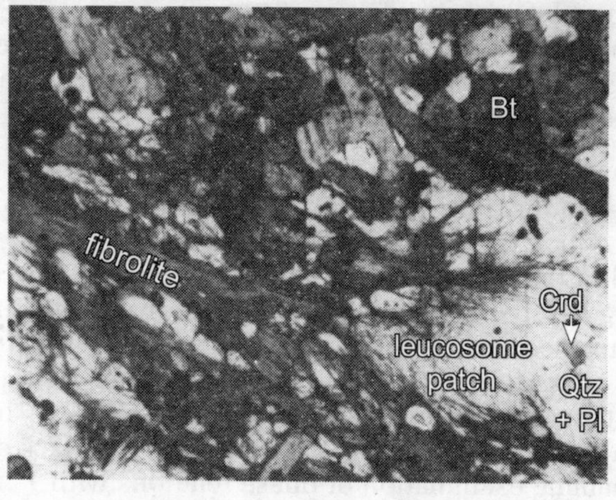

(c)

Fig. 2c. Quartzofeldspathic leucosome patch containing pinitized peritectic cordierite within andalusite-free metapelites. Fibrolite is concentrated in seams that terminate against the leucosome patches by dissipating into a spray of fibrolite.

The fine-grained (0.1-0.5 mm) mesosome consists of biotite + plagioclase + quartz + fibrolite ± andalusite, muscovite, cordierite, ilmenite, and tourmaline. It contains small (approximately 1 mm across) leucocratic patches that are coarser grained than the matrix of the mesosome, and which comprise quartz, muscovite and/or plagioclase with small (<0.1 mm) euhedral grains of cordierite that exhibit varying degrees of pinitization (Fig. 2c). The leucocratic patches are associated with discontinuous mats and seams of fibrolite (Fig. 2c) that are rich in fibrolitized biotite and disseminated ilmenite. Macroscopic podiform discordant leucosomes (Fig. 2b) comprise quartz, plagioclase, muscovite and tourmaline. Euhedral cordierite grains may or may not be present.

The fibrolitized biotite and ilmenite within the fibrolite seams are consistent with base cation leaching by an acidic volatile phase (e.g. Wintsch, 1975; Vernon, 1979; Nabelek, 1997, McLelland *et al.*, 2002); exsolution of ilmenite reflects the low solubility of Ti (and Fe). The fibrolite seams are interpreted to record passage of a high a_{H+} volatile phase through the system. Quartz-sillimanite veins are inferred to represent thin metapsammite horizons that concentrated and were metasomatized by sub-horizontal volatile phase flow. Discordant leucosomes and small leucocratic patches within mesosome are interpreted to represent frozen melt that was variably fluxed by the H_2O-rich volatile phase.

We interpret field and petrographic features with reference to an NCKFMASH pseudosection that is drawn for an average subaluminous Silverton Formation metapelite (cf. Johnson *et al.*, 2003b). Fig. 3 shows a projection into T-XH_2O space contoured for molar melt proportion isopleths (as mole fraction) that illustrates the influence of the H_2O volatile phase.

The prograde subsolidus evolution in the interval 640-650°C is characterized by a major dehydration step (segment I). The pulse of water liberated along this segment potentially infiltrated overlying rocks and influenced melting of suprasolidus rocks. Segment (I) exhausts first muscovite then sillimanite from the modelled bulk composition by the reactions,

$$\text{(Pl)} + \text{(Crd)} + \text{Ms} + \text{Qtz} \Leftrightarrow \text{Kfs} + \text{(Bt)} + \text{Sil} + H_2O, \quad (1)$$

$$\text{(Pl)} + \text{Bt} + \text{Sil} + \text{Qtz} \Leftrightarrow \text{Kfs} + \text{Crd} + H_2O \quad (2)$$

(in reactions, minor participant phases are shown in parentheses). Together, reactions 1 and 2 liberate about 0.7 mol. % of H_2O, and less than 0.1 mol. % of H_2O is produced by a trivial amount of reaction in the interval 650-660°C along segment II.

Upon reaching the solidus (black dot in Fig. 3), melt production is governed by both increasing T and the potential supply of H_2O from underlying rocks. An increase in T with no external supply of H_2O (segment III) will result in the formation of small quantities (<1 mol. %) of melt with peritectic cordierite and K-feldspar by the reaction,

$$\text{Pl} + \text{Bt} + \text{Qtz} \Leftrightarrow \text{Kfs} + \text{Crd} + \text{L}. \quad (3)$$

The aluminium required for peritectic garnet production is supplied by the major reactant, biotite, which becomes progressively depleted in the eastonite component by Tschermak exchange. If at some fixed T along segment (III) an influx of H_2O occurs (i.e. segment IV) additional melt will be produced by the congruent reaction,

$$\text{Kfs} + \text{Pl} + \text{Qtz} + \text{(Bt)} + \text{(Crd)} + H_2O \Leftrightarrow \text{L}, \quad (4)$$

which results in a five-fold increase in the melt fraction to 3.5-4 mol. % for an addition of 0.7 mol. % H_2O, the amount released over the dehydration segment I (by reactions 1 and 2). Melts (and leucosomes) produced by reaction 4 will not be associated with any peritectic reaction products.

The major pulse of H_2O from underlying rocks by reactions 1 and 2 occurs over a temperature interval of about 10°C. Consequently, the prograde suprasolidus path is most likely to have followed a vector combining both increasing T and H_2O-supply. Segment (V) shows the vector for a rock absorbing a 0.7 mol. % pulse of water over 10°C, the equivalent of absorbing the water produced over the temperature interval of a prograde evolution along segment I. Such a vector is near orthogonal to the melt proportion isopleths and will be efficient at producing melt; in excess of 4 mol. % melt is predicted to form by reaction 4. The overall abundance and coarse-grained nature of discontinuous leucosomes suggest that they may have resulted from congruent melting following such H_2O influx.

In nature, the prograde path may follow any vector with a positive slope in Fig. 3, with the slope depending on the rate of temperature increase and the rate of supply of H_2O. The rate of supply of H_2O to any particular volume of rock (as well as the absolute quantities involved) will be strongly influenced by the heterogeneity of volatile phase distribution, i.e. whether transfer is pervasive or, more likely, channelled. The vectors for suprasolidus cordierite growth or consumption are shown in Fig. 3. Peritectic cordierite is produced only if the prograde vector is inclined more steeply ($>dT/dX_{H_2O}$) than the diagonal vector of constant cordierite molar proportion (e.g. along segment VI), whereas existing cordierite is consumed (by reaction 1) if the prograde vector is more shallow (e.g. along segment V).

Small volume leucocratic patches within mesosome are ubiquitously associated with peritectic cordierite, suggesting limited H_2O-fluxed melting, assisted by any available B (cf. Acosta-Vigil *et al.*, 2001). This conclusion is consistent with the common discontinuity of fibrolite mats associated with the leucocratic patches, and implies only local (millimetre-scale) redistribution of H_2O within mesosome. Decimetre-scale discordant leucosome segregations are commonly not associated with peritectic cordierite, implying that the larger volume of volatile phase derived from the dehydrating rocks beneath was channelled into the quartz-sillimanite veins and/or fibrolite seams to flux congruent melting. However, fluid flow (both volatile and melt phases) occurred only on an outcrop scale (Johnson *et al.*, 2003b), consistent with the limited immobility of high aH_2O melts.

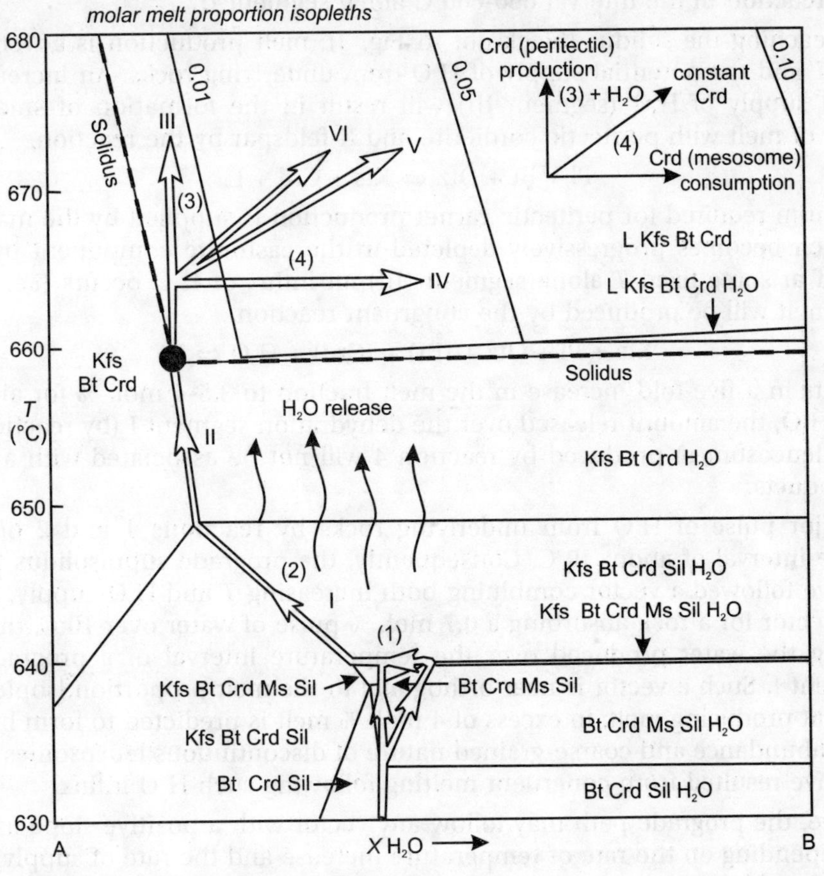

Fig. 3. T-X_{H_2O} pseudosection calculated for the mean composition of Silverton Formation metapelites. The white arrows show the potential prograde paths, numbered segments (I-VI) of which are discussed in the text. The black arrows show vectors of changing cordierite proportions as suprasolidus reaction progresses. With no H_2O influx and a rise in T (vertical vector), peritectic cordierite is produced. With H_2O -influx and no rise in T (horizontal vector), mesosome cordierite is consumed. The diagonal vector marks no change in the cordierite content. The length of the arrows marks a change of 0.25 mol.% cordierite. The normalized bulk composition of the left and right hand edges of the diagram in terms of the components H_2O:SiO_2:Al_2O_3:CaO:MgO:FeO:K_2O:Na_2O are 4.10:64.99:13.44.0.55:5.52.:8.12:2.11:1.17 and 6.37:63.46:13.12:0.53:5.39:7.93:2.06:1.14 respectively. Numbers in parentheses refer to reactions discussed in the text that occur across the relevant multivariant field. The darkest fields are quinivariant.

Moderate and low aH_2O (volatile phase-absent) melting

Muscovite consuming reactions

We take as our example of low aH_2O melting at P>'IP1' (Fig. 1a & b) metapelitic migmatites from the Central Maine belt, which were deformed and metamorphosed during the Acadian orogeny (c. 404 Ma) in the northern Appalachians (e.g. Solar & Brown, 2001). The migmatites represent the metamorphic high within a NE-SW oriented belt of greenschist to upper-amphibolite facies regionally metamorphosed rocks. Within the anatectic domains, metatexites and diatexites are variably developed in fibrolite- and biotite-rich metapelitic rocks that are now depleted in a granitic component with respect to presumed protolith compositions; up to 20 vol. % bulk melt depletion is implied (Solar & Brown 2001). At subsolidus grades up to the migmatite front, primary fabric forming muscovite is abundant. At grades beyond the migmatite front, primary muscovite is absent although retrograde, skeletal development of this phase is ubiquitous. Fibrolite and biotite are the primary (fabric forming) phases in the residual mesosome. Although K-feldspar may comprise up to 5 vol. % of leucosome, it is usually absent from the rocks. There is only limited evidence for extensive incongruent volatile phase-absent, biotite-consuming melting, and cordierite, orthopyroxene and/or spinel do not occur in migmatites. Rare garnet-bearing leucosomes in the cores of the anatectic domains suggest granulite facies conditions may have been reached locally.

We interpret the evolution of the Central Maine belt migmatites with reference to pseudosections in the MnNCKFMASH system that are based on an average subaluminous metapelite considered to be representative of the presumed protolith (Solar & Brown, 2001; Johnson *et al.*, 2003a). The addition of Mn expands garnet stability to lower-P, consistent with field observations (e.g. Mahar *et al.*, 1997); at near solidus T and before the onset of incongruent biotite breakdown, garnet proportions and reaction coefficients are small. Fig. 4 is a P-T projection showing the predicted stable parageneses and molar melt proportion isopleths (in mol. %). The presence of primary muscovite at the migmatite front and the abundance of skeletal retrograde muscovite imply that both the prograde and retrograde segments of the P-T path crossed the solidus at P>'IP1'. Subsolidus rocks preserve evidence for thickening that continued after the regional thermal peak, so that peak T preceded peak P along the P-T evolution and the gross P-T path was counterclockwise (e.g. Holdaway *et al.*, 1982; Johnson *et al.*, 2003a). The schematic P-T path is broadly consistent with petrological constraints (e.g. Solar & Brown, 2001; Johnson *et al.*, 2003a).

The principal melting reaction is,

$$\text{Ms} + \text{Qtz} + \text{Pl} \Leftrightarrow \text{Kfs} + \text{Sil} + (\text{Bt}) + \text{L}, \quad (5)$$

which produces 8-10 mol. % melt over a small temperature interval (Fig. 4). At the metamorphic peak (asterisk in Fig. 5a), the rocks should contain abundant K-feldspar but lack muscovite. The general lack of K-feldspar in the migmatites and abundance of retrograde muscovite requires explanation.

Fig. 5a shows an isobaric (at P = 4.5 kbar) T-X_{melt} diagram that illustrates the effect of melt loss and/or influx into the migmatites (White & Powell, 2002). At the inferred metamorphic peak (asterisk), the modelled bulk composition contains about 10 mol. % melt (Fig. 4). Upon cooling to the temperature of the solidus, complete K-feldspar consumption is only predicted if almost all (>90%) of the 10 mol. % melt remains within the equilibration

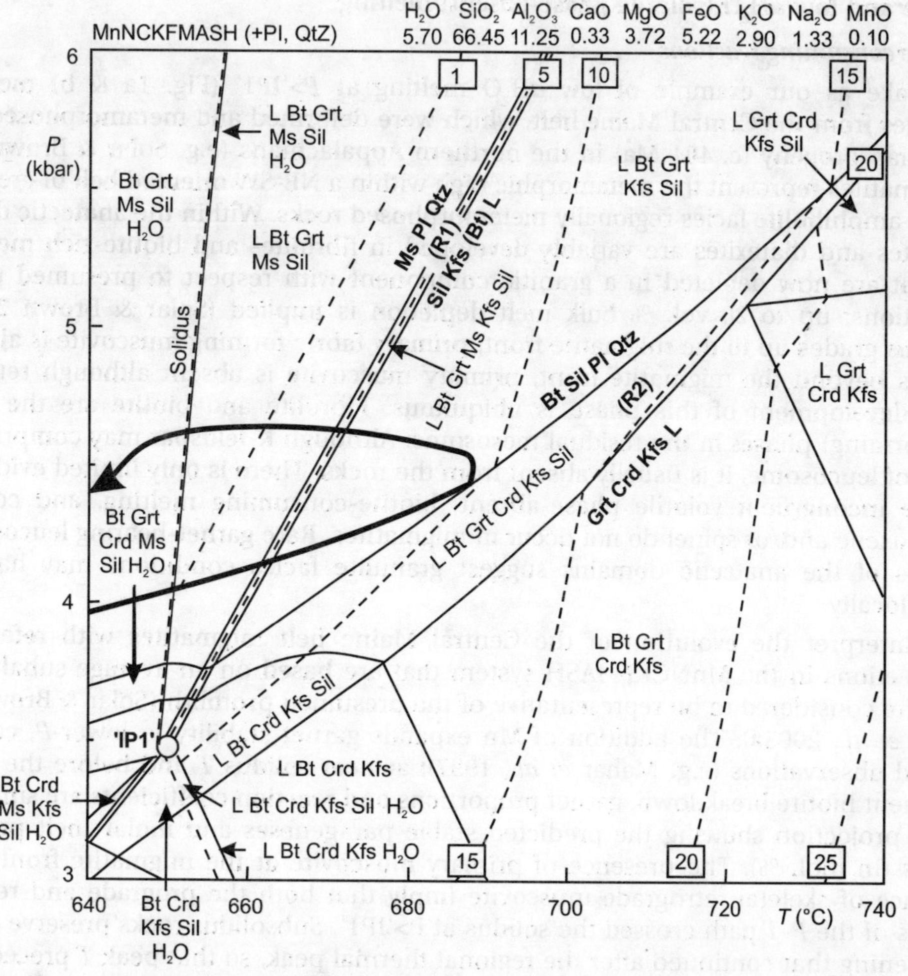

Fig. 4. Suprasolidus *P-T* pseudosection in the MnNCKFMASH model system for a representative composition of a Central Maine Belt metapelite. The dashed lines show contours of modal melt proportions. The two narrow trivalent fields show the range over which the major volatile phase-absent melting reactions occur; R1 consumes muscovite; R2 consumes biotite and sillimanite and results in peritectic garnet production. Significant quantities of melt (>2 mol. %) will not be produced until the 'effective' solidus (R1) is encountered. The darkest field is quadrivariant. The bulk composition contains sufficient H_2O (~6 mol. %) to just saturate the solidus at high-*P*.

volume (e.g. at point C in Fig. 5a). If some (5-90%) of the melt remains within the system on cooling, both K-feldspar and muscovite should be stable (e.g. point B). If >95% of the 10 mol. % melt is lost, only K-feldspar is predicted. Geochemical arguments (Solar & Brown, 2001) suggest that the rocks lost up to 20 vol. % melt only, and consequently K-feldspar should be present in these rocks.

Fig. 5b shows an isobaric (at $P = 4.5$ kbar) T-X_{H_2O} diagram that illustrates the effect of volatile influx into the migmatites. The bulk composition for which Fig. 5b is drawn is a residual migmatite from which all the 10 mol. % melt has been removed (i.e. the right-hand

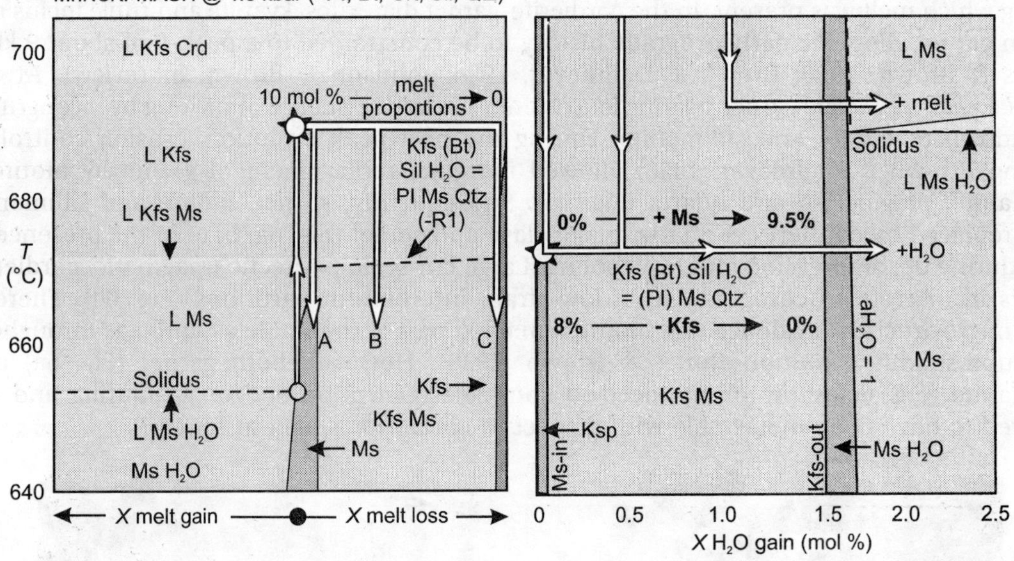

Fig. 5. (a) *T-Xmelt* pseudosection showing the effect of melt loss and melt gain on a mean metapelite composition (see Fig. 4) heated to 690°C (asterisk). (b) $T\text{-}X_{H_2O}$ pseudosection showing the effect of the addition of H_2O to a residual bulk composition. The absence of K-feldspar implies in excess of 1.5 mol. % addition of H_2O.

side of Fig. 5a). The left-hand side of Fig. 5b represents the modelled residual bulk composition with no influx of H_2O and shows that, on cooling to subsolidus T, K-feldspar should be abundant. A subsequent evolution to the right shows the effect of adding up to 2.5 mol. % of H_2O to the rocks. As H_2O is added, K-feldspar and sillimanite are consumed, and muscovite is produced by the reaction,

$$\text{Kfs + Sil + (Bt) + } H_2O \Leftrightarrow \text{Ms + Qtz + Pl.} \tag{6}$$

With the addition of just 1.5 mol. % of H_2O, total K-feldspar consumption is predicted, consistent with observations. Further addition of H_2O will produce melt (if T>655°C) or a free volatile phase (if T<655°C).

We interpret the lack of K-feldspar and the abundance of retrograde muscovite in the Central Maine belt migmatites to have been the combined result of melt loss followed by H_2O-rich volatile phase influx at near solidus *T*. We suggest that a supply of H_2O is likely given the abundance of crystallizing granite magma within the system (Solar & Brown, 2001).

Biotite-consuming reactions

We take as examples of volatile phase-absent, biotite consuming reactions migmatites from the highest-grade part of the Bushveld Complex aureole and regional migmatites from Southern Brittany. These two terrains differ significantly in their *P-T* evolution, degree of melt loss and subsequent retrograde history.

Southern Brittany. Metapelitic migmatites in Southern Brittany occur as both metatexites and cordierite-garnet diatexites that record a complex suprasolidus *P-T*-deformation evolution during which melt was present. In the cordierite-garnet diatexites, kyanite and rutile inclusions within garnet allow the early prograde history to be constrained to a peak-P of about 9 kbar (Jones & Brown, 1990; Brown & Dallmeyer, 1996; Johnson & Brown, in review). At the thermal peak, peritectic garnet-bearing leucosomes preserve evidence of melting by incongruent breakdown of biotite and sillimanite. During the post-peak evolution, erosion controlled cooling (Brown & Dallmeyer, 1996) allowed marginal replacement of garnet by biotite ± sillimanite, plagioclase and quartz (Fig. 6a). Subsequently, garnet, biotite and sillimanite were replaced by cordierite ± quartz, plagioclase and spinel (Fig. 6a, b, c) in the presence of melt during decompression and fast cooling. Large cm-scale peritectic (magmatic) cordierite occurs in discrete leucosomes within low-strain interboudin partitions (Fig. 6d). There is good microstructural evidence that biotite remained part of the stable assemblage throughout the suprasolidus evolution (Jones & Brown, 1990). However, both garnet (Fig. 6a) and sillimanite (Fig. 6c) show pronounced to complete resorption and/or armouring and are inferred to have been metastable with respect to cordierite + melt at lower-P.

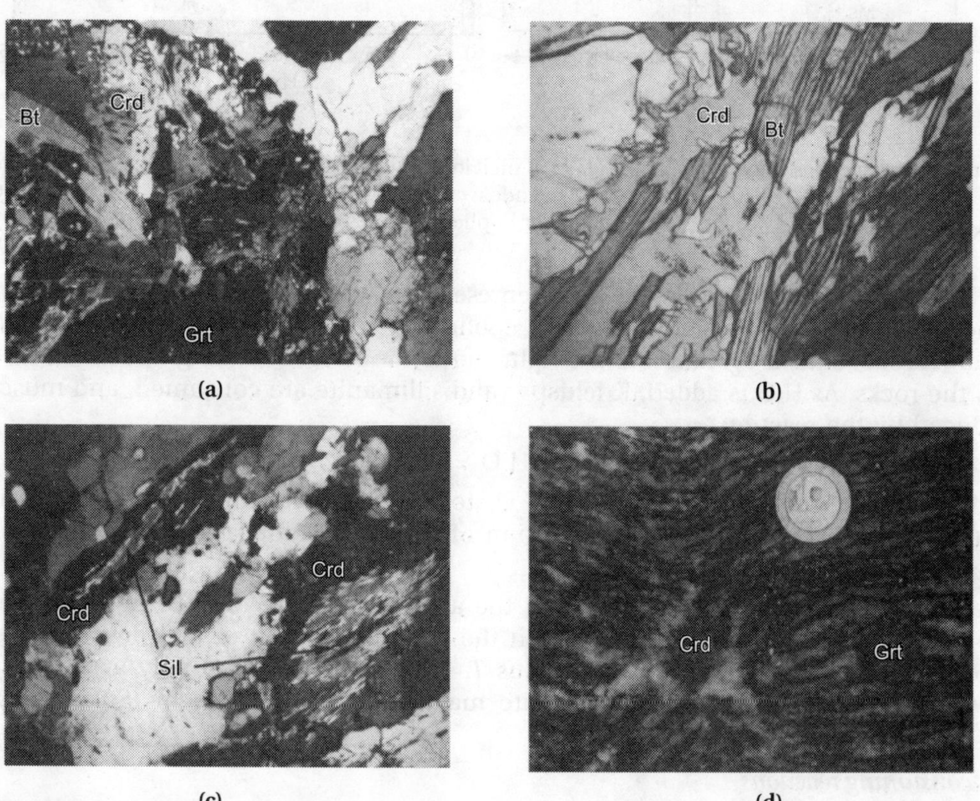

Fig. 6. Photomicrographs to show petrographic aspects of the Southern Brittany migmatites. (a) Garnet porphyroblast surrounded by complex intergrowth of biotite, cordierite, quartz and plagioclase that isolates garnet from coarse leucosome (width of field of view is 3.5 mm). (b) and (c) Cordierite replacing biotite and sillimanite respectively, the result of decompression. Biotite is still inferred to be a stable part of the assemblage. Sillimanite is always surrounded by cordierite and is inferred to be metastable (width of field of view is 1.5 mm and 3 mm, respectively). (d) Garnet-cordierite gneiss. Peritectic cordierite commonly occurs within interboudin partitions.

There are no exposed lower-grade (unmigmatized) equivalents of the Southern Brittany metapelitic migmatites. Thus, to place constraints on the evolution of these rocks, we use the *P-T* pseudosection constructed for an average subaluminous metapelitic rock in the NCKFMASH model system. The inferred *P-T* path for the Southern Brittany migmatites is shown in Fig. 7; this *P-T* path has been modified after Jones & Brown (1990) and Brown & Dallmeyer (1996) to take into account new field observations (Johnson & Brown, in review). However, the modelled equilibria remain in good agreement with the petrological and thermobarometric constraints provided by these authors.

Segment A along the prograde path (Fig. 7) exhausts muscovite by reaction 5, producing about 10 mol. % melt. Continued heating produces an additional 10 mol. % melt by the volatile phase-absent, biotite-consuming incongruent reaction,

$$\text{Bt} + \text{Sil} + \text{Pl} + \text{Qtz} \Leftrightarrow \text{Grt} + \text{Kfs} + \text{L}. \quad (7)$$

At the thermal peak (B), around 25 mol. % garnet is predicted, consistent with the abundance of this phases in most metapelitic rocks (Fig. 6a, d).

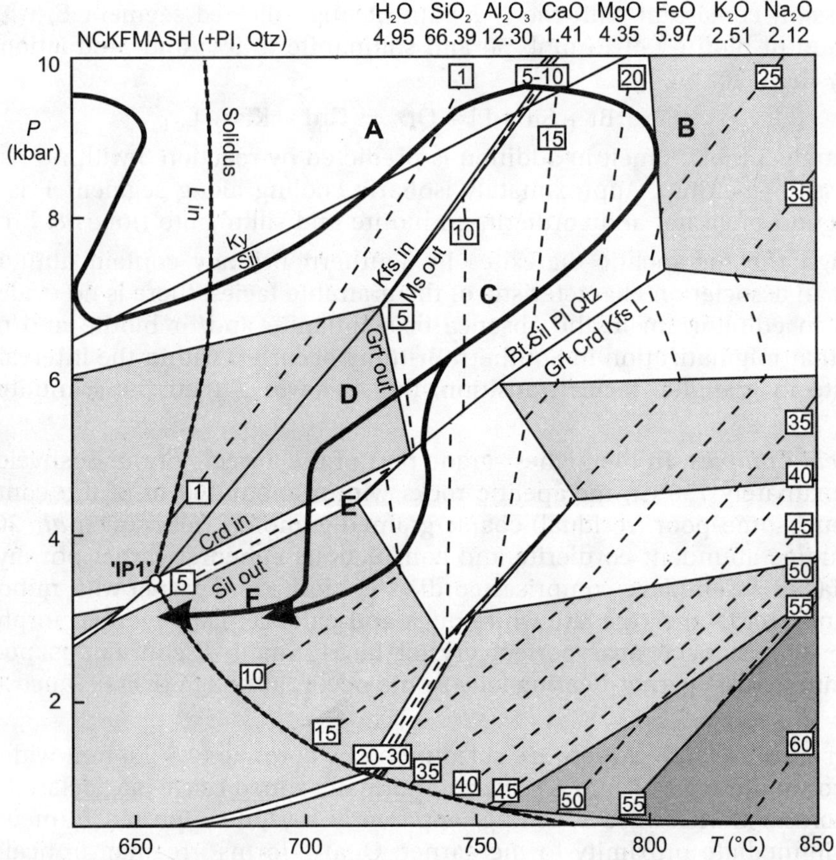

Fig. 7. *P-T* pseudosection calculated for the average metapelite composition used in Fig. 1a & b. Molar melt proportion isopleths are shown as dashed lines. The *P-T* path shown is consistent with pre-existing data. The letters refer to reactions discussed in the text.

Retrograde reaction along the cooling segment C causes resorption of garnet and liquid by the reverse of reaction 7, although stoichiometric coefficients must change. Reaction microstructures support replacement of garnet by biotite ± plagioclase and sillimanite. Such retrograde replacement is only possible if melt is available as a reactant, implying its retention within the system. Continued retrograde evolution along segment D should result in complete garnet replacement (Fig. 7). Although complete replacement may occur (Jones & Brown, 1990), garnet commonly remains as cores surrounded by wide reaction coronae; it is clearly metastable.

There is no evidence for the *P-T* path having crossed the narrow divariant field that marks the transition into the granulite field by the reaction (Fig. 7),

$$\text{Bt} + \text{Sil} + \text{Pl} + \text{Qtz} \Leftrightarrow \text{Grt} + \text{Crd} + \text{Kfs} + \text{L}. \qquad (8)$$

Reaction 8 produces about 5 mol. % melt and peritectic cordierite on decompression, but is also associated with a second phase of peritectic garnet production, for which there is no evidence.

Microstructures show that cordierite replaces biotite and sillimanite (Fig. 6a, b) implying decompression consistent with the *P-T* path having followed segment E, which results in consumption of biotite (30-15 mol. %) and sillimanite (9-0%) and production of cordierite (0-25%) by the reaction,

$$\text{Bt} + \text{Sil} + \text{Pl} + \text{Qtz} \Leftrightarrow \text{Crd} + \text{Kfs} + \text{L}. \qquad (9)$$

Approximately 5 mol. % melt in addition is predicted by reaction 9 with no rise in *T* (i.e. by decompression melting). Approximately isobaric cooling along segment F is suggested by minor reaction of garnet and cordierite to biotite and sillimanite (Jones & Brown, 1990).

Although the metapelitic diatexites in Southern Brittany contain abundant garnet + cordierite, an association characteristic of the granulite facies, there is no evidence that they co-existed in equilibrium in the absence of sillimanite and/or biotite and/or quartz. We conclude that migmatization in Southern Brittany occurred within the interval of the upper amphibolite to granulite facies transition, but at lower *T* than the granulite facies *sensu stricto*.

Bushveld Complex. In the highest-grade part of the aureole of the Bushveld Complex (at *T* of approximately 750°C), metapelitic rocks within about 100 m of the contact comprise massive leucosome-poor (residual) coarse-grained granofels (Johnson *et al*., 2003b; Fig. 8a), which contains abundant cordierite and conspicuous euhedral garnet porphyroblasts. The granulite facies assemblage comprises cordierite, biotite and garnet with minor plagioclase, ilmenite and quartz and rare late white mica and chlorite. Large garnet porphyroblasts (up to 10 mm across) have near perfect crystal faces. Small discontinuous pods of smaller (about 1 mm across) garnet-bearing leucosome occur, although overall leucosome fractions are <5 vol. %.

In thin section, large garnets are surrounded by coronae (0.5-2.0 mm wide) of euhedral, pinitized cordierite (up to 1 mm across) and normally-zoned calcic plagioclase (approximately 0.1 mm across) that form a touching, apparently crystal-supported framework (Fig. 8b), except in immediate proximity to the garnet. Quartz forms irregular, optically continuous patches that have pronounced cuspate margins with low dihedral angles that are interstitial to the matrix phases. Such microstructures suggest melting and subsequent melt loss.

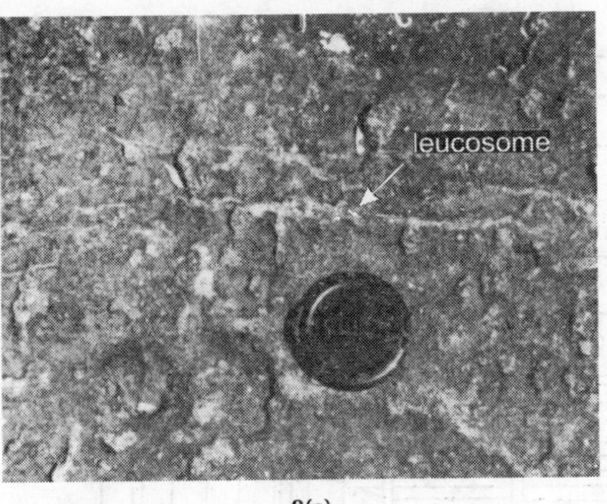

8(a)

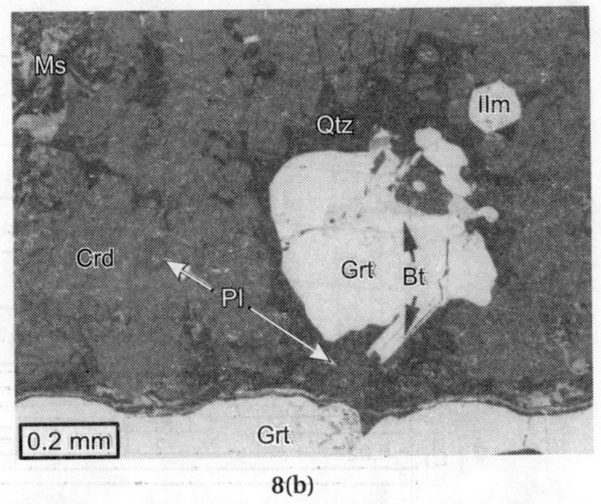

8(b)

Fig. 8. (a) Cordierite-garnet granofels approximately 100 m beneath the ultramafic-country rock contact in the aureole of the Bushveld Complex. Thin leucosome veins are conspicuous. (b) Cordierite-rich mantle around large garnet (bottom). The compacted microstructures suggest melt loss. The cuspate, interstitial quartz patch is interpreted to pseudomorph a trapped melt pocket.

From Fig. 1a, b it can be seen that a prograde suprasolidus evolution up to 725°C at 3 kbar is characterized by the formation of peritectic cordierite and melt (by reaction 3), followed by peritectic garnet and melt by the reaction,

$$\text{Bt} + \text{Qtz} + \text{Pl} + (\text{Crd}) \Leftrightarrow \text{Grt} + \text{Kfs} + \text{L}. \qquad (10)$$

The aluminium required for peritectic garnet production is supplied by biotite, which continues to be progressively depleted in the eastonite component. Melt fractions of 15-20 mol. % are predicted by reaction 10 (Fig. 1a).

Fig. 9 is an isobaric (P = 3 kbar) T-X_{melt} diagram illustrating the effect of potential melt loss and melt gain (cf. White *et al.*, 2001; White & Powell, 2002) from the average Silverton Formation bulk composition heated to 740°C (point X in Fig. 9). Within the trivariant field {L + Pl + Kfs + Bt + Crd + Grt + Qtz}, reaction 10 results in the formation of about 6 mol. % peritectic garnet, a quantity similar to that observed in the granofels. As melt is lost from the equilibration volume, the bulk composition will be displaced towards the right-hand edge of Fig. 9.

If all of the melt is lost from the equilibration volume before significant cooling (point A), the rock will encounter the quadrivariant subsolidus field {Kfs + Pl + Bt + Crd + Grt + Qtz} with no resorption of garnet. The subvertical garnet molar proportion isopleths (Fig. 9) indicate that subsequent cooling will have little or no effect on garnet molar proportions, particularly as diffusion will be greatly decreased since most of the melt is lost from the equilibration volume. If some melt remains trapped within the equilibration volume, cooling results in garnet resorption, i.e. garnet molar proportion isopleths are sub-horizontal within the melt-bearing field (Fig. 9). Point B shows a rock in which 60% of the 12 mol. % melt is lost. The remaining 40% (about 5 mol. %) reacts with garnet on cooling by a reversal of reaction 10. As the last melt crystallizes, 2-3 mol. % garnet remains. Garnet in such rocks would be expected to exhibit extensive reaction coronas (containing biotite and/or cordierite) and/or pronounced corrosion. If >60% melt remains (point C), complete garnet resorption is predicted (Fig. 9).

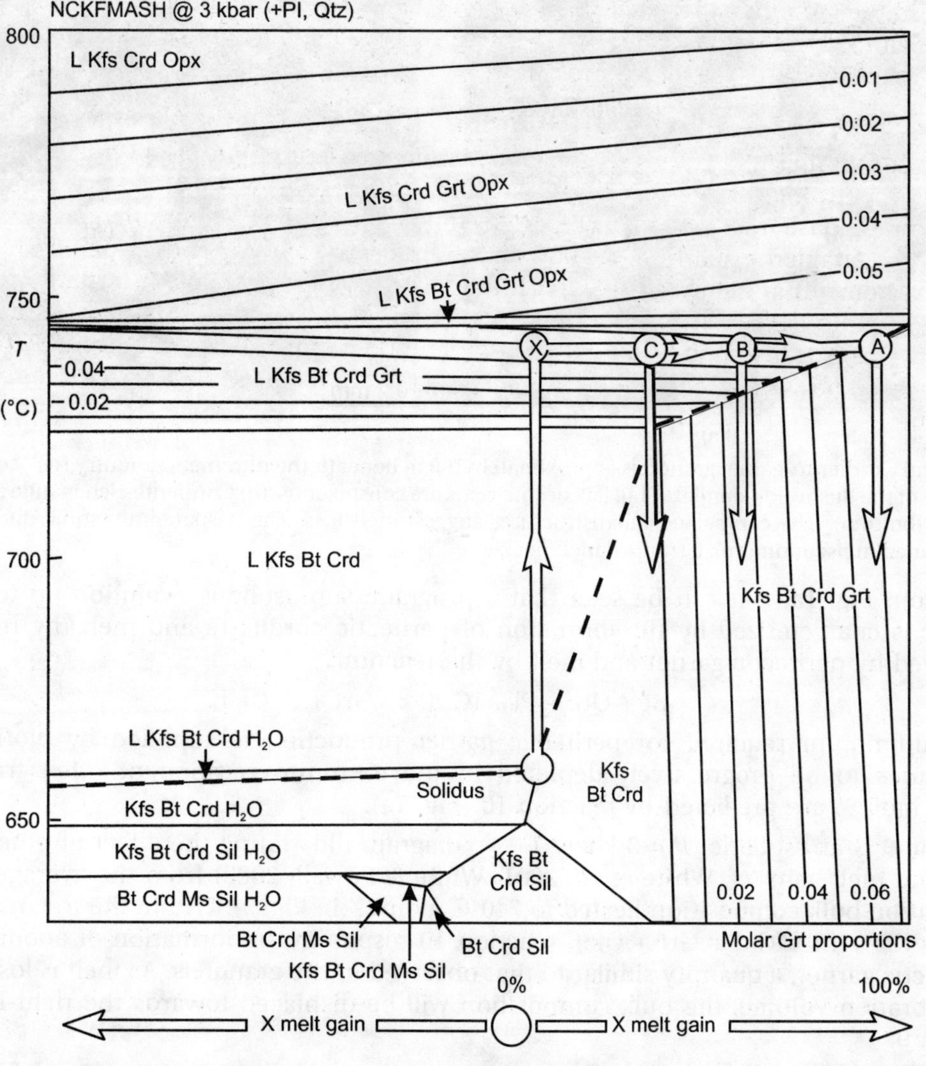

Fig. 9. T-X_{melt} pseudosection calculated for the mean composition of Silverton Formation metapelites. The garnet-bearing fields are contoured for molar proportions of garnet. The diagram considers an average regional metapelite heated to 740°C (up-T white arrow) at which point it contains about 12 mol. % melt. A subsequent evolution to the right involves variable degrees (0-100%) of melt loss; to the left, the effect of melt gain is shown. The normalized bulk composition of the left and right hand edges of the diagram in terms of the components H_2O:SiO_2:Al_2O_3:CaO:MgO:FeO:K_2O:Na_2O are 14.91:70.22:6.98:0.19:0.27:1.20:3.62:2.61 and 2.92:63.77:14.39:0.60:6.34:9.19:1.86:0.93 respectively. The darkest fields are quinivariant.

Although a limited degree of retrogression is evident in thin section, the granofels contains garnets with perfect crystal faces and exhibits texturally equilibrated microstructures that suggest retrogression was minimal. X-ray composition maps show that garnets of all sizes are unzoned, apart from at their extreme edges, consistent with this conclusion. Garnet should have been strongly resorbed if a significant amount of melt remained trapped within the equilibration volume (Fig. 9). The evidence suggests that the modal quantities of garnet present in the granofels approximates that predicted at the metamorphic peak and that most of the melt (>90%) was removed. Microstructures preserving evidence for extreme melt loss (Fig. 8) and the scarcity of leucosome in these rocks attest to the efficacy of this process.

DISCUSSION AND CONCLUSIONS

The continuing refinement of thermodynamic data and appropriate mixing models for phases of petrological interest allows a quantitative or semi-quantitative treatment of equilibrium phase relations. The incorporation of additional components into complex phases has enabled expansion of the commonly used KFMASH model pelite system into larger systems that more closely resemble natural rocks. Using internally consistent thermodynamic data sets to model phase relations with e.g. THERMOCALC is a powerful tool enabling extrapolation of limited experimental data into semi-continuous P-T-X space. The suprasolidus NCKFMASH model system is limited in that it cannot account for additional components, particularly Ti and Fe^{3+}, which are known be an important control on the stability of e.g. biotite and spinel; small quantities of these components also partition into melt. However, accepting these caveats, an equilibrium pseudosection approach to petrological problems commonly yields results that are consistent with field and petrographic constraints on natural rocks.

High aH_2O anatexis gives rise to low volume, essentially immobile melts and there may be only limited macroscopic evidence for melting in the rocks. Volatile-fluxed melting is likely to be rare in nature given the low porosity of high-grade metamorphic rocks, although granite crystallizing within the system may locally supply an H_2O-rich volatile phase to drive retrogression. At higher T, around the amphibolite to granulite facies transition, evidence for high aH_2O melting will be obliterated. Larger volumes of low aH_2O melt may potentially segregate and migrate out of the system, leaving behind desiccated, dense residua. Deformation is a key driving force for melt segregation, although ultimately melt will tend to move upward under buoyancy; thus, there is a feedback relation between melting, deformation and melt segregation, ascent and emplacement (e.g. Brown & Solar, 1999; Brown, 2001b). If significant volumes of melt stagnate within the system retrogression will result. It is intriguing to speculate that limited occurrence of UHT granulites within common granulite terrains may reflect retrogression of formerly more extensive UHT mineral assemblages facilitated by retention of some melt within the system (e.g. Moraes *et al.*, 2002), although we acknowledge the parallel issue of the rarity of the high-Mg bulk compositions necessary to record a mineralogical imprint of UHT conditions (e.g. Kelsey *et al.*, 2003).

REFERENCES

Acosta-Vigil, A., Pereira, M.D., Shaw, D.M. & London, D., 2001. Contrasting behaviour of boron during crustal anatexis. *Lithos*, **56**, 15-31.

Ashworth, J.R., 1985. Introduction. In: Ashworth, J.R. (ed.) *Migmatites*. Glasgow: Blackie, 1-35.

Ashworth, J.R. & McLennan, E.L., 1985. Textures. In: Ashworth, J.R. (ed.) *Migmatites*. Glasgow: Blackie, 180-203.

Atherton, M.P. & Brotherton, M.S., 1982. Major element composition of the pelites of the Scottish Dalradian. *Geological Journal*, **17**, 185-221.

Ayres, M. & Vance, D., 1997. A comparative study of diffusion profiles in Himalayan and Dalradian garnets; constraints on diffusion data and the relative duration of the metamorphic events. *Contributions to Mineralogy and Petrology*, **128**, 66-80.

Brown, M., 1973. The definition of metatexis, diatexis and migmatites. *Proceedings of the Geologists' Association*, **84**, 371-82.

Brown, M., 1994a. Melt segregation mechanism controls on the geochemistry of crustal melts. *V.M. Goldschmidt Conference, Mineralogical Magazine*, **58A**, 124-25.

Brown, M., 1994b. The generation, segregation, ascent and emplacement of granite magma: The migmatite-to-crustally-derived granite connection in thickened orogens. *Earth-Science Reviews*, **36**, 83-130.

Brown, M., 2001a. From microscope to mountain belt: 150 years of petrology and its contribution to understanding geodynamics, particularly the tectonics of orogens. *Journal of Geodynamics*, **32**, 115-64.

Brown, M., 2001b. Orogeny, migmatites and leucogranites: a review. *Proceedings of the Indian Academy of Sciences (Earth and Planetary Science)*, **110**, 313-36.

Brown, M., 2002. Retrograde processes in migmatites and granulites revisited. *Journal of Metamorphic Geology*, **20**, 25-40.

Brown, M., 2003. Hot orogens, tectonic switching, and creation of continental crust: Comment and Reply. *Geology: On-Line Forum.* June 2003, DOI 10.1130/0091-7613(2003)31<e9:HOTSAC>2.0.CO;2.

Brown, M. & Dallmeyer, D., 1996. Rapid Variscan exhumation and the role of magma in core complex formation: southern Brittany metamorphic belt, France. *Journal of Metamorphic Geology*, **14**, 361-79.

Brown, M. & Raith, M., 1996. First evidence of ultrahigh-temperature decompression from the granulite province of Southern India. *Journal of the Geological Society, London*, **153**, 819-22.

Brown, M. & Solar, G.S., 1999. The mechanism of ascent and emplacement of granite magma during transpression: a syntectonic granite paradigm. *Tectonophysics*, **312**, 1-33.

Cann, J.R., 1970. Upward movement of granitic magma. *Geological Magazine*, **107**, 335-40.

Cenki, B., Kriegsman, L.M. & Braun, I., 2002. Melt-producing and melt-consuming reactions in the Achankovil cordierite gneisses, South India. *Journal of Metamorphic Geology*, **20**, 543-61.

Clemens, J.D. & Droop, G.T.R., 1998. Fluids, P-T paths and the fates of anatectic melts in the Earth's crust. *Lithos*, **44**, 21-36.

Dalrymple, D.J., 1995. *Contact anatexis of Dalradian metapelites from the Huntly-Knock area, Aberseenshire, N.E. Scotland.* Unpublished PhD Thesis, University of Manchester.

Galán, G. & Marcos, A., 2000. The metamorphic evolution of the high pressure mafic granulites of the Bacariza formation (Cabo Ortegal Complex, Hercynian Belt, NW Spain). *Lithos*, **54**, 139-71.

Harley, S.L., 1998. On the occurrence and characterization of ultrahigh-temperature crustal metamorphism. In: Treloar, P.J., O'Brien, P.J. (Eds.), *What drives metamorphism and metamorphic reactions?* Geological Society, Special Publication, **138**, 81-107.

Harley, S.L. & Motoyoshi, Y., 2000. Al zoning in orthopyroxene in sapphirine quartzite: evidence for > 1120°C UHT metamorphism in the Napier Complex, Antarctica, and implications for the entropy of sapphirine. *Contributions to Mineralogy and Petrology*, **138**, 293-307.

Holdaway, M.J., Guidotti, C.V., Novak, J.M. & Henry, W.E., 1982. Polymetamorphism in medium- to high-grade pelitic metamorphic rocks, west-central Maine. *Geological Society of America Bulletin*, **93**, 572-84.

Holland, T.J.B. & Powell, R., 1998. An internally consistent thermodynamic data set for phases of petrological interest. *Journal of Metamorphic Geology*, **16**, 309-43.

Holtz, F. & Johannes, W., 1994. Maximum and minimum water contents of granitic melts; implications for chemical and physical properties of ascending magmas. *Lithos*, **32**, 149-59.

Johnson, T. & Brown, M. (in review). Quantitative Constraints on Metamorphism in the Variscides of Southern Brittany — a Complementary Pseudosection Approach. *Journal of Petrology.*

Johnson, S. & Vernon, R.H., 1995. Stepping stones and pitfalls in the determination of an anticlockwise *P-T-t* deformation path: The low-*P*, high-*T* Cooma Complex, Australia. *Journal of Metamorphic Geology*, **13**, 165-83.

Johnson, T.E., Hudson, N.F.C. & Droop, G.T.R., 2001. Partial melting in the Inzie Head gneisses: the role of water and a petrogenetic grid in KFMASH applicable to anatectic migmatites. *Journal of Metamorphic Geology*, **19**, 99-118.

Johnson, T.E., Brown, M. & Solar, G.S., 2003a. Low-pressure subsolidus and suprasolidus phase equilibria in the MnNCKFMASH system: Constraints on conditions of regional metamorphism in western Maine, northern Appalachians. *American Mineralogist*, **88**, 624-38.

Johnson, T.E., Gibson, R.L., Brown, M. & Buick, I.S., 2003b. Partial melting of metapelitic rocks beneath the Bushveld Complex, South Africa. *Journal of Petrology*, **44**, 789-813.

Jones, K.A. & Brown, M., 1990. High-temperature 'clockwise' *P-T* paths and melting in the development of regional migmatites: an example from southern Brittany, France. *Journal of Metamorphic Geology*, **8**, 551-78.

Kelsey, D.E., White, R.W. & Powell, R., 2003. Orthopyroxene-sillimanite-quartz assemblages: distribution, petrology, quantitative *P-T-X* constraints and *P-T* paths. *Journal of Metamorphic Geology*, **21**, 439-53.

Kohn, M.J., Spear, F.S. & Valley, J.W., 1997. Dehydration-melting and fluid recycling during metamorphism: Rangeley Formation, New Hampshire, USA. *Journal of Petrology*, **38**, 1255-77.

Kretz, R.M., 1983. Symbols for rock forming minerals. *American Mineralogist*, **68**, 277-79.

Kriegsman, L.M., 2001. Partial melting, partial melt extraction and partial back reaction in anatectic migmatites. *Lithos*, **56**, 75-96.

Kriegsman, L.M. & Hensen, B.J., 1998. Back reaction between restite and melt; implications for geothermobarometry and pressure-temperature paths. *Geology*, **26**, 1111-14.

Laporte, D., Rapaille, C. & Provost, A., 1997. Wetting angles, equilibrium melt geometry, and the permeability threshold of partially molten crustal protoliths. In: Bouchez, J.L., Hutton, D. & Stephens, W.E. (Eds.), *Granite: From melt segregation to emplacement fabrics*, 31-54. Kluwer Academic Publishers, Dordrecht, The Netherlands.

McGregor, V.R. & Friend, C.R.L., 1997. Field recognition of rocks totally retrogressed from granulite facies: An example from Archean rocks in the Paamiut region, south-west Greenland. *Precambrian Research*, **86**, 59-70.

Mahar, E.M., Baker, J.M., Powell, R., Holland, T.J.B. & Howell, N., 1997. The effect of Mn on mineral stability in metapelites. *Journal of Metamorphic Geology*, **15**, 223-38.

Marchildon, N. & Brown, M., 2002. Grain-scale melt distribution in two contact aureole rocks: implications for controls on melt localization and deformation. *Journal of Metamorphic Geology*, **20**, 381-96.

McLelland, J., Morrison, J., Selleck, B., Cunningham, B., Olson, C. & Schmidt, K., 2002. Hydrothermal alteration of late- to post-tectonic Lyon Mountain Granitic Gneiss, Adirondack Mountains, New York: Origin of quartz-sillimanite segregations, quartz-albite lithologies, and associated Kiruna-type low-Ti Fe-oxide deposits. *Journal of Metamorphic Geology*, **20**, 175-90.

Mehnert, K.R., 1968. *Migmatites and the origin of granitic rocks*. Amsterdam: Elsevier.

Mezger, K., Bohlen, S.R. & Hanson, G.N., 1990. Metamorphic history of the Archean Pikwitonei granulite domain and the Cross Lake Subprovince, Superior Province, Manitoba, Canada. *Journal of Petrology*, **31**, 483-517.

Moraes, R., Brown, M., Fuck, R.A., Camargo, M.A. & Lima, T.M., 2002. Characterization and P-T Evolution of Melt-bearing Ultrahigh-temperature Granulites: an Example from the Anápolis-Itauçu Complex of the Brasilia Fold Belt, Brazil. *Journal of Petrology*, **43**, 1673-1705.

Morand, V., 1990. Low-pressure regional metamorphism in the Omeo Metamorphic Complex, Victoria, Australia. *Journal of Metamorphic Geology*, **8**, 1-12.

Nabelek, P.I., 1997. Quartz-sillimanite leucosomes in high-grade schists, Black Hills, South Dakota: A perspective on the mobility of Al in high-grade metamorphic rocks. *Geology*, **25**, 995-98.

O'Brien, P.J., 2001. The fundamental Variscan problem: high-temperature metamorphism at different depths and high-pressure metamorphism at different temperatures. In: Franke, W., Haak, V., Onken, O. & Tanner, D. (Eds.), *Orogenic processes: quantification and modelling in the Variscan Belt*. Geological Society, Special Publication, **179**, 369-86.

O'Brien, P.J. & Rötzler, J., 2003. High-pressure granulites: Formation recovery of peak conditions and implications for tectonics. *Journal of Metamorphic Geology*, **21**, 3-20.

Pattison, D.R.M. & Tracy, R.J., 1991. Phase equilibria and thermobarometry of metapelites. In: *Reviews in Mineralogy* (ed. Kerreck, D.), **26**, 105-206.

Pattison, D., Chacko, T., Farquhar, J. & McFarlane, C.R.M., 2003. Temperatures of granulite-facies metamorphism based on experimental phase equilibria combined with garnet-orthopyroxene thermobarometry corrected for retrograde exchange. *Journal of Petrology*, **44**, 867-900.

Powell, R. & Holland, T.J.B., 1988. An internally consistent thermodynamic dataset with uncertainties and correlations. III. Application methods, worked examples and a computer program. *Journal of Metamorphic Geology*, **6**, 173-204.

Raith, M., Karmakar, S. & Brown, M., 1997. Ultrahigh-temperature metamorphism and multi-stage decompressional evolution of sapphirine granulites from the Palni Hill Ranges, Southern India. *Journal of Metamorphic Geology*, **15**, 379-99.

Rushmer, T., 2001. Volume change during partial melting reactions; implications for melt extraction, melt geochemistry and crustal. *Tectonophysics*, **342**, 389-405.

Sawyer, E.W., 1998. Formation and evolution of granite magmas during crustal reworking: the significance of diatexites. *Journal of Petrology*, **39**, 1147-67.

Sawyer, E.W., 2001. Melt segregation in the continental crust: Distribution and movement of melt in anatectic rocks. *Journal of Metamorphic Geology*, **19**, 291-309.

Schumacher, J.C., Hollacher, K.T., Robinson, P. & Tracy, R.J., 1990. Progressive reactions and melting in the Acadian metamorphic high of central Massachusetts in southwestern New Hampshire, USA. In: Ashworth, J.R. & Brown, M. (Eds.) *High-temperature Metamorphism and Crustal Anatexis*, Unwin Hyman, New York, 198-234.

Senior, A. & Leake, B.E., 1978. Regional metasomatism and the geochemistry of the Dalradian metasediments of Connemara, western Ireland. *Journal of Petrology*, **19**, 585-625.

Shaw, D.M., 1956. Geochemistry of pelitic rocks. Part III: Major elements and general geochemistry. *Bulletin of the Geological Society of America*, **67**, 919-34.

Solar, G.S. & Brown, M., 2001. Petrogenesis of migmatites in Maine, USA. Possible source of peraluminous granite in plutons. *Journal of Petrology*, **42**, 789-823.

Spear, F.S., Kohn, M.J. & Cheney, J.T., 1999. P-T paths from anatectic pelites. *Contributions to Mineralogy and Petrology*, **134**, 17-32.

Stüwe, K., 1997. Effective bulk composition changes due to cooling; a model predicting complexities in retrograde reaction textures. *Contributions to Mineralogy and Petrology*, **129**, 43-52.

Thompson, A.B., 2001. Clockwise P-T paths for crustal melting and H_2O recycling in granite source regions and migmatite terrains. *Lithos*, **56**, 33-45.

Thompson, A.B. & Connolly, J.A.D., 1995. Melting of the continental crust; some thermal and petrological constraints on anatexis in continental collision zones and other tectonic settings. *Journal of Geophysical Research, B, Solid Earth and Planets*, **100**, 15, 565-79.

Tinkham, D.K., 2002. MnNCKFMASH metapelite phase equilibria, garnet activity modeling, and garnet Sm-Nd chronology: applications to the Waterville Formation, Maine and south-central Nason terrene, Washington. Unpublished PhD thesis, University of Alabama.

Vernon, R.H., 1979. Formation of late sillimanite by hydrogen metasomatism (base-leaching) in some high-grade gneisses. *Lithos*, **12**, 143-52.

Walraven, F., Armstrong, R.A. & Kruger, F.J., 1990. A chronostratigraphic framework for the north-central Kaapvaal Craton, the Bushveld Complex and Vredefort structure. *Tectonophysics*, **171**, 23-48.

Waters, D.J., 1988. Partial melting and the formation of granulite facies assemblages in Namaqualand, South Africa. *Journal of Metamorphic Geology*, **6**, 387-404.

Waters, D.J. & Lovegrove, D.P., 2002. Assessing the extent of disequilibrium and overstepping of prograde metamorphic reactions in metapelites from the Bushveld aureole. *Journal of Metamorphic Geology*, **20**, 135-49.

Westphal, M., Schumacher, J.C. & Boschert, S., 2003. High-temperature metamorphism and the role of magmatic heat sources at the Rogaland anorthosite complex in Southwestern Norway. *Journal of Petrology*, **44(6)**, 1145-62.

White, R.W. & Powell, R., 2002. Melt loss and the preservation of granulite facies mineral assemblages. *Journal of Metamorphic Geology*, **20**, 621-32.

White, R.W., Powell, R. & Holland, T.J.B., 2001. Calculation of partial melting equilibria in the system Na_2O-CaO-K_2O-FeO-MgO-Al_2O_3-SiO_2-H_2O (NCKFMASH). *Journal of Metamorphic Geology*, **19**, 139-53.

Wintsch, R.P., 1975. Solid-fluid equilibria in the system $KAlSi_3O_8$-$NaAlSi_3O_8$-Al_2SiO_5-H_2O-HCl. *Journal of Petrology*, **16**, 57-79.

2

EVOLUTION OF EARLY PROTEROZOIC METAMORPHISM WITHIN TIM-YASTREBOVSKAYA PALEORIFT, VORONEZH CRYSTALLINE MASSIF, EAST-EUROPEAN PLATFORM: METAPELITE SYSTEMATICS, PHASE EQUILIBRIUM, AND P-T CONDITIONS

TATYANA N. POLYAKOVA*, KONSTANTIN A. SAVKO*, VYACHESLAV YU. SKRYABIN*

ABSTRACT

The results of investigation of phase relations and mineral compositions as well as of the metamorphic conditions of the different types of metapelites in Tim-Yastrebovskaya rift structure have been presented. Mineral assemblages corresponding to greenschist facies where chlorite-potassium feldspar and biotite-muscovite grades of biotite subfacies are identified. The staurolite-chlorite-muscovite and andalusite biotite subfacies are recognized within staurolite facies at medium grade metamorphism. The cordierite-biotite grade rocks occur in the high temperature part of staurolite facies. The garnet of the rocks can be subdivided into two generations: early kinematic with respect to the D1 stage and related to the folding and late kinematic which occurs near intrusive bodies. The greenschist facies metapelites was related to regional metamorphism; the staurolite facies mineral assemblages developed locally and are spatially restricted to compositionally variable intrusive bodies. The greenschist facies temperatures of metamorphism are 220-250°C for biotite subfacies, 300-500°C for garnet subfacies and 500-580°C for staurolite facies. The Tim-Yastrebovskaya structure was affected by two metamorphic episodes: early regional metamorphism in the greenschist facies conditions at 2250-2086 Ma and younger more high-temperature contact metamorphism at 2086±5 Ma. The latter event produced the metamorphic zoning observable within the structure.

KEY WORDS

Metapelite; Phase equilibrium; P-T conditions; Paleorift; Tim-Yastrebovskaya and Evolution.

INTRODUCTION

The advances achieved over the past two decades in precise techniques for the assaying the physicochemical conditions of lithospheric mineralizing processes made it possible to model geodynamic environments in correlations with various types of

* Department of Geology, Voronezh State University, Voronezh, 394006, Russia.
Corresponding author. E-mail: gfkig304@main .vsu.ru

physicochemical metamorphic conditions (De Yoreo *et al.*, 1991; England and Thompson, 1984 and others).

Metamorphic zoning is commonly interpreted as reflecting either the spatial distribution of local heat fluxes (thermal dome model) or the regional P-T gradients during the subsidence of crustal blocks (gradient model), with the temperature of the latter model depending exclusively on the burial depth of the rocks. However, intruded magmatic bodies (contact regional metamorphism) can also induce metamorphic zoning.

Detailed data obtained on the metapelites of the Tim-Yastrebovskaya structure demonstrate that its metamorphic zoning was caused by the effect of intrusions of various compositions.

GEOLOGY

Within the basement structure of the East European Platform, the Voronezh crystalline massif (VCM) is a large shallow-buried uplift, 600 × 700 km size. It is limited in the north-east and southwest by Pachelmsky and Dniepr-Donetsky aulacogens respectively; in the southeast by Precaspian depression and in the northwest — by Prypyat depression. In the structure of Precambrian VCM basement two blocks are recognized: the eastern one composed Early Proterozoic formations, and the western block in which the rocks of Archean foot prevails. Within the western block Early Proterozoic deposits occur only in the narrow linear zones from southeast to northwest direction. They are intracontinental rifts, formed in Early Proterozoic on the Late Archean protoplatform and transformed later to the syncline structures as a result of folding and block movement.

One of the largest synforms of that type is Tim-Yastrebovskaya paleorift that is situated in the central part of VCM and extends from the northwest to the southeast 110 km length, its width being up to 30 km (Fig. 1). At the present time it is a zonal metamorphic complex composed from Early Proterozoic iron-rich rocks of banded iron formation and terrigenous-carbonaceous deposits of the ancient post-Archean protoplatform sedimentary cover and volcanoceous rocks of the proper riftogenesis stage which are folded and intruded of granite and granodiorite-diorite massifs.

The Tim-Yastrebovskaya structure includes the Early Proterozoic Kurskaya and Oskol'skaya groups, which are extensively folded and metamorphosed. All of the stratified rocks are characterized by a systematic transitions from terrigenous-sedimentary to terrigenous-carbonate and terrigenous-volcanic rocks with increasing concentrations of carbonaceous matter and volcanic material.

The rocks of the Kurskaya Group occur in the southwestern flank of the Tim-Yastrebovskaya structure (Fig. 2) and are dominated by quartz-mica schists and jaspilites. The Oskol'skaya Group is characterized by cyclical intercalations of conglomerates, gritstones, and metasandstones, which upward grade into carbonate-micaceous schists with bands of amphibolites and metamorphosed limestones and dolomites and abundant rocks with carbonaceous matter and volcanics. The group consists of two conformable formations: the lower Rogovskaya terrigenous-carbonate and the upper Timskaya terrigenous-volcanic formations.

The rocks of the Rogovskaya Formation are exposed in the southwestern flank and the southern structural closure of the Tim-Yastrebovskaya structure (Fig. 2). These are quartz-biotite, quartz-muscovite, quartz-garnet-biotite, and quartz-biotite-staurolite schists with thin beds of carbonate rocks, metasandstones, and amphibolites.

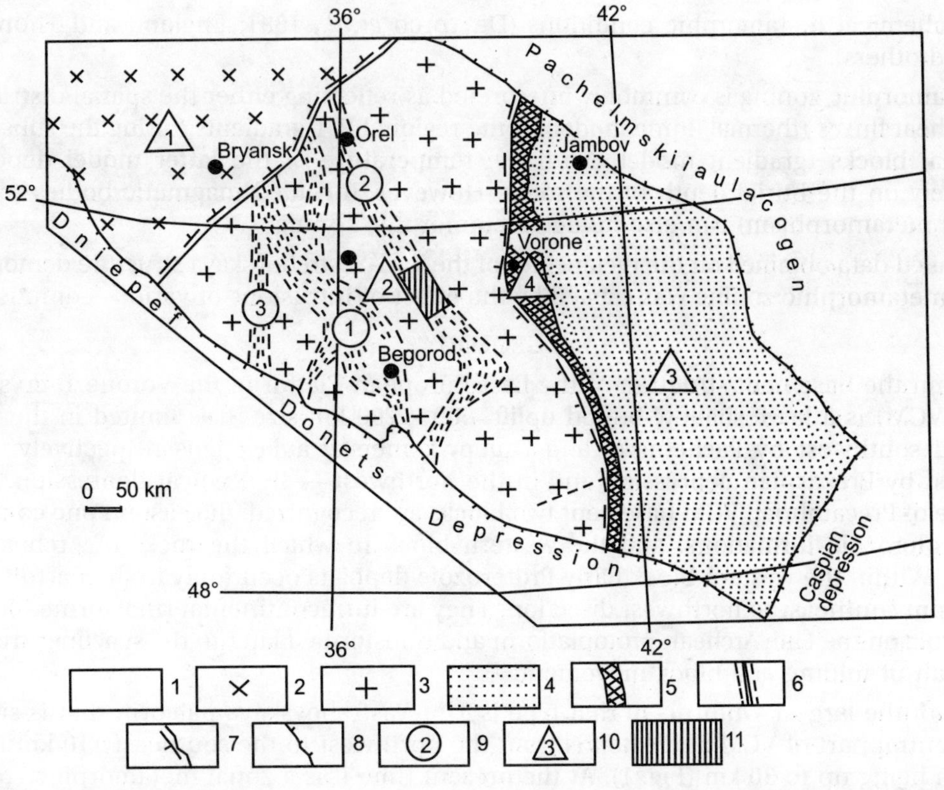

Fig. 1. Schematic structural-geological map of the Voronezh crystalline massif (modified after Chernyshov *et al.*, 1997) (1-6) Tectono-stratigraphic complexes: (1) Riphean and Phanerozoic platform complexes, (2-6) pre-Riphean complexes (2 and 3-of the Bryansk and Kursk megablocks of the Early Archean consolidation, 3-Voronstovskii pericraton trough, 5-Livny-Bogucharskaya suture, 6-second-order Riphean structures); (7-8) geologic boundaries; (7) between megablocks, (8) of the Voronezh crystalline massif; (9) second-order riftogenic structures (circled numbers: 1-Belgorod-Mikhailovskaya, 2-Orel-Timskaya, 3-Krupetsko-Krivoi Rog); (10) first-order structures (numbers in triangles: 1-Bryansk megablock, 2-Kursk megablock, 3-Voronezh megablock, (4) Livny-Bogucharskaya suture); (11) the Tim-Yastrebovskaya structure (study area).

Widespread carbonaceous sulfidized schists with subordinate amounts of beds and strata of metasandstones, quartzites, calc-silicate rocks, marbles, and volcanic rocks characterize the Timskaya Formation, which makes up the central part of the structure.

The rocks of the Oskol'skaya Group were dated by U-Pb techniques on zircon from acid volcanics and syngenetic metapelites (Artemenko *et al.*, 1992). The Timskaya metadacites were dated in the northern portion of the structure. Dykes of quartz metaporphyries and two-mica granite porphyries were studied in the drill core from the eastern part of the Tim-Yastrebovskaya structure. The U-Pb isochron plots with concordia date the volcanic event at 2167 ± 288 Ma for the metadacite and 2254 Ma for the quartz metaporphyries. The relatively high errors of the former isochron date are explained by the small spread of the U-Pb isotopic ratios for distinct zircon fractions from this sample. Hence, the isotopic age of the sedimentary-volcanic rocks of the Oskol'skaya Group is 2250 Ma.

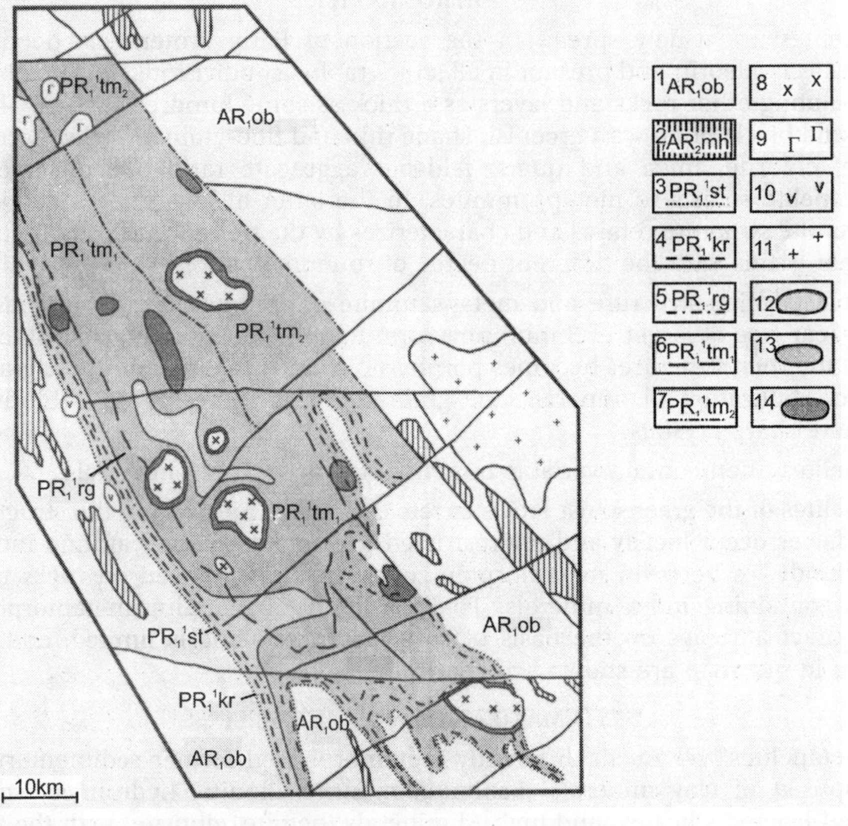

Fig. 2. Schematic geological map of the Tim-Yastrebovskaya structure and metamorphic zoning in its metapelites. 1 – Oboyanskaya Group, 2 – Mikhailovskaya Group, 3-7 – Geologic boundaries: 3-4 – Kursk Group: 3 – Stoilenskaya Formation, 4 – Korobkovskaya Formation, 5-7 – Oskol Group: 5 – Rogovskaya Formation, 6 – Timskaya Formation, lower subformation, 7 – Timskaya Formation, upper subformation, 8-11 – intrusive complexes: 8-Stoilo-Nikolaevskii, 9 – Zolotukhinskii, 10 – Oskoletskii, 11 – Atamanskii, 12-14 – Metamorphic facies: 12 – Greenschist facies, biotite subfacies, 13 – Greenschist facies, garnet subfacies, 14 – Staurolite subfacies.

The igneous rocks of the structure belong to the premetamorphic Atamanskii Complex of moderately alkaline granites (plagioclase-microcline biotite granites) and the Oskoletskii Complex of migmatitic plagiogranites (plagiogranite, granodiorite, adamellite, and granite) and to the postmetamorphic Stoilo-Nikolaevskii dyke complex of the gabbro-diorite-granodiorite association (granodiorite, diorite, quartz diorite) and the Zolotukhinskii peridotite-gabbronorite complex (Fig. 2) (gabbronorite, gabbro, diorite, peridotite, and serpentinite).

The intrusive rocks of the Stoilo-Nikolaevskii Complex were isotopically dated on biotite-hornblende mesodiorite and granodiorite (Artemenko *et al.*, 1992). Because the U-Pb isotopic ratios of zircon from these samples were characterized by very insignificant distinctions, a single concordia plot was calculated for both of them, which determined an isotopic age of 2086 ± 5 Ma.

PETROGRAPHY

Metapelites are widely spread in the section of Early Proterozoic deposits of Tim-Yastrebovskaya paleorift and present in all stratigraphic subdivisions, forming both separate interbeds among other rocks and layers as a thick as some hundred meters. They are grey, dark grey and black, often with greenish shade thin and fine-grained rocks. According to the grain size, chlorite, mica and quartz-feldspar aggregate ratio one observes transition metaaleuropelite schists to metapsammites. In the latter at low-grade metamorphism the relict psammite structure retains and characterizes by the presence of primary clastic quartz and feldspar grains with the different degree of roundness.

The schists microstructure and metapsammite groundmass are lepidogranoblastic. In case of appearance of larger (2-3 mm) new forming minerals (garnet, staurolite, andalusite, cordierite) the rock structures becomes porphyroblastic. If the metamorphic grade increases due to recrystallization of minerals, the grain size also increases and chlorite and mica acquire more sharp crystals.

Metapelites often contain variable quantity of carbonaceous materials.

Metapelites of the greenschist facies covers the largest area within the structure, and the staurolite facies occur locally and are restricted to contact aureoles around intrusions (Fig. 2). The boundaries between metamorphic zones are drawn based on rocks recovered by holes and containing index minerals. The possibility of mapping metamorphic zones in narrow contact aureoles on the basis of core materials alone is limited, and, thus, some boundaries in our map are shown schematically.

SYSTEMATIZATION OF METAPELITES

By "metapelites" we mean differently metamorphosed former sedimentary rocks, that were composed of clay minerals (kaolinite, montmorillonite, hydromica, smectites and other mixed-layered silicates) and high-Al minerals (bemite, gibbsite) with the admixture of terrigenous quartz and feldspar. The boundary between epigenesis and metamorphism of such rocks is considered to be the moment of complete physico-chemical equilibrium, that is accompanied by disappearance of metastable assemblages, equilibrium component distribution among the mineral phases and ordering of the layered silicates structures (Dobretsov *et al.,* 1970; Korikovskii, 1979). During low-grade metamorphism muscovite-phengitic mica, paragonite, chlorite, stilpnomelane, pyrophyllite and diaspore are formed, terrigenous plagioclase being replaced by albite. Characteristic feature of low-grade metapelites is the presence of small amount of calcite, ankerite-dolomite and magnesite-siderite carbonates. This fact may probably be explained by rather high CO_2/H_2O ratio in fluid phase at low-grade metamorphism. The calcite and dolomite amount in metapelites should be limited by the ratio $Al_2O_3/(CaO + Na_2O + K_2O) \geq 1$ in them, that would provide complete transition of calcium into plagioclase when the metamorphic grade increases. With increase in the metamorphic grade ankerite-dolomite and magnesite-siderite carbonates in metapelites disappear from parageneses with quartz and result in producing amphiboles cummingtonite-grunerite series; with potassium feldspar in the biotite formation and with diaspore in the spinel formation.

Thus assuming that the only calcium-bearing phase in metapelites is plagioclase, and subtracting the portion of silica ($SiO_2 = 2CaO$) and alumina ($Al_2O_3 = CaO$) contained in it, the compositions of these rocks may be considered within calcium free system. This system becomes simpler when subtracting equivalent quantity of Fe_2O_3 from the sum FeO + MgO, contained in magnetite, that is commonly present in metapelites in small amount. Proceeding from this fact and taking into account maximum possible concentrations of carbonates and mineral phases that originate at the initial metamorphic stages it is possible to determine the metapelites composition area in the system (SiO_2-2CaO) – (Al_2O_3-CaO) – (FeO + MgO-Fe_2O_3) – (Na_2O + K_2O). It will be limited by assemblages including diaspore, spinel, chlorite, muscovite, paragonite, pyrophyllite, biotite, stilpnomelane, potassium feldspar, albite, amphiboles cummingtonite-grunerite series and quartz (Fig. 3). Compositions of paragonite-free metapelites may be considered in the sodium-free system subtracting equivalent amount Na_2O contained in plagioclase from silica and alumina (SiO_2-2CaO-6Na_2O) – (Al_2O_3-CaO-Na_2O) – (FeO + MgO-Fe_2O_3) – K_2O.

In both sodium-bearing and sodium-free systems the fields of parageneses Chl + Ms(Pg) + Prl and Chl + Bt + Ms(Pg) divide the areas of metapelites composition into rocks oversaturated with silica and ones oversaturated with alumina.

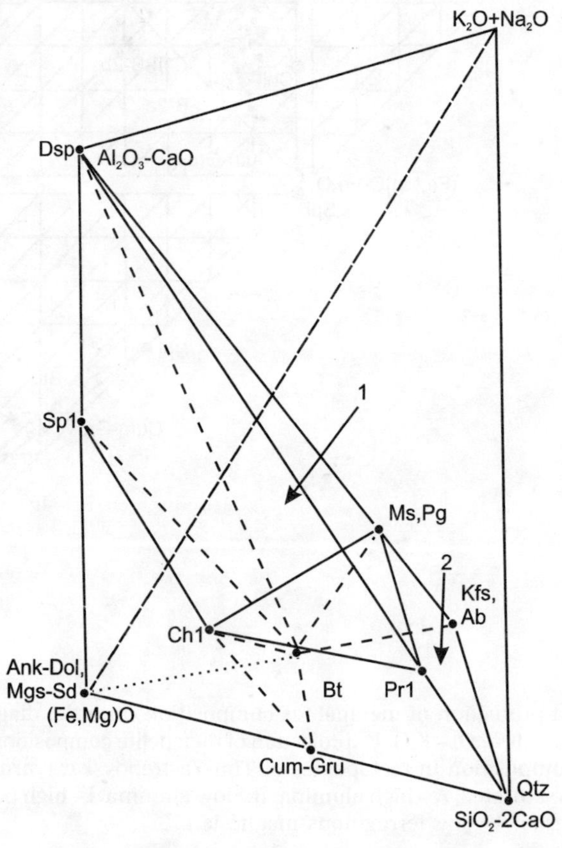

Fig. 3. Compositions of metapelites in the system: (Al_2O_3-CaO) – (Fe, Mg)O – (SiO_2-2CaO) – (K_2O+Na_2O). 1 – alumina oversaturated metapelites, 2 – silica oversaturated metapelites. Mineral symbols after Kretz (1983).

Metapelites of Tim-Yastrebovskaya structure do not contain paragonite; in the system (SiO_2-2CaO-6Na_2O) – (Al_2O_3-CaO-Na_2O) – (FeO + MgO-Fe_2O_3) – K_2O the points of their composition lie in the area of rocks oversaturated with silica (Fig. 4). This fact is confirmed by stable presence of quartz in them. In such metapelites SiO_2 is an excess component and they may be considered in the system (Al_2O_3-CaO-Na_2O) – (FeO + MgO-Fe_2O_3) – K_2O.

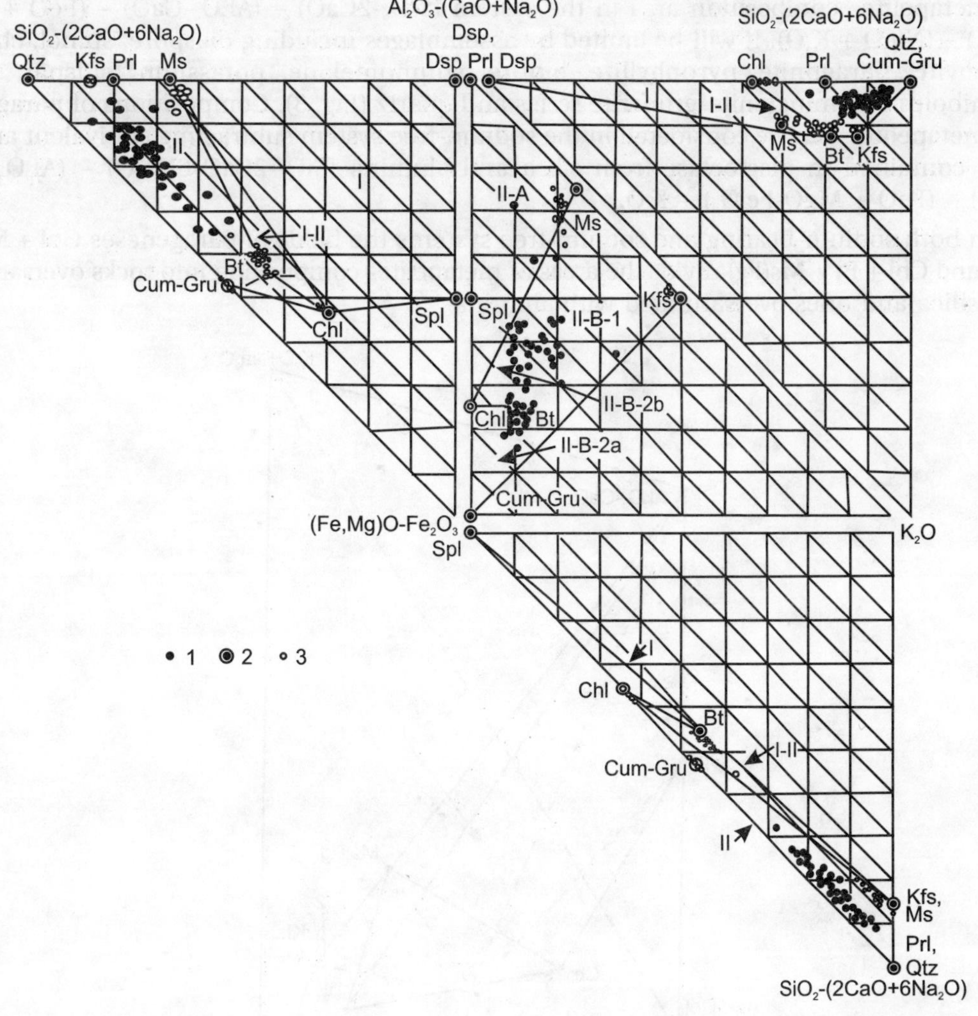

Fig. 4. Systematization and protection of metapelites composition onto the diagrams Al_2O_3-(CaO + Na_2O) – (Fe,Mg) O-Fe_2O_3 – SiO_2-(2CaO + 6Na_2O) – K_2O. 1 – projection of metapelite composition; 2 – theoretical composition of minerals; 3 – mineral composition in metapelites of Tim-Yastrebovskaya structure. Metapelites: I – Al_2O_3 oversaturated, II – SiO_2 oversaturated, A – high alumina, B – low alumina, 1 – high potassium, 2 – low potassium, a – high ferruginous-magnesian, b – low ferruginous-magnesian.

In this system conode Chl-Ms divides the oversaturated with silica rocks composition area into two fields, corresponding to high- and low-alumina metapelites (Fig. 5). In high alumina rocks at the low grade metamorphism pyrophyllite is present in addition to chlorite and muscovite. With the metamorphic grade increasing chloritoid, andalusite or kyanite form in them within greenschist facies. Low alumina metapelites in addition to muscovite and chlorite may contain potassium feldspar, stilpnomelane and iron-magnesian carbonates.

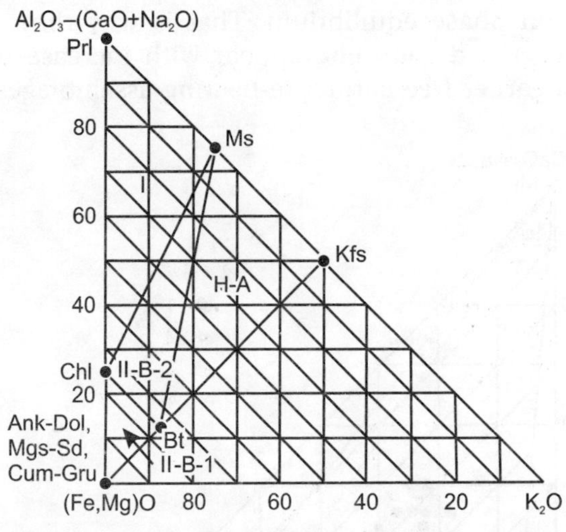

Fig. 5. Mineralogical-petrochemical systematization of SiO_2 oversaturated metapelites. I – High alumina, II – low alumina, A – high potassium, B – low potassium, 1 – high magnesian-ferruginous, 2 – low high magnesian-ferruginous.

To refer a rock to high or low alumina on the basis of chemical analysis the molecular ratio of $Al_2O_3/(FeO + MgO + K_2O)$ in it may be used. As Ms-Chl conode that in the system (Al_2O_3-CaO-Na_2O) – (FeO + MgO-Fe_2O_3) – K_2O divides the field of oversaturated with silica metapelites into high- and low-alumina ones is a straight-line, it is described by linear regression equation $y = a + bx$. To resolve this equation molecular ratios $Al_2O_3/(FeO + MgO + K_2O) = y$ and $K_2O/(FeO + MgO + K_2O) = x$ may be used as coordinates. In chlorite with the composition $(Fe,Mg)_9Al_3[Si_5Al_3O_{20}](OH)_6$ similar to the one observed in most metapelites the ratio $Al_2O_3/(FeO + MgO + K_2O) = 1/3$ when $K_2O/(FeO + MgO + K_2O) = 0$. In muscovite with theoretical composition $KAl_2[Si_3AlO_{10}](OH)_2$ at $K_2O/(FeO + MgO + K_2O) = 1$ the ratio $Al_2O_3/(FeO + MgO + K_2O) = 3$. Using the two points and solving linear regression equation, we obtain that along the conode Chl-Ms $Al_2O_3 = 1/3\ (FeO + MgO) + 3K_2O$. The ratio of molecular amount of Al_2O_3 in the rock to this expression may be considered as a alumina modulus (a). Its values on the conode will be equal to unity; for high alumina metapelites they will be more than unity and for low-alumina metapelites they will be less than unity. Taking into consideration the amount of Al_2O_3 equivalent to CaO and Na_2O contained in plagioclase and amount FeO equivalent to Fe_2O_3 contained in magnetite, the value of alumina modulus will be as follow: $a = 3(Al_2O_3\text{-}CaO\text{-}Na_2O)/(FeO\text{-}Fe_2O_3 + MgO + 9K_2O)$.

Low alumina metapelites are divided into high potassium and low potassium by the Bt-Ms conode. In high potassium (but no oversaturated) metapelites after reactions Stp + Kfs = Bt + Qtz + H_2O or Chl + Kfs = Bt + Ms + Qtz + H_2O excess potassium feldspar remains. Chlorite and muscovite remains in low potassium metapelites. Chl-Bt conode allows subdividing low potassium rocks into high- and low-magnesian-ferruginous ones. In high magnesian-ferruginous biotite and chlorite are stable with stilpnomelane and magnesian-ferruginous carbonates at the expense of which amphibole of cummingtonite-grunerite series with the metamorphic grade increasing. In low magnesian-ferruginous metapelites assemblage chlorite and biotite with muscovite is possible.

In both high- and low-magnesian-ferruginous metapelites the ratio of FeO and MgO in rocks has a great effect on phase equilibrium. That is why they should be divided into ferruginous in which garnet and staurolite appear with increase in the temperature and magnesian ones in which garnet-free cordierite-bearing assemblages may be stable (Fig. 6).

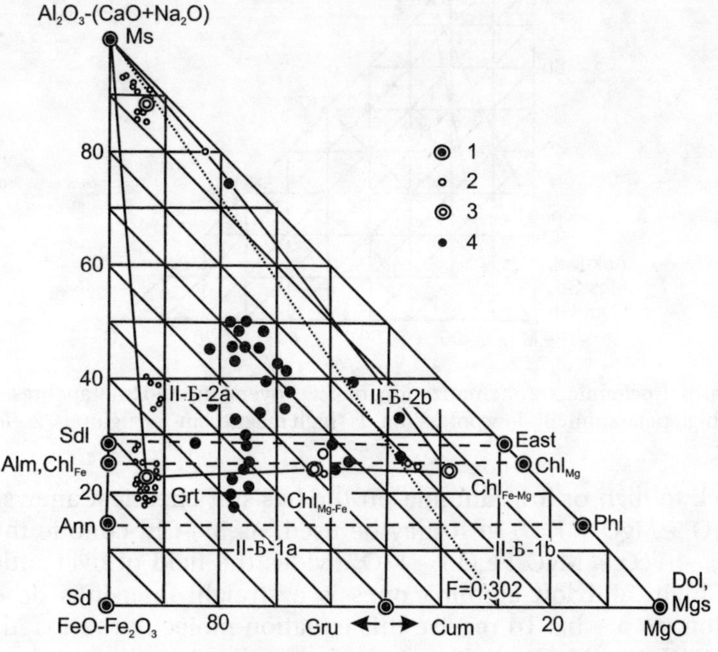

Fig. 6. Division of low potassium metapelites of Tim-Yastrebovskaya structure on the basis iron content. 1. Theoretical mineral composition, 2. Mineral composition from metapelite of Tim-Yastrebovskaya structure, 3. Average mineral composition from metapelites of Tim-Yastrebovskaya structure, 4. Low potassium metapelites of Tim-Yastrebovskaya structure (II- Б -1a – high ferruginous-magnesian ferruginous, II- Б -1b – high ferruginuous-magnesian, II- Б -2a – low ferruginous-magnesian ferruginous, II- Б -2b – ferruginuous-magnesian magnesian.

Thus, the supposed metapelite systematics based on mineral paragenesis changing with increase in the metamorphic grade is attributed to the hierarchy shown in Fig. 7.

Within Tim-Yastrebovskaya structure high alumina metapelites are represented by chlorite-muscovite and garnet-chlorite-muscovite schists only. They are most abundant in the foot of the section of Post-Archean sedimentary cover. This fact is due to erosion and redeposition of ancient weathering crusts. In proper rifting formations high-alumina rocks play subordinate role and here low-alumina high-potassium metapelites prevailed. In their formation, migmatized and granitized Archean rocks probably are the source of the wash out. Among low-potassium both high- and low-magnesian-ferruginous varieties magnesian metapelites are of primary importance that is indicative of participation of basic volcanism products of proper rift-stage in their formation.

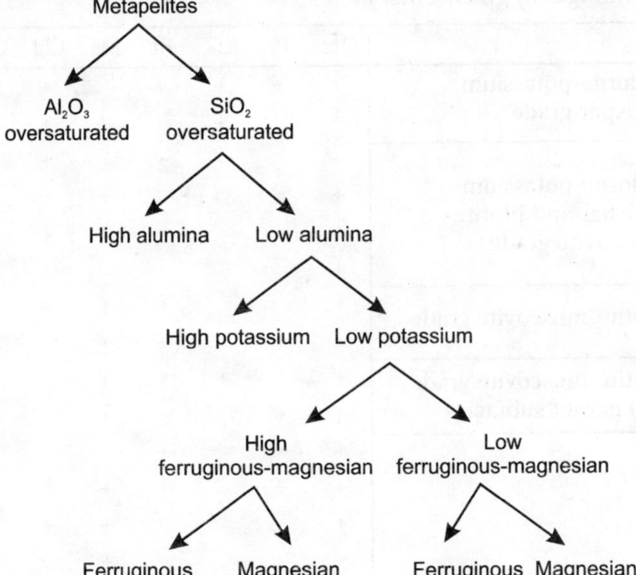

Fig. 7. Hierarchical relation of rock groups by mineralogical-petrochemical systematization of metapelites.

PHASE EQUILIBRIUM

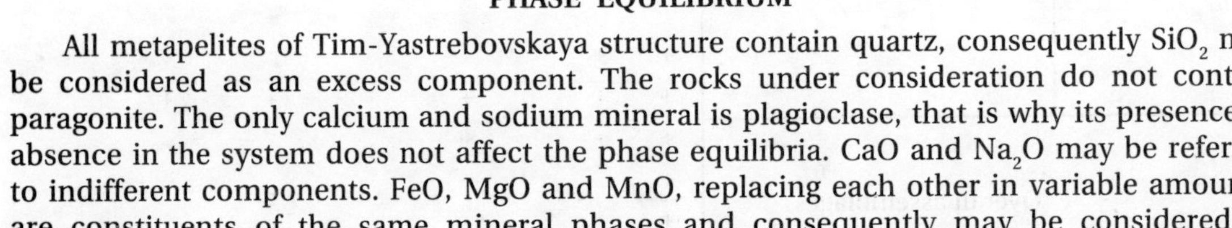

All metapelites of Tim-Yastrebovskaya structure contain quartz, consequently SiO_2 may be considered as an excess component. The rocks under consideration do not contain paragonite. The only calcium and sodium mineral is plagioclase, that is why its presence or absence in the system does not affect the phase equilibria. CaO and Na_2O may be referred to indifferent components. FeO, MgO and MnO, replacing each other in variable amounts, are constituents of the same mineral phases and consequently may be considered as isomorphic components. Concentration of TiO_2 and Fe_2O_3 in rocks is commonly small and they either scatter as isomorphic admixtures in rock-forming minerals or form accessory minerals that do not greatly affect phase equilibria in the system. Volatile compounds (water, carbon dioxide, fluor etc.) easily migrate by means of diffusion and infiltration and consequently are quite mobile components of metamorphic system (Kerrick, 1974; Ferry, 1991 and others). Because of this phase equilibria in metapelites of Tim-Yastrebovskaya structure should be analyzed in the system Al_2O_3-(Fe, Mn, Mg)O-K_2O.

GREENSCHIST FACIES

Biotite subfacies

Chlorite-potassium feldspar grade. The lowest grade assemblages in the studied low alumina low ferruginous-magnesian metapelites of Tim-Yastrebovskaya structure are Chl + Kfs + Bt and Chl + Kfs + Ms (Table 1). Presence of Bt + Kfs assemblage is indicative of metamorphic grade exceeding the temperature of biotite formation in the result of the reaction ferruginous-magnesian carbonates with potassium feldspar: Mgs-Sd (Ank-Dol) + Kfs + H_2O = Bt(±$CaCO_3$) + CO_2 (Fig. 8a). Stability of Chl + Kfs assemblage allows to estimate the lowest temperature of metamorphism as corresponding to the lower part of biotite

Table 1: *Mineral assemblages of greenschist facies*

		Qtz	Pl	Kfs	Bt	Ms	Chl	Grt	Cum-Gru	Gph
Biotite subfacies	Chlorite-potassium feldspar grade	+	+	+	+		+			
		+		+		+	+			+
	Chlorite-potassium feldspar and biotite-muscovite grade	+			+		+			+
		+				+	+			
		+	+			+	+			+
		+					+			+
	Biotite-muscovite grade	+			+	+	+			
		+	+	+	+	+		+(Sps)		+
Biotite and garnet subfacies	Biotite-muscovite grade and garnet subfacies	+		+	+	+				+
		+	+		+	+				
Garnet subfacies	–	+			+		+	+		
		+	+		+	+	+	+		
		+				+	+	+		+
		+			+	+		+		
		+	+		+			+		
		+				+		+		
		+	+	+	+			+		
		+					+	+		
		+	+		+				+	
		+	+		+			+	+	
Overall assemblages		+	+		+					
		+			+					
		+	+			+				
		+		+		+				+

subfacies of greenschist facies (chlorite-potassium feldspar grade). In some cases (boreholes 35. and 992) stability of Chl + Kfs assemblage coincides with that of interlayered metapelites of different composition of assemblage Bt + Ms is observed. Formation of the latter at this metamorphic grade is only possible in the rock that initially contained stilpnomelane and potassium feldspar besides muscovite. Appearance of biotite in this case is due to reaction Stp + Kfs = Bt + Qtz + H_2O.

Thus, the range of the lowest temperature conditions is limited on the one hand by disappearance of paragenesis Stp + Kfs and biotite appearance at their expense and by stability of Chl + Qtz assemblage on the other hand, that corresponds to the upper part of chlorite-potassium feldspar grade of biotite subfacies.

Biotite-muscovite grade. In more high-temperature rocks Chl + Kfs paragenesis disappears, being replaced by assemblage biotite and muscovite: Chl + Kfs = Bt + Ms + Qtz + H_2O (Fig. 8b). In the low alumina low ferruginous-magnesian metapelites assemblage either with excess potassium feldspar (Bt + Kfs + Ms) or with chlorite (Bt + Chl + Ms) appears. The

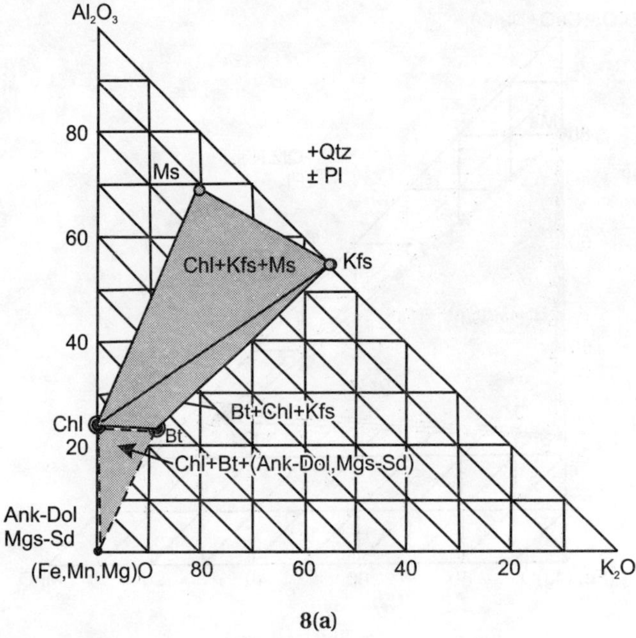

8(a)

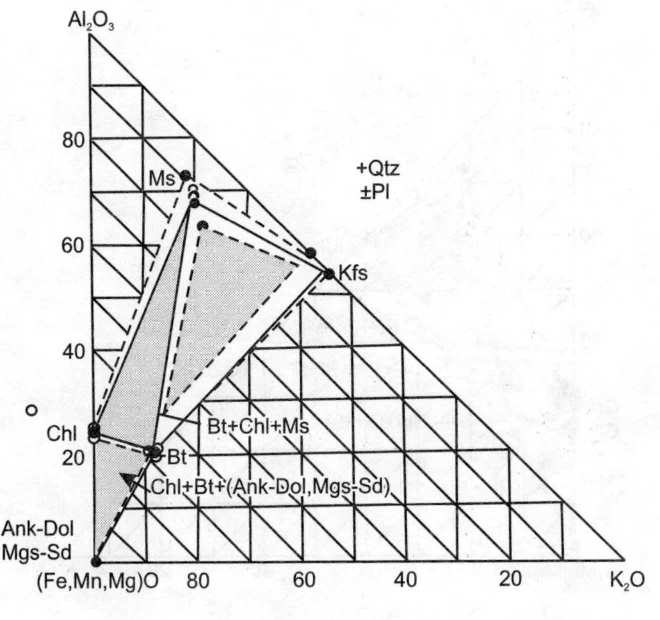

8(b)

Fig. 8. Phase relations in low alumina metapelites of greenschist facies of metamorphism within Tim-Yastrebovskaya structure. A-C – biotite subfacies: (a) chlorite-potassium feldspar grade; (b) biotite-muscovite grade.

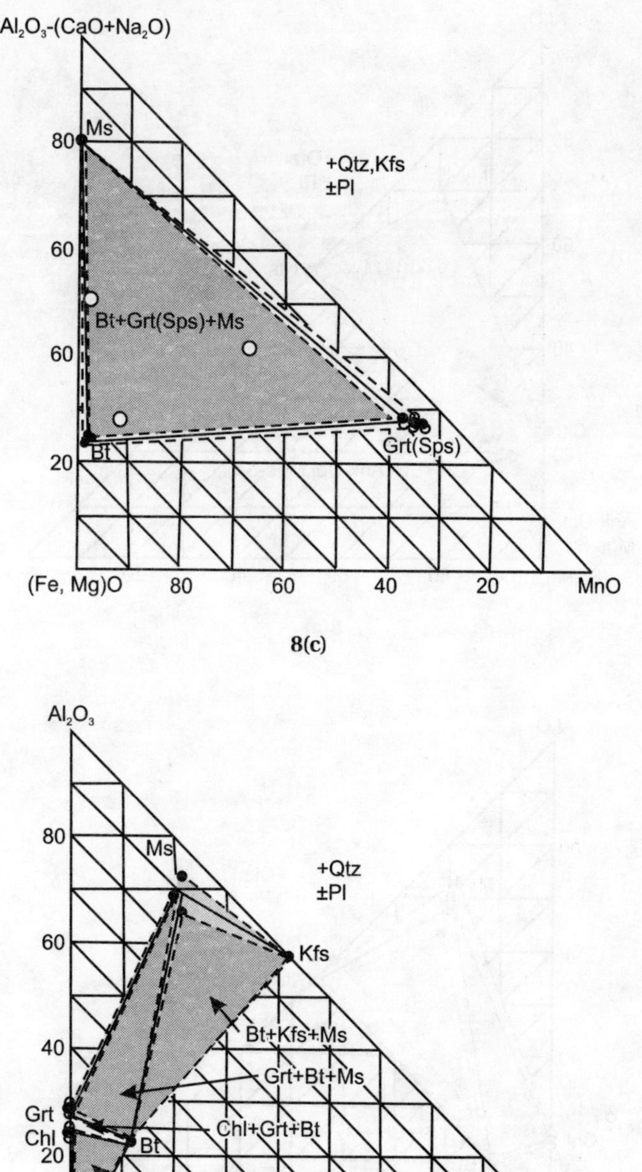

8(c)

8(d)

Fig. 8. Phase relations in low alumina metapelites of greenschist facies of metamorphism within Tim-Yastrebovskaya structure. (c) biotite subfacies: biotite-muscovite grade, (d) garnet subfacies in high potassium high manganese metapelites. Component relation in high manganese metapelites is shown by light colour circles.

presence of which is indicative of metamorphic conditions, corresponding to the upper part of biotite subfacies that we refer to as the biotite-muscovite grade.

In some case in metapelites with excess potassium feldspar the spessartine garnet ($Sps_{75-80}Grs_{13-14}Prp_{4-6}Alm_{1-5}$) is present. Its appearance in biotite and even pre-biotite subfacies described previously by (Brown, 1967, 1969). It follows from the analysis of phase relations (Fig. 8c, e), that the formation of spessartine is only possible in rocks with the MnO/(Fe, Mg)O bulk ratio higher than in typical for metapelites silicate minerals existing in these conditions. According to this value high potassium metapelites may be divided into high manganese that contain spessartine and low manganese – spessartine-free ones. Compositions of some studied rocks according to the bulk ratio Al_2O_3/(Fe, Mg)O/MnO are beyond the limits of Bt + Kfs + Sps field, that is indicative of the presence of other manganese-bearing minerals together with spessartine. So, V.I. Myasnyankin *et al.* (pers. comm.) described braunite, hausmannite, alabandite, ferruginous-manganese sphalerite, magnesian-manganese carbonates, manganocalcite and rhodochrosite in manganese rocks of Tim-Yastrebovskaya structure. Clearly, spessartine garnet appears in biotite and pre-biotite subfacies not due to the result of chlorite and quartz reaction but at the expense of other manganese-bearing low temperature minerals.

Garnet subfacies

In low potassium metapelites garnet appears at higher grade, corresponding to the upper part of the greenschist facies.

At the top of the greenschist facies, garnet crystallization is due to the reaction Chl + Qtz = Grt + H_2O. Here spessartine is formed at the expense of Mn-component in chlorite and

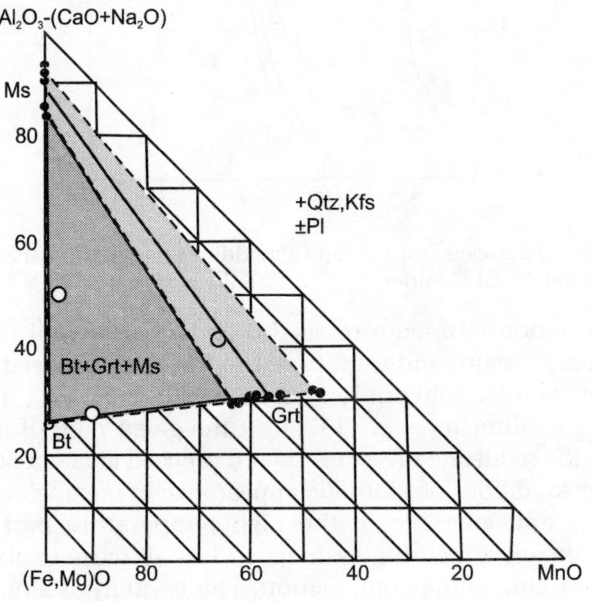

Fig. 8(e). Phase relations in low alumina metapelites of greenschist facies of metamorphism within Tim-Yastrebovskaya structure. (e) garnet subfacies in high potassium high manganese metapelites. Component relation in high manganese metapelites is shown by light coloured circles.

almandine — at the expense of Fe-component ones. These phase equilibria have been studied experimentally. The lowest temperature reaction of spessartine appearance proceeds in the pressure interval from 1 to 3 kbar at T°C = 400 (Fig. 9). Its stability does not depend on variations of oxidation-reduction conditions commonly observed during metamorphism. Unlike spessartine, almandine crystallization at the expense of ferruginous chlorite is facilitated by relatively reduction environment. The lowest temperatures of ferruginous garnet formation at iron-wüstite buffer, is 560-570°C. These oxidation-reduction conditions of almandine crystallization are probably the most similar to natural ones, because at oxygen fugacity corresponding to QFM buffer, the temperature field of its stability sharply decrease (Spear and Cheney, 1989), that contradicts widespread occurrence of ferruginous garnets in high-grade metamorphic facies.

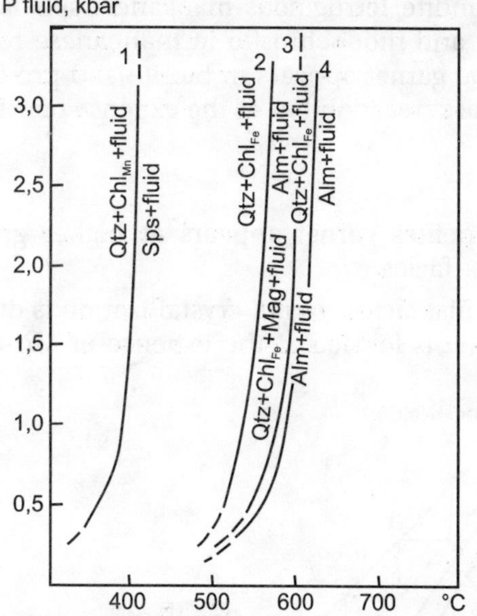

Fig. 9. Phase relations by producing spessartine (1) and almandine (2-4) after Spear and Cheney (1989). Oxygen fugacity: 2 - iron-wustite buffer, 3 - QFM buffer, 4 - Ni-bunsenite buffer.

The almandine formation temperature according to Korikovskii (1979) decreases when the ratio of partial water pressure and total pressure in the system reduces (Fig. 10). On the basis of calculations of mineral dehydration curves mean value P_{H_2O} in greenschist facies is estimated as 0.7-0.8 P_{tot} (Dobretsov *et al.*, 1970). At the given ratio of fluid at total pressure, almandine formation at the moderate pressures should proceed at about 500°C. This temperature boundary is close to the conditions of staurolite appearance in metapelites. Thus, crystallization of almandine garnet is only possible in the high temperature part of garnet subfacies. However in natural manganese-bearing systems within all garnet subfacies, garnets are of intermediate spessartine-almandine composition. The content of almandine component in them increasing as the metamorphic grade rises (Brown, 1969; Atherton, 1968; Cygan and Lasaga, 1982; Loomis, 1983). In this connection the staurolite formation in metapelites is the upper temperature boundary of garnet subfacies. In staurolite-free assemblages it is based

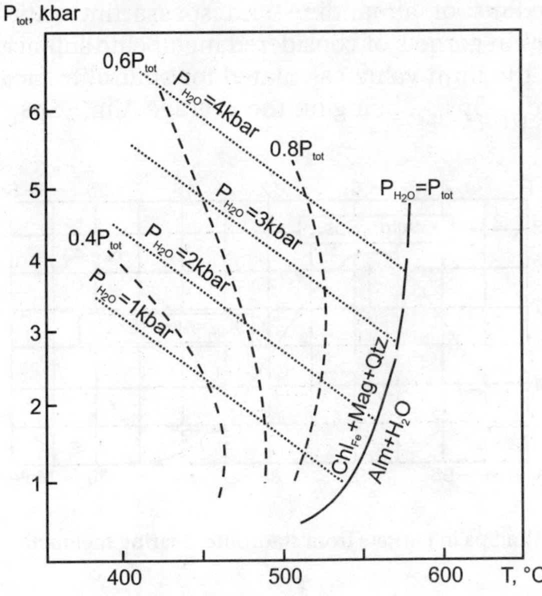

Fig. 10. Reaction curves of almandine producing (Chl_{Fe} + Mag + Qtz = Alm + H_2O) by various relations $P_{tot.}$ and P_{H_2O} after Korikovskii (1979).

on garnet composition and is mainly empirical character. So, Dobretsov *et al.* (1972) separate it according to MnO (5 wt. %) concentration in garnet. Similar MnO concentrations (5-6 wt. %) at almandine isograd are observed in Barrow zones (Scotland) (Atherton, 1968). However in metamorphic complex Abukuma (Japan) even at the lower amphibolite facies MnO contents in garnet is 16 wt. % (Miyashiro, 1953 & 1958).

Because of this we have made an attempt to determine the garnet composition limit at the upper temperature boundary of garnet subfacies on the basis statistic methods. The maximum permissible at the given level of probability garnet composition with student's criteria was considered to be the basic parameter. Absence of andradite component in the studied garnets (max. up to 2 mol. %) as well low concentrations of pyrope (up to 10 mol. %) and grossularite (up to 23 mol. %) allows to analyze only the ratio of almandine and spessartine concentrations. Because of high dispersion of the garnet composition in the garnet subfacies under consideration, the distribution of the ratio of these concentrations from conjugate staurolite facies has been analyzed. The garnet compositions from staurolite-bearing assemblages from different regions of the world (Perchuk, 1971) as well as our own analytical data on garnets from metapelites of Tim-Yastrebovskaya structure have been used for statistic treatment.

In staurolite-bearing rocks from different regions the average ratio of almandine and spessartine in garnets is $Alm_{92.6}Sps_{7.4}$, its maximum value (with probability 0.95) being estimated as $Alm_{85.6}Sps_{14.4}$ (Fig. 11). In staurolite-bearing metapelites and interbedded with them garnet-biotite and cummingtonite-garnet-biotite schists of Tim-Yastrebovskaya structure the average value of this ratio asymptotically approaches pure almandine — its limit value with probability 0.95 being $Alm_{91.2}Sps_{8.8}$ (Fig. 12). Discrepancy in the estimation of garnet composition limit in metapelites of different regions of the world and Tim-Yastrebovskaya structure is insignificant at the observed dispersions of almandine and spessartine ratio in the utilized arrays. Maximum Alm/Sps values in garnets of considered metapelite subfacies of Tim-Yastrebovskaya structure do not exceed the limit value calculated for staurolite facies assemblages and vary from $Alm_{13.9}Sps_{86.1}$ to $Alm_{87.6}$-$Sps_{12.4}$, being on the average $Alm_{58.0}Sps_{42.0}$.

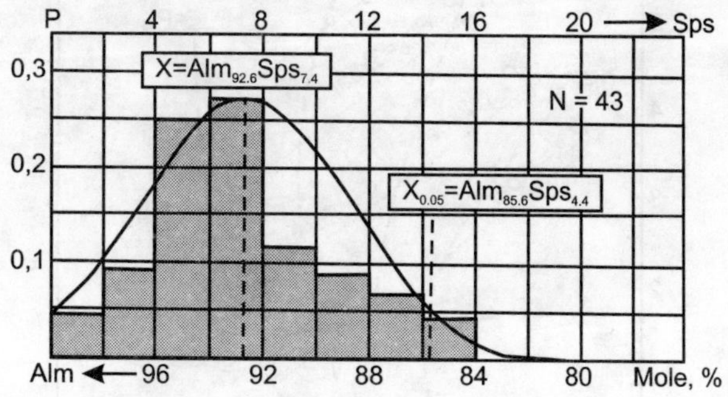

Fig. 11. Distribution of ratio Alm:Sps in garnets from staurolite-bearing metapelites from different regions of the world.

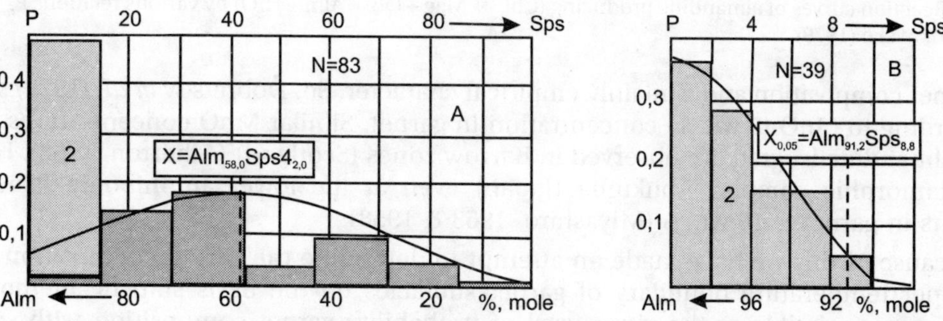

Fig. 12. Distribution of ratio Alm:Sps in garnets from Tim-Yastrebovskaya structure (A) and enlarged part for the garnets of staurolite facies (B). 1 - greenschist facies, 2 - staurolite facies.

The garnet subfacies conditions being reach, crystallization of garnet leads to appearance of assemblages Bt + Grt, Chl + Grt, Bt + Chl + Grt, Bt + Grt + Ms, Bt + Chl + Grt + Ms in low alumina low potassium metapelites of Tim-Yastrebovskaya structure and Grt + Ms, Chl + Grt + Ms — in high alumina metapelites. Beginning from high temperature part of greenschist facies mineral assemblages stability in low potassium metapelites is greatly influenced by Fe/Mg ratio of coexisting minerals. So, in case of high magnesian chlorite composition the assemblages Bt + Chl + Ms and Chl + Ms remain stable up to high temperature part of staurolite facies. Because of this, phase equilibria of staurolite facies for low potassium rocks will be also considered in the system Al_2O_3-(Fe, Mn)O-MgO.

In high, low potassium ferruginous-magnesian metapelites within garnet subfacies, amphiboles of cummingtonite-grunerite series appear as a result of reaction Mgs-Sid + Qtz + H_2O = Cum-Gru + CO_2 and Cum-Bt and Cum-Bt-Grt assemblages form (Fig. 8d).

Mineral assemblages in high potassium rocks from previous metamorphic grade remain stable in garnet subfacies. In high manganese metapelites, enrichment of spessartine garnet with almandine component take place.

Staurolite facies

Further changes of metamorphism conditions are fixed within Tim-Yastrebovskaya structure on the basis staurolite appearance in metapelites. In the field of staurolite-bearing assemblages stability in silica oversaturated rocks, according to Dobretsov *et al.* (1972) the facies muscovite-staurolite schists begins and the appearance and complete decomposition of staurolite with quartz are considered to be its low and high temperature boundaries respectively. The low grade of staurolite facies is the staurolite formation with chlorite and chloritoid — the upper one being disappearance of assemblage Ms + Qtz + St or the staurolite decomposition in the presence of quartz in muscovite free rocks (Korikovskii, 1979).

For the first time staurolite crystallizes at the expense of chloritoid, chlorite and andalusite in high alumina metapelites (Cld + Chl + And = St + Qtz + H_2O). In low alumina, low potassium rocks, it is formed at the expense of garnet, chlorite and muscovite (Chl + Grt + Ms = Bt + St + Qtz + H_2O). According to calculation data (Spear and Cheney, 1989) it proceeds at moderate pressure conditions at 500-550°C. Its temperature conditions may be influenced by spessartine and grossular components present in garnet. So, some investigators consider that plagioclase and ilmenite take part in the given reaction as well (Yardley *et al.*, 1980). They provide the redistribution of calcium and manganese among coexisting minerals. The staurolite crystallization in metapelites is also possible without the garnet as a result of muscovite and chlorite reaction. It is reproduced experimentally by Hoshek (1969) in conditions of QFM buffer and water fluid. According to the results, the reaction occurs in the pressure conditions 4 kbar at 540°C, and 7 kbar at 565°C.

In the staurolite-free metapelites the garnet appearance with maximum Alm/Sps ratio ($Alm_{91.2}Sps_{8.8}$) may be considered as a low boundary of staurolite facies (we touched it previously in upper boundary of garnet subfacies). The staurolite breakdown reaction in muscovite bearing rocks Ms + St + Qtz = Al_2SiO_5 + Bt + Grt + H_2O at pressure 1-4 kbar proceeds according to the calculation (Yardley, 1981; Spear and Cheney, 1989) in the temperature interval 550–640°C. On the basis of experimental studies, the staurolite with quartz decomposition proceeds at 660–680°C at 5 kbar (Richardson, 1966), 643–658°C at 3.25 kbar (Dutrow and Holdaway, 1989).

Available experimental and calculation data are obtained for water saturated systems and they may be considered as maximum possible. Taking into account that in natural occurrences partial water pressure in metamorphic systems is less than total pressure it follows that temperature boundaries of stability of staurolite bearing mineral assemblages probably should be slightly shifted into lower temperature area. So, on geothermometry evidence staurolite appears as a result of the reaction Chl + Grt + Ms = Bt + St + Qtz + H_2O at 450-470°C, maximum temperature of staurolite stability with muscovite and quartz being 550-570°C, that of staurolite with quartz approximately 600°C (Korikovskii, 1979). In low pressure conditions, the staurolite stability reduces due to the reaction St + Qtz = Al_2SiO_5 + Crd + H_2O.

Staurolite-chlorite-muscovite subfacies

Within Tim-Yastrebovskaya structure, metapelites containing assemblage And + Chl + Cld are absent. The staurolite appearance in high alumina and low alumina low potassium rocks at the expense of chlorite, muscovite and garnet is the evidence of staurolite-chlorite-muscovite subfacies conditions. In high alumina metapelites the reaction Cld + Qtz = St + Chl + Grt + H_2O is low temperature boundary. In low alumina rocks and metapelites bearing Chl + Ms and Chl + Ms + Grt assemblages the stability of staurolite with muscovite and chlorite is of great importance in subfacies temperature range. That is why we propose to call the subfacies under consideration a staurolite-chlorite-muscovite one.

After staurolite crystallization in metapelites of Tim-Yastrebovskaya structure the assemblages Bt + Grt + Ms + St, Bt + Grt + St, Bt + Ms + St, Bt + Chl + Grt + St are formed (Fig. 13a, b). Within entire staurolite facies the mineral assemblages of high potassium and high ferruginous-magnesian metapelites that formed at the earlier metamorphic stages as well as most mineral assemblages of low potassium rocks (Bt + Grt, Chl + Grt, Bt + Chl + Grt, Bt + Grt + Ms) remain stable. As the metamorphism grade increases, only change in mineral compositions at the components redistribution among coexisting mineral phases take place.

Andalusite-biotite subfacies

In some occurrences in staurolite facies conditions andalusite-bearing assemblages are observed. The earliest andalusite crystallization occurs at the expense of pyrophyllite (Prl = And(Ky) + Qtz + H_2O) in high alumina rocks. According to the experimental data the upper limit of its stability in silica oversaturated rocks 360-400°C. The decreasing of the ratio of partial water pressure and total pressure in the systems, moves this reaction to the even lower temperatures. According to calculations (Korikovskii, 1979) the pyrophyllite decomposition at $P_{H2O} = 0.8P_s$ proceeds in the 350-380°C. These results are very similar to natural ones because andalusite (kyanite) is observed in biotite subfacies conditions (greenschist facies). In the high temperature part of staurolite facies in both high alumina and for the first time in low alumina metapelites andalusite (kyanite) crystallizes at the expense of chlorite and

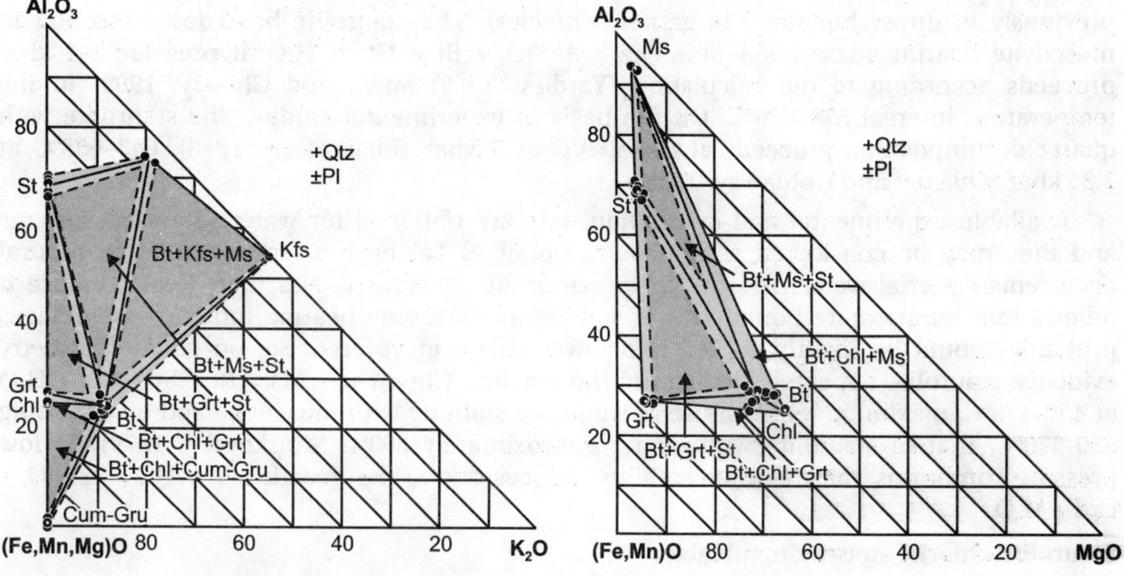

Fig. 13(a) Fig. 13(b)

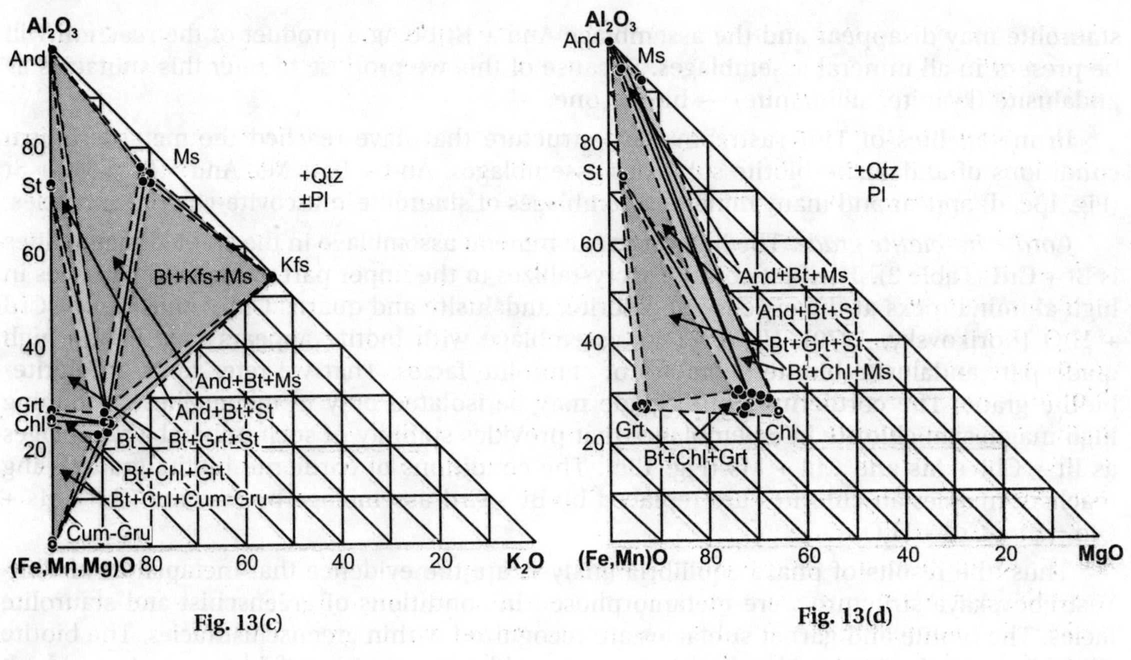

Fig. 13(c)

Fig. 13(d)

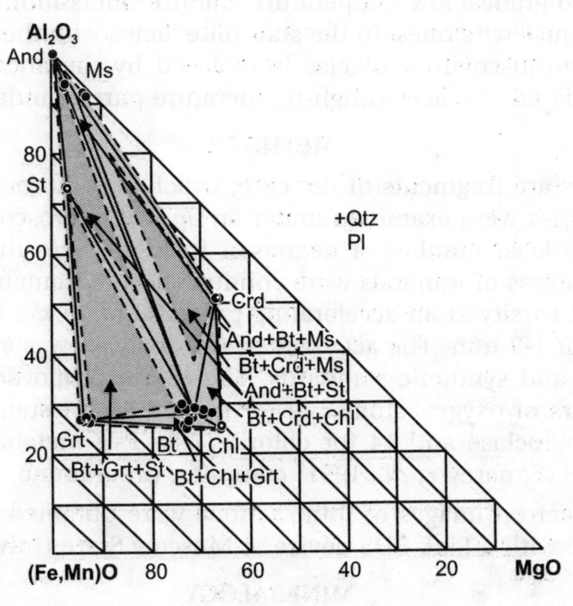

Fig. 13(e)

Fig. 13. Phase relation of staurolite facies metapelites of Tim-Yastrebovskaya structure: (a, b) staurolite-chlorite-muscovite subfacies, (c, d) andalusite-biotite subfacies, (e) cordierite-biotite grade.

muscovite and chlorite partial decomposition: St + Chl + Ms = And + Bt + Qtz + H_2O, that is the low boundary temperature of andalusite (kyanite, sillimanite)-biotite-staurolite subfacies (Korikovskii, 1979). However, depending on the mineral phase number during this reaction staurolite may disappear and the assemblage And + Bt being a product of the reaction will be present in all mineral assemblages. Because of this we propose to refer this subfacies as andalusite (kyanite, sillimanite) — biotite one.

In metapelites of Tim-Yastrebovskaya structure that have reached the metamorphism conditions of andalusite-biotite subfacies assemblages, And + Bt + Ms, And + Bt + Ms + St (Fig. 13c, d) appear and many mineral assemblages of staurolite-muscovite-chlorite subfacies.

Cordierite-biotite grade. The highest-grade mineral assemblage in the studied metapelites is Bt + Crd (Table 2). The first cordierite crystallizes in the upper part of greenschist facies in high alumina rocks at the expense of chlorite, andalusite and quartz: Chl + And + Qtz = Crd + H_2O (Korikovsky, 1979). However its assemblage with biotite appears only in the high grade part andalusite-biotite subfacies of staurolite facies. That we refer to as cordierite-biotite grade. The cordierite-biotite grade may be isolated only in metapelites containing high magnesian chlorite in assemblages that provides stability of such mineral assemblages as Bt + Chl + Ms and Chl + Ms (Fig. 13e). The conditions of cordierite-biotite grade being reached, magnesian chlorites are replaced by Bt + Crd assemblage by reaction Chl + Ms + Qtz = Bt + Crd + H_2O.

Thus, the results of phase equilibria analysis are the evidence that metapelites of Tim-Yastrebovskaya structure were metamorphosed in conditions of greenschist and staurolite facies. The biotite and garnet subfacies are recognized within greenschist facies. The biotite subfacies includes two grades: low temperature chlorite-potassium feldspar grade and high temperature biotite-muscovite ones. In the staurolite facies with the temperature increasing the staurolite-chlorite-muscovite subfacies is replaced by the andalusite-biotite one. The cordierite-biotite grade takes place at high-temperature part of andalusite-biotite subfacies.

METHODS

All of our samples are fragments of the core, which was thoroughly studied during the field works. The samples were examined under an optical microscope, and the minerals of assemblages with the least number of degree of freedom were then analyzed on a laser microprobe. Spot analyses of minerals were conducted on a Camebax SX 50 microprobe at the Moscow State University at an accelerating potential of 15 kV, beam current of 1-2 nA, and beam diameter of 1-2 mm. The accuracy of the analyses was systematically controlled by analyzing natural and synthetic standards. The cation proportions were normalized to the following numbers of oxygen atoms: 12 for garnet, 46 for staurolite, 11 for muscovite and biotite, 8 for plagioclase and 14 for chlorite. The P-T metamorphic conditions were calculated by the TPF (Fonarev *et al.*, 1991) computer programme.

Back-scattered electron images of thin sections were obtained on a CamScan electron microscope equipped with a Link EDS device at Moscow State University.

MINERALOGY

The metapelites from the Tim-Yastrebovskaya structure contain quartz, plagioclase, biotite, muscovite, and chlorite. In high grade rocks garnet, staurolite, andalusite, cordierite and cummingtonite-grunerite series amphiboles are present.

The plagioclase is present in most assemblages. The plagioclase composition from biotite subfacies varies from oligoclase ($Ab_{89}An_{11}$) to andesine ($Ab_{59-65}An_{35-41}$) (Table 3) that is not typical for the greenschist facies, where it is commonly represented by albite

Table 2: *Mineral assemblages of staurolite facies*

		Qtz	Pl	Kfs	Bt	Ms	Chl	Grt	St	And	Crd	Cum-Gru	Gph
Staurolite-chlorite-muscovite and andalusite-biotite subfacies	Staurolite-chlorite-muscovite and andalusite-biotite subfacies	+	+		+			+	+				
		+	+		+	+			+				
		+	+		+		+	+	+				
		+	+		+	+		+	+				
andalusite-biotite subfacies	–	+	+		+	+			+	+			
		+	+		+	+				+			
	Cordierite-biotite grade	+	+		+						+		+
Overall assemblages		+	+		+	+		+					
		+	+		+			+					
		+	+		+							+	
		+	+		+			+				+	
		+		+		+							+
		+		+	+	+							+
		+	+		+								
		+			+								
		+	+		+	+							
		+	+			+							

Table 3: *Representative microprobe analyses of plagioclase and potassium feldspar*

	Greenschist facies					Staurolite facies							
	Biotite subfacies			Garnet subfacies		Staurolite-chlorite-muscovite and andalusite-biotite subfacies			Andalusite-biotite subfacies				
	Chlorite-potassium feldspar grade		Biotite-muscovite grade										
Assemblage	Bt+Chl+Kfs		Bt+Cal+Grt(Sps)+Kfs+Ms	Bt+Grt+Kfs		Bt+Ms+St	Bt+Cum		And+Bt+Ms	Bt			
Sample no.	992/174.0		3063/590.0-595.0	3678/256.0		3068/611.0	3700/234.9		4317/614.5	4317/397.0			
Analysis no.	Pl-21	Pl-26	Pl-6	Pl-19	Pl-22	Pl-44	Pl-99	Pl-100	Pl-47	Pl-25	Pl-26	Pl-31	Pl-32
SiO_2	59.12	57.68	66.60	53.53	56.46	62.34	63.21	63.56	53.22	48.92	51.65	52.21	52.50
TiO_2	0.03	n.d.	0.01	n.d.	n.d.	n.d.	n.d.	n.d.	0.04	0.07	0.17	n.d.	0.07
Al_2O_3	26.58	26.70	21.31	30.27	28.20	23.60	22.99	22.85	30.29	32.00	30.48	30.06	30.06
FeO	0.12	0.16	0.03	0.14	0.15	n.d.	0.22	0.23	0.04	0.20	0.05	0.27	0.26
MnO	0.05	n.d.	n.d.	0.06	0.01	n.d.	0.13	n.d.	n.d.	0.20	0.04	n.d.	n.d.
MgO	0.01	0.01	0.02	0.02	n.d.	n.d.	n.d.	n.d.	0.01	n.d.	n.d.	0.01	0.04
CaO	7.32	8.48	2.04	11.86	9.95	4.93	4.50	4.34	12.41	15.92	13.48	13.08	12.63
Na_2O	7.45	6.73	9.83	4.08	5.18	8.93	8.85	8.78	4.58	2.56	4.02	4.36	4.37
K_2O	0.04	0.05	0.10	0.09	0.04	0.03	0.01	0.02	0.05	0.13	0.10	0.01	0.08
Total	100.72	99.81	99.94	100.05	99.99	99.83	99.91	99.78	100.64	100.00	99.99	100.00	100.01
Oxygen	8.00	8.00	8.00	8.00	8.00	8.00	8.00	8.00	8.00	8.00	8.00	8.00	8.00
Si	2.62	2.59	2.94	2.43	2.55	2.76	2.81	2.83	2.39	2.24	2.35	2.37	2.37
Al^{IV}	1.38	1.41	1.11	1.62	1.50	1.23	1.20	1.20	1.61	1.73	1.63	1.61	1.61
SR^{IV}	4.00	4.00	4.05	4.05	4.05	3.99	4.01	4.03	4.00	3.97	3.98	3.98	3.98
Fe^{2+}	0.00	0.01	0.00	0.01	0.01	0.00	0.01	0.01	0.00	0.01	0.00	0.01	0.01
Mn	0.00	0.00	0.00	0.00	0.00	0.00	0.01	0.00	0.00	0.01	0.00	0.00	0.00
Ca	0.35	0.41	0.10	0.58	0.48	0.23	0.21	0.21	0.60	0.78	0.66	0.64	0.61
Na	0.64	0.59	0.84	0.36	0.45	0.77	0.76	0.76	0.40	0.23	0.35	0.38	0.38
K	0.00	0.00	0.01	0.01	0.00	0.00	0.00	0.00	0.00	0.01	0.01	0.00	0.01
SR^{VI}	0.99	1.01	0.95	0.96	0.94	1.00	0.99	0.98	1.00	1.04	1.02	1.03	1.01
Ab	0.65	0.59	0.89	0.38	0.48	0.77	0.78	0.79	0.40	0.22	0.35	0.38	0.38
An	0.35	0.41	0.10	0.62	0.52	0.23	0.22	0.21	0.60	0.78	0.65	0.62	0.61
Or	0.00	0.00	0.01	0.00	0.00	0.00	0.00	0.00	0.00	0.00	0.00	0.00	0.01

(Contd.)

	Staurolite facies andalusite-biotite subfacies					Greenschist facies							
						Biotite subfacies Garnet							Subfacies
			Cordierite-biotite grade			Chlorite-potassium feldspar grade	Biotite-muscovite grade						
Assemblage	Bt+Cum+Grt		Bt+Crd			Bt+Chl+Kfs	Bt+Cal+Grt(Sps)+Kfs+Ms						Bt+Grt+Kfs
Sample no.	4317/370.1		3129/311.0			992/174.0	3063/590.0-595.0						3678/256.0
Analysis no.	Pl-54	Pl-56	Pl-37	Pl-35	Pl-36	Kfs-22	Kfs-2	Kfs-14	Kfs-17	Kfs-18	Kfs-23	Kfs-19	Kfs-25
SiO_2	58.87	58.72	53.94	62.50	62.68	65.33	64.41	64.03	66.02	64.06	63.61	64.36	64.53
TiO_2	0.20	0.06	0.03	n.d.	n.d.	0.01	n.d.	n.d.	n.d.	0.02	n.d.	n.d.	n.d.
Al_2O_3	25.69	25.70	31.22	23.98	23.43	19.39	18.57	19.27	17.79	19.18	19.85	19.22	19.37
FeO	0.35	0.23	0.07	0.22	0.09	0.11	n.d.	n.d.	0.08	n.d.	0.03	n.d.	0.20
MnO	0.07	n.d.	n.d.	n.d.	0.11	n.d.	0.20	0.03	n.d.	0.14	n.d.	n.d.	n.d.
MgO	n.d.	n.d.	n.d.	n.d.	0.04	0.01	0.01	n.d.	n.d.	n.d.	n.d.	0.01	0.09
CaO	7.36	7.56	12.10	4.94	4.76	0.02	0.01	n.d.	0.02	n.d.	0.47	n.d.	1.34
Na_2O	7.40	7.54	4.72	8.24	8.84	0.02	0.64	0.55	0.59	0.48	1.63	0.72	0.79
K_2O	0.08	0.05	0.02	0.12	0.05	15.07	14.55	15.04	13.81	15.08	13.34	14.71	12.82
Total	100.02	99.86	102.10	100.00	100.00	99.96	98.39	98.92	98.31	98.96	98.93	99.02	99.14
	8.00	8.00	8.00	8.00	8.00	8.00	8.00	8.00	8.00	8.00	8.00	8.00	8.00
Si	2.63	2.62	2.39	2.78	2.78	3.04	3.03	2.99	3.12	2.99	2.95	3.00	2.96
Al^{IV}	1.35	1.35	1.63	1.26	1.22	1.06	1.03	1.06	0.99	1.06	1.09	1.06	1.05
ΣR^{IV}	3.98	3.97	4.02	4.04	4.00	4.10	4.06	4.05	4.11	4.05	4.04	4.06	4.01
Fe	0.01	0.01	0.00	0.01	0.00	0.00	0.00	0.00	0.00	0.00	0.00	0.00	0.01
Mn	0.00	0.00	0.00	0.00	0.00	0.00	0.00	0.00	0.00	0.01	0.00	0.00	0.00
Ca	0.35	0.36	0.57	0.24	0.23	0.00	0.00	0.00	0.00	0.00	0.02	0.00	0.07
Na	0.64	0.65	0.41	0.71	0.76	0.00	0.06	0.05	0.05	0.04	0.15	0.07	0.07
K	0.00	0.00	0.00	0.01	0.00	0.89	0.87	0.90	0.83	0.90	0.79	0.88	0.75
ΣR^{VI}	1.00	1.01	0.98	0.97	0.99	0.89	0.93	0.95	0.88	0.95	0.96	0.95	0.90
Ab	0.64	0.64	0.41	0.75	0.77	0.00	0.06	0.05	0.06	0.05	0.16	0.07	0.08
An	0.36	0.36	0.59	0.25	0.23	0.00	0.00	0.00	0.00	0.00	0.02	0.00	0.07
Or	0.00	0.00	0.00	0.00	0.00	1.00	0.94	0.95	0.94	0.95	0.82	0.93	0.85

n.d. — not detected

(Dobretsov *et al.*, 1970; Korikovskii, 1979; Miyashiro, 1973). Some researchers describe oligoclase or simultaneous presence of albite and oligoclase in greenschist facies. It may be related with miscibility gap in peristerite solvus area. In these cases oligoclase commonly produce peripheral rims with distinct boundaries around the albite cores. However, there are no sufficient evidence of albite-oligoclase miscibility gap and the change of plagioclase composition may be the result of progressive temperature increasing during metamorphism. In garnet subfacies assemblages the content of anorthite component in plagioclase is ever greater and it is represented by labrador ($Ab_{38-48}An_{51-62}$). Its composition in greenschist facies conditions may be presumable explained by extremely low partial CO_2 pressure regulating the process of deanorthization and albite with calcite producing or calcium including into plagioclase structure. However to solve this problem additional studies should be carried out.

In different assemblages of staurolite facies plagioclase is characterized by wide range of compositions from $Ab_{77-79}An_{21-23}$ to $Ab_{22}An_{78}$ (Table 3).

The potassium feldspar is represented in metapelites of Tim-Yastrebovskaya structures by microcline that is often latticed and rather relatively fixed composition $Or_{82-100}Ab_{0-15}An_{0-7}$ (Table 3).

The micas of the metapelitic rocks are muscovite, biotite, and chlorite. The early-kinematic generation of micas defines the schistosity of the rock matrix and pressure shadows around garnet grains in the greenschist facies. The biotite and muscovite of the late-kinematic generation occur as elongated sub-hedral laths, which cut the schistosity and fragmentary replace the rims of garnet grains in the staurolite facies.

The biotite produces fine isometric xenomorphic flakes in the lowest grade rocks (chlorite-potassium feldspar of biotite subfacies). They are pleochroism in yellowish-brownish or brown-green shades. When the grade of metamorphism increases its size become larger and crystallographic outlines begin to observe. In metapelites of biotite subfacies biotite is characterized by great variations in X_{Fe} values from 0.06 to 0.55 (Table 4), its composition change from typical biotites to phlogopites (Fig. 14). Within garnet subfacies X_{Fe} values of biotites slightly varies from 0.33 to 0.35. X_{Fe} values of biotites increase to 0.55 in staurolite facies assemblages. In transition from greenschist to staurolite facies the amount of Ti

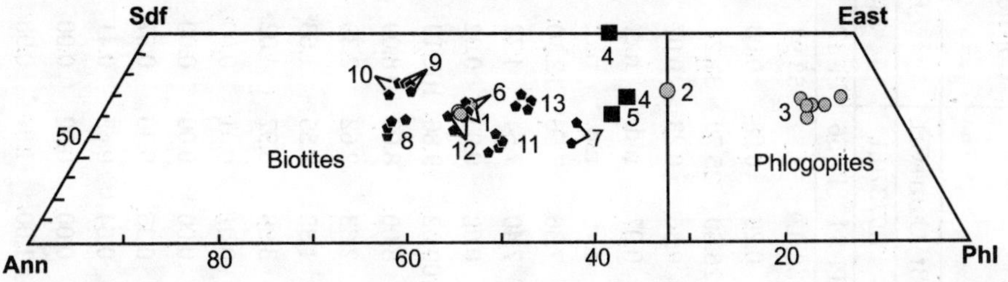

Fig. 14. Composition of biotites and phlogopites from metapelites of Tim-Yastrebovskaya structure. 1-5 - greenschist facies: 1-3 biotite subfacies: 1 - chlorite - potassium feldspar grade (Bt+Chl+Kfs), 2,3 - biotite-muscovite grade (2 – Bt + Cl + Ms, 3 - Bt + Cal + Kfs + Ms + Sps), 4, 5,- garnet subfacies (4 – Bt + Grt + Kfs, 5 – Bt + Chl + Grt), 6-13 - staurolite facies: 6-9 - staurolite-chlorite-muscovite and andalusite-biotite subfacies (6 – Bt + Grt + St, 7 – Bt + Ms + St, 8 – Grt + Bt, 9 – Bt + Cum), 10-13 - andalusite-biotite subfacies (10 – And + Bt + Ms + St, 11 – Bt, 12 – Bt + Cum + Grt), 13 – cordierite-biotite grade (Bt + Crd).

Table 4: *Representative microprobe analyses of biotite*

	Greenschist facies												
	Biotite subfacies										Garnet subfacies		
	Chlorite-potassium feldspar grade			Biotite-muscovite grade							–		
Assemblage	Bt+Chl+Kfs			Bt+Chl+Ms	Bt+Cal+Grt(Sps)+Kfs+Ms						Bt+Grt+Kfs		Chl+Bt+Grt
Sample no.	992/174.0			3697/292.0	3063/590.0-595.0						3678/256.0		3070/180.4
Analysis no.	Bt-20	Bt-5	Bt-25	Bt-28	Bt-5	Bt-10	Bt-24	Bt-27	Bt-20	Bt-22	Bt-21	Bt-23	Bt-40
SiO_2	34.89	37.62	35.47	34.78	41.20	40.27	40.22	41.36	40.92	40.87	44.99	38.90	37.35
TiO_2	1.53	1.06	1.57	1.59	0.84	1.07	0.79	0.74	1.14	0.74	1.13	1.15	1.65
Al_2O_3	19.15	20.40	18.77	19.05	17.83	18.16	17.87	17.74	18.43	18.36	16.85	18.76	19.00
FeO*	19.26	10.92	20.59	20.29	3.68	3.95	4.38	4.04	2.93	1.01	11.69	12.75	13.69
MnO	0.18	0.08	0.20	0.17	0.70	0.76	0.70	0.73	0.78	1.33	0.26	0.38	0.42
MgO	9.40	15.76	9.44	9.46	21.42	20.10	21.30	21.39	21.88	21.57	12.50	14.38	14.09
CaO	n.d.	0.09	0.04	0.03	n.d.**	0.18	0.01	n.d.	n.d.	0.07	0.46	0.12	0.07
Na_2O	0.14	0.22	0.14	0.14	0.08	0.20	0.06	0.09	0.04	0.19	0.28	0.16	0.24
K_2O	8.43	8.41	8.51	8.46	9.80	9.66	9.65	9.77	10.10	9.94	8.03	9.57	9.17
Total	92.98	94.56	94.73	93.97	95.55	94.35	94.98	95.86	96.22	94.08	96.19	96.17	95.68
Oxygen	11.00	11.00	11.00	11.00	11.00	11.00	11.00	11.00	11.00	11.00	11.00	11.00	11.00
Si	2.80	2.83	2.80	2.76	2.97	2.95	2.92	2.97	2.92	2.98	3.41	2.91	2.82
Al^{IV}	1.20	1.17	1.20	1.24	1.03	1.05	1.08	1.03	1.08	1.02	0.59	1.09	1.18
ΣR^{IV}	4.00	4.00	4.00	4.00	4.00	4.00	4.00	4.00	4.00	4.00	4.00	4.00	4.00
Al^{VI}	0.60	0.64	0.55	0.55	0.48	0.52	0.45	0.48	0.47	0.53	0.91	0.57	0.51
Ti	0.09	0.06	0.09	0.10	0.05	0.06	0.04	0.04	0.06	0.04	0.06	0.07	0.09
Fe	1.29	0.69	1.36	1.35	0.22	0.24	0.27	0.24	0.17	0.06	0.74	0.80	0.86
Mn	0.01	0.01	0.01	0.01	0.04	0.05	0.04	0.04	0.05	0.08	0.02	0.02	0.03
Mg	1.12	1.77	1.11	1.12	2.30	2.19	2.30	2.29	2.33	2.33	1.41	1.60	1.59
ΣR^{VI}	3.11	3.17	3.12	3.13	3.09	3.06	3.10	3.09	3.08	3.04	3.14	3.06	3.08
Ca	0.00	0.01	0.00	0.00	0.00	0.01	0.00	0.00	0.00	0.01	0.04	0.01	0.01
Na	0.02	0.03	0.02	0.02	0.01	0.03	0.01	0.01	0.01	0.03	0.04	0.02	0.04
K	0.86	0.81	0.86	0.86	0.90	0.90	0.89	0.90	0.92	0.92	0.78	0.91	0.88
ΣR^{VIII}	0.88	0.85	0.88	0.88	0.91	0.94	0.90	0.91	0.93	0.96	0.86	0.94	0.93
Sdf	0.36	0.20	0.34	0.34	0.06	0.08	0.07	0.07	0.05	0.04	0.35	0.23	0.22
East	0.31	0.51	0.28	0.28	0.57	0.58	0.51	0.56	0.55	0.64	0.65	0.45	0.39
Ann	0.18	0.08	0.21	0.31	0.04	0.04	0.05	0.04	0.04	0.02	0.00	0.11	0.14
Phl	0.15	0.21	0.17	0.17	0.33	0.30	0.37	0.33	0.36	0.30	0.00	0.21	0.25
Fe/(Fe+Mg)	0.54	0.28	0.55	0.55	0.10	0.12	0.12	0.11	0.09	0.06	0.34	0.33	0.35

(Contd.)

	Staurolite facies														
	Staurolite-chlorite-muscovite and andalusite-biotite subfacies					Andalusite-biotite subfacies									
												Biotite-cordierite grade			
Assemblage	Bt+Grt+St		Bt+Grt			Bt+Cum+Grt				And+Bt+Ms+St		Bt+Crd			
Sample no.	3055/421.0		3706/286.6			4317/370.1				3716/608.2-608.6		3129/311.0			
Analysis no.	Bt-80	Bt-82	Bt-65	Bt-66	Bt-58	Bt-59	Bt-52	Bt-55	Bt-18*	Bt-5*	Bt-6*	Bt-37	Bt-29	Bt-31	Bt-34
SiO_2	35.17	35.91	35.02	35.01	34.99	34.21	36.15	35.73	36.74	35.88	35.63	36.30	36.12	36.67	36.62
TiO_2	1.93	2.11	1.92	1.98	1.99	1.85	2.00	1.99	1.69	2.34	2.30	2.69	2.86	2.55	2.76
Al_2O_3	19.97	20.15	18.41	18.22	18.94	18.61	17.06	17.26	17.94	20.45	19.98	19.98	19.72	20.56	20.07
FeO	19.75	19.76	24.00	24.10	23.23	25.65	22.00	22.32	21.73	21.76	22.55	16.97	17.22	16.26	16.55
MnO	n.d.	0.01	n.d.	n.d.	0.03	0.14	0.05	0.08	n.d.	0.06	0.08	0.12	0.09	0.24	0.11
MgO	9.40	9.39	7.75	7.67	7.99	8.21	9.85	9.66	10.12	7.15	7.13	10.81	10.35	10.27	10.94
CaO	n.d.	n.d.	0.17	0.09	0.12	0.22	0.02	n.d.	0.15	0.03	0.05	0.09	0.03	0.11	0.17
Na_2O	0.48	0.30	0.12	0.33	0.27	0.23	0.48	0.21	0.26	0.18	0.20	0.41	0.29	0.18	0.25
K_2O	9.10	8.79	8.45	8.66	8.41	7.23	8.32	8.41	8.01	8.52	8.54	9.13	9.29	9.25	8.73
Total	95.65	96.42	95.84	96.06	95.97	96.35	95.93	95.66	96.64	96.37	96.46	96.50	95.97	96.09	96.20
Oxygen	11.00	11.00	11.00	11.00	11.00	11.00	11.00	11.00	11.00	11.00	11.00	11.00	11.00	11.00	11.00
Si	2.72	2.78	2.78	2.77	2.76	2.70	2.82	2.81	2.85	2.82	2.80	2.77	2.78	2.82	2.81
Al^{IV}	1.28	1.22	1.22	1.23	1.24	1.30	1.18	1.19	1.15	1.18	1.20	1.23	1.22	1.18	1.19
ΣR^{IV}	4.00	4.00	4.00	4.00	4.00	4.00	4.00	4.00	4.00	4.00	4.00	4.00	4.00	4.00	4.00
Al^{VI}	0.55	0.61	0.50	0.46	0.52	0.44	0.40	0.41	0.49	0.71	0.65	0.57	0.57	0.68	0.62
Ti	0.11	0.12	0.11	0.12	0.12	0.11	0.12	0.12	0.10	0.14	0.14	0.15	0.17	0.15	0.16
Fe	1.28	1.27	1.59	1.59	1.53	1.69	1.44	1.46	1.41	1.43	1.48	1.08	1.11	1.04	1.06
Mn	0.00	0.00	0.00	0.00	0.00	0.01	0.00	0.01	0.00	0.00	0.01	0.01	0.01	0.02	0.01
Mg	1.09	1.08	0.92	0.90	0.94	0.97	1.15	1.13	1.17	0.84	0.84	1.23	1.19	1.18	1.25
ΣR^{VI}	3.03	3.09	3.12	3.07	3.11	3.22	3.11	3.13	3.17	3.12	3.12	3.04	3.05	3.07	3.10
Ca	0.00	0.00	0.02	0.01	0.01	0.02	0.00	0.00	0.01	0.00	0.00	0.01	0.00	0.01	0.01
Na	0.07	0.04	0.02	0.05	0.04	0.04	0.07	0.03	0.04	0.03	0.03	0.06	0.04	0.03	0.04
K	0.90	0.87	0.85	0.87	0.85	0.73	0.83	0.84	0.79	0.85	0.86	0.89	0.91	0.91	0.85
ΣR^{VIII}	0.97	0.92	0.89	0.93	0.90	0.79	0.90	0.87	0.84	0.88	0.89	0.96	0.95	0.95	0.90
Sdf	0.33	0.36	0.37	0.35	0.36	0.32	0.28	0.29	0.32	0.47	0.45	0.30	0.31	0.35	0.31
East	0.28	0.31	0.21	0.20	0.22	0.18	0.22	0.22	0.27	0.28	0.25	0.34	0.33	0.38	0.37
Ann	0.21	0.18	0.27	0.29	0.26	0.32	0.28	0.28	0.22	0.16	0.19	0.17	0.17	0.13	0.15
Phl	0.18	0.15	0.15	0.16	0.16	0.18	0.22	0.21	0.19	0.09	0.11	0.19	0.19	0.14	0.17
Fe/(Fe+Mg)	0.54	0.54	0.64	0.64	0.62	0.64	0.56	0.57	0.55	0.63	0.64	0.47	0.48	0.47	0.46

*FeO measured as total Fe

**n.d. – not detected

cations in the structure of biotite increases on the average from 0.07 to 0.13 reaching maximum values (average 0.16) in conditions of the cordierite-biotite grade. When the X_{Fe} values and Ti content rises in biotites one observes decreasing in Al^{VI} concentration on the average from 0.67 in greenschist to 0.60 in staurolite facies. Such change in biotite composition is probably related with heterovalent replacement in the structure $2Al^{3+} \rightarrow Fe^{2+}Ti^{4+}$ in assemblages with high alumina minerals (staurolite, andalusite, cordierite).

The muscovite occurs in the metapelites of the greenschists facies as small flakes. As the metamorphic grade increases, the size of the flakes also increases, so that the rocks of the staurolite facies contain relatively large (up to 1 mm) platelets. Muscovite in greenschist facies metapelites is characterized by rather fixed composition $Ms_{68-82}Pg_{4-25}Phn_{6-14}$ (Table 5) and by absence of interrelation changes in the paragonite and phengite contents. In transition to staurolite facies the phengite contents in muscovite decrease on the average from 9.0 to 5.5% (Fig. 15).

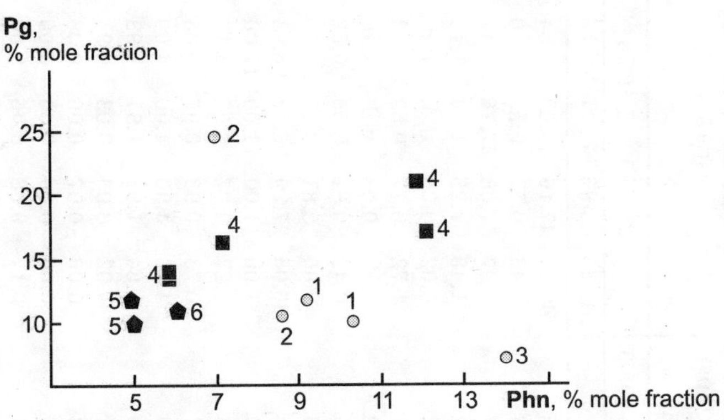

Fig. 15. Ratio of paragonite and phengite components in muscovites. 1 - Chlorite-potassium feldspar and biotite-muscovite grade (Chl + Ms), 2, 3 - biotite-muscovite grade (2 – Bt + Chl + Ms, 3 – Bt + Cal + Kfs + Ms + Sps), 4 - garnet subfacies (Grt + Ms), 5, 6 - staurolite facies, andalusite-biotite subfacies (5 – And + Bt + Grt + Ms + St, 6 – And + Bt + Ms).

The chlorite within the greenschist facies according to the amount Si cations (5.06-5.57 p.f.u.) is referred to sheridanite-ripidolite series (Fig. 16) with fixed Al content in R^{VI} coordination (Fig. 17). In biotite subfacies assemblages X_{Fe} values in chlorites varies in wide range (0.20-0.52) (Table 6).

The garnet of the early-kinematic generation occurs in metapelites of the Tim-Yastrebovskaya structure only within the greenschist facies and consists of porphyroblasts with the size from 0.33 to 1.70 mm of equant or irregular shapes with clear pressure shadows and contains abundant oriented inclusions of quartz, plagioclase, and carbonaceous matter, which compose trails and characteristic "petal" patterns (Fig. 18, 1-6). Practically always they are of S-like, sometimes spiral-like, snowball internal structure indicative of crystals rotation during their growth. In staurolite facies metapelites most garnets are represented by poikiloblasts with the size 0.5-2.1 mm (Fig. 18, 7-9). Within biotite subfacies of greenschist facies garnet has spessartine composition $Sps_{76-81}Grs_{13-15}Prp_{5-7}Alm_{1-3}$. It is characterized by low X_{Fe} = 0.34-0.47 (Table 7). X_{Fe} slightly increases from garnet cores to rims.

Table 5: *Representative microprobe analyses of muscovite*

	Greenschist facies										Staurolite facies			
	Biotite subfacies					Garnet subfacies					Andalusite-biotite subfacies			
	Chlorite-potassium feldspar and biotite-muscovite grade		Biotite-muscovite grade			–					–			
Assemblage	Chl+Ms		Bt+Chl+Ms		Bt+Cal+Grt(Sps) +Kfs+Ms	Grt+Ms					And+Bt+Ms+St		And+Bt+Ms	
Sample no.	1045/240.0		3697/292.0		3063/590.0-595.0	3615/304.2-304.8					3716/608.2-608.6		4317/614.5	
Analysis no.	Ms-8	Ms-13	Ms-6	Ms-7	Ms-9	Ms-45	Ms-46	Ms-47	Ms-48	Ms-49	Ms-7	Ms-8	Ms-39	Ms-41
SiO_2	45.89	45.82	45.58	47.48	47.18	49.23	47.16	46.02	47.64	45.40	47.25	47.62	45.44	46.75
TiO_2	0.43	0.35	0.28	0.22	0.28	0.43	0.52	0.57	0.60	0.71	0.33	0.39	1.15	0.69
Al_2O_3	35.40	35.66	35.04	34.54	32.77	32.97	37.08	37.16	36.41	35.87	38.42	37.45	36.56	37.12
FeO	2.72	2.48	1.91	1.83	0.82	1.48	1.25	1.04	1.13	2.31	1.20	1.03	0.54	0.49
MnO	0.05	n.d.	0.01	n.d.	0.03	0.02	0.05	n.d.	n.d.	n.d.	0.20	n.d.	n.d.	0.08
MgO	0.92	0.93	0.91	0.66	2.71	1.72	0.77	0.63	0.63	1.07	0.39	0.43	1.01	1.12
CaO	n.d.	0.02	0.11	1.59	0.08	n.d.	0.03	n.d.	n.d.	0.06	0.06	0.02	0.01	0.03
Na_2O	0.67	0.70	0.78	1.74	0.33	0.42	0.62	0.59	0.62	0.67	0.79	0.88	0.74	0.74
K_2O	9.61	9.32	9.23	7.97	10.35	9.80	9.81	10.21	9.90	9.44	8.64	9.46	9.36	9.05
Total	95.69	95.28	93.85	96.03	94.55	96.08	97.29	96.22	96.93	95.53	97.28	97.28	94.81	96.07
Oxygen	11.00	11.00	11.00	11.00	11.00	11.00	11.00	11.00	11.00	11.00	11.00	11.00	11.00	11.00
Si	3.05	3.06	3.08	3.13	3.13	3.27	3.07	3.03	3.12	3.01	3.08	3.11	3.04	3.08
Al^{IV}	0.95	0.94	0.92	0.87	0.87	0.73	0.93	0.97	0.88	0.99	0.92	0.89	0.96	0.92
ΣR^{IV}	4.00	4.00	4.00	4.00	4.00	4.00	4.00	4.00	4.00	4.00	4.00	4.00	4.00	4.00
Al^{VI}	1.83	1.86	1.88	1.82	1.69	1.84	1.92	1.91	1.93	1.81	2.04	1.99	1.91	1.97
Ti	0.02	0.02	0.01	0.01	0.01	0.02	0.03	0.03	0.03	0.04	0.02	0.02	0.06	0.03
Fe^{2+}	0.15	0.14	0.11	0.10	0.03	0.08	0.07	0.06	0.06	0.13	0.07	0.06	0.03	0.03
Mn	0.00	0.00	0.00	0.00	0.00	0.00	0.00	0.00	0.00	0.00	0.01	0.00	0.00	0.00
Mg	0.09	0.09	0.09	0.07	0.27	0.17	0.08	0.06	0.06	0.11	0.04	0.04	0.10	0.11
ΣR^{VI}	2.09	2.11	2.09	2.00	2.00	2.11	2.10	2.07	2.08	2.09	2.18	2.11	2.10	2.14
Ca	0.00	0.00	0.01	0.11	0.01	0.00	0.00	0.00	0.00	0.00	0.00	0.00	0.00	0.00
Na	0.09	0.09	0.10	0.22	0.04	0.05	0.08	0.08	0.08	0.09	0.10	0.11	0.10	0.09
K	0.81	0.79	0.80	0.67	0.88	0.83	0.82	0.86	0.83	0.80	0.72	0.79	0.80	0.76
ΣR^{VIII}	0.90	0.88	0.91	1.00	0.93	0.88	0.90	0.94	0.91	0.89	0.82	0.90	0.90	0.85
Pg	0.10	0.11	0.11	0.25	0.04	0.17	0.16	0.14	0.14	0.21	0.12	0.10	0.11	0.11
Phn	0.10	0.09	0.09	0.06	0.14	0.12	0.07	0.06	0.06	0.11	0.05	0.05	0.06	0.06
Ms	0.80	0.80	0.80	0.69	0.82	0.71	0.77	0.80	0.80	0.68	0.83	0.85	0.83	0.83

n.d. – not detected

Table 6: *Representative microprobe analyses of chlorite*

	Greenschist facies										
	Biotite subfacies							Garnet subfacies			
	Chlorite-potassium feldspar grade		Chlorite-potassium feldspar grade and Biotite-muscovite grade			Biotite-muscovite grade		–			
Assemblage	Bt+Chl+Kfs		Chl+Ms			Bt+Chl+Ms		Chl+Grt			
Sample no.	992/174.0		1045/240.0			3697/292.0		2215/9			
Analysis no.	Chl-23	Chl-24	Chl-10	Chl-12	Chl-14	Chl-2	Chl-3	Chl-2	Chl-3	Chl-4	Chl-5
SiO_2	22.91	23.40	26.96	27.82	27.42	25.62	25.92	26.97	25.86	26.09	26.67
TiO_2	0.06	0.08	0.05	0.08	0.02	0.03	0.07	n.d.	0.06	0.05	0.08
Al_2O_3	22.50	23.10	24.09	23.71	23.41	25.11	24.90	24.14	21.88	24.98	25.18
FeO	24.85	25.60	10.38	10.47	10.82	14.42	14.13	27.11	26.17	27.93	26.76
MnO	0.36	0.30	0.16	0.25	0.09	0.17	0.14	0.11	0.05	0.09	0.11
MgO	12.88	13.23	24.07	23.48	23.84	21.04	22.12	15.45	13.14	14.60	14.48
CaO	0.11	0.10	0.07	0.08	0.08	0.05	0.03	0.13	0.15	0.13	0.13
Na_2O	0.05	0.01	0.03	0.02	0.01	0.04	0.02	0.01	0.05	0.02	0.01
K_2O	0.07	0.04	0.05	0.02	0.05	0.12	0.02	0.03	0.02	0.05	0.03
Total	83.79	85.86	85.86	85.93	85.74	86.60	87.35	93.95	87.38	93.94	93.45
Oxygen	28.00	28.00	28.00	28.00	28.00	28.00	28.00	28.00	28.00	28.00	28.00
Si	5.07	5.06	5.37	5.57	5.49	5.17	5.17	5.32	5.47	5.17	5.29
Al^{IV}	2.93	2.94	2.63	2.43	2.51	2.83	2.83	2.68	2.53	2.83	2.71
ΣR^{IV}	8.00	8.00	8.00	8.00	8.00	8.00	8.00	8.00	8.00	8.00	8.00
Al^{VI}	2.93	2.95	3.03	3.15	3.02	3.15	3.01	2.93	2.92	2.99	3.18
Ti	0.01	0.01	0.01	0.01	0.00	0.01	0.01	0.00	0.01	0.01	0.01
Fe^{2+}	4.59	4.62	1.73	1.75	1.81	2.43	2.35	4.46	4.62	4.62	4.43
Mn	0.07	0.06	0.03	0.04	0.01	0.03	0.02	0.02	0.01	0.02	0.02
Mg	4.24	4.26	7.15	7.00	7.11	6.33	6.57	4.54	4.14	4.31	4.28
Ca	0.03	0.02	0.02	0.02	0.02	0.01	0.01	0.03	0.03	0.03	0.03
Na	0.02	0.00	0.01	0.01	0.00	0.02	0.01	0.00	0.02	0.01	0.00
K	0.02	0.01	0.01	0.01	0.01	0.03	0.01	0.01	0.01	0.01	0.01
ΣR^{VI}	11.91	11.93	11.99	11.99	11.98	12.01	11.99	11.99	11.76	12.00	11.96
Fe/(Fe+Mg)	0.52	0.52	0.20	0.20	0.20	0.28	0.27	0.50	0.53	0.52	0.51

n.d. — not detected

Table 7: *Average composition of garnet*

	Greenschist facies																			
	Biotite subfacies Biotote-muscovite grade										Garnet subfacies									
Assemblage	Bt+Cal+Grt(Sps)+Kfs+Ms										Bt+Grt+Kfs		Bt+Chl+Grt		Chl+Grt				Grt+Ms	
Sample no.	3063/590.0-595.0										3678/256.0		3070/180.4		2215/9				3615/304.2-304.8	
Grain No.	2		5		3		1		4		1		1		1		2		1	
Position	core	rim	core	rim	core	rim	core	rim	core	rim	core	rim	core	rim	core	rim	core	rim	core	rim
Number of analysis	4	2	1	1	1	2	8	2	1	1	2	2	1	2	2	2	2	2	9	2
SiO_2	36.75	36.56	36.66	36.86	37.00	36.75	36.54	36.60	36.77	36.80	37.07	37.03	36.83	36.90	35.99	35.80	35.52	36.46	36.49	36.08
TiO_2	0.42	0.18	0.34	0.14	0.50	0.39	0.41	0.20	0.38	0.27	0.54	0.03	0.15	0.12	0.08	0.04	0.08	0.05	0.14	0.02
Al_2O_3	20.80	20.83	20.83	20.74	20.94	20.73	20.82	21.30	20.78	20.74	21.70	22.12	20.95	21.01	21.72	21.70	21.73	21.50	20.54	20.75
FeO	1.23	1.76	1.10	1.86	1.59	2.48	1.47	2.06	2.07	2.47	6.64	15.42	12.43	16.07	21.27	30.98	29.18	31.28	23.68	32.60
MnO	34.28	34.08	34.96	33.58	33.53	33.09	34.20	33.75	33.46	32.72	25.25	19.18	24.52	20.58	14.50	5.50	5.57	4.58	11.74	6.14
MgO	1.11	1.34	1.20	1.31	1.17	1.61	1.23	1.40	1.39	1.56	0.67	1.42	0.87	1.34	0.90	1.59	1.48	1.62	0.95	2.36
CaO	4.70	4.94	4.62	4.90	4.63	4.70	4.76	4.65	4.87	4.83	7.90	4.48	4.08	3.72	5.00	3.86	6.30	4.39	6.32	1.86
Total	99.29	99.69	99.71	99.39	99.36	99.75	99.53	99.96	99.72	99.39	99.77	99.78	99.83	99.74	99.46	99.47	99.86	99.88	99.86	99.81
Oxygen	12.00	12.00	12.00	12.00	12.00	12.00	12.00	12.00	12.00	12.00	12.00	12.00	12.00	12.00	12.00	12.00	12.00	12.00	12.00	12.00
Si	2.99	2.96	2.97	2.99	3.01	2.97	2.97	2.95	2.97	2.98	2.99	3.00	2.99	3.00	2.93	2.90	2.87	2.96	2.96	2.93
Al	0.01	0.04	0.03	0.01	0.00	0.03	0.03	0.05	0.03	0.02	0.01	0.00	0.01	0.00	0.07	0.10	0.13	0.04	0.04	0.07
ΣR^{IV}	3.00	3.00	3.00	3.00	3.01	3.00	3.00	3.00	3.00	3.00	3.00	3.00	3.00	3.00	3.00	3.00	3.00	3.00	3.00	3.00
Al	1.98	1.95	1.96	1.97	2.00	1.94	1.96	1.97	1.95	1.96	2.01	2.10	2.00	2.01	2.01	1.98	1.94	2.01	1.92	1.91
Ti	0.03	0.01	0.02	0.01	0.03	0.02	0.03	0.01	0.02	0.02	0.03	0.00	0.01	0.01	0.00	0.00	0.00	0.00	0.01	0.00
Fe^{3+}	0.00	0.07	0.03	0.02	0.00	0.04	0.02	0.06	0.02	0.02	0.00	0.00	0.00	0.00	0.06	0.11	0.18	0.02	0.10	0.16
ΣR^{VI}	2.01	2.03	2.01	2.00	2.03	2.00	2.01	2.04	1.99	2.00	2.04	2.10	2.01	2.02	2.07	2.09	2.12	2.03	2.03	2.07
Fe^{2+}	0.08	0.04	0.05	0.11	0.11	0.13	0.08	0.08	0.12	0.15	0.45	1.04	0.84	1.09	1.39	2.01	1.78	2.08	1.50	2.06
Mn	2.36	2.34	2.40	2.31	2.31	2.26	2.35	2.30	2.29	2.25	1.72	1.31	1.69	1.41	1.00	0.37	0.38	0.31	0.80	0.42
Mg	0.13	0.16	0.14	0.16	0.14	0.19	0.15	0.17	0.17	0.19	0.08	0.17	0.11	0.16	0.11	0.19	0.18	0.19	0.11	0.29
Ca	0.41	0.43	0.40	0.43	0.40	0.41	0.41	0.41	0.42	0.42	0.68	0.39	0.36	0.32	0.44	0.33	0.54	0.38	0.55	0.16
ΣR^{VIII}	2.98	2.97	2.99	3.01	2.96	2.99	2.99	2.96	3.00	3.01	2.93	2.91	3.00	2.98	2.94	2.90	2.88	2.96	2.96	2.93
Alm	0.03	0.01	0.02	0.04	0.04	0.04	0.03	0.03	0.04	0.05	0.15	0.36	0.28	0.37	0.47	0.69	0.62	0.70	0.51	0.70
Sps	0.79	0.79	0.80	0.77	0.78	0.76	0.78	0.78	0.76	0.75	0.59	0.45	0.56	0.47	0.34	0.13	0.13	0.11	0.27	0.14
Prp	0.04	0.05	0.05	0.05	0.05	0.06	0.05	0.06	0.06	0.06	0.03	0.06	0.04	0.05	0.04	0.06	0.06	0.06	0.04	0.10
Grs	0.14	0.14	0.13	0.14	0.13	0.14	0.14	0.13	0.14	0.14	0.23	0.13	0.12	0.11	0.15	0.11	0.17	0.13	0.17	0.05
Adr	0.00	0.01	0.00	0.00	0.00	0.00	0.00	0.00	0.00	0.00	0.00	0.00	0.00	0.00	0.00	0.01	0.02	0.00	0.01	0.01
Fe/(Fe+Mg)	0.38	0.42	0.34	0.44	0.43	0.46	0.40	0.45	0.46	0.47	0.85	0.86	0.89	0.87	0.93	0.92	0.92	0.92	0.93	0.89

(Contd.)

Table 7: *(Contd.)*

	Staurolite facies						
	Staurolite-chlorite-muscovite and andalusite-biotite subfacies						Andalusite-biotite subfacies
Assemblage	Bt+Grt+St		Bt+Grt				Bt+Cum+Grt
Sample no.	3055/421.0		3706/286.6				4317/370.1
Grain No.	1		1		2		1
Position	core	rim	core	rim	core	rim	rim
Number of analysis	2	2	1	2	1	2	2
SiO_2	36.67	36.79	36.94	36.56	36.52	36.50	35.99
TiO_2	0.12	0.03	0.09	0.02	n.d.	0.20	0.13
Al_2O_3	21.03	20.82	20.87	20.89	20.59	20.63	21.51
FeO	36.19	37.01	34.73	36.77	35.43	36.67	36.95
MnO	1.48	0.72	2.50	0.88	1.92	0.94	0.38
MgO	2.16	2.18	1.49	1.71	1.56	1.64	2.14
CaO	2.12	2.00	3.18	3.04	3.83	3.26	2.68
Total	99.74	99.63	99.80	99.85	99.85	99.83	99.77
Oxygen	12.00	12.00	12.00	12.00	12.00	12.00	12.00
Si	2.98	2.99	3.01	0.98	2.97	2.97	2.92
Al	0.02	0.01	0.00	0.02	0.03	0.03	0.08
ΣR^{IV}	3.00	3.00	3.01	3.00	3.00	3.00	3.00
Al	2.00	1.98	2.00	1.98	1.94	1.95	1.97
Ti	0.01	0.00	0.01	0.00	0.00	0.02	0.01
Fe^{3+}	0.03	0.02	0.00	0.06	0.09	0.06	0.09
ΣR^{VI}	2.04	2.00	2.01	2.03	2.03	2.02	2.07
Fe^{2+}	2.43	2.35	2.36	2.44	2.31	2.44	2.41
Mn	0.10	0.05	0.17	0.06	0.13	0.07	0.03
Mg	0.26	0.26	0.18	0.21	0.19	0.20	0.26
Ca	0.19	0.18	0.28	0.27	0.33	0.29	0.24
ΣR^{VIII}	2.97	2.99	2.99	2.97	2.96	2.98	2.93
Alm	0.84	0.85	0.81	0.84	0.80	0.84	0.84
Sps	0.03	0.02	0.05	0.02	0.04	0.02	0.01
Prp	0.08	0.08	0.06	0.06	0.06	0.06	0.08
Grs	0.06	0.05	0.08	0.08	0.10	0.08	0.07
Fe/(Fe+Mg)	0.90	0.90	0.93	0.93	0.93	0.93	0.91

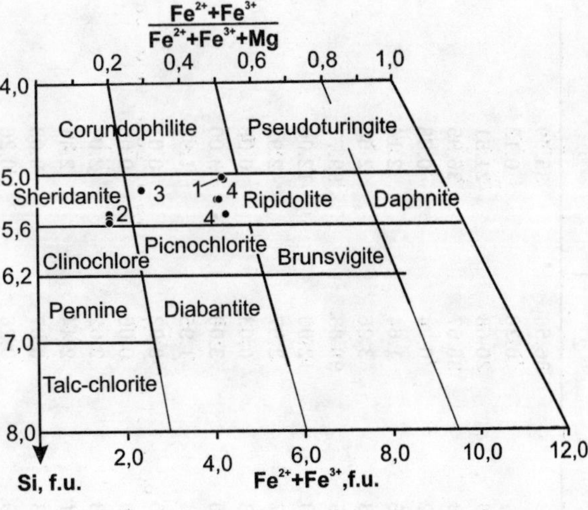

Fig. 16. Nomenclature of chlorites from greenschist facies metapelites after Hey (1954). 1-3 - Biotite subfacies: chlorite-potassium feldspar grade (Bt + Chl + Kfs), 2 - chlorite-potassium feldspar and biotite-muscovite grade (Chl + Ms), 3 - biotite-muscovite grade (Bt + Chl + Ms), 4 - garnet subfacies (Bt + Chl + Grt).

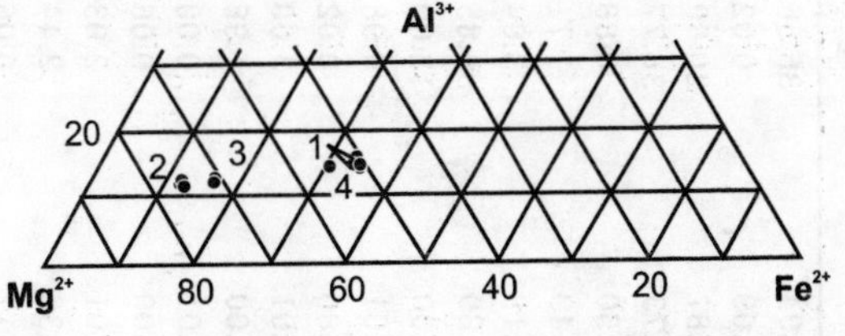

Fig. 17. Ratio of cations Fe^{2+}, Mg^{2+} and Al^{3+} in R^{VI} coordination in chlorites of greenschist facies. 1-3 - biotite subfacies: 1 - chlorite-potassium feldspar grade (Bt + Chl + Kfs), 2 - chlorite-potassium feldspar and biotite-muscovite grade (Chl + Ms), 3 - biotite-muscovite grade (Bt + Chl + Ms), 4 - garnet subfacies (Bt + Chl + Grt).

In metapelites of garnet subfacies there are both almandine-spessartine ($Sps_{49-62}Alm_{10-30}Grs_{11-25}Prp_{3-7}$) and spessartine-almandine ($Alm_{41-75}Sps_{10-38}Grs_{7-18}Prp_{4-7}$) garnets. They sharply distinguish from garnets of biotite subfacies by X_{Fe} = 0.84-0.93. X_{Fe} decreases from garnet cores to rims.

The late-kinematic garnet generation comprises sub-hedral grains of this mineral without pressure shadows (Fig. 18, 7-9). Garnet grains of this generation can be readily identified by the presence of randomly oriented inclusions, with some of these porphyroblasts containing no inclusions and noted for their low Mn concentrations. Their outlines clearly cut across the schistosity. Postkinematic garnet occurs in the metapelites within the staurolite facies. Here garnets is characterized by sufficient almandine composition ($Alm_{80-87}Sps_{0-5}Grs_{4-10}Prp_{6-10}$). With the rise of metamorphic grade the iron mole fraction of garnet slightly increases (X_{Fe} = 0.88-0.93), but as distinct from greenschist garnets there is no regular change from cores of crystals to rims. Because of this no clear zoning was detected.

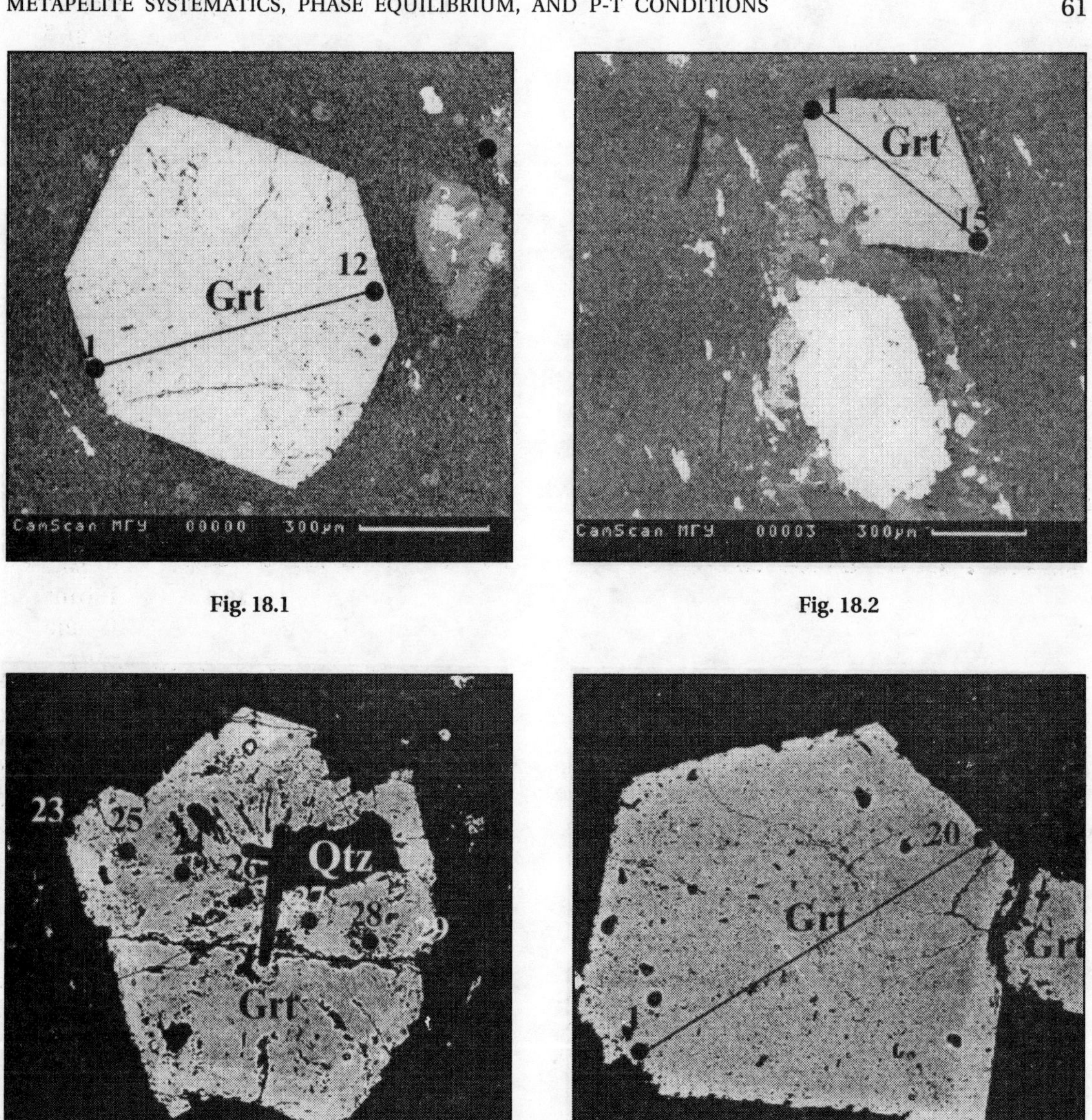

Fig. 18.1

Fig. 18.2

Fig. 18.3

Fig. 18.4

Garnets of greenschist facies display clear prograde growth zoning in terms of Ca, Mg, Fe, and Mn concentrations. The CaO and MnO concentrations decrease, whereas those of FeO and MgO increase from the cores to the rims of crystals (Fig. 19). The prograde zoning is related with the temperature increasing in the process of garnet crystallization.

At the present time there are two models of such progressive decreasing of Mn content in garnet with the temperature rise (Hollister, 1966; Miyashiro and Shido, 1973). One of

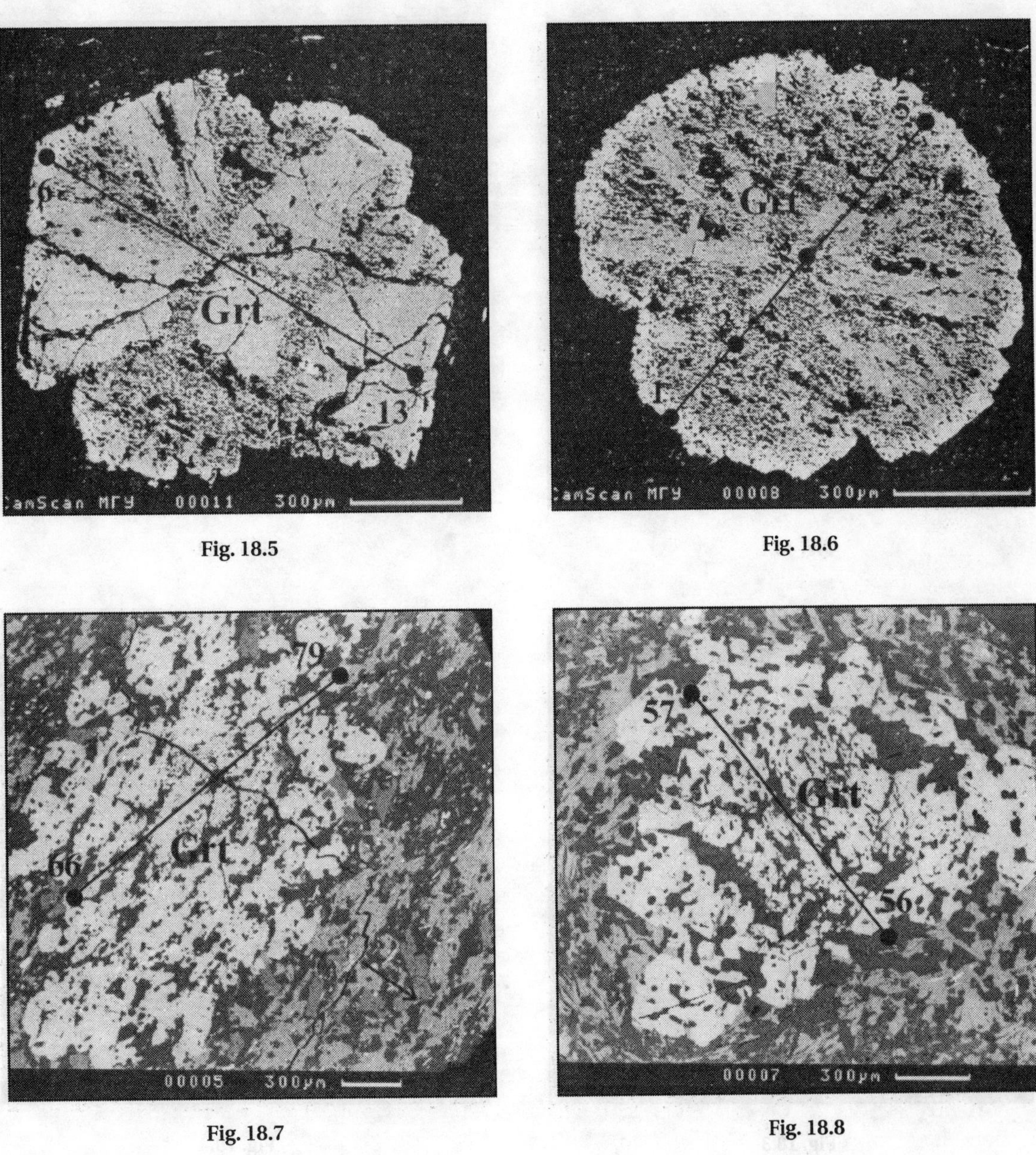

Fig. 18.5

Fig. 18.6

Fig. 18.7

Fig. 18.8

them is based on the absence or extremely insignificant role of diffusion in garnet. It is assumed that only the outermost crystal rim is in equilibrium with surrounding minerals. Their core largely enriched in Mn being excluded from the further reaction. In this model complete fractionating of Mn proceed in the garnet crystallization. The second model assumes complete equilibrium of garnet and surrounding minerals. In this case the decrease of Mn concentration is due to the increase in the amount of crystallizing garnet. Though the above

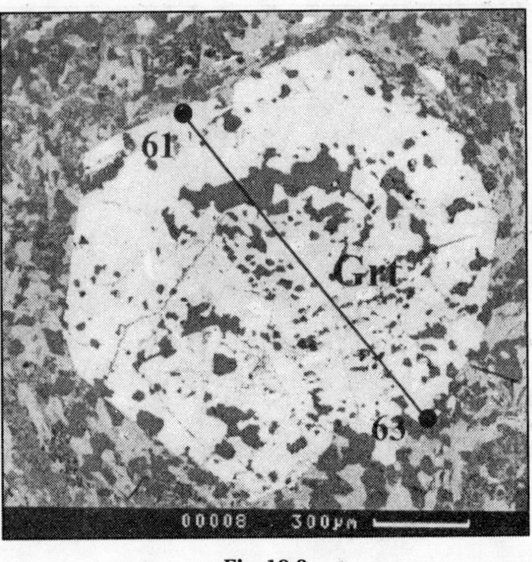

Fig. 18.9

Fig. 18. Morphological types (images in back-scattered electrons) of garnets from metapelites of Tim-Yastrebovskaya structure. 1-6 – Greenschist facies: 1, 2 – biotite subfacies (Bt + Cal + Kfs + Ms + Sps), 3-6 – garnet subfacies (3 – Bt + Grt + Kfs, 4 – Grt + Ms, 5, 6 – Grt + Chl), 7-9 – staurolite facies (7 – Bt + Grt + St, 8, 9 – Bt + Grt).

mentioned models are opposite, the calculated values and tendencies of changes in MnO concentrations in garnet are similar (Fig. 20).

In garnet of greenschist facies MnO concentrations decrease, whereas those of FeO increase from the cores to rims of crystals and character of zoning is subject to the degree dependence. In higher temperature staurolite facies this character changes and corresponds to exponential dependence. This distinction between garnets zoning is the evidence of different mechanisms of concentration zoning producing. In garnets of greenschist facies at low rate of component diffusion the growth zoning produces that is accompanied by MnO fractionating in metamorphic system. In conditions of staurolite facies with the rise of temperature during metamorphism the rate of diffusion greatly increases that provides leveling off component concentration in rims (Yardley, 1977). The increase in the width of MnO diffusion rim in garnets from staurolite facies assemblages in comparison with garnets from greenschist facies (Fig. 21) is the evidence of the increase in diffusion rate with the temperature rise. On this basis we can say that in natural conditions both models of concentration zoning producing in garnets are feasible and realization of any of them depends on temperature of metamorphism.

Amphiboles of cummingtonite-grunerite series in the studied metapelites are represented be cummingtonite (X_{Fe} = 0.44-0.47) and grunerite (X_{Fe} = 0.53-0.58) and have been analyzed in rock of staurolite facies only (Table 8).

Staurolite produces prismatic crystals with the size 0.1-0.2 mm with pleochroism in gold-yellow shades (Fig. 22). Staurolite has X_{Fe}=0.83-0.89 and contains notably amounts of ZnO (0.42-1.84 wt %) (Table 9). The ZnO concentrations in staurolite depend on its amount in the rock, because in other minerals its concentrations are present in trace amounts. The

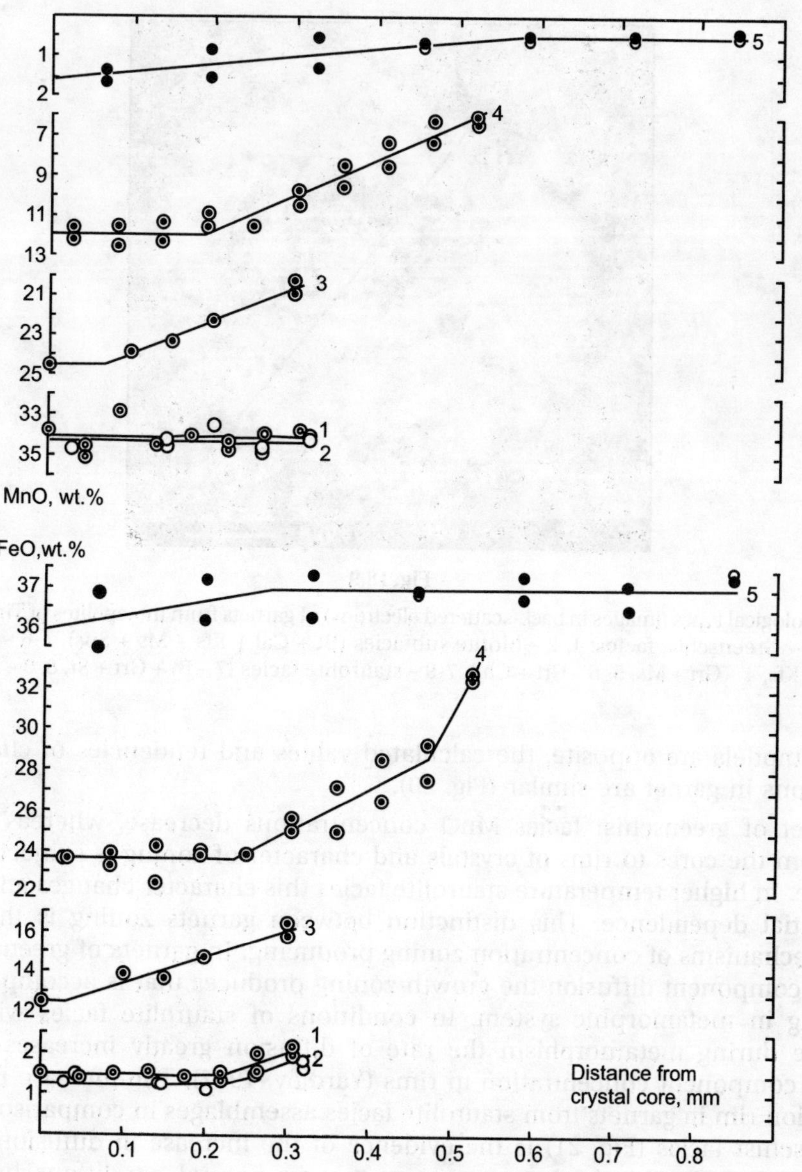

Fig. 19. Microprobe profiles across garnet grains. 1, 2 – Biotite subfacies, biotite-muscovite grade (high manganese metapelite (Bt + Cal + Kfs + Ms + Sps), 3, 4 – garnet subfacies (3 – Bt + Chl + Grt, 4 – Grt + Ms), 5 – staurolite facies (Bt + Grt + St).

high ZnO concentration is caused by a decrease in the staurolite content toward the high-temperature part of the facies, with Fe-rich staurolite being the first to decompose, Zn-enriched varieties surviving to higher temperatures, and the last remaining crystals becoming strongly enriching in Zn. In the studied metapelites staurolites enriched in ZnO are observed in both high-grade and low-grade area of staurolite facies. Because of this variations in the ZnO concentrations are conditioned by rock chemistry.

Table 8: *Representative microprobe analyses of cummingtonite*

	Staurolite facies									
	Staurolite-chlorite-muscovite and andalusite-biotite subfacies				Andalusite-biotite subfacies					
Assemblage	Bt+Cum				Bt+Grt+Cum					
Sample no.	3700/234.9				4317/370.1					
Analysis no.	Cum-98	Cum-101	Cum-102	Cum-104	Cum-51	Cum-53	Cum-13	Cum-14	Cum-16	Cum-17
SiO_2	54.74	55.35	55.29	54.48	52.78	53.12	55.08	53.77	53.58	53.34
TiO_2	0.10	n.d.	n.d.	n.d.	0.10	0.17	0.04	0.07	0.06	0.04
Al_2O_3	0.27	0.12	0.25	1.03	1.48	0.72	0.75	1.00	1.10	0.89
FeO	26.09	26.08	26.07	25.63	31.66	32.17	28.59	29.43	31.56	31.05
MnO	0.60	0.53	0.62	0.64	0.10	0.02	0.11	0.08	0.07	0.03
MgO	17.07	17.00	17.01	17.29	12.87	13.00	14.16	14.06	12.80	12.56
CaO	0.47	0.63	0.61	0.57	0.50	0.47	0.50	0.56	0.57	0.45
Na_2O	0.23	0.10	n.d.	0.21	0.38	0.20	0.04	0.03	0.04	0.05
K_2O	0.03	n.d.	n.d.	n.d.	0.05	0.01	0.01	n.d.	0.01	n.d.
Total	99.60	99.81	99.85	99.85	99.92	99.88	99.28	99.00	99.79	98.41
Oxygen	23.00	23.00	23.00	23.00	23.00	23.00	23.00	23.00	23.00	23.00
Si	7.91	7.99	7.98	7.83	7.79	7.87	8.13	7.94	7.94	8.01
Al^{IV}	0.05	0.02	0.04	0.17	0.26	0.13	0.13	0.17	0.19	0.16
ΣR^{IV}	7.96	8.01	8.02	8.00	8.05	8.00	8.26	8.11	8.13	8.17
Ti	0.01	0.00	0.00	0.00	0.01	0.02	0.00	0.01	0.01	0.00
Fe^{3+}	0.19	0.03	0.00	0.23	0.26	0.15	0.00	0.01	0.00	0.00
Fe^{2+}	2.96	3.12	3.14	2.84	3.64	3.82	3.52	3.62	3.90	3.89
Mn	0.07	0.06	0.08	0.08	0.01	0.00	0.01	0.01	0.01	0.00
Mg	3.67	3.65	3.66	3.70	2.83	2.87	3.11	3.09	2.83	2.81
Ca	0.07	0.10	0.09	0.09	0.08	0.07	0.08	0.09	0.09	0.07
Na	0.06	0.03	0.00	0.06	0.11	0.06	0.01	0.01	0.01	0.01
K	0.01	0.00	0.00	0.00	0.01	0.00	0.00	0.00	0.00	0.00
ΣR^{VI}	7.04	6.99	6.97	7.00	6.95	6.99	6.73	6.84	6.85	6.78
Mg/(Mg+Fe)	0.53	0.53	0.53	0.56	0.42	0.42	0.47	0.46	0.42	0.42

n.d. — not detected

Table 9: *Representative microprobe analyses of staurolite*

	Staurolite facies								
	Staurolite-chlorite-muscovite and andalusite-biotite subfacies					Andalusite-biotite subfacies			
Assemblage	Bt+Ms+St			Bt+Grt+St		And+Bt+Ms+St			
Sample no.	3068/611.0			3055/421.0		3716/608.2-608.6			
Position	Profile across crystal			core	rim	Profile across crystal			
Analysis no.	St-39	St-40	St-41	St-81	St-83	St-1	St-2	St-3	St-4
SiO_2	26.92	26.07	26.21	26.96	27.59	26.61	27.20	27.63	27.45
TiO_2	0.29	0.57	0.54	0.33	0.36	0.44	0.44	0.42	0.43
Al_2O_3	53.32	53.10	52.97	52.87	53.11	54.69	54.91	54.45	54.27
FeO	14.08	14.31	14.36	15.55	14.86	13.10	12.89	12.84	12.71
MnO	0.17	0.28	0.36	n.d.	n.d.	0.23	0.16	0.18	0.22
MgO	1.43	1.41	1.23	1.52	1.70	0.89	0.86	0.93	0.94
CaO	n.d.	0.02	0.05	n.d.	0.09	0.02	n.d.	n.d.	n.d.
Na_2O	0.22	0.67	0.33	0.23	n.d.	0.05	0.07	0.05	0.07
K_2O	n.d.	0.05	0.07	n.d.	0.03	n.d.	n.d.	0.01	n.d.
ZnO	1.55	1.45	1.84	0.47	0.41	1.79	1.77	1.72	1.50
Total	97.98	97.93	97.96	97.93	98.15	97.82	98.30	98.22	97.59
Oxygen	23.00	23.00	23.00	23.00	23.00	23.00	23.00	23.00	23.00
Si	3.83	3.70	3.73	3.83	3.91	3.79	3.85	3.92	3.92
Al^{IV}	8.93	8.87	8.89	8.85	8.87	9.18	9.17	9.10	9.12
Ti	0.03	0.06	0.06	0.04	0.04	0.05	0.05	0.05	0.05
ΣR^{IV}	8.96	8.93	8.95	8.89	8.91	9.23	9.22	9.15	9.17
Fe^{2+}	1.67	1.70	1.71	1.85	1.76	1.56	1.53	1.52	1.51
Mn	0.02	0.03	0.04	0.00	0.00	0.03	0.02	0.02	0.03
Mg	0.30	0.30	0.26	0.32	0.36	0.19	0.18	0.20	0.20
Ca	0.00	0.00	0.01	0.00	0.01	0.00	0.00	0.00	0.00
Na	0.06	0.18	0.09	0.06	0.00	0.01	0.02	0.01	0.02
K	0.00	0.01	0.01	0.00	0.01	0.00	0.00	0.00	0.00
Zn	0.16	0.15	0.19	0.05	0.04	0.19	0.19	0.18	0.16
ΣR^{VI}	2.21	2.37	2.31	2.28	2.18	1.98	1.94	1.93	1.92
Fe/(Fe+Mg)	0.85	0.85	0.87	0.85	0.83	0.89	0.89	0.89	0.89

n.d. — not detected

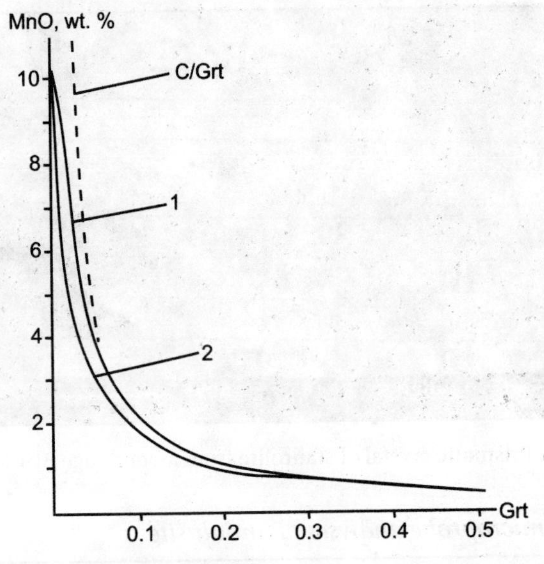

Fig. 20. Dependence of MnO content on garnet weight concentration (Grt) for models of total fractionating (1) and total equilibrium (2) after Atherton (1968). C – MnO average content (wt. %) in a total of ferruginous-magnesian minerals including garnet.

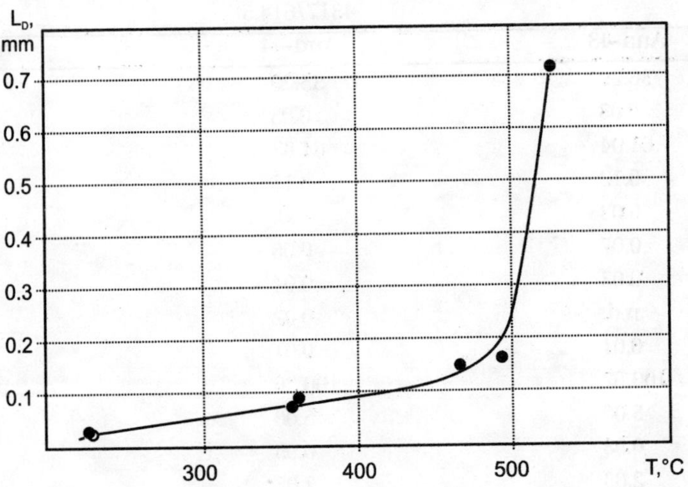

Fig. 21. Dependence of diffusion zone width (L_D) MnO on temperature of garnet crystallization in metapelites of Tim-Yastrebovskaya structure.

Andalusite in staurolite facies metapelites is characterized by the composition similar to pure Al_2SiO_5 with small amount of Fe_2O_3=0.05-0.15 wt. % (Table 10).

Cordierite occurs as isometric rounded grains and contains abundant biotite and muscovite inclusions. In some cases cordierite had sectorial trillings and sixlings. Magnesian composition of cordierite (X_{Fe}=0.30-0.31) (Table 11) is typical for metamorphism conditions of high temperature area of staurolite facies.

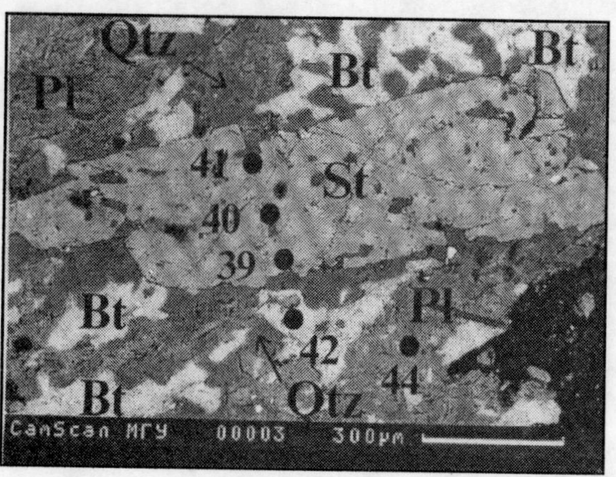

Fig. 22. Prismatic crystal of staurolite from assemblage Bt + Ms + St

Table 10: *Representative microprobe analyses of andalusite*

	Staurolite facies Andalusite-biotite subfacies		
Assemblage	And+Bt+Ms		
Sample no.	4317/614.5		
Analysis no.	And-43	And-44	And-45
SiO_2	36.29	35.20	35.63
TiO_2	0.03	0.05	0.07
Al_2O_3	64.04	64.83	64.77
FeO	0.12	0.15	0.05
MnO	0.03	n.d.	0.02
MgO	0.07	0.06	0.07
CaO	0.03	0.04	0.04
Na_2O	n.d.	0.02	0.01
K_2O	0.01	0.01	n.d.
Total	100.62	100.36	100.66
Oxygen	5.00	5.00	5.00
Si	0.98	0.95	0.96
Al	2.03	2.06	2.06
Ti	0.00	0.00	0.00
Fe	0.00	0.00	0.00
Mn	0.00	0.00	0.00
Mg	0.00	0.00	0.00
Ca	0.00	0.00	0.00
Na	0.00	0.00	0.00
K	0.00	0.00	0.00

n.d. — not detected

Table 11: *Representative microprobe analyses of cordierite*

	Staurolite facies, Andalusite-biotite subfacies, Cordierite-biotite grade			Staurolite facies, Andalusite-biotite subfacies, Cordierite-biotite grade	
Assemblage	Bt+Crd		Assemblage	Bt+Crd	
Sample no.	3129/311.0		Sample no.	3129/311.0	
Analysis no.	Crd-28	Crd-30	Analysis no.	Crd-28	Crd-30
SiO_2	49.99	49.78	Oxygen	18.00	18.00
TiO_2	n.d	0.01	Si	5.02	5.00
Al_2O_3	33.65	33.42	Al^{IV}	0.98	1.00
FeO	6.76	7.09	ΣR^{IV}	6.00	6.00
MnO	0.22	0.28	Al^{VI}	3.00	2.96
MgO	9.10	9.08	ΣR^{VI}	3.00	2.96
CaO	0.03	0.05	Fe	0.57	0.59
Na_2O	0.25	0.28	Mn	0.02	0.02
K_2O	0.02	n.d	Mg	1.36	1.36
Total	100.02	99.99	Na	0.05	0.05
			ΣR^{VIII}	2.00	2.02
			Mg/(Fe+Mg)	0.70	0.69

P-T METAMORPHIC CONDITIONS

The P-T metamorphic conditions of the Tim-Yastrebovskaya metapelites were determined by two methods. First, we compared the mineral assemblages observed in the rocks with known petrogenetic grids (Spear and Cheney, 1989; Powell and Holland, 1990; Korikovskii, 1979 and others) and the experimentally established boundaries of the stability fields of minerals and mineral assemblages (for example, alumina silicates, staurolite with quartz, and others). The advantage of this method is its capability of determining the overall position of the assemblages in a P-T space and the character of metamorphism (facies series). This technique can be particularly successfully applied to medium-temperature metapelites, in which a relatively narrow temperature interval may involve the appearance and decomposition of several minerals (for example, chlorite, staurolite, muscovite, etc.), which systematically vary in composition. At the same time, this technique is disadvantageous because it often cannot provide precise quantitative P-T estimates.

The second method, mineralogical thermobarometry, make use of the composition of chemically zoned minerals to reconstruct P-T coordinates and paths. These techniques can be more successfully utilized in calculating the P-T trajectories of high-temperature rocks, because the careful interpretation of thermobarometric results potentially offers valuable information on the evolution of the metamorphic complexes. The disadvantage of this method is the possibility of errors caused by the poorly known thermodynamic constant of some mineral solid solutions and the affect of some components (such as MnO in garnet or ZnO in staurolite) on the properties of the minerals, as well as by difficulties in determining the equilibrium compositions of minerals utilized in thermobarometry.

In order to estimate the temperatures of the earlier metamorphic episode we used the compositions of the early-kinematic garnet and the garnet-biotite (Perchuk, 1986; Thompson, 1976; Ferry and Spear, 1978), garnet-chlorite (Perchuk, 1989; Patrik and Evans, 1989), and garnet-muscovite (Hanes and Forest, 1988; Krogh and Raheim, 1978) thermometers (Table 12). We utilized the compositions of the marginal parts of garnet grains with prograde zoning and the compositions of adjacent biotite, muscovite, and chlorite in the rock matrix.

The pressure was quantified by the garnet-plagioclase-andalusite-quartz geobarometer (Aranovich and Podlesskii, 1980) with the use of the compositions of garnet in the margins of grains with prograde zoning and the matrix plagioclase.

Our results indicate that the Tim-Yastrebovskaya metapelites underwent metamorphism to the greenschist facies during the early stage. The temperature values for assemblages of biotite-muscovite grade vary in the interval 228-241°C. For the central part of crystals from the chlorite-potassium feldspar grade they are the lowest, being 217-231°C. The metamorphic conditions corresponded to the high-grade greenschist facies with temperatures of 300-500°C (Table 12) and pressures of 3.0-3.1 kbar.

The thermodynamic parameters of the higher temperature metamorphic episode were quantified based on the compositions of late-kinematic garnet and the garnet-biotite (Perchuk, 1986; Thompson, 1976; Ferry and Spear, 1978), garnet-staurolite (Perchuk, 1989), and biotite-muscovite (Hoisch, 1989) geothermometers. The compositions of the minerals were measured in the margins of unzoned garnet grains and biotite in the rock matrix (Sample 4317/370.1; 3055/421.0; 3706/286.6). The estimation of temperature values does not allow characterizing temperature conditions of staurolite-chlorite-muscovite and andalusite-biotite subfacies separately. So we can discuss staurolite facies temperature conditions in general.

We have determined that the metapelites were metamorphosed to the staurolite facies at 500-580°C (Table 12). We failed to evaluate the pressure with the use of known geobarometers because of the rarity of suitable mineral assemblages. Because of this, our pressure estimates for the staurolite facies metapelites were based on the results of paragenetic analysis. The presence of the staurolite plus andalusite assemblage in the metapelites suggests pressures of lower 3-4 kbar (Korikovskii, 1979). According to calcite-dolomite thermobarometry for calcite-dolomite-pargasite-spinel-forsterite hornfelses from contact aureole near intrusion of granodiorites of the Stoilo-Nikolaevskii Complex — the pressure is evaluated 1.7 kbar at temperature 570°C.

THE ORIGIN OF THE TIM-YASTREBOVSKAYA STRUCTURE AND THE METAMORPHIC EVOLUTION OF METAPELITES

In the general geodynamic evolutionary model for the Voronezh crystalline massif, the Tim-Yastrebovskaya structure is currently regarded as a second-order rift-related structure within the Proterozoic continent of the Kursk Magnetic Anomaly.

Early during Proterozoic time and the most active uplift of the Kursk and Voronezh-Samara continent, the Kursk Magnetic Anomaly was a dry land. An inner continental sea basin existed at the site of the structure in the earliest Proterozoic. In lagoons of this basin, terrigenous sediments and the clayey redeposition products of weathering crusts were accumulated. This time period was also responsible for the deposition of chemogenic carbonate-iron formations (the Kursk Group and the Rogovskaya Formation of the Oskol'skaya Group). Changes in the steady-state thermal regime brought about the reactivation of the

Table 12: *Metamorphic temperatures calculated for metapelites of the Tim-Yastrebovskaya structure*

Facies	Grade	Assemblage	Sample no.	Grain no.	Analysis no. Grt	Analysis no. Bt, Chl, Ms	Position	Thermometer	Temperature °C
Biotite subfacies	Chlorite-potassium feldspar (?) and biotite-muscovite grade	Bt+Cal+Kfs+Ms+Sps	3063/590,0-595,0	2	4, 6, 9, 10	5. 10 20–22, 24, 27	core	P	217
					1, 12		rim		228
				5	1		core		221
					2		rim		228
				3	4		rim		221
					3, 5		rim		241
				1	3–9, 13		core		223
					1, 15		rim		231
				4	8		core		231
					6		rim		240
				6	13		Small grain		236
Garnet subfacies		Bt+Kfs+Grt	3678/256,0	1	24, 26	21, 23	core	P	301
					23, 29		rim		364
		Chl+Bt+Grt	3070/180,4	1	33	40	core		317
					34, 37		rim		359
		Chl+Grt	2215/9	1	9, 10	2–5	core	P	403
					6, 13		rim		473
					9, 10		core	P&E	436
					6, 13		rim		468
				2	2, 3		core	P	471
					1, 5		rim		480
					2, 3		core	P&E	494
					1, 5		rim		481
		Grt+Ms	3615/304,2-304,8	1	7–15	45–49	core	K&R	345
					1, 20		rim		409
					7–15		core	H&F	453
					1, 20		rim		495
Staurolite facies		Bt+Cum+Grt	4317/370,1	1	2–4, 7, 8	18, 52, 55	rim	P	521
		Bt+Grt+St	3055/421,0	1	72, 73	80, 82	core	P	511
					66–70		rim		526
		Bt+Grt	3706/286,6	1	60	58, 59, 65, 66	core	P	514
					61, 63		rim		535
				2	54		core		521
					56, 57		rim		528

Note: Thermometers: P (Perchuk, 1989), P&E (Patric and Evans, 1989), K&R (Krogh and Raheim, 1978), H&F (Hanes and Forest, 1988).

Kursk block and the onset of a prerift stage, when an extensive uplift started to develop in place of the shallow-water basin. Its growth was coupled with active volcanism and conglomerate deposition on the slopes. A still-shallow northwest-trending rift trough started to develop during the early rifting stage in the axial weakened zone of the uplift. As the trough evolved, the subsidence became progressively more intense and thick masses of fine-grained and graywacke silty metapelites began to fill the structure. The sedimentation process was accompanied by active volcanism, which extended outside the Tim-Yastrebovskaya structure. Sedimentation and volcanism of the late rifting stage were concentrated only within the ever-deepening trough. The closing rifting stages of the structure were marked by the transition from fine terrigenous and chemogenic sediments to continental molasses. The closure of the Tim-Yastrebovskaya rift was induced by collision between the Kursk and Voronezh-Samara geoblocks. The onset of compression, which was oriented from northeast to southwest, caused extensive imbrication and overthrusting within the structure. The growth of a folded orogen at the site of the rift trough was accompanied by the first episode of regional metamorphism.

The U-Pb age of volcanism that produced the Timskaya Formation of the Oskol'skaya Group is 2254 (it was determined on zircon from acid volcanics that rest conformably with the rock of the formation; Artemenko *et al.*, 1992). It can be hypothesized that regional metamorphism of the Tim-Yastrebovskaya deposits proceeded slightly later than 2250 Ma but earlier than the intrusion of the postmetamorphic granodiorites of the Stoilo-Nikolaevskii Complex, which was dated at 2086 Ma (Artemenko *et al.*, 1992).

The compressional environment predetermined the appearance of a number of crustal melting zones, which caused the intrusion of granodiorites of the Stoilo-Nikolaevskii Complex, after the closure of the Tim-Yastrebovskaya rift. The igneous rocks served as the sources of heat for younger contact-regional metamorphism, which occurred locally and produced the staurolite-andalusite, staurolite-sillimanite, and sillimanite-muscovite metamorphic zones.

Intrusions of this complex were dated at 2086 ± 5 Ma (Artemenko *et al.*, 1992). The metamorphic event that produced the metamorphics, including the metapelites, should have been coeval with these magmatic processes. Hence, a fragment of a young platform appeared in the Tim-Yastrebovskaya rift structure late in the Early Proterozoic.

CONCLUSIONS

The results of our investigation of different types of metapelites in Tim-Yastrebovskaya rift structure, of phase relations and mineral compositions as well as of the metamorphic conditions allow to make following conclusions:

The presence of oversaturated with silica metapelites and rare occurrence of their high alumina varieties is the evidence of low maturity and small amount of weathering products of Archean basement rocks because Early Proterozoic metapelites were formed at their re-deposition. Rather vast occurrence of high potassium metapelites is the sign of the fact that Archean rocks had undergone the processes of migmatization and granitization.

The lowest temperature grade conditions of metamorphism corresponds to greenschist facies where chlorite-potassium feldspar and biotite-muscovite grades of biotite subfacies are isolated. The staurolite-chlorite-muscovite and andalusite biotite subfacies are recognized within staurolite facies at medium grade metamorphism. The cordierite-biotite grade isolated in the high temperature part of staurolite facies.

The mineral composition study has shown that the increase of metamorphic grade leads to the rise of iron mole fraction and the amount of Ti, Al^{VI} in biotite being reduced. Moreover Fe concentrations increase; MnO simultaneously decreases from the cores to rims of garnet crystals from low to high metamorphic grade. The investigation of compositional zoning in garnets in greenschist facies conditions indicates growth zoning which was due to low rate of the component diffusion. The fractionating process governs FeO enrichment of garnet rims. With the metamorphic grade increasing and accordingly diffusion rate in garnets from staurolite facies, the diffusion zoning is characterized by leveling off component concentrations at the garnet rims.

The garnet of the rocks can be subdivided into two generations: early kinematic with respect to the D1 stage and related to the folding and late kinematic which occurs near intrusive bodies.

The greenschist facies metapelites was related to regional metamorphism; the staurolite facies mineral assemblages developed locally and are spatially restricted to compositionally variable intrusive bodies.

The greenschist facies temperatures of metamorphism are 220-250°C for biotite subfacies, 300-500°C for garnet subfacies and 500-580°C for staurolite facies.

The Tim-Yastrebovskaya structure was affected by two metamorphic episodes: early regional metamorphism in the greenschist facies conditions at 2250-2086 Ma and younger more high-temperature contact metamorphism at 2086±5 Ma. The latter event produced the metamorphic zoning observable within the structure.

ACKNOWLEDGMENT

This study was supported by the Integration Federal Programme (project E-0348), Universities of Russia Programme (project UR.09.01.038), a grant from the President of the Russian Federation (project 00-15-99397), a grant of the Russian Ministry of Education (project E00-9.0-2), and the Russian Foundation for Basic Research (project no. 00-05-64522, 02-05-79023).

REFERENCES

Aranovich, L.Ya. & Podlesskii, K.V., 1980. Garnet-plagioclase geobarometer. *Doklady Akademii Nauk SSSR*, **251**, 1216-19.

Artemenko, G.V., Bartnitskii, E.N. & Myasnyankin, V.I., 1992. The U-Pb age of igneous rocks from Orlovsko-Timskii greenstone structure of the Voronezh Crystalline Massif. *Doklady Akademii Nauk Ukrainy*, **7**, 113-17.

Atherton, M.P., 1968. The variation in garnet, biotite and chlorite composition in medium grade pelitic rocks from the Dalradian, Scotland, with particular reference to the zonation in garnet. *Contributions to Mineralogy and Petrology*, **18**, 347-71.

Brown, E.H., 1967. The greenschist facies in part of eastern Otago, New Zealand. *Contributions to Mineralogy and Petrology*, **14**, 259-92.

Brown, E.H., 1969. Some zoned garnets from the greenschist facies. *American Mineralogist*, **54**, 1662-72.

Chernyshov, N.M., Nenakhov, V.M. & Lebedev, I.P., *et al.*, 1997. A Model for the Geodynamic Development of the Voronezh Crystalline Massif. *Geotektonika*, **3**, 21-30.

Cygan, R.T. & Lasaga, A.C., 1982. Crystal growth and the formation of chemical zoning in garnets. *Contributions to Mineralogy and Petrology*, **79**, 187-200.

De Yoreo, J.J., Lux, D.R. & Guidotti, C.V., 1991. Thermal modeling in low pressure/High Temperature Metamorphic Belts. *Tectonophysics*, **188**, 253-68.

Dobretsov, N.L., Reverdatto, V.V. & Sobolev, V.S. *et al.*, 1970. *Fatsii Metamorfizma (Facies of Metamorphism).* Moscow: Nedra, 432.

Dobretsov, N.L., Sobolev, V.S. & Khlestov, V.V., 1972. *Fatsii Regionalnogo Metamorfizma Umerennyh Davlenii (Facies of Regional Metamorphism at moderate pressures).* Moscow: Nedra, 286.

Dutrow, B.L. & Holdaway, M.J., 1989. Experimental determination of the upper thermal stability of Fe-staurolite at medium pressures. *Journal of Petrology*, **30**, 239-48.

England, P.C. & Thompson, A.B., 1984. Pressure-temperature-time paths of regional metamorphism: I. Heat transfer during the evolution of regions of thickened continental crust. *Journal of Petrology*, **25**, 894-928.

Ferry, J.M., 1991. Dehydration and decarbonation reactions as a record of fluid infiltration. *Mineralogical Society of America Reviews in Mineralogy*, **26**, 351-93.

Ferry, J.M. & Spear, F.S., 1978. Experimental calibration of the partitioning of Fe and Mg between biotite and garnet. *Contributions to Mineralogy and Petrology*, **66**, 113-17.

Fonarev, V.I., Graphchikov, A.A. & Konilov, A.N., 1991. A consistent system of geothermometers for metamorphic complexes. *International Geological Review*, **33**, 743-83.

Hanes, A. & Forest, R.C., 1988. Empirical garnet-muscovite geothermometry in low-grade metapelites, Selwyn Range (Canadian Rockies). *Journal of Metamorphic Geology*, **6**, 297-309.

Hey, M.H., 1954. A new review of the chlorites. *Mineralogical Magazine*, **30**, 277-92.

Hoisch, T.D., 1989. A muscovite-biotite geothermometer. *American Mineralogist*, **74**, 565-72.

Hollister, L.S., 1966. Garnet zoning, an interpretation based on the Rayleigh fractionation model. *Science*, **154**, 1647-51.

Hoschek, G., 1969. The stability of staurolite and chloritoid and their significance in metamorphism of pelitic rocks. *Contributions to Mineralogy and Petrology*, **22**, 208-32.

Kerrick, D.M., 1974. Review of metamorphic mixed-volatile (H_2O-CO_2) equilibria. *American Mineralogist*, **59**, 729-62.

Korikovskii, S.P., 1979. *Fatsii Metamorfizma Metapelitov (Metamorphic Facies of Metapelites).* Moscow: Nauka, 260.

Kretz, R., 1983. Symbols for rock-forming minerals. *American Mineralogist*, **68**, 277-79.

Krogh, E.J. & Raheim, A., 1978. Temperature and pressure dependence of Fe-Mg partitioning between garnet and phengite, with particular reference to eclogite. *Contributions to Mineralogy and Petrology*, **66**, 75-80.

Loomis, T.P., 1983. Compositional zoning of crystals: a record of growth and reaction history// In: Saxena, S.K (ed.) *Kinetics and Equilibrium in Mineral Reactions*, 1-61.

Miyashiro, A., 1953. Calcium-poor garnet in relation to metamorphism. *Geochimica et Cosmochimica Acta*, **4**, 179-208.

Miyashiro, A., 1958. Regional metamorphism of the Gosaisyo-Takanuki district in the central Abukuma Plateau. *Tokyo University Faculty Science Journal*, **11**, 219-72.

Miyashiro, A., 1973. *Metamorphism and metamorphic belts.* New York: Halsted Press, J. Wiley & Sons. N.Y., 492.

Miyashiro, A. & Shido, F., 1973. Progressive compositional change of garnet in metapelites. *Lithos*, **6**, 13-20.

Patric, B.E. & Evans, B.W., 1989. Metamorphic evolution of the Seaward Peninsula Blueschist Terrane. *Journal of Petrology*, **30**, 351-555.

Perchuk, L.L., 1971. *Sosushchestvuyushchiye Mineraly (Coexisting minerals).* Leningrad: Nedra, 424.

Perchuk, L.L., 1986. Evolution of metamorphism. In Zharikov (ed.), *Eksperiment v reshenii aktual'nykh zadach geologii (Experiment in Solving Topical Problems of Geology).* Moscow, 151-73.

Perchuk, L.L., 1989. Mutual coordination of some Fe-Mg geothermometers on the basis of the Nernst law: A Revision. *Geokhimiya*, **5**, 611-22.

Powell, R. & Holland, T., 1990. Calculated mineral equilibria in the pelite system, KFMASH (K_2O-FeO-MgO-Al_2O_3-SiO_2-H_2O). *American Mineralogist*, **75**, 367-80.

Richardson, S.W., 1966. Staurolite stability in a part of the system Fe-Al-Si-O-H. *Journal of Petrology*, **9**, 467-88.

Spear, F.S. & Cheney, J.T., 1989. A petrogenetic grid for pelitic schists in the system SiO_2-Al_2O_3-FeO-MgO-K_2O-H_2O. *Contributions to Mineralogy and Petrology*, **101**, 149-64.

Thompson, A.B., 1976. Mineral reactions in pelitic rocks: 2. Calculation of some P-T-X (Fe-Mg) phase relations. *American Journal of Science*, **276**, 425-54.

Yardley, B.W.D., 1977. Homogenization of zoned garnets from pelitic schists. *American Mineralogist*, **62**, 793-800.

Yardley, B.W.D., 1981. A note on the composition and stability of Fe-staurolite. *Neues Jahrbuch Miner. Monath*, **3**, 127-32.

Yardley, B.W.D., Leake, B.E. & Farrow, C.M., 1980. The metamorphism of Fe-rich pelites from Connemara, Ireland. *Journal of Petrology*, **21**, 365-99.

3

METAMORPHOSED CARBONATE-EVAPORITIC ROCKS AT TRANSITION OF HIGH-PRESSURE AMPHIBOLITE/ECLOGITE FACIES CONDITIONS: A CASE STUDY FROM THE SARE SANG LAPIS-LAZULI DEPOSIT (AFGHANISTAN)

SHAH WALI FARYAD*

ABSTRACT

Volatile-bearing minerals with CO_2, H_2O, halogens, sulfur and boron are widespread in the lapis-lazuli deposits at Sare Sang (Western Hindukush). The rocks hosting these minerals represent marbles, different varieties of calc-silicate rocks and whiteschists that derived from carbonate-evaporite layers within volcanic and psammopelitic sequence of the Sare Sang Series. Textural relations and mineral compositions indicate a high-pressure amphibolite facies metamorphism of these rocks with various stages of retrogression. The main volatile-bearing phases in the rocks are carbonates, sodalite-group minerals, scapolite, amphibole and biotite, however most of these minerals have retrograde origin. Minerals stable at peak P-T conditions were carbonates, graphite, scapolite, apatite, amphibole, phlogopite, talc, clinohumite, pyrothite and pentlandite. Calculated log (X_F/X_{OH}) and log (X_{Cl}/X_{OH}) in coexisting biotite and amphibole indicates preferential partitioning of Cl into amphibole and F into biotite. Retrogression that was accompanied by granite formation led to recrystallization of these minerals as well as to fluid infiltration and metasomatic reactions in the surrounding rocks. These processes led to formation of the main lapis-lazuli deposit in the Sare Sang locality.

KEY WORDS

Amphibolite facies metamorphism, carbonate-evaporite sequence, Sare Sang Afghanistan.

INTRODUCTION

Metamorphic fluids in the lowermost crust are derived either from original pore water, when the rocks were at shallower water level, and progressively modified by the release of volatiles by metamorphic breakdown, or they have mantle origin and introduced from below, by crystallization of magma in the lower crust (Yardley *et al.*, 2000). Carbonate-evaporitic rocks are the best reservoir of fluids (CO_2, F, Cl, S and B) of various composition

* Institute of Petrology and Structural Geology, Charles University, Albertov 6, 12843 Prague 2, Czech Republic. e-mail: faryad@natur.cuni.cz

that enter mineral structure during metamorphism. As the volatiles infiltrate hosting silicate rocks and migrate to higher crustal levels, their origin in high-grade terrains is not always clear. Evidences of evaporitic protoliths were found even in ultra-high-pressure rocks (Schertl *et al.*, 1991; Chopin, 1984). The Sare Sang locality, which is known for the presence of lapis lazuli deposits and whiteschists, is the best example, where rocks containing minerals with various volatile contents form lenses and layers reflecting primary sedimentary sequences. The Sare Sang and nearby areas are characterized by sillimanite ± kyanite + garnet ± cordierite gneiss, amphibolite, calc-silicate rocks and marbles metamorphosed under upper amphibolite facies conditions (Blaise and Cesbron, 1966). Granitoid rocks with biotite, amphibole and rare pyroxene are associated with migmatized gneisses. The lapis lazuli deposits occur at contacts with granites or form lenses or layers in marbles. According to Yurgenson and Sukharev (1984), the lapis lazuli deposits were result from interaction between granite-related hydrothermal fluids (Na-metasomatism) on marbles with high amounts of pyrite.

An alternative interpretation on metamorphosed sedimentary origin for the Sare Sang rocks was proposed by Kulke (1976) and Schreyer and Abraham (1976). On the basis of relict anhydrite and gypsum (Kulke, 1976; Blaise and Cesbron, 1966) they assumed that the lapis lazuli and whiteschists are part of a metamorphosed evaporite deposit that was extensively modified by melting and volatilization. Detailed petrological studies (Grew, 1988; Faryad, 1999, 2002) indicated that most of volatiles are incorporated in silicate minerals in metasedimentary rocks. This work is aimed to summary mineralogy of both metamorphosed carbonate and evaporite rocks that are part of volcano-sedimentary sequences of the Sare Sang series. This is to confirm or otherwise the postulated origin of volatiles from carbonate-evaporite source rocks and implications for their infiltration to the host silicate rocks during metamorphism. Variations of halogen contents and their preferential partitioning in coexisting silicate minerals are also discussed.

GEOLOGICAL BACKGROUND

The Sare Sang locality with lapis lazuli deposit is part of the South Badakhshan block, which occurs at boundary between Afghanistan in western Hindukush and Tajikistan in southwest Pamir (Fig. 1). According to Boulin (1991), the South Badakhshan block together with Ukraine-Turan and Kirghizian blocks belongs to the Baykalian Proterozoic basement. It consists of 6.7-9.2 km thick terrigenous carbonate and volcanic rocks, which are subdivided into two series: the lower-Sare Sang and the upper-Kohe Lal (Dronov and Feoktistov, 1980). The present position of the South Badakhshan block and nearby areas results from Neogene and Quaternary tectonic activity. The Sare Sang series with metaevaporite-carbonate rocks is based on U-Pb-isotopic data (2700-2400 Ma) from Pamir (Horeva *et al.*, 1971) assumed to be Achaean in age (Dronov and Feoktistov, 1980). Hubbard *et al.* (1999) recently obtained a Neogene cooling age ($^{40}Ar/^{39}Ar$ data from biotite, phlogopite and hornblende) for upper amphibolite facies whiteschists in the Pamir (Tajikistan). The high-grade rocks in Tajik side in Pamir called Goran series are represented by 3400-4000 m thick metasediments, mainly metapelite or semipelite, marbles and calc-silicate rocks, with intercalation of quartzite and amphibolite (Kiselyov and Budanov, 1986). Kiselyov and Budanov, (1986) assumed earlier eclogite/granulite facies metamorphism, based on mineral assemblages in metabasites and charnockites.

The best exposures of the Sang series are along the north-flowing Kokcha river valley [ca. 100 km SSE from Paiz Abad in the Badakhshan Province (Fig. 1)]. The common rocks of the Sare Sang series are calcite and dolomite marbles and schists intercalated with biotite-amphibole (± garnet, ±diopside) gneisses, amphibolites and quartzites. Granites and pegmatites that cross cut the surrounding metamorphic rocks are also present. The Sare Sang lapis lazuli deposits, formed mainly by lazurite with various amounts of haüyne, sodalite, scapolite, afghanite and phlogopite, are the main reservoir of S and halogens. They form lenses or layers in marbles and occur at contacts with granites and pegmatite. Blaise and Cesbron, (1966) reported tourmaline-rich quartz + albite rocks that occur immediately under the marbles.

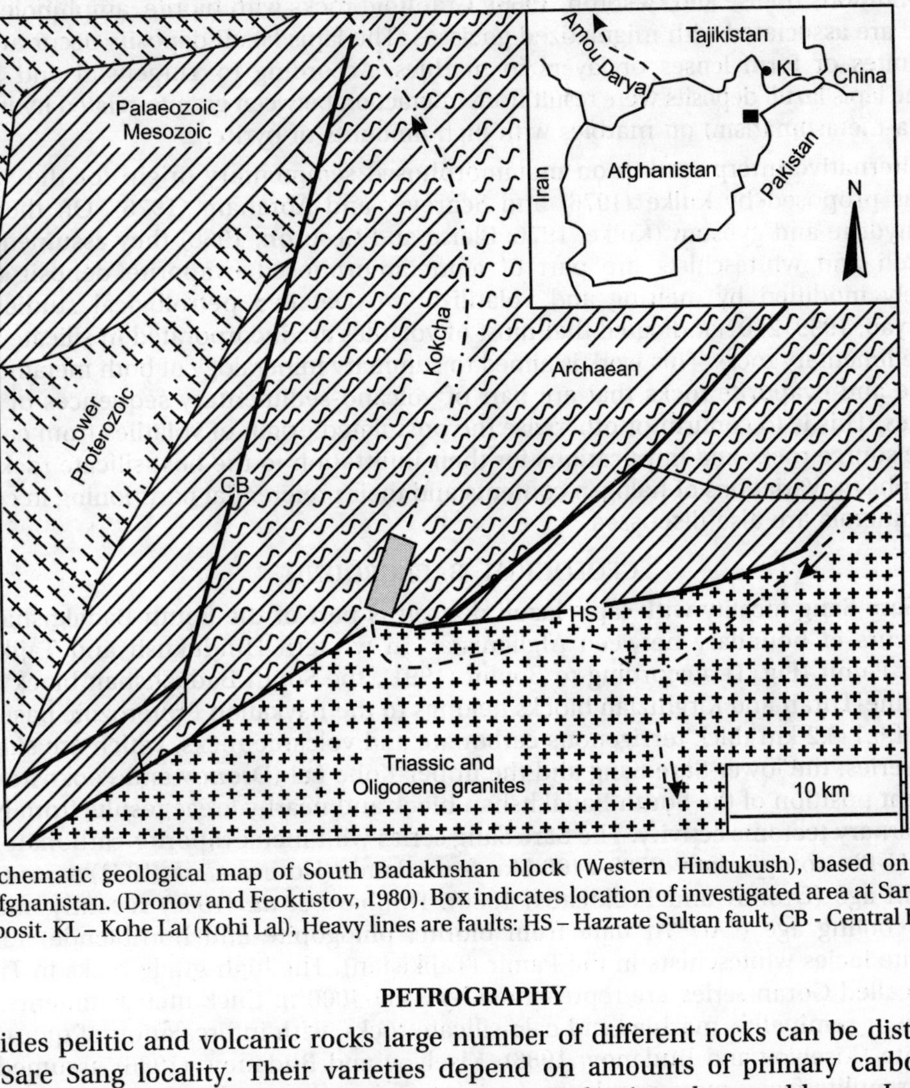

Fig. 1. Schematic geological map of South Badakhshan block (Western Hindukush), based on Geological Map of Afghanistan. (Dronov and Feoktistov, 1980). Box indicates location of investigated area at Sare Sang lapis lazuli deposit. KL – Kohe Lal (Kohi Lal), Heavy lines are faults: HS – Hazrate Sultan fault, CB - Central Badakhshan fault.

PETROGRAPHY

Besides pelitic and volcanic rocks large number of different rocks can be distinguished in the Sare Sang locality. Their varieties depend on amounts of primary carbonate and evaporite mixture that result formation of variegated mineral assemblages. In addition, metasomatic reactions between granite/pegmatite and adjacent carbonate-evaporite rocks led to formation of lapis lazuli deposit. The Sare Sang and nearby areas are largely underlain by sillimanite ± kyanite + garnet ± cordierite gneiss, amphibolite, calc-silicate rocks and dolomitic marble metamorphosed under upper amphibolite facies conditions (Blaise and Cesbron, 1966). Besides marbles (calcite, dolomite and rarely magnesite), the metacarbonatic/ metaevaporitic rocks can be based on mineral assemblages subdivided into several types (Fig. 2):

1. Calc-silicate rocks that includes three varieties: forsterite-bearing, garnet-bearing and garnet-hyperstene-bearing
2. Kyanite-talc whiteschists and ultra-magnesian rocks
3. Na-Ca (K)-rich rocks with various amounts of halogen-, S- and CO_2-bearing phases
4. Lapis lazuli deposits.

Calc-silicates

Forsterite-bearing calc silicates consist mainly of forsterite and phlogopite, but they contain also one or more of the minerals: diopside, magnesiohornblende, pyrothite, apatite, rutile, clinohumite and sapphirine. Textural relations (olivine and phlogopite enclosed by dolomite and inclusions of pyrothite, diopside and of clinohumite in olivine) suggest equilibrium conditions for the assemblage forsterite + phlogopite + diopside + clinohumite (Table 1). Minerals clearly formed by retrogression are antigorite, spinel, lazurite and partly calcite. Spinel penetrates grain boundary of magnesioamphibole and phlogopite and it is partly replaced by carbonate.

Table 1. *Peak P-T and retrograde minerals in metamorphic rocks from Sare Sang*

	Peak P-T minerals	Retrograde minerals	
Rock	1 stage	2 stage	3 stage
Ca-silicate	Fo, Phl, Pyth, Di, Cln, Rt	Spr, Mg-Hbl, Sp, Cc	Ant, Laz, Cc
	Di, Scp_1, Pl_1, Kf, Grt, Ap, Tnt, Do(Cc)	Scp_2, Pl_2, Hbl, Ep, Cc	Sod, Cc
	Di, En, Grt, Pyth, Do (Cc), Scp_1, Pl_1, Gph	Hbl, Scp_2, Pl_2 (An_{62})	Tr, Bt
Whiteschists	Als, Tc, Ged, Phl, Tur, Qtz	Anth, Spr, Crd, Krn, Crn	
	Fo, Di (En), Phl, Pyth	Spr, Sp	Laz
Na-Ca(K)-	Ab, Cpx, Fo, Phl, Tur	Neph, Kf, Dnb	Ap, Ep, Sod, Pyr
rich rocks	Di+Phl	Hbl	Ap, Sod
Metabasite	Cpx, Grt, Hbl, Pl	Act, Ab	
Metapelite	Ky, Grt, Bt, Pl, Ms	Sil, Ms	

Ab = albite, Act = actinolite, Als = aluminosilicates (kyanite/sillimanite), Ap = apatite, Ant = antigorite, Anth = anthophyllite, Bt = biotite, Cc = calcite, Cln = clinohumite, Crn = Corundum, Cor = cordierite, Cpx = clinopyroxene, Di = diopside, Dnb = danburite, Do = dolomite, En = enstatite, Ep = epidote, F = forsterite, Ged = gedrite, Grt = garnet, Hbl = hornblende, Kf = K-feldspar, Krn = Kornerupine, Ky = kyanite, Laz = lazurite, Ms = muscovite, Neph = nepheline, Phl = phlogopite, Pl = plagioclase, Pyr = pyrite, Pyth = pyrothyte, Rt = rutile, Scp = scapolite, Sil = sillimanite, Sod = sodalite, Sp = spinel, Spr = sapphirine, Tc = talc, Tnt = titanite, Tr = tremolite, Tur-turmaline

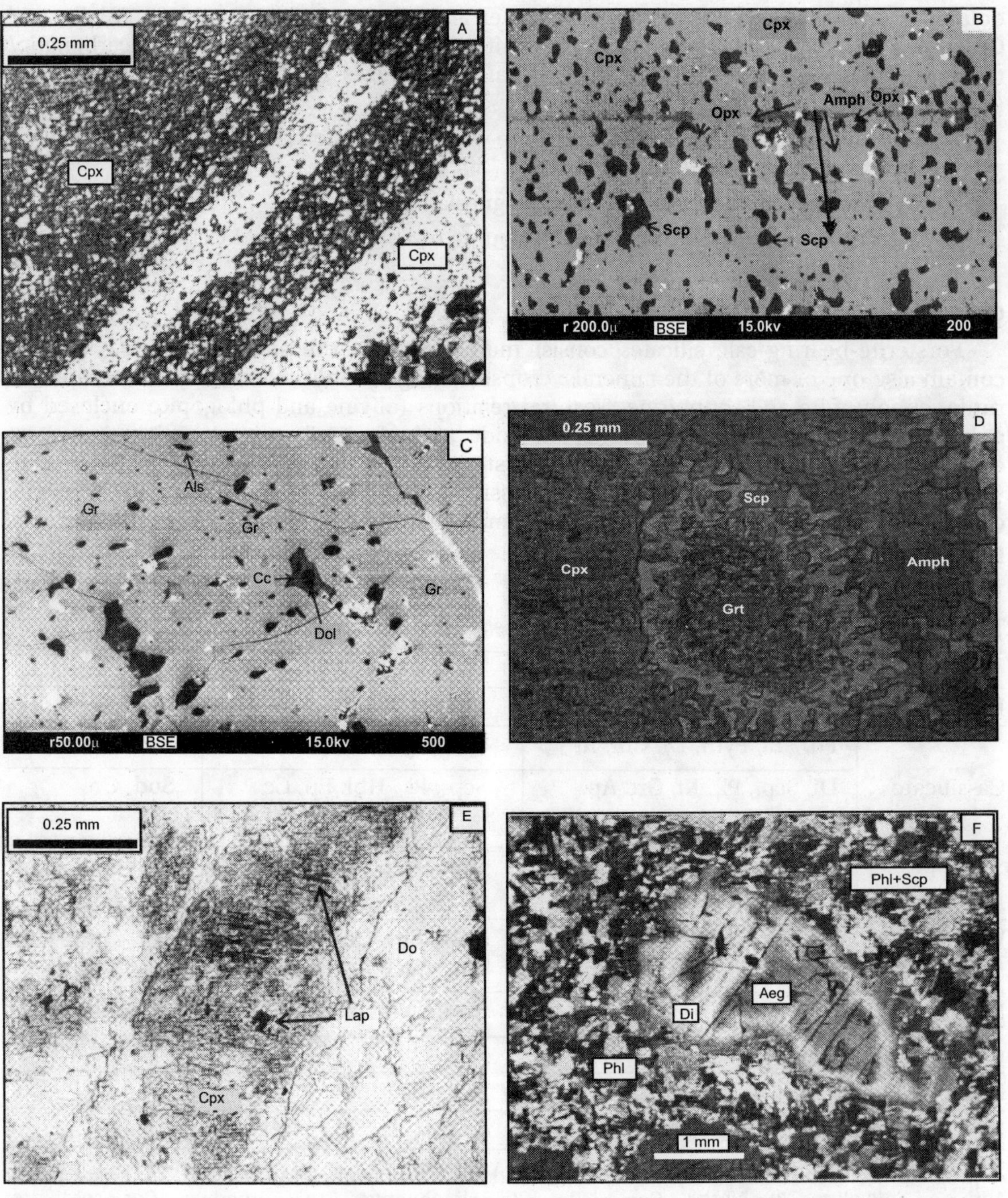

Fig. 2. Photomicrographs and back-scattered electron images depicting reaction textures in calc-silicate rocks: A-B. Clinopyroxene with inclusions of scapolite, enstatite and amphibole, C-garnet with inclusions of dolomite rimmed by calcite and aluminosilicate, D. Scapolite + amphibole corona around garnet at contact with clinopyroxene, E-lazurite replacing diopside in marble, F. Zoned porphyroblast of clinopyroxene in the phlogopite+scapolite matrix.

Garnet-bearing calc-silicates are represented by garnet, clinopyroxene, scapolite, quartz, calcite, plagioclase, epidote and accessory titanite, apatite and K-feldspar. Clinopyroxene contains inclusions of quartz, calcite and rarely of scapolite (Scp_1). Garnet encloses scapolite (Scp_1) or clinopyroxene with overgrowing scapolite, quartz and calcite. Amphibole replaces pyroxene or forms individual grains in contact with pyroxene and scapolite. Epidote occurs in plagioclase-rich parts or forms mono- or bimineralic aggregates with coexisting calcite; however, some calcite grains contain inclusions of epidote. Scapolite (Scp_2) enclosing garnet, pyroxene and epidote is a retrograde phase. At least three textural types of calcite are present (Table 1). The texturally early type forms inclusions in garnet and clinopyroxene, the latter type is intergrown with scapolite and epidote and the last variety of calcite forms veins in garnet.

The main constituents of garnet-enstatite-bearing calc-silicate rocks are enstatite, diopside, plagioclase, garnet, hornblende, biotite and scapolite. Both pyroxenes form mostly relatively large crystals upto 5 mm in size, which may contain inclusions of scapolite (Scp_1), graphite, dolomite and pyrothite. Garnet contains very fine-grained aluminosilicate (Al_2SiO_5) with plagioclase corona, formed by the reaction Grs + Als = Pl. Intergranular spaces between pyroxene grains are filled by retrograde minerals: plagioclase, scapolite (Scp_2), biotite, amphibole, quartz and by symplectites of Pl + Cpx + Amph. Coronas between garnet and pyroxene and garnet and amphibole are comprised of plagioclase and scapolite. Some garnet grains have irregular boundaries that are overgrown by plagioclase, amphibole and biotite. Besides symplectites with plagioclase, amphibole may form individual grains that contain inclusions of plagioclase, clinopyroxene and quartz. Accessory phases are apatite, rutile, ilmenite, pentlandite, chalcopyrite and hematite. Rutile rimming ilmenite occurs with amphibole, biotite, scapolite and plagioclase.

Kyanite whiteschists and ultramagnesian rocks

These rocks are strongly foliated with preferred orientation of kyanite porphyroblasts (upto 5 cm in length). In some cases kyanite is replaced by sillimanite. The kyanite crystals are mostly rimmed by cordierite. Most samples contain gedrite, which is similar to kyanite replaced by cordierite. The gedrite crystals are partly rimmed by anthophyllite. According to Schreyer and Abraham (1975) and Kulke and Schreyer (1973), the mineral assemblages in whiteschists from Sare Sang indicate three stages during retrogression. Kyanite represents a peak pressure mineral in all assemblage and it is usually replaced by the reactions:

kyanite + talc = cordierite + corundum (gedrite + quartz + H_2O)

kyanite + Al-anthophyllite = cordierite + Al-sapphirine

kyanite + Al-sapphirine = cordierite + corundum

kyanite + gedrite = cordierite + corundum

The gedrite-bearing assemblage may have formed under conditions of higher temperature or—more likely the same temperature under conditions of lower water fugacity.

The ultramagnesian type of rocks, first reported by Blaise and Cesbron (1966) and later investigated by Grew (1988) is represented by the equilibrium assemblage of kornerupine + phlogopite + magnesite + enstatite + sapphirine. Kornerupine (upto 1 mm) forms symplectites with magnesite and encloses sapphirine. Retrograde phase in the rocks are spinel, anthophyllite. Grew (1988) assumed replacement of kyanite in whiteschist by the reactions similar to that reported by Schreyer and Abraham (1976):

$$Ky\ (Ged) + Mgs = En+Spr + (Phl)$$
$$Ky+Tur+Ged + H_2O = Krn+Phl+Crd$$

The breakdown of tourmaline to kornerupine is predicted to have been a hydration reaction in the Sare Sang rocks. As a result, kornerupine-tourmaline relations in the Sare Sang rocks differ from those expected in chemically more complex metapelites, in which these breakdown reactions are probably dehydration reactions and are partly controlled by metamorphic fluids or anatectic melts.

Na-Ca (K)-rich rocks

In general these rocks are free of garnet, enstatite and olivine, however the latter was found as accessory phase. Most of these rocks have various amounts of tourmaline, Na-scapolite, phlogopite, sodalite and nepheline. Faryad (2002) described a thin-layered rock that changes composition from Na-rich to Ca-rich varieties. At least four layers with different constituents were observed in a single thin section. All of them contain Cpx ± Scp ± Phl but they differ each from other by the presence of one or more of the minerals: sodalite, albite, K-feldspar, calcite, quartz and accessory olivine, dolomite, nepheline, diopside, danburite and pyrite (Table 1). Besides occurring as individual grains, clinopyroxene also forms inclusions in phlogopite. Sodalite usually occurs between feldspar grains, which rims phlogopite and scapolite. Some phlogopite-rich parts contain aggregates of scapolite and calcite with accessory quartz and dolomite.

Some Ca-rich varieties have phlogopite, clinopyroxene, Ca-scapolite and calcite or plagioclase. Clinopyroxene encloses calcite, titanite, phlogopite, K-feldspar and rarely albite. The colorless clinopyroxene is overgrown or rimmed by phlogopite, scapolite and partly by calcite. Phlogopite is usually fine-grained, but in calcite- and scapolite-rich parts may form upto 1 mm large crystals which enclose clinopyroxene. At least two or three textural varieties of scapolite are present in plagioclase-bearing variety: scapolite enclosed by clinopyroxene, occurring as fine-grained aggregates in the matrix and forming porphyroblasts upto 3 cm in maximum diameter. Their porphyroblasts have inclusions of phlogopite, plagioclase, calcite, quartz and rectangular-shaped pseudomorphs of radial chlorite after phlogopite. The large scapolite crystals are partly crossed by calcite veins.

Haüyne and flourapatite veinlets were observed in clinopyroxene-phlogopite- and pargasite-bearing rock (Table 1). The bluish pale green haüyne replaces pyroxene and form corona around apatite.

Lapis lazuli deposit

The lapis lazuli form lenses (length of upto 40 m and thickness of 2 m) or layers in marbles and occur at contacts with granite and pegmatite. They mostly contain minerals formed by metasomatic reactions and during retrogression. The lenses are usually zoned and from core to rim they contain (Yurgenson & Sukharev, 1984): (1) alkali feldspar + quartz (calcic plagioclase) and accessory clinopyroxene, titanite, phlogopite, pyrite, epidote, titanite, apatite and scapolite, (2) diopside (40-90 wt %) + forsterite + scapolite + phlogopite + tremolite + titanite + zoisite + garnet and rare nepheline, haüyne and lazurite and (3) calcite + diopside + lazurite + haüyne + afghanite + nepheline. Large amounts of lapis lazuli deposits come from the layers, which are thick ca 6 m and reach a length of several hundred m (maximum 450 m). In most cases a band of calcite-diopside-lazurite is rimmed by diopside-rich zone from one side and by phlogopite-diopside-calc-silicate zone from other side. Such bands or zones may alternate several times in a vertical section and they remain a rhythmic sedimentation sequence. Minerals present in the layers are similar to that in the lenses. Lazurite together with scapolite, nepheline, afghanite and haüyne replace diopside and amphibole. Lazurite may indicate zoning character with blue purple or blue core and dark blue rim. The last variety of lapis lazuli deposit is developed at contact between marble and granite or pegmatite. They have lens shaped forms of upto, 2 × 0.30 m in size and they are formed mainly by lazurite and haüyne, rarely by afghanite, diopside, calcite, tremolite, phlogopite, pyrite, forsterite, nepheline and scapolite. In some cases the lapis lazuli bodies are separated from granite by a 0.4-1 cm zone of diopside (75 wt %), plagioclase, forsterite and muscovite. Diopside forms symplectites with plagioclase and it is partly replaced by tremolite. Pyroxene is usually replaced by lazurite and haüyne. On the outer side of lapis lazuli bodies at contact with marble, a zone formed by phlogopite, diopside, albite, sodalite and nepheline can be present. In calc-silicate rocks, lazurite forms veins and aggregates that replace olivine, clinopyroxene and calcite, but phlogopite remains as relicts in lazurite. The lazurite aggregates are cross cut by thin veins of calcite.

MINERAL COMPOSITIONS

Volatile-bearing minerals

CO_2-bearing phases

The main sources of CO_2 are carbonate marbles (calcite, dolomite and magnesite). Some CO_2 is involved in scapolite as it exhibits a wide range of composition from $Eq_{An} = 0.07$, $X_{Cl} = 0.99$ to $Eq_{An} = 0.61$, $X_{Cl} = 0.07$ (Fig. 3). Composition of scapolite depends on the whole rock composition as well as on its origin during prograde or retrograde history. The presence of graphite in enstatite could be interpreted either as result of reduction of oxidized carbon-bearing species, such as CO_2, or inheritance from a former carbonate (dolomite, calcite).

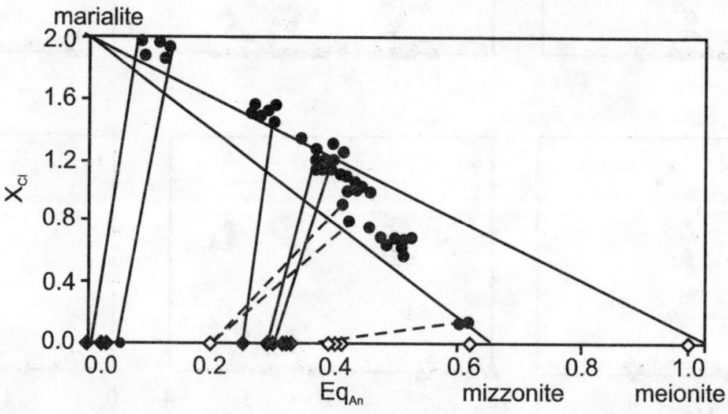

Fig. 3. Compositional range of scapolite from carbonate-evaporitic rocks at Sare Sang. Lines bracket the stoichiometric range of scapolites proposed by Evans *et al.* (1969). Plagioclase compositions plot on the X-axis where $X_{Cl} = 0$. Solid lines connect coexisting and dashed non-coexisting scapolite and plagioclase.

F-bearing phases

Biotite. Phlogopite ($X_{Mg} = 0.96$-1.00) from calc-silicate rocks has usually high F = 0.6-1.55 atoms per formula unit (a.f.u.), calculated based on 22 oxygens; it is poor in Cl (Fig. 4). The high-F phlogopite was found in rocks with flourapatite and haüyne veins. Low F < 0.5 a.f.u.

occurs in biotite from metabasites and metapelites. F contents indicate a positive correlation with Mg and a negative correlation with Al in phlogopite and biotite (Fig. 4).

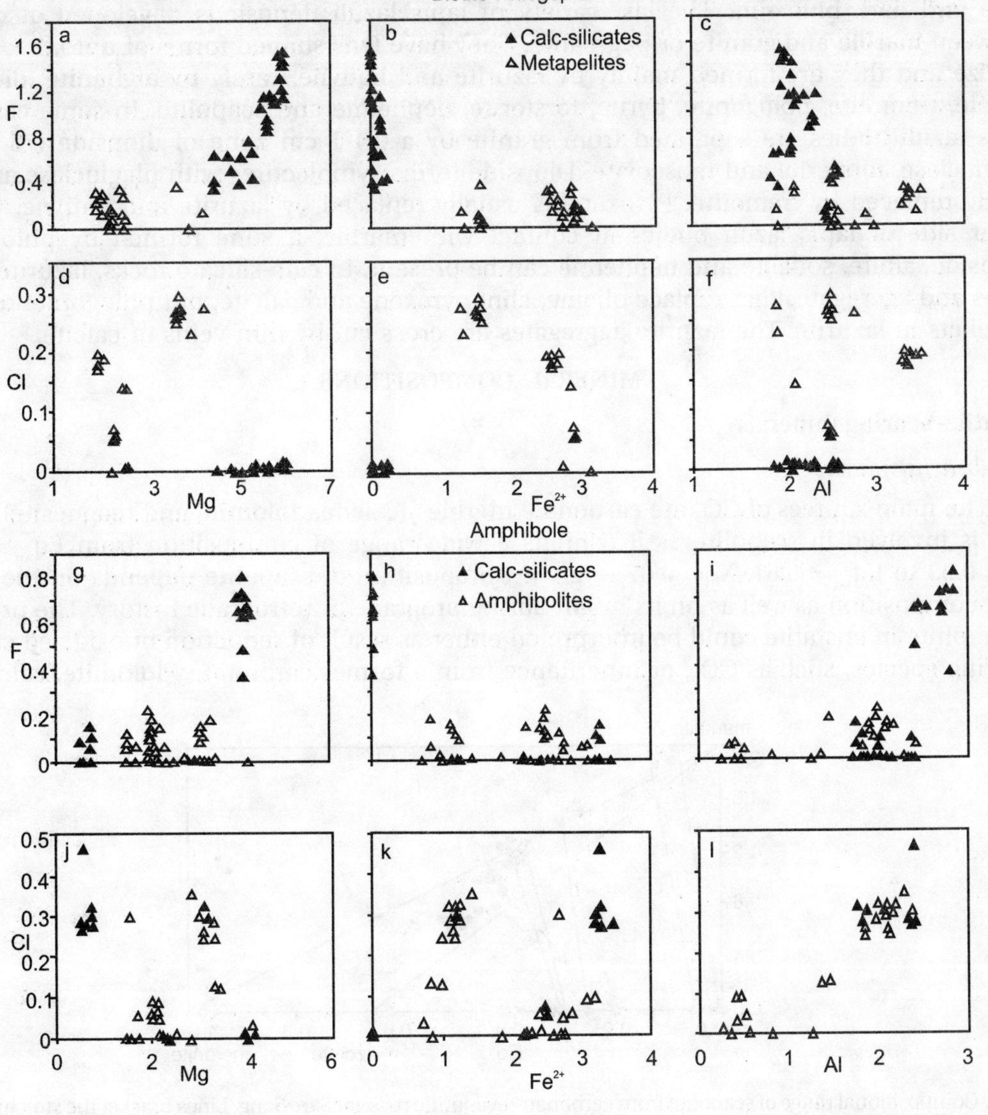

Fig. 4. F and Cl [a.f.u.] contents vs. Mg, Fe^{2+} and Al [a.f.u.] in biotite/phlogopite (a-f) and amphibole (g-l) from calc-silicates and nearby metapelites and metabasites of the Sare Sang locality.

Amphibole. Ca-amphibole is mostly magnesiohornblende (X_{Mg} = 0.96-0.98) in calc-silicate rocks and contain F < 0.2 a.f.u. similar to tschermackite or ferrotschermackite in metabasites. Maximum fluorine content (F = 0.65 a.f.u.) was analyzed in pargasitic amphibole (Na_A = 0.9 a.f.u., X_{Mg} = 0.99) cross cutting the haüyne veins in Na-Ca(K)-rich rock. No Cl was detected in gedrite (X_{Mg} = 0.96) and anthophyllite (X_{Mg} = 0.93-0.98), but F content may reach 0.36 a.f.u. Similar to biotite and phlogopite there is correlation between F, Mg and Al (Fig. 4). No regular variation of F occurs in amphibole from metabasites.

Apatite. Analyzed apatite has generally high F (3.0 – 4.9 wt %) contents. Large apatite crystals reveal irregular zoning which can be easily observed in back-scattered electron images. The F contents generally decrease towards the rims. In contrast, low F (< 0.5 wt %), but high chlorine contents (6.85 wt %) was analyzed in enstatite-bearing calc-silicate.

Titanite. Variation of Al, Fe and F in titanite ranges between 1.0-3.5 wt % Al_2O_3, 0.5-1.3 wt % Fe_2O_3 and F = 0.6-1.4 wt %. Generally, there is a good correlation between $(Al+Fe^{3+})/(Ti+Al+Fe^{3+})$ and F contents, the former indicating substitution of $(Al, Fe^{3+}) + (F,OH)^- = Ti^{4+} + O^{2-}$ (Coombs *et al.*, 1976; Markl & Piazole, 1999). Some large titanite crystals show a weak zonation with a decrease of Al and F and increase of Ti towards the rim.

Clinohumite. Clinohumite enclosed in olivine has Mg = 8.76 a.f.u., Ti = 0.15 a.f.u., Fe = 0.09 a.f.u., Si = 3.99 a.f.u. and F = 1.1 a.f.u.

Cl-bearing phases

Scapolite. Meionite-poor scapolites with Eq_{An} = 0.07-0.13 come from Na-Ca(K) rocks. Some optically zoned scapolite porphyroblasts with Cl = 0.31, Na = 1.43, Ca = 2.03 a.f.u. are rimmed by relatively marialite-rich scapolite having Cl = 0.40, Na =1.67, Ca = 1.69 a.f.u. Scapolite from single sample is generally homogenous within the expected analytical errors. However, scapolite inclusions in clinopyroxene have somewhat lower Eq_{An} (0.27-0.44) compared to retrograde scapolite in the matrix (Eq_{An} = 0.42-0.46).

Sodalite-group minerals. Analyzed sodalite has a composition close to the ideal end-member ($Na_4Al_6Si_6O_{12}Cl$), with Cl = 1.0 a.f.u. and Ca < 0.015 a.f.u., calculated on the basis of 25 oxygens. Small amounts of Cl is present in haüyne (Cl = 0.76 wt %) and lazurite (Cl = 0.28 wt %).

Biotite. As mentioned above, phlogopite is poor in Cl, but biotite (X_{Mg} = 0.7) from enstatite-bearing rock has a little fluorine, but considerable chlorine (0.29 a.f.u.). Cl content is low (<0.1 a.f.u.) in biotite from metabasite, but biotite in metapelites has high Cl = 0.15 – 0.22 a.f.u. contents. Excepting phlogopite from calc-silicates, Cl reveals a negative correlation with Mg and a positive correlation with Fe^{2+} (Fig. 4).

Amphibole. Amphibole of parasite-edenite and tschermakite composition (X_{Mg} = 0.64-0.80) from enstatite-bearing rock has high Cl (up to 0.39 a.f.u.), but low F (<0.2 a.f.u.). A potassium hastingsite (0.47 a.f.u. K) with X_{Mg} = 0.13-0.19 from garnet-bearing calc-silicate has little F (< 0.2 a.f.u.) and high chlorine content (0.32 a.f.u.). It occurs as retrograde phase with scapolite and replaces garnet. Cl contents in amphibole from metabasites is low (<0.1 a.f.u.). There is a correlation between Cl, Mg and Fe^{2+} in amphibole (Fig. 4). Excepting F-rich amphibole in haüyne-bearing calc-silicate, there is a weak positive correlation between F and Al.

Apatite. Most analyzed apatites from calc-silicate rocks have low chlorine content (<0.5 wt %). Cl-rich apatite (Cl = 6.85, F < 0.5 wt %) was analyzed in enstatite-bearing calc-silicate.

Boron-bearing phases

The most common borone phase is tourmaline, which is present in accessory amounts in various lithologies, but forms also lenses of the so-called "tourmalinite" with albite and quartz. Kornerupine, forming symplectite with magnesite was described by Blaise and Cesbron (1966) and Grew (1988). The intergrowth of magnesite, kornerupine and phlogopite is interpreted as result from coeval crystallization. Both tourmaline and kornerupine are rich in Mg with minor FeO < 0.33 wt %. Accessory danburite ($CaB_2Si_2O_8$) was reported by Faryad (2002) from Na- and K-rich calc-silicates.

S-bearing phases

Most sulfur from the Sare Sang locality is involved in lazurite ($Na_{5.5}Ca_{0.9}Al_{5.8}Si_6S_{1.7}Cl_{0.1}O_{29}$), haüyne ($Na_{5.1}Ca_{1.2}Al_6Si_6S_{1.6}Cl_{0.2}O_{29}$) and afghanite $[(Na,Ca,K)_8(Si,Al)_{12}O_{24}(SO_4,Cl,CO)_3\cdot(H_2O)]$. Maximum SO_4 value of 0.1 wt % was analysed in sodalite. In addition sulfide minerals (pyrite, pyrothite, pentlandite, chalcopyrite and galena) are widespread. They form aggregates, veins, but may form inclusion in other minerals. Pyrothite inclusions were found in pyroxenes from enstatite-garnet-bearing calc-silicate.

Some important volatile-free minerals

Pyroxene. The most common clinopyroxene in the investigated rocks is diopside, but diopside-hedenbergite solid solutions and aegirine augite was also found. Accessory manganoan diopside (ca 7% MnO) and omphacite (Jd = 29 mol %) were found in albite+ K-feldspar layer in Na-Ca(K)-rich rocks. Clinopyroxene with exsolution lamellae of enstatite is augite. Most clinopyroxene analyses from the symplectites or that enclosing enstatite have a relatively large jadeite content (maximum 17 mol %). Enstatite (X_{Mg} = 0.76-0.78) has two compositional varieties. The first variety, with abundant graphite inclusions, has less Al_2O_3 (<0.9), CaO (<0.3), Cr_2O_3 (<0.07 wt %.) than the second, more common variety with Al_2O_3 = 1.5 - 2.7, CaO = 1.2-1.5, Cr_2O_3 = 0.3-0.5 wt %.

Garnet. Garnet composition is quite homogeneous in every single sample. In scapolite-bearing calc-silicates, it is rich in grossularite (45-96 mol %) with small pyrope (0.1-2.5 mol %) and spessartine (0.5-7.9 mol %) contents. Almandine content is low (< 0.9 mol %) with exception of garnet from enstatite-bearing rocks with Alm_{41}. In epidote-bearing samples, garnet has relatively high andradite (25 mole %) content. High-pyrope garnet (Py_{36-43} Alm_{41-50} Grs_{10-15} Sps_{1-2} Adr_{0-3}) is present in enstatite-bearing rock.

Feldspars. Na-rich feldspar (Ab_{69} Or_{30}) is associated with perthite (Ab_{30} Or_{70}) in Na-Ca(K)-rich rocks. Albite and orthoclase inclusions are often in clinopyroxene. In garnet-bearing calc-silicate plagioclase has a composition of An_{32}. Plagioclase associated with scapolite in enstatite-bearing rock has An_{39-41}, but retrograde plagioclase (An_{95-65}) forms coronas around aluminosilicate inclusions in garnet or occurs with scapolite and amphibole along the margins of garnet.

Other volatile-free phases. Nepheline occurring with sodalite has about 6 wt % K_2O and Na/(Na+K) ca 70 %. Olivine from all analysed samples corresponds to forsterite with Mg/($Mg+Fe^{2+}$) = 0.98 – 1.00. Ilmenite has low MnO (0.39 wt %), but relatively high MgO (2.32 vol. %) contents, indicating 8.5 mol % geikielite solid solution. It is in contact with rutile containing 0.52 wt % Cr_2O_3 and 0.11 wt % Al_2O_3. Because of relatively high Al contents, sapphirine was identified by Schreyer and Abraham (1976) as peraluminous phase. It is free of iron, which follow the substitution of $MgSiAl_{-1}A_{-1}$. Sapphirine of similar composition was described by Grew *et al.* (1994) from Kohe Lal in Tajikistan.

P-T CONDITIONS

Textural relations and chemical equilibrium from varies lithologies (Kulke and Schreyer, 1973; Schreyer and Abraham, 1976; Grew, 1988; Faryad, 1999, 2002) indicated that the Sare Sang rocks underwent extensive retrogression. Several stages of mineral growing and extensive retrogression, which led to subsequent relatively low-grade assemblages are distinguished (Table 1). Kiselyova and Budanov (1986), assumed and earlier eclogite/granulite facies metamorphism, based on the mineral assemblages in metabasites and charnockites in the Tajik part of the South Badakhshan block. This was confirmed by Schreyer & Abraham (1976) for whiteschists of the Sare Sang locality, where they assumed pressure and temperature of 1.1 GPa and ca 800°C for the equilibrium conditions of kyanite, talc gedrite in whiteschists (Fig. 5). Pressures of ca. 0.7 GPa and 650-680°C were estimated by Massonne (1989) and Grew (1988) for chlorite-talc-cordierite- and kornerupine-phlogopite-sapphirine-bearing assemblages that represent decompression and destabilization of kyanite and gedrite in the whiteschists. Mineral representing maximum P-T conditions in calc-silicate rocks are forsterite, clinohumite, enstatite, dolomite, magnesite, pyrothite and diopside (Table 1). Combination of maximum Jd content (29%) in clinopyroxene that associates with albite and composition of alkali feldspar pairs with Or_{70} Ab_{30} and Or_{30}, Ab_{69}, $An_{0.5}$ or Or_{15} Ab_{83} $An_{1.3}$ (Fuhrman & Lindsley, 1988), indicated pressure of 1.18 GPa at 750°C (Faryad, 2002). P-T calculations, using Fe-Mg exchange thermometry between clinopyroxene and garnet (Ganguly *et al.*, 1996) and garnet-clinopyroxene-plagioclase barometry (Powell & Holland, 1988; Eckert *et al.*, 1991) used for different sample yielded temperatures of 690 - 780°C and 1.2-1.4 GPa. Textural relations, mainly the presence of small inclusions of scapolite in diopside and enstatite from calc-silicates suggest that scapolite was formed during prograde metamorphism in amphibolite facies conditions. This was confirmed also by the equilibrium reactions involving scapolite, calculated using thermodynamic data of Berman (1996) and the TWQ programme, which yielded P-T conditions of 700-800°C/1.0-1.4 GPa (Faryad, 2002).

The first stage of retrogression is well documented in the enstatite-bearing rock, where diopside + scapolite and diopside + amphibole symplectites and anorthite coronas around aluminosilicate inclusions in garnet were formed. Schreyer and Abraham (1976) indicated the peak-pressure minerals (Ky, talc, gedrite and phlogopite) were destabilized during decompression and temperature increase in whiteschists. The new minerals formed in this stage are cordierite, sapphirine, plagioclase and some dravite and various kinds of myrmekitic intergrowth within cordierite rims around gedrite or phyllosilicate symplectite.

Appearance of amphibole and biotite, mainly at the margins of some garnet from enstatite-bearing calc-silicates, suggest addition of H_2O to the system (Faryad, 2002). If it is assumed that plagioclase (An_{40} and An_{95}) coronas around aluminosilicate enclosed in garnet were formed due to breakdown of aluminosilicate and garnet, then a pressure decrease from 1.3 GPa (An_{28}) through 1.2 GPa (An_{40}) to 1.0 GPa (An_{95}) at 750°C can be calculated. Temperatures calculated on the bases of amphibole-plagioclase thermometry of Holland & Blundy (1994) are 740-750°C at 0.5 – 1.0 GPa for plagioclase An_{40}. The most common retrograde textures in calc-silicates are the replacement of clinopyroxene and garnet by calcite and amphibole. In some scapolite-free calc-silicate rocks clinopyroxene is replaced by lazurite and olivine by antigorite.

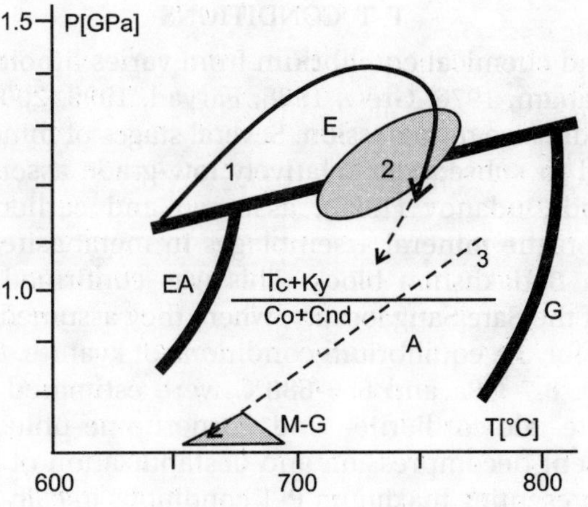

Fig. 5. Summary of estimated P-T data for the Sare Sang locality in Hindukush. Numbers in circles are results of P-T calculations for various lithologies and mineral reactions: 1- indicate P-T paths obtained for the scapolite-bearing calc-silicates (Faryad, 2002), 2- maximum P-T conditions calculated for metabasites and metapelites (Faryad, 1999), 3- retrograde P-T path from stage 2 to 3 (Schreyer and Abraham, 1976) inferred on the basis of the breakdown reaction kyanite + talc to cordierite + corundum. M-G: metamorphic conditions after Massonne (1989) and Grew (1988) for kyanite-cordierite- and kornerupine-sapphirine-bearing whiteschists in Sare Sang. Fields of metamorphic facies (AE-epidote-amphibolite, A-amphibolite, E-eclogite and G-granulite) are after Spear (1993).

DISCUSSION AND CONCLUSIONS

The presence of NaCl-rich scapolite in metamorphosed rocks is generally thought to imply an evaporitic source with the NaCl provided from slate-rich layers (Mora & Valley, 1989; Oliver *et al.*, 1994; Ellis, 1978). The best argument for sedimentary origin of NaCl and scapolite formations in the Sare Sang locality is the presence of relict anhydrite and gypsum (Blaise and Cesbron, 1966; Kulke, 1976) and stratigraphic control on the occurrence of carbonate and Na-, Al- or Mg-rich minerals that follow bedding in the rocks. Presumably the scapolite and other volatile-bearing phases crystallized during regional metamorphism in the rock layers that originally contained halite, limestone and dolomite. The occurrence of phlogopite or Mg- and Al-rich biotite with only minor K-feldspar may also indicate metamorphism of evaporitic argillites (Moine *et al.*, 1989). Similarly the presence of lenses of albite + tourmalinite + quartz, the so-called tourmalinite or albite (Blaise and Cesbron, 1966) may suggest sedimentary source of boron in these rocks.

Metamorphic rocks recrystallized to very low porosities under mid-crustal conditions expel most of the saline pore water in granulite facies (Yardley and Valley, 1997). Textural and/or chemical relations indicated that the Sare Sang carbonate-evaporitic rocks reached high-pressure amphibolite facies and significant amounts of CO_2, halogens, B and S were conserved in mineral structure, mainly in calcite, dolomite, magnesite, F-rich phlogopite, tourmaline, scapolite, apatite, titanite, pyrothite and graphite. The high F content in phlogopite is consistent with experimental data on the stability of F-phlogopite and with natural data from ultra-high T and P rocks (Harlov and Melzer, S., 2002; Motoyoshi, and Hensen, 2001). Low H_2O activity fluids (CO_2 and concentrated Cl, F brines) play a significant role during the high-grade amphibolite facies conditions. Some irregular variations of F and Cl in biotite and amphibole may result from different fluid composition in original rocks and infiltrated to surrounding rocks.

Retrogression and partial melting led to mobilization of S and halogens and formation of sodalite-group minerals. Beside lithology, variation of volatile in minerals was controlled by avoidance criteria of elements in mineral structure. Correlation of F and Mg in phlogopite and biotite (Fig. 4) is in agreement with preferentially incorporation of F into biotite with high Mg/Fe ratios and Fe-F avoidance as it has been extensively documented by experimental works and in natural systems (Ekstrom, 1972; Valley *et al.*, 1982). According to Markl & Piazolo (1998), no correlation between Cl in biotite and amphibole is found in most calc-silicates. In our samples Cl is incorporate in biotite and amphibole structure by factor near 1:1, however, F indicates preferential partitioning into biotite by factor of 9 : 4 (Fig. 6). Calculated log (X_F/X_{OH}) and log (X_{Cl}/X_{OH}) in coexisting biotite and amphibole indicates preferential partitioning of Cl into amphibole and F into biotite (Fig. 7). This is consistent with data from Nijland & Maijer (1993) from amphibolite to granulite facies transition rocks at Arendal, Norway. F-OH and Cl-OH partitioning coefficients for apatites showed preferential partitioning of F into apatite with respect to biotite (Faryad, 2002). The extent of preference is dependent on temperature and Fe concentration in biotite (Zhu & Sverjensky, 1992; Nijland & Maijer, 1993).

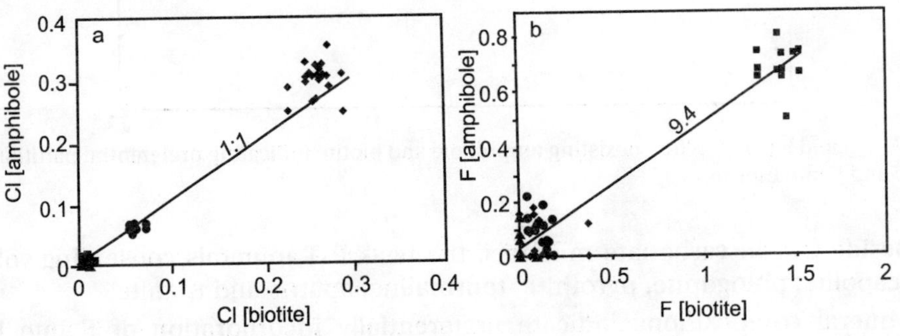

Fig. 6. Correlations of Cl [a.f.u.] (a) and F [a.f.u.] (b) in biotite/phlogopite vs. amphibole from different verities of the Sare Sang rocks. The regression equation for F contents (b) is y = 0.45x + 0.024.

The presence of apatite, scapolite and sodalite-group minerals in veins suggests prograde devolatization of high-grade metaevaporite layers during decompression and retrogression. Infiltration of extremely F-, S- and Cl-rich fluids led to formation of lapis-lazuli deposits. In addition, it is reasonable to assume that the granitoid magma also contributed at least some fluids that infiltrated the surrounding rocks. The crystallization of the numerous minor discordant intrusions of post-metamorphic granite would have released water, which permeated the carbonate-evaporite sequence and enriched it in S, Cl and CO_2. The best evidence for this model is accumulation of lapis lazuli near contact of granite.

Combining stratigraphic control and mineral chemical composition of rocks, the following conclusions are drawn:

1. Precursor of the Sare Sang NaCl-, F, S, CO_2 and B-bearing rocks and whiteschists was a carbonate-evaporite sequence that underwent high-pressure amphibolite facies metamorphism.

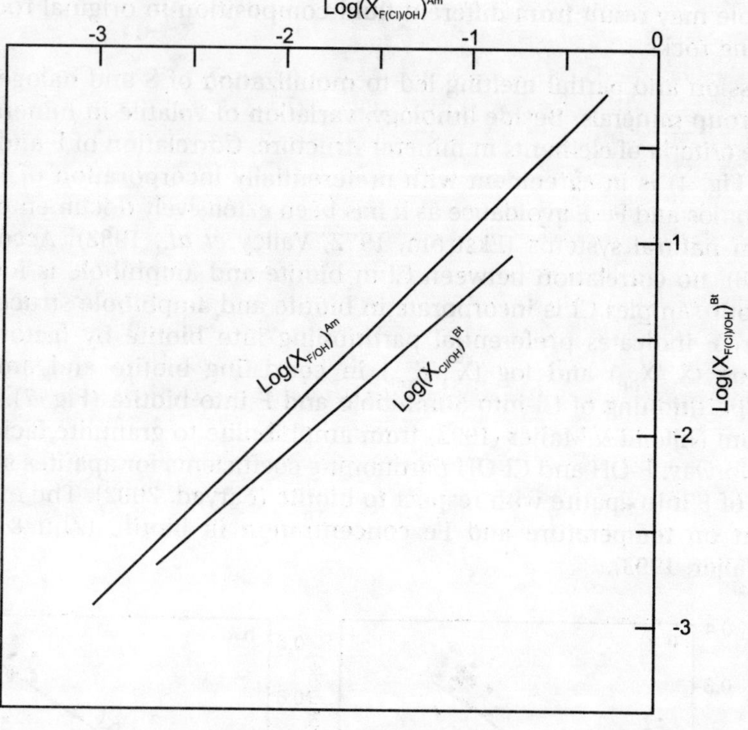

Fig. 7. $log(X_F/_{OH})$ and $log(X_{Cl}/_{OH})$ for coexisting amphibole and biotite indicating preferential partitioning of Cl into amphibole and F into biotite.

2. Beside CO_2 in carbonate minerals, the peak P-T minerals conserving volatiles were scapolite, phlogopite, pyrothite, tourmaline, apatite and titanite.
3. Mineral compositions indicate preferentially incorporation of F into biotite and amphibole with high Mg/Fe. Cl correlates with Fe and Al contents in amphibole. F incorporates preferentially into apatite with respect to biotite.
4. F-, S- and Cl-rich phases forming veins and replacing high-grade minerals were results from devolatization and infiltration of fluids into surrounding rocks during retrogression.
5. Water released by crystallization of post-metamorphic granite and infiltrated the carbonate-evaporite also played significant role in formation of scapolite and sodalite group minerals.

REFERENCES

Berman, R., 1996. TWQ program for calculation of mineral-fluid equilibria, *TWQ201 (version). http://www.gis.nrcan.gc.ca/twq.html*

Blaise, J. & Cesbron, F., 1966. Données minéraliques et pétrographiques sur ke gisment de lapis-lazuli de Sare Sang, Hindou-Kouch, Afghanistan. *Bulletin de la Société Française de Minéralogie et de Cristallographie*, **89**, 333-43.

Boulin, J., 1991. Structures in Southwest Asia and evolution of the eastern Tethys. *Tectonophysics*, **196**, 211-68.

Chopin, C., 1984: Coesite and pure pyrope in high grade pelitic blueschists of the Western Alps: a first record and some consequences. *Contributions to Mineralogy and Petrology*, **86**, 107-18.

Coombs, D.S., Nakamura, Y. & Vugnat, M., 1976. Pumpellyite – actinolite facies schists of the Taveyanne formation near Loche Valais, Switzerland. *Journal of Petrology*, **17**, 440-71.

Dronov, V.I., Abdulah, Sh. & Chmirev, V.M., 1980. Geology and mineral deposits of Afghanistan book 1-Geology, Nedra, Moscow, 418-49 (in Russian).

Dronov, V.I. & Feoktistov, V.P., 1980. Stratigraphic evolution of Archaean group in Afghanistan in: Geology and mineral deposits of Afghanistan book 1-Geology (Dronov, V.I., Abdulah, Sh., Chmirev, V.M eds.), Nedra, Moscow, 46-52 (in Russian).

Eckert, J.O. Jr., Newton, R.C. & Kleppa, O.J., 1991. The ΔH of reaction and recalibration of garnet-pyroxene-plagioclase-quartz geobarometers in the CMAS system by solution calorimetry. *American Mineralogist*, **76**, 148-60.

Ekstrom, T.K., 1972. The distribution of fluorine among some coexisting minerals. *Contributions to Mineralogy and Petrology*, **34**, 192-200.

Ellis, D.M., 1978. Stability and phase equilubria of chloride- and carbonate-bearing scapolites at 750°C and 4000 bar. *Geochimica et Cosmochimica Acta*, **42**, 1271-81.

Evans, B.W., Shaw, D.M. & Haughton, D.R., 1969. Scapolite stoichiometry. *Contributions to Mineralogy and Petrology*, **24**, 293-305.

Faryad, S.W., 1999. Metamorphic evolution of the Precambrian South Badakhshan block, based on mineral reactions in metapelites and metabasites associated with whiteschists from Sare Sang (Western Hindu Kush, Afghanistan). *Precambrian Research*, **98**, 223-41.

Faryad, S.W., 2002. Metamorphic conditions and fluid compositions of scapolite-bearing rocks from the lapis lazuli deposit at Sare Sang Afghanistan. *Journal of Petrology*, **43**, 725-47.

Fuhrman, M.L. & Lindsley, D.H., 1988. Ternary-feldspar modeling and thermometry. *American Mineralogist*, **73**, 201-15.

Ganguly, J., Cheng, W. & Tirone, M., 1996. Thermodynamics of aluminosilicate garnet solid solution: new experimental data, an optimized model, and thermodynamic application. *Contributions to Mineralogy and Petrology*, **126**, 137-51.

Grew, E.S., 1988. Kornerupine at the Sare Sang, Afghanistan, whiteschist locality: Implications for tourmaline-kornerupine distribution in metamorphic rocks. *American Mineralogist*, **73**, 345-57.

Grew, E., Pertsev, N.N., Yates, M.G., Christy, A.G., Marquez, N. & Chernosky, J.V., 1994. Sapphirine + forsterite and sapphirine + humite-group minerals in an ultra-magnesian lens from Kuhi Lal, SW Pamirs, Tajikistan: Are these assemblages forbidden? *Journal of Petrology*, **35**, 1275-93.

Harlov, D.E. & Melzer, S., 2002. Experimental Partitioning of Rb and K between Phlogopite and a Supercritical (K, Rb) Cl Brine: Implications for the Role of Supercritical KCl Brines in the Depletion of Rb in Phlogopite and the Stability of Phlogopite During Charnockite Genesis. *Lithos*, **64**, 15-28.

Holland, T. & Blundy, J., 1994. Nonideal interactions in calcic amphiboles and their bearing on amphibole-plagioclase thermometry. *Contributions to Mineralogy and Petrology*, **116**, 433-47.

Horeva, B.Ja., Iskanderova, A.D., & Shergina, Ju.P., 1971. Age of protoliths from metamorphic series of SW Pamir, according to the Pb isotope method (in Russian). *Izvestia AN SSSR, ser. geologicheskaya*, **8**, 40-46.

Hubbard, M.S., Grew, E.S., Hodges, K.V., Yates, M.G. & Pertsev, N.N., 1999. Neogene cooling and exhumation of upper-amphibolite facies whiteschists in the southwest Pamir Mountains, Tajikistan. *Tectonophysics*, **305**, 325-37.

Kiselyov, V.I. & Budanov, V.I., 1986. Precambrian magnesian-skarn deposits of the Southern Pamirs (in Russian). Dushanbe, Donish, 224.

Kulke, H. & Schreyer, W., 1973. Kyanite-Talc schist from Sare Sang, Afghanistan. *Earth Planetary Science Letters*, **18**, 324-28.

Markl, G. & Piazolo, S., 1998. Halogen-bearing minerals in syenites and high-grade marbles of Dronning Maud Land, Antarctica: monitors of fluid compositional changes during late-magmatic fluid-rock interaction processes, *Contributions to Mineralogy and Petrology*, **132**, 246-68.

Markl, G. & Piazolo, S., 1999. Stability of high-Al titanite from low-pressure calcsilicates in light of fluid and host-rock composition, *American Mineralogist*, **84**, 692-92.

Massonne, H.J., 1989. The upper thermal stability of chlorite + quartz: an experimental study in the system MgO-Al_2O_3-SiO_2-H_2O. *Journal of Metamorphic Geology*, **7**, 567-81.

Moine, B., Fortune, J.P., Moreu, Ph & Viguer, F., 1989. Comparative Mineralogy, Geochemistry and conditions of formation of two metasomatic talc and chlorite deposits: Trimouns (Pyrenees France) and Rabenwald (Eastern Alps Austria). *Economic Geology*, **84**, 1398-1416.

Mora, C.I. & Velley, J.W., 1989. Halogen-rich scapolite and biotite: implication for metamorphic fluid-rock interaction. *American Mineralogist*, **74**, 721-37.

Morimoto, N., 1988. Nomenclature of pyroxenes. *Schweiz. Mineral. Petrogr. Mitt.*, **68**, 95-111.

Motoyoshi, Y. & Hensen, B.J., 2001. F-rich phlogopite stability in ultra-high-temperature metapelites from the Napier Complex, East Antarctica. *Amer. Mineralogists*, **86**, 1404-13.

Nijland, T.G. & Maijer, C., 1993. The regional amphibolite to granulite facies transition at Arendal, Norway: Evidence for a thermal dome. *Neues Jahrbuch für Mineralogie Abhandlung*, **165**, 191-221.

Oliver, N.H.S., Rawling, T.J., Cartwright, I. & Pearson, P.J., 1994. High-temperature fluid-rock interaction and scapolitization in an extension-related hydrothermal system, Mary Kathleen, Australia. *Journal of Petrology*, **35**, 1455-91.

Powell, R. & Holland, T.J.B., 1988. An internally consistent dataset with uncertainties and correlations: 3. Applications to geobarometry, worked examples and a computer program. *Journal of Metamorphic Geology*, **6**, 173-204.

Schertl, H.-P., Schreyer, W. & Chopin, C., 1991. The pyrope-coesite rocks and their country rocks at Parigi, Dora Maira Massif, Western Alps: detailed petrography, mineral chemistry and P-T path. *Contrib. Mineral. Petrol.*, **108**, 1-21.

Schreyer, W. and Abraham, K. 1975. Peraluminous sapphirine as a metastable reaction product in kyanite-gedrite-talc schist from Sare Sang, Afghanistan. *Mineralogical Magazine*, **40**, 171-80.

Schreyer, W. and Abraham, K. 1976. Three stage metamorphic history of whiteschist from Sare Sang, Afghanistan, as part of a former evaporite deposit. *Contributions to Mineralogy and Petrology*, **59**, 111-30.

Spear, F.S., 1993. Metamorphic phase equilibria and pressure-temperature-time paths. *Mineralogical Society of America*, 799.

Valley, J.W., Bowman, J.R., Essene, E.J. & Petersen, E.U., 1982. Fluorphlogopite and fluortremolite in Adirondack marbles and calculated C-O-H-F fluid compositions. *American Mineralogist*, **167**, 545-57.

Yardley, B., Gleeson, S., Bruce, S. & Banks, D., 2000. Origin of retrograde fluids in metamorphic rocks. *Journal of Geochemica Exploration*, **69**, 281-85.

Yardley, B.W.D. & Valley, J.W., The petrologic case for a dry lower crust. *Journal of Geophysical Research*, **102**, 12173-85.

Yurgenson, G.A & Sukharev, B.P., 1984. Localization of Lapis Lazuli bodies and their mineral zonation in Badakhshan (in Russian). *Zapiski Vsesoyuznogo Mineralogicheskogo Obshchestva*, **113**, 498-505.

Zhu, C. & Sverjensky, D.A., 1992. F-Cl-OH partitioning between biotite and apatite. *Geochimica et Cosmochimica Acta*, **56**, 3435-67.

4

PETROGENESIS AND EVOLUTION OF PEÑA NEGRA, AN ANATECTIC COMPLEX IN THE SPANISH CENTRAL SYSTEM

M. DOLORES PEREIRA GÓMEZ*

ABSTRACT

The Peña Negra complex, crops out in the central-north area of the Sierra de Gredos, Avila batholith, central Spain. It is a perfect example of partial melting process, representing Hercynian events, and this can be followed through the outcrops, step by step. The Migmatites and anatectic granitoids were generated during the Upper Carboniferous from a Neoproterozoic-Lower Cambrian schist-greywacke complex with minor interlayered orthogneisses. Granitoids appear as subhorizontal sheets of megacrystic granodiorite and as lenticular bodies of leucogranite. These formed by *in situ* anatexis and correspond to melt/restite segregations under two different conditions, namely high melt-fraction in the case of the granodiorite, and low melt-fraction in the case of the leucogranites. The leucogranites display large enclaves of sillimanite-rich composition, representing the restitic fraction. Granodiorite was generated as consequence of crustal thickening under conditions of $T \approx 750$ to 800°C and $P \approx 4$ kb, coinciding with the second phase of the Hercynian orogeny. After this event, an extensional episode took place involving smaller degrees of decompression melting, generating small batches of cordierite leucogranites from an already depleted source. Anatexis was enhanced, in both cases, by shear structures and by the existence of a fertile protolith.

KEY WORDS

Peña Negra, Avila Batholith, Spain, Hercynian, anatexis.

INTRODUCTION

The Avila batholith consists of a large number of granitic bodies, and most of them may be classified chemically and petrographically as S-type granites. The Peña Negra anatectic complex could represent the level at which these granites were produced. The Peña Negra anatectic complex (PNAC: Pereira, 1992; Pereira and Bea, 1994; Pereira and Rodríguez-Alonso, 2000) is located in the middle part of the Gredos sector (Fig. 1). The PNAC comprises high-grade metamorphic rocks, mostly migmatites, and autochthonous or sub-autochthonous granitic bodies. These bodies occur either as small, lens-shaped bodies or as sub-horizontal sheets interlayered with the migmatites. The PNAC has a sub-horizontal orientation due to

* Depto. de Geología (Petrología y Geoquímica), Facultad de Ciencias, Universidad de Salamanca, 37008, Salamanca, SPAIN. e-mail mdp@gugu.usal.es

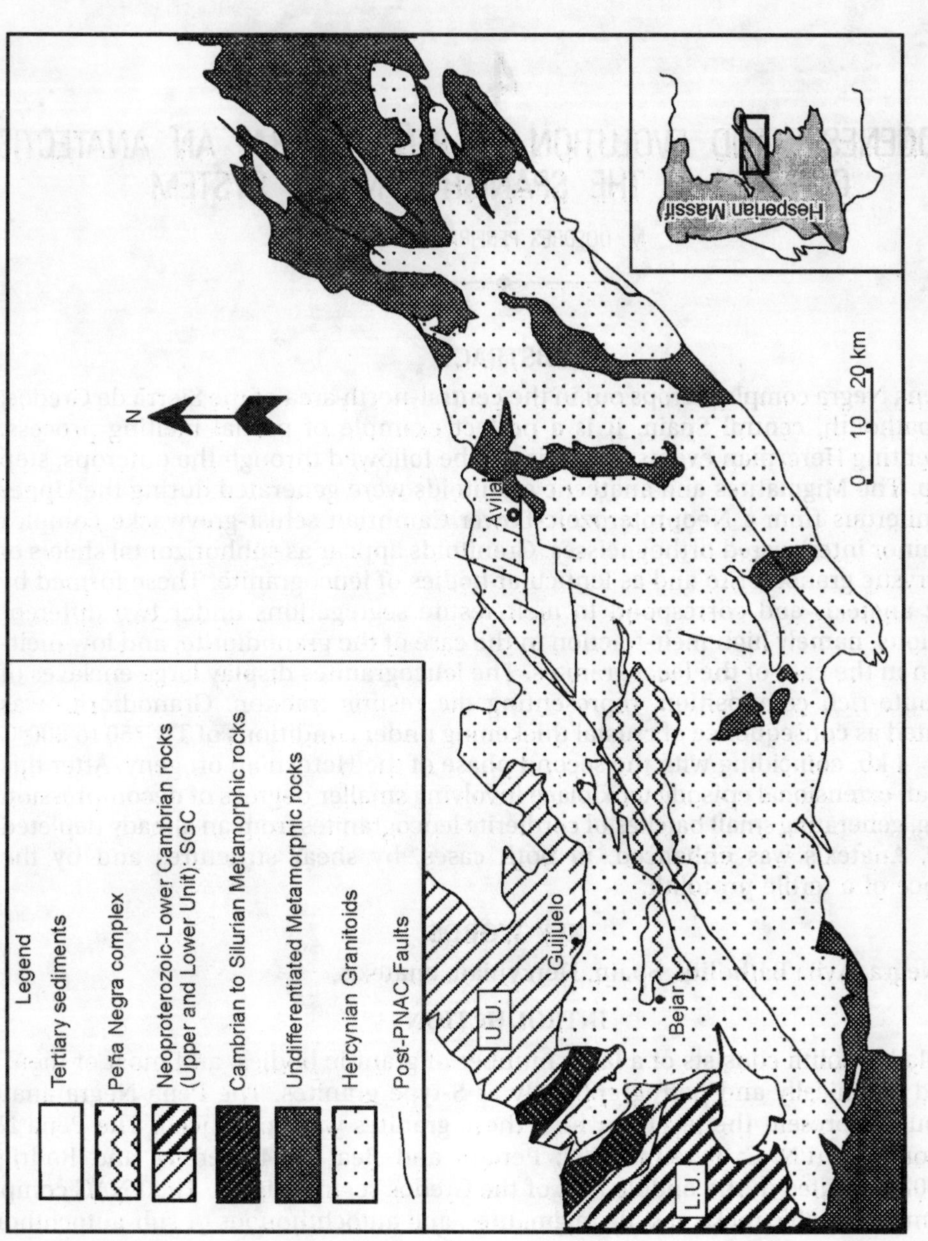

Fig. 1. Location of the Peña Negra anatectic complex within the Avila batholith. From Pereira and Rodriguez-Alonso, 2000.

the second phase of Hercynian deformation, and exhibits sub-horizontal structures such as isoclinal folds, foliation and shear zones. This complex is surrounded by several granite plutons with contacts ranging from concordant to intrusive (Fig. 2).

The Schist-Greywacke Complex (SGC) has been documented (e.g. Beetsma, 1995) in the northwestern part of the Iberian Central System. It comprises Neoproterozoic-Lower Cambrian metasedimentary rocks considered as a possible protolith of the Peña Negra complex, as well as of many other granitic complexes in central Iberia (Ramírez and Grundvig, 2000; Pereira and Rodríguez-Alonso, 2000).

In this contribution is presented the contrasting geochemical evolution of the two anatectic granitic rocks, granodiorites and leucogranites, that form part of the Peña Negra complex, together with a synthesis of the compositional characteristics of the rocks involved in the study. Specifically, major, trace element and oxygen isotope compositions of the anatectic rocks and their potential source rocks are compared. An attempt is made to identify the processes involved in the generation of these granitic rocks and the evidence for postulating a metasedimentary protolith. At the same time, a petrogenetic path for the generation of the complex is suggested here.

GEOLOGICAL BACKGROUND

Regional Setting

The PNAC covers an area of more than 350 km^2 in the Central Iberian Zone, within the confines of the Avila batholith (Fig. 1). This batholith crops out over a section of 13,000 km^2 in the middle axial part of the Hercynian Massif, in the western part of the Iberian Peninsula (Julivert *et al.*, 1972). It has been divided into two sectors according to lithology and age, Gredos (western part) and Guadarrama (eastern part: Fuster and Villaseca, 1979; Bea, 1985; Ibarrola *et al.*, 1987). The Gredos sector is composed of coalesced plutons of strongly peraluminous, porphyritic granodiorites and monzogranites, together with large anatectic complexes (e.g. PNAC). Most of the granitoids are syn-kinematic with the second phase of the Hercynian deformation, and their age is around 310 Ma. This age also dates the episode of the volumetrically most important anatexis (see below).

The PNAC consists mainly of migmatites, granodiorites, leucogranites, orthogneisses and sillimanite-rich restite. Gabbros and diorites can be found as small outcrops in a nearby area, cut by the granodiorites. It seems there is no interaction between them, but the scarce existence of intermediate and basic rocks within the PNAC cannot be neglected in the interpretation of its generation and evolution (Castro *et al.*, 1999; Bea *et al.*, 1999). However, these materials will not be discussed in this contribution.

Field relations, petrography and mineralogy

Peña Negra rock types. In terms of volume, migmatites represent the most important rock type in the PNAC (Pereira, 1989; Bea, 1991; Pereira, 1992). In outcrop, they appear as mesocratic rocks composed of a granodioritic leucosome (quartz, plagioclase, potassic feldspar, cordierite and biotite), and a melanosome, made up of sillimanite, biotite and cordierite, locally with a high content of ilmenite, which shows up as small enclaves or schlieren interlayered with the leucosome. Depending on the mesoscopic aspect of the mesocratic migmatites we have described three different facies: (1) schlieren, (2) nebulitic and (3) transitional. Each facies has the same mineralogy and geochemistry, but differs in texture. These rocks were affected by the same petrogenetic process but with different degrees of

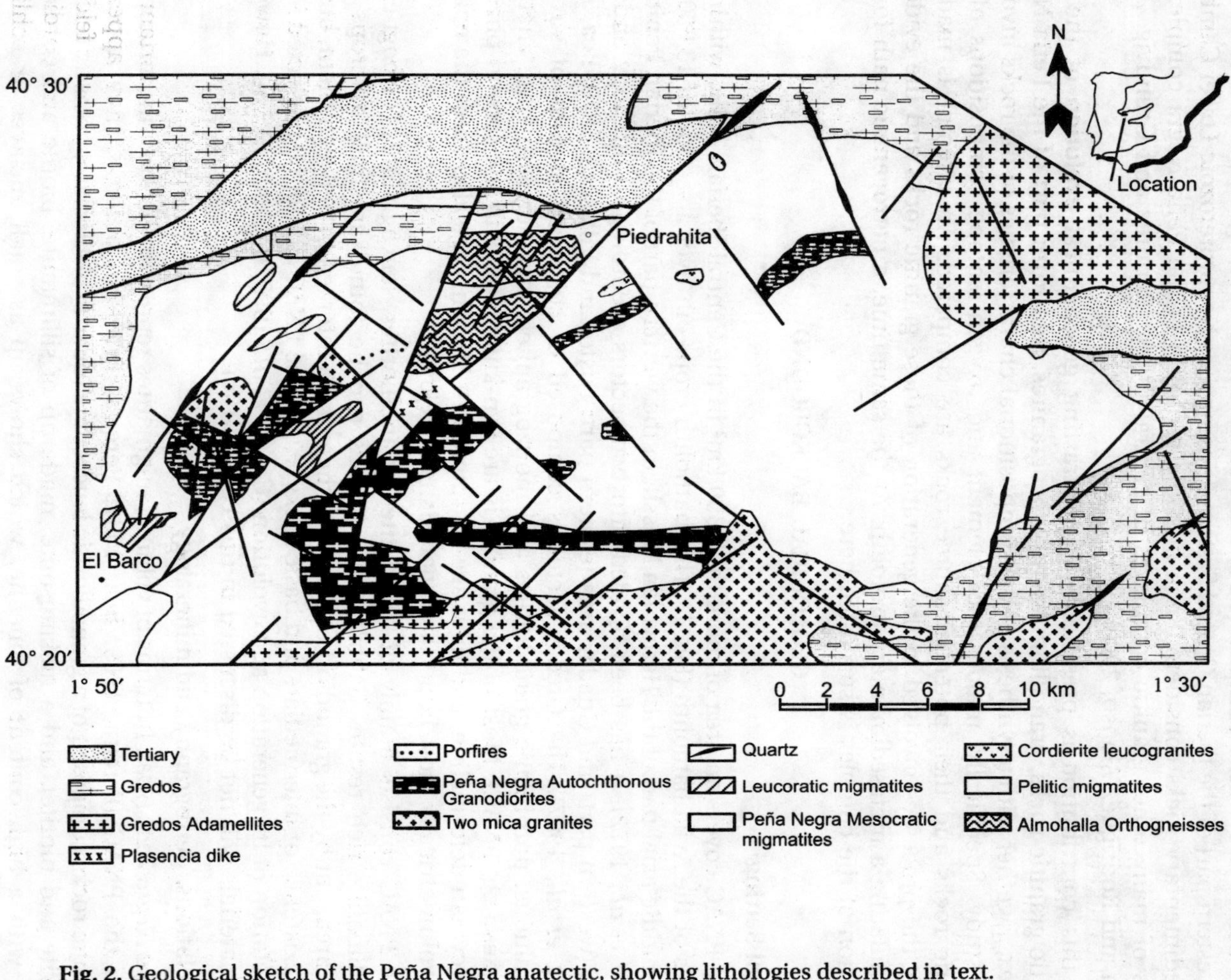

Fig. 2. Geological sketch of the Peña Negra anatectic, showing lithologies described in text.

segregation and textural evolution. The schlieren facies has centimetric compositional layering defined by melanocratic and leucocratic layers. In places the schlieren display complex folding, as "wild migmatite", which exhibits sub-horizontal deformation attributed to the second phase of the Hercynian Orogeny. Nebulitic and transitional facies reveal the sub-horizontal structures *albeit* less pronounced because the higher intensity of anatexis has caused a partial obliteration. Transitional facies show up in the transition between migmatites and granodiorites and the contact is invariably gradational. This facies has some feldspar phenocrysts, which is the main petrographic feature of the granodiorites, and ultimately it is difficult to determine whether they should be called migmatites or diatexitic granodiorites. Relative textural maturity established on the different facies of migmatites (McLellan, 1983); the results indicate that the degree of anatexis increases from the schlieren to the nebulitic and transitional facies.

Porphyritic granodiorite crops out as medium- to coarse-grained rocks, with abundant tabular potassic feldspar megacrysts and prismatic, commonly chloritized cordierite. The fabric of the granodiorites is sub-horizontal, concordant with the migmatites and the general structure of the PNAC, and displays a strong sub-horizontal foliation defined by feldspar megacrysts, biotite and prismatic cordierite.

Mineralogical compositions of the migmatites and granodiorites are similar. Both contain quartz, plagioclase (migmatites: core An_{33-18}, rim An_{18-12}; granodiorite: core An_{30-29}, rim An_{17-12}), alkali feldspar, biotite, cordierite and sillimanite as principal phases. The leucosomes of migmatites have hypidiomorphic textures, composed of quartz + plagioclase + cordierite + biotite + alkali feldspar. The melanosome is composed of alternating granoblastic (cordierite ± quartz ± plagioclase ± alkali feldspar) and schistose (sillimanite + biotite + ilmenite ± cordierite) layers. All these rocks contain accessory apatite, zircon, ilmenite, tourmaline, monazite and xenotime. Garnet is as a minor phase in one outcrop of the schlieren migmatites. This outcrop is located in a narrow band, coinciding with the largest outcrop of cordierite leucogranite generated from melting of the migmatites. Specific studies on this outcrop are needed to understand this anomaly.

Leucogranites are commonly spatially related to the mesocratic migmatites. They consist of cordierite-bearing leucocratic rocks that outcrop as small bodies within the migmatites, with contacts ranging from transitional to intrusive (Figs. 4 a, b: Bea, 1991). These rocks are made up of quartz, alkali feldspar, plagioclase (core An_{25-20}, rim An_{17-13}) and cordierite, with the same accessories as in the granodiorites and migmatites. Cordierite is commonly euhedral to subhedral. Locally, very small (less than 1 m diameter) intrusive bodies of a garnet leucogranite occur and are unrelated to the anatexis.

Orthogneisses occur as narrow bands interlayered with the migmatites. They are leucocratic, strongly foliated, and show the structures produced by the different phases of the Hercynian deformation (Díez Balda *et al.*, 1990). Orthogneisses have been affected by several stages of migmatization (Pereira, 1992) and augen structures are common. These rocks consist of quartz, plagioclase (core An_{37}, rim An_{33}), alkali feldspar and biotite, with zircon, apatite, monazite and ilmenite as accessory phases. Muscovite and chlorite occur as secondary products.

Sillimanite-rich restite forms irregular, small bodies typically in contact with the granodiorites, and are characterized by conspicuous sillimanite crystals 3-4 cm long. These restites have granoblastic textures, and the modal content of sillimanite ranges between 30 and 60%. The mineral assemblage is as in the melanosomes. In addition to sillimanite, biotite and cordierite occur as principal phases, augmented by quartz and feldspar. Ilmenite (locally abundant), pyrite, muscovite, zircon and apatite represent accessory phases that are commonly hosted by biotite crystals.

Basic and intermediate rocks (i.e. gabbros and tonalites) have been found in areas nearby the complex. Small enclaves have been studied for geochemistry, oxygen isotope analyses and geochronology. Results obtained until now show that there is no connection between the generation of granitoids and these rocks. However, a more precise study is needed to be more conclusive and to get an agreement between the study presented here and experimental results obtained by other authors (Castro *et al.*, 1999).

The protolith of the complex. Carbonate and calc-silicate lenses within the migmatitic series indicate that the migmatites of the PNAC derived from a metasedimentary source (Pereira, 1992). Field relationships suggest that the Neoproterozoic-Lower Cambrian metasedimentary rocks within the extensive antiforms (between very narrow Ordovician synclines at the north and western part of the PNAC area) are the potential source for the migmatites and other Hercynian granites. These metasedimentary rocks, known as the Schist Greywacke Complex (SGC), are characterized by a very thick sandstone-mudstone sequence with some conglomeratic and carbonate intercalations. This complex grades upward into a Lower Cambrian sandy and carbonate facies (Tamames Sandstone and Tamames Limestone Fms) with trilobites, archaeocyaths and ichnofossils. All are unconformably overlain by Ordovician-Silurian metasedimentary rocks.

Two different main units, described in the SGC, can also be distinguished throughout the Central-Iberian Zone. The Lower Unit is lithologically monotonous and occupies the western part of the PNAC area (containing low-grade metamorphic assemblages), whereas the Upper Unit outcrops mainly toward the northwestern part and constitutes a varied lithological succession that includes a more intensely metamorphosed and tectonized nucleus (Martinamor-Castellanos Antiform) in which orthogneiss and granites intruded the pre-Ordovician formations. The whole stratigraphic sequence is interpreted as representing continuous siliciclastic sedimentation with the development of a turbiditic facies, which evolved upward into a mixed siliciclastic-carbonate slope-platform sedimentation, within an area of relative tectonic instability associated with volcanic activity (Vidal *et al.*, 1994; Rodríguez Alonso and Palacios, 1994).

The Lower Unit is generally devoid of marker beds and is composed of alternating sandstones and mudstones, with some intercalation of conglomerates and disordered levels. The Upper Unit comprises a predominantly gray, muddy succession with several levels of black mudstones, conglomerates and sandstones as well as some discontinuous siliciclastic-carbonate strata and olistostromes. Other rock units described from the Upper Unit include phosphate-rich layers, volcaniclastic, calc-silicate and amphibolitic strata, each showing conspicuous lateral variations in facies. This sequence is enriched in quartz and feldspar phases toward the northwestern part of the PNAC area, where two informal lithostratigraphic units have been described: Monterrubio and Aldeatejada Formations. This area was affected by intensive Hercynian deformation, with the development of a medium- to high-grade metamorphism that induced anatexis in some areas.

The Monterrubio Fm outcrops in the core of phase I and III Variscan antiforms, and consists of sandy mudstones and laminated carbonaceous black mudstones, with some quartz-arenite and conglomerate intercalations. Some feldspathic conglomerates have been named porphyroids (Pereira and Rodríguez Alonso, 2000), which imply a derivation from an andesite-rhyodacite provenance, in order to indicate their genesis from volcanic rocks, together with intercalations of andesites and rhyodacites. Amphibolites and calcsilicate rocks are also present. Phosphate-clasts are common and constitute good marker-horizons. The estimated thickness of this formation is 2,000 m. It represents sedimentation in a siliciclastic platform environment (sensu lato) with repeated changes in sea level and evidence of wave action. The boundary with the overlaying Aldeatejada Fm is gradual (Pereira and Rodríguez Alonso, 2000).

The Aldeatejada Fm, which has an estimated thickness of more than 2,000 m, consists of siltstones and mudstones with fine intercalations of sandstones, carbonates (limestones, breccias, conglomerates and nodules) and calc-silicate rocks, together with black laminated carbonaceous to gray mudstone interbeds. The latter constitute a good marker-horizon, with some nodules or thin layers rich in phosphate. They have been interpreted as contourite deposits in a platform environment, which evolved to near-shore sandstone facies in the Tamames Sandstone Fm (Pereira and Rodríguez Alonso, 2000), straddling the Precambrian-Cambrian boundary.

Mudstones and their metamorphic equivalents are the most abundant group, augmented by minor sandstone, conglomerate and carbonate intercalations. Phosphate-rich rocks and cherts are also present. In areas of low metamorphic grade, original textures can be recognized, although most rocks are recrystallized and deformed. Evidence of magmatism contemporaneous with the sedimentation is provided by volcanogenic lithic fragments and crystals mixed in different proportions with siliciclastic constituents within volcanoclastic and calc-silicate rocks. Outcrops with amphibolites, quartziferous amphibolites and rhyolites also occur. Textures, mineralogy and chemical composition of these low-grade rocks confirm their relationship with calc-alkaline magmatism that involved rhyolite, andesites and basalts.

Mudstones comprise a fine groundmass of phylosilicates (chlorite, muscovite) with varied proportions of silt and fine-grained quartz sand. Accessory minerals are feldspar, biotite, opaque, tourmaline, zircon and apatite. Some samples show a fine lamination, containing abundant carbonaceous material, giving a darker color to the sample. Occasionally, we can find phosphate-rich horizons or nodules.

Among the sandstones we include quartz-arenites, lithic sandstones, quartz-, lithic- and feldspathic-graywackes and volcanoclastic sandstones. Conglomerates are both grain- and matrix-supported, and also oligo- and polymictic in composition. The mineralogy of the sandstones and conglomerates reveal different provenance areas for the rock fragments and quartz grains: sedimentary (siliciclastic), metamorphic and igneous (plutonic and volcanic).

Calc-silicate rocks are considered volcanoclastic, with clastic textures in the sand-size grains, made up mainly of quartz, plagioclase and intermediate igneous rock fragments (with microcrystalline texture and composition of quartz+plagioclase+tremolite±actinolite) similar to the quartziferous amphibolite rocks. They have abundant interstitial matter of the same composition of the fragments, where irregular crystals of plagioclase can be found interacting with grains of quartz and feldspar.

Among the carbonate materials are crystalline laminated limestones and dolomites with very few allochemical components and small quantities of quartz and pyrite together with pelitic and acid volcanic rock fragments. Other accessory phases are feldspar, tourmaline, zircon, apatite, muscovite, biotite, chlorite, opaque, titanite and clinozoisite.

Quartz-amphibolite rocks present a microcrystalline laminated texture. Opaque grains, calcite and epidote are disposed as fine and aligned aggregates. The microcrystalline groundmass is made up of a feldspathic aggregate with scarce quartz in which amphibole (actinolite-tremolite) shows a fasciculate texture and clinozoisite is found as idiomorphic or irregular aggregates together with epidote and opaque phases.

WHOLE-ROCK GEOCHEMISTRY

Analytical methods

Major elements in whole-rock samples were analyzed by optical emission spectrometry (ICP-OES) and Na and K by atomic absorption spectrometry (AA) at the University of Salamanca (Spain). Analytical precision was about ± 0.25 % rel., ± 0.75 % rel. and ± 2.5 % rel. for analytic concentrations of 10 wt %, 1 wt % and 0.1 wt % respectively. Trace elements and rare-earth elements (REE) were determined by mass spectrometry (ICP-MS) at the laboratory of Perkin-Elmer in Uberlingen (Germany) and at the analytical laboratories of Granada University. Precision was about ± 2% rel. and ± 5% rel. for analyte concentrations of 50 and 5 ppm respectively. For analytical comparison, some of the samples were analyzed by instrumental neutron activation analysis (INAA) at the nuclear reactor of McMaster University (Canada). Boron analyses were done by prompt gamma neutron activation analysis (PGNAA) at McMaster University (Canada). Description of the method can be found in Pereira and Shaw (1997).

Oxygen isotope analyses were performed at the University of Saskatchewan (Canada). The compositions were measured using the BrF_5 technique of Clayton and Mayeda (1963).

Results

Major-element compositions of mesocratic migmatites and sub-autochthonous granodiorites are almost identical (Table 1). Their compositions plot between the granodiorite and granite fields on the An-Ab-Or diagram. The migmatites are particularly heterogeneous. The SiO_2 content of the migmatites and granodiorites ranges from 60.56 to 67.08%. Al_2O_3 contents range between 10.02 and 16.72%. The average value of CaO, Na_2O and K_2O are 1.51% (2.57% for granodiorites), 2.59% and 3.92% (4.36% for granodiorites) respectively. The rocks are all peraluminous: (mean ASI is 1.4 for migmatites and 1.3 for granodiorites). The ratio Fe/(Fe + Mg) is quite constant, with an average of 0.73. TiO_2 concentration is locally high, in the range 0.25 - 1.25%, and MnO, between 0.03 and 0.11%. P_2O_5 can reach values of 0.55%. Cordierite leucogranites have a SiO_2 content between 62.98 and 76.32%. They are highly peraluminous (ASI = 1.3) and tend to be grouped on the cordierite field of the AFM diagram. Restites are depleted in SiO_2 and enriched in TiO_2, Al_2O_3 and CaO. Orthogneisses have a composition very close to that of the migmatites, except for their enrichment in CaO, in which they resemble the granodiorites, and in Na_2O. Fig. 5 shows the compositional variations. There is a trend of enrichment for TiO_2, Al_2O_3, FeO, MgO and content of H_2O (measured as Loss on Ignition) from the most evolved materials to the restites.

The major element composition of the postulated protolith (SGC) is close to that for average migmatite and is enriched in FeO and MgO, but depleted in Na_2O and CaO. Major-element compositions of the orthogneisses are between those of migmatites and the SGC.

Table 1: ***Major element composition of PNAC and the protolith rock types.***

Ref.	SiO_2	TiO_2	AL_2O_3	FeOT	MgO	MnO	CaO	Na_2O	K_2O	P_2O_5	H_2O	Lithol
B-304	62.71	0.69	18.32	5.27	2.08	0.06	1.60	2.91	4.29	0.33	1.56	M
B-305	71.06	0.42	14.90	2.63	1.01	0.04	1.35	3.26	4.40	0.29	0.34	M
B-307	60.94	0.77	18.29	6.75	3.08	0.11	1.49	2.77	3.55	0.17	1.84	M
B-320	66.21	0.62	17.35	4.72	1.90	0.06	1.40	2.30	3.92	0.23	0.84	M
B-339	65.01	0.87	18.60	5.69	2.40	0.09	1.07	2.37	3.49	0.18	0.00	M
B-341	70.12	0.61	15.07	3.67	1.64	0.08	1.37	2.63	3.13	0.18	1.07	M
B-342	68.00	0.71	16.05	3.94	1.97	0.09	1.12	2.37	3.67	0.14	1.60	M
B-343	67.82	0.65	16.50	3.49	1.55	0.06	1.08	2.89	4.29	0.11	1.33	M
B-344	68.89	0.73	14.76	2.11	0.78	0.05	0.99	2.39	6.57	0.55	1.86	M
B-345	65.22	0.79	17.22	4.79	1.95	0.11	1.32	2.82	4.02	0.11	1.45	M
LO-14	72.67	0.25	14.38	1.31	0.28	0.03	0.62	3.01	5.10	0.42	1.79	M
LO-16	60.56	1.25	17.27	5.64	2.31	0.06	3.79	3.19	2.98	0.38	1.69	M
LO-18	66.15	0.77	17.16	4.64	1.57	0.06	1.94	2.48	3.34	0.16	0.98	M
LO-25	64.64	0.89	17.03	5.77	1.87	0.07	1.65	1.97	3.54	0.16	1.98	M
LO-3	64.50	0.82	15.64	4.63	1.65	0.06	2.23	2.85	4.07	0.34	2.51	M
LO-4	78.75	0.34	10.02	2.10	0.99	0.04	0.94	3.18	2.31	0.15	1.19	M
B-516	63.03	0.95	16.46	5.00	1.77	0.06	2.95	2.96	4.07	0.19	1.76	GD
LO-17	63.41	0.96	15.95	4.57	1.57	0.00	4.24	3.07	4.15	0.21	1.56	GD
B-381	68.01	0.62	15.02	2.98	0.80	0.03	2.02	3.02	4.53	0.34	1.63	GD
B-382	65.17	1.01	15.41	5.43	1.82	0.05	1.91	2.51	3.50	0.30	1.29	GD
B-400	65.45	0.95	15.69	4.72	1.72	0.06	2.83	2.98	3.32	0.35	1.17	GD
LO-23	64.63	0.59	17.74	4.14	1.47	0.06	0.53	1.65	4.86	0.21	3.56	GD
B-306	71.58	0.22	14.90	1.78	0.54	0.04	1.27	3.57	3.96	0.30	1.47	CL
B-309	65.96	0.84	16.58	5.45	2.04	0.07	1.24	1.89	3.49	0.32	1.87	CL
B-311	72.00	0.19	14.34	1.46	0.42	0.10	1.04	2.20	7.07	0.29	0.71	CL
B-347	72.14	0.39	13.97	1.67	0.74	0.07	1.08	2.58	5.47	0.25	1.35	CL
B-373	68.33	0.85	16.03	5.13	2.00	0.07	0.62	1.79	3.99	0.18	0.70	CL
LO-24	71.98	0.08	14.27	0.69	0.38	0.02	0.12	2.72	6.48	0.14	2.51	CL
LO-28	76.34	0.19	13.11	1.23	0.53	0.02	0.54	2.73	3.48	0.20	1.23	CL
LO-30	71.29	0.35	14.58	2.05	0.62	0.04	1.23	2.35	5.32	0.26	1.19	CL
LO-32	70.36	0.37	14.76	2.00	0.63	0.04	1.45	2.65	6.26	0.39	0.96	CL
LO-33	62.98	0.75	17.30	4.84	2.09	0.08	2.25	2.71	4.30	0.23	1.89	CL
LO-5	71.47	0.33	14.54	2.44	0.57	0.04	1.37	2.82	5.14	0.19	0.85	CL
B-370	57.25	1.20	22.45	5.75	1.46	0.08	4.22	1.10	3.44	0.21	1.66	R
LO-7	41.20	1.49	31.52	10.69	4.10	0.14	0.01	0.50	5.82	0.13	3.09	R
LO-9	49.57	1.50	21.78	10.00	4.16	0.12	0.67	1.92	6.18	0.09	2.94	R
B-312	66.41	0.40	16.80	2.88	0.60	0.04	3.91	4.31	3.50	0.42	0.47	O
B-321	67.12	0.37	17.46	2.58	0.63	0.04	3.37	3.95	3.56	0.20	0.65	O
B-322	69.37	0.51	15.57	3.56	1.41	0.06	1.01	3.74	3.80	0.14	0.23	O
B-374	66.10	0.74	17.94	3.72	0.76	0.06	3.31	3.66	3.03	0.18	0.33	O
B-375	65.17	0.56	19.03	2.84	0.57	0.05	2.90	3.93	4.06	0.18	0.31	O
B-376	69.58	0.47	16.57	2.32	0.46	0.04	2.55	3.47	3.67	0.16	0.34	O

(Contd...)

Table 1: *Contd...*

Ref.	SiO_2	TiO_2	AL_2O_3	FeOT	MgO	MnO	CaO	Na_2O	K_2O	P_2O_5	H_2O	Lithol
B-377	68.74	0.42	16.57	2.24	0.42	0.04	2.81	3.58	3.62	0.16	0.97	O
B-378	70.30	0.42	15.62	2.16	0.40	0.05	2.61	3.48	3.65	0.15	0.65	O
LO-26	68.01	0.69	16.04	4.13	1.61	0.06	1.62	2.64	4.31	0.16	0.34	O
LO-27	67.93	0.68	15.42	4.62	1.74	0.07	1.39	2.15	3.65	0.16	1.59	O
LR-379	61.71	0.48	17.39	5.17	2.96	0.05	0.67	1.01	3.80	0.31	6.54	Pel.+Ph
LR-350	58.32	0.85	20.18	8.91	1.79	0.05	0.17	0.87	3.90	0.32	4.98	Pel.+Ph
LR-458	66.49	0.72	14.95	9.93	0.65	0.03	0.22	0.07	3.45	0.16	3.19	Con.+Ph
LR-396	62.13	0.73	16.94	4.80	2.47	0.08	3.29	1.18	3.99	0.32	4.04	Con.+Ph
B-617	60.38	0.73	18.95	6.16	3.30	0.30	0.80	1.86	3.55	0.16	3.32	C.P.
B-618	60.86	0.68	18.83	6.02	3.14	0.27	0.73	1.77	3.79	0.13	3.08	C.P.
B-619	60.62	0.71	18.91	6.35	3.33	0.30	0.75	1.80	3.74	0.15	3.01	C.P.
LR-506	51.81	0.23	15.55	7.16	9.14	0.50	13.10	0.39	1.38	0.25	0.89	ANF
LR-84	68.59	0.56	14.65	5.47	1.89	0.12	2.22	4.05	0.35	0.12	1.87	SAND.
LR-392	87.15	0.11	3.68	6.77	0.06	0.03	0.03	0.07	0.18	0.08	0.74	SAND.
LR-45	70.31	0.38	12.03	8.67	2.57	0.08	0.21	1.18	1.33	0.08	2.82	SAND.
LR-198	75.28	0.59	11.82	1.56	0.64	0.02	0.20	1.33	2.74	0.08	5.71	SAND.
LR-190	80.36	0.59	6.02	3.76	1.93	0.10	4.31	0.57	0.55	0.25	1.18	VOL. S.
LR-140	71.46	0.83	12.12	5.21	1.85	0.08	1.26	3.58	1.42	0.15	1.32	VOL. S.

PNAC lithologies: M = Migmatite; GD = Granodiorite; CL = Cordierite leucogranite; R = Restite; Protolith lithologies: SAND. = Sandstone; VOL. S. = Volcanoclastic sandstone; ANF = Quartz-Amphibolite rock; C.P. = Carbonaceous pelite; Con.+Ph = Conglomerate with phosphate-rich nodules; Pel.+Ph = Black Pelite with phosphatic nodules; O = Orthogneiss

The trace element variation is portrayed in spider diagrams (Fig. 3) with data from this study normalized to the upper continental crust (Taylor and McLennan, 1985). The granodiorites are very similar to the migmatites, and reveal a depletion of B, Li, Sc, V and S compared to them. The depletion of volatiles is attributed to consumption of these components in the generation of melt. Enrichment of Zr, Mo, Pb, Th and Sn has to do with the retention of these elements in accessory minerals, hosted in restite that remained as enclaves in the generation of granodiorite anatectic melts. Restites concentrate Li, Rb, Sc, V, Ni, Cu, Y, Zr, Nb, Mo, Hf, Ta, Th, U, Sn and Ga. This enrichment is compensated by the depletion of the same elements in the leucogranites. Boron, as in the case of granodiorites, was lost in the melting process and it is depleted in both restites and leucogranites (Acosta *et al.*, 2000).

The total REE content (Table 2, Fig. 4) in the mesocratic migmatites is rather high for this type of lithology. Chondrite-normalized patterns are very similar for all samples; they are characterized by a relatively strong fractionation of LREE against HREE (La_N/Lu_N = 11-36). All show a large negative anomaly for Eu (Eu/Eu^* = 0.43-0.73). A similarity of the REE spectrum of migmatites and of Post-Archean Australian Shales (PAAS, McLennan, 1989) supports the interpretation of a sedimentary precursor for the migmatites.

Table 2: *Trace element composition on selected samples from PNAC.*

	LO-18	LO-25	LO-16	LO-3	B-516	LO-17	LO-23	LO-28	LO-29	LO-30	LO-5	LO-7	LO-8	LO-9
	Migmatites				Granodiorites			Cordierite Leucogranites				Restites		
B	4.2	5.8	5.2	124.3	15.9	6.9	16.9	13.8	6.2	11.2	4.5	21.7	25.8	8.5
Li	63.09	63.93	52.58	137.59	54.31	63.61	17.00	40.35	42.47	41.15	34.28	139.43	390.00	134.57
Rb	83.81	124.82	202.07	169.62	197.65	200.97	109.64	233.12	232.14	166.78	174.92	120.31	510.29	356.91
Be	2.24	1.13	1.38	3.16	3.86	1.05	2.70	3.04	1.51	2.49	0.56	1.66	3.17	2.36
Sr	143.74	115.86	274.33	140.26	182.53	263.05	107.55	172.95	189.42	152.59	123.36	8.32	22.54	66.42
Ba	482.22	272.01	418.10	546.60	767.55	866.99	513.64	928.00	544.86	537.28	885.56	178.63	373.32	492.61
Sc	13.18	26.87	14.06	16.09	4.81	19.69	2.78	3.26	3.81	49.23	7.32	26.16	n.d.	51.17
V	99.41	99.40	84.81	79.47	93.86	72.98	57.47	23.79	25.57	28.75	27.22	259.57	318.31	207.95
Co	60.01	50.57	55.95	46.62	74.20	45.75	40.80	57.92	61.88	46.30	62.01	22.65	39.34	46.97
Ni	15.70	18.78	10.52	13.72	13.95	11.05	10.85	5.75	5.71	6.86	5.87	34.85	38.19	24.42
Cu	28.71	26.73	18.36	25.83	27.78	17.68	16.73	5.15	4.40	7.18	5.01	106.37	27.34	25.09
Zn	148.59	121.14	117.28	96.63	104.03	98.70	83.62	43.74	40.44	43.34	57.86	156.24	302.74	175.63
Y	13.07	11.19	28.55	26.81	29.24	22.77	10.35	21.61	12.47	14.54	21.23	11.17	54.39	44.74
Zr	56.91	33.64	20.59	63.82	49.43	57.69	75.17	86.68	78.90	109.45	91.24	223.39	280.54	130.41
Nb	13.09	20.73	20.66	14.25	16.23	15.99	11.60	7.74	9.02	9.05	13.69	42.71	56.37	43.85
Mo	0.91	0.43	1.03	0.97	1.00	4.54	0.10	0.58	0.25	0.28	0.09	103.76	0.37	0.36
Hf	2.26	1.56	1.36	2.36	2.20	2.17	2.62	3.03	2.85	3.61	3.35	6.07	7.10	4.18
Ta	1.08	1.30	1.32	1.16	1.02	1.15	0.64	0.54	0.78	0.68	0.89	3.66	4.39	3.50
Pb	16.76	16.85	15.72	22.24	20.85	23.07	29.64	32.63	33.45	39.17	29.23	7.09	8.38	14.22
Th	6.94	10.53	9.17	13.10	14.97	20.49	9.05	12.12	12.53	11.75	12.11	10.88	19.27	19.98
U	3.10	2.94	1.75	2.75	2.45	2.89	2.23	3.81	3.48	3.50	2.41	6.36	7.22	7.20
Sn	7.48	3.44	2.15	7.50	4.43	3.67	64.80	2.72	3.20	4.21	6.88	9.87	16.11	10.81
Ga	25.12	29.77	28.49	23.41	27.45	25.28	19.59	18.64	17.72	18.74	20.55	67.17	61.47	49.33
S	2008	659	6594	2199	1717	1852	337	126	108	350	294	20.75	n.d.	38.22
La	27.06	36.32	47.75	46.07	62.05	70.68	27.40	39.40	38.14	28.75	7.00	34.25	108.58	90.82
Ce	70.34	85.37	108.24	110.19	141.77	163.52	67.55	92.83	90.03	73.54	17.22	54.97	245.94	200.30
Pr	6.18	7.80	10.01	10.38	13.26	15.34	6.15	8.48	8.02	6.84	2.23	5.72	22.22	18.82
Nd	25.96	32.06	42.16	43.94	54.78	64.06	25.82	34.14	31.53	27.48	9.13	26.20	90.02	78.10
Sm	5.30	5.80	7.85	9.07	10.60	12.21	5.18	7.03	6.70	6.03	2.35	5.73	17.14	15.26
Eu	0.83	0.76	1.59	1.37	1.64	1.89	0.73	1.12	1.17	0.89	0.26	0.30	0.91	1.18
Gd	4.62	5.20	7.55	7.87	9.83	11.09	4.42	6.71	5.75	4.96	1.88	4.77	17.16	14.41
Tb	0.65	0.62	0.99	1.01	1.13	1.22	0.54	0.88	0.65	0.67	0.31	0.67	1.99	1.79
Dy	2.90	2.76	4.88	4.87	5.02	4.97	2.53	3.93	2.76	3.14	1.54	3.42	9.14	8.27
Ho	0.50	0.45	0.90	0.88	1.00	0.78	0.40	0.63	0.32	0.44	0.28	0.57	1.77	1.55
Er	1.33	1.22	2.37	2.33	2.31	1.99	1.06	1.72	1.05	1.41	0.66	1.61	4.50	3.67
Tm	0.14	0.11	0.31	0.31	0.30	0.22	0.10	0.20	0.10	0.15	0.08	0.18	0.64	0.46
Yb	0.92	0.79	1.61	2.00	1.85	1.26	0.68	1.23	0.65	0.98	0.54	1.18	3.42	2.62
Lu	0.15	0.10	0.23	0.27	0.29	0.19	0.12	0.19	0.10	0.15	0.07	0.19	0.49	0.38
Rb/Sr	0.58	1.08	0.74	1.21	1.08	0.76	1.02	1.35	1.23	1.09	1.42	14.46	22.64	5.37

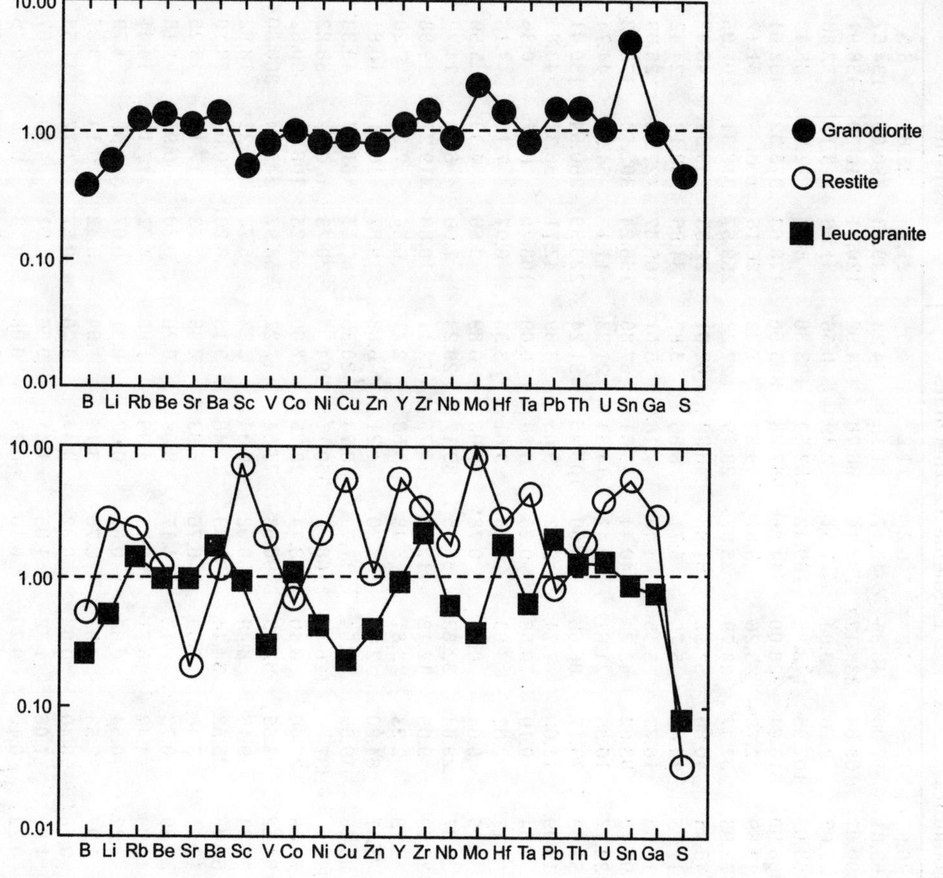

Fig. 3. Average trace element concentration in PNAC lithologies, normalized to average trace element data for the Schist-Greywacke complex.

For the anatectic products, different REE chondrite-normalized patterns are observed for granodiorites compared to the leucogranites. Total REE contents of granodiorites are high, matching those for the migmatites, and chondrite-normalized patterns show the same features. Fractionation of LREE versus HREE is strong ($La_N/Lu_N = 26$), and they also have a significant Eu anomaly ($Eu/Eu^* = 0.53$). For the leucogranites, the total REE contents are lower than in the granodiorites and mesocratic migmatites. Chondrite-normalized patterns show significant differences, and most plot beneath the PAAS spectrum (Bea, 1991; Pereira, 1992).

The orthogneisses reveal enrichment in almost all trace elements analyzed, compared to the migmatites. The most striking depletions are observed in B, Li, V, Co and Zn (Table 3). There is enrichment in light rare earth elements, and small Eu anomaly ($Eu/Eu^* = 0.86$) (Fig. 4).

Sillimanite-rich restites have very high REE contents; fractionation is again very high ($La_N/Lu_N = 33$) and the negative Eu anomaly is pronounced ($Eu/Eu^* = 0.17$) (Fig. 4). REE are controlled mainly by the biotite, which enclose many of the trace element-bearing accessories.

We have compared the trace element composition of the PNAC rock units with those of the SGC. The migmatites define a similar trend, but there is a conspicuous enrichment in U

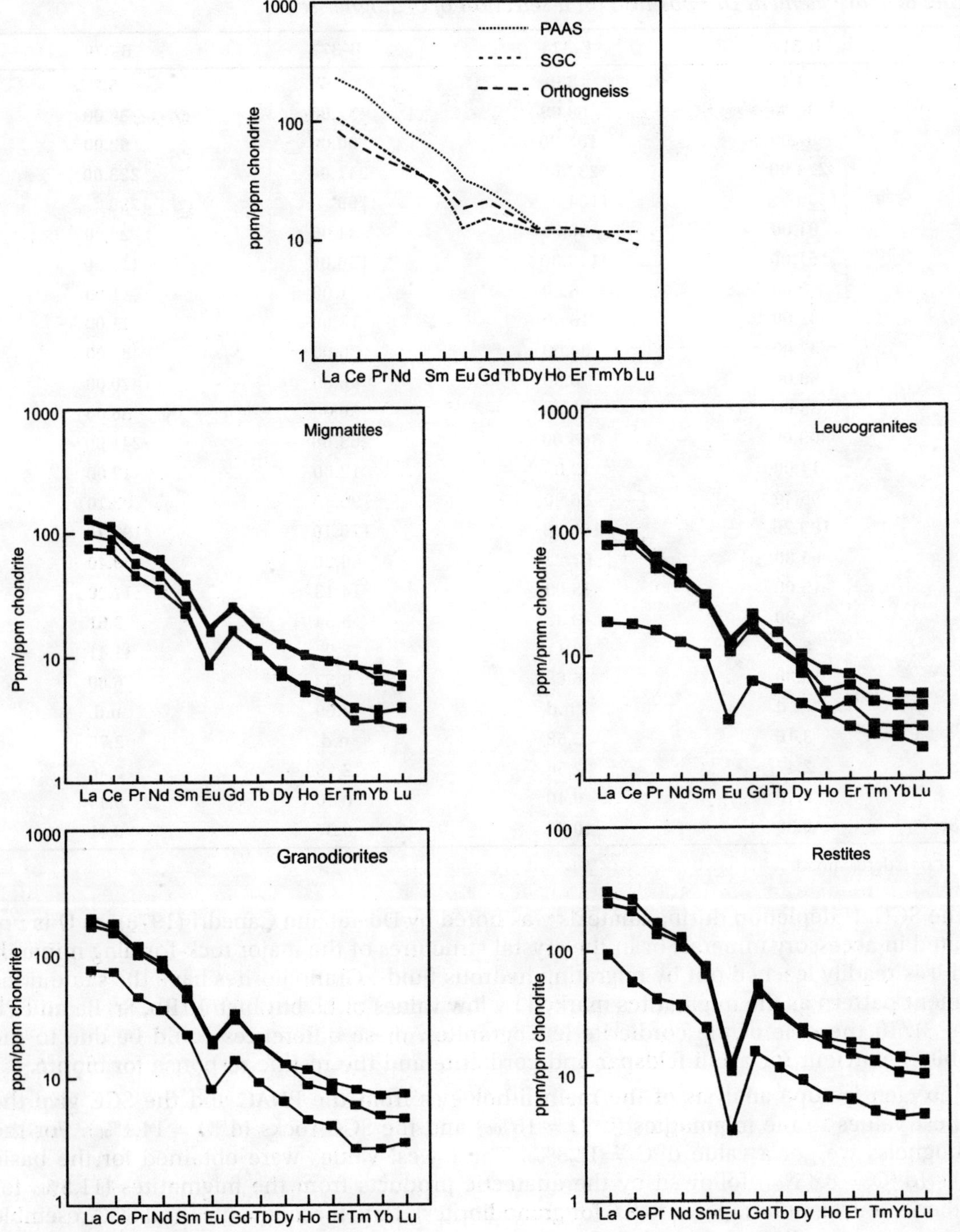

Fig. 4. Average concentration of rare-earth elements in Peña Negra lithologies and Schist-Greywacke complex, normalized to chondritic abundances. The protolith pattern is compared to the PAAS pattern.

Table 3: *Trace element composition on a selection of orthogneisses.*

	B-312	B-374	B-377	B-378
B	4.4	3.9	5.5	5.7
Li	36.00	68.00	37.00	36.00
Rb	87.00	106.00	90.00	92.00
Sr	224.00	237.00	241.00	223.00
Ba	1539	1104	1799	1790
V	81.00	50.00	34.00	26.00
Cr	151.00	113.00	108.00	111.00
Co	8.00	6.00	1.00	1.00
Ni	32.00	16.00	13.00	11.00
Cu	37.00	34.00	36.00	32.00
Zn	68.00	78.00	46.00	70.00
Y	38.00	31.00	36.00	36.00
Zr	309.00	365.00	263.00	241.00
Nb	14.00	19.00	12.00	12.00
La	95.40	86.90	90.30	83.10
Ce	183.20	180.28	170.10	193.10
Nd	65.80	67.22	60.20	59.40
Sm	15.00	15.09	14.13	17.20
Eu	3.90	3.32	3.54	3.61
Gd	8.95	10.11	12.03	11.41
Dy	8.36	6.69	6.52	6.80
Ho	n.d.	n.d.	2.69	n.d.
Er	3.16	2.68	n.d.	2.75
Yb	2.47	2.54	2.93	2.46
Lu	0.40	0.40	0.40	0.44
Rb/Sr	0.39	0.45	0.37	0.41

n.d. – not determined

in the SGC. U depletion during anatexis was noted by Dostal and Capedri (1978), as U is not located in accessory minerals or in the crystal structures of the major rock-forming minerals and it is readily leached out by migrating hydrous fluids. Granodiorites have the same trace element pattern as the migmatites marked by low values of Li, but high K/Rb, Sr, Ba and Pb (Fig. 3). In the case of the cordierite leucogranites, these differences could be due to the mobile behaviour for alkali feldspar and cordierite and the restitic response for biotite.

Oxygen isotope analysis of the main lithologies from the PNAC and the SGC give the highest values to the migmatites ($d^{18}O$ = 16‰) and the SGC rocks ($d^{18}O$ = 14.1‰). For the orthogneiss we got a value $d^{18}O$ = 12.3‰. The lowest values were obtained for the basic rocks ($d^{18}O$ = 8.45‰) followed by the anatectic products from the migmatites (11.5‰ for cordierite leucogranites, down to 9.6 for granodiorites). Data for migmatites and SGC resemble those for metasedimentary rocks (Kempton and Harmon, 1992) and suggest a pelite-rich source. The values obtained for the Peña Negra rocks indicate a heterogeneous evolution in the granitic production path, from a metasedimentary source to pure melts, where residue from partial melting has an important role, as bulk composition and hence, mineralogy, determine variations in the isotopic composition.

GEOTHERMOBAROMETRIC STUDY

There is a large collection of references dedicated to the calibration of geothermometers based on the same mineralogical association that is found in the materials used for the present study (*e.g.* Perchuk, 1969, 1970a, 1970b; Hensen and Green, 1973; Currie, 1974; Holdaway and Lee, 1977; Perchuk, 1977; Ferry and Spear, 1978; Hodges and Spear, 1982; Perchuk and Lavrent'eva, 1983; Ganguly and Saxena, 1984; Indares and Martignole, 1985; Bhattacharya *et al.*, 1988; Spear and Peacock, 1989). However, only few were used to calculate P and T conditions for the PNAC (Table 4) avoiding those that gave aberrant results. More specifically, for the pair garnet-cordierite the algorithm of Bhattacharya *et al.* (1988) was used, and for the pair garnet-biotite: Ferry and Spear (1978), Hodges and Spear (1982), Ganguly and Saxena (1984), Perchuck and Lavrent'eva (1983), Indares and Martignole (1985) and Spear and Peacock (1990). The geobarometer used was the one by Hodges and Spear (1992), using the association garnet-sillimanite-plagioclase-quartz.

Table 4: *Application of geothermometers for an invariable P of 3 kb.*

Thermometer	core	middle	rim	Observation
1	741°C	622°C	513°C	
2	785°C	643°C	488°C	
3	795°C	652°C	496°C	
4	764°C	658°C	535°C	dWMn=3000 cal
5	678°C	610°C	525°C	
6	676°C	555°C	430°C	
7	782°C	644°C	494°C	

Cordierite-garnet pair: 1 = Bhattacharya *et al.* (1988). Biotite-garnet pair: 2 = Ferry & Spear (1978); 3 = Hodges & Spear (1982); 4 = Ganguly & Saxena (1984); 5 = Perchuck and Lavrent'eva (1983); 6 = Indares and Martignole (1985); 7 = Ferry & Spear (1990).

Chemistry of the garnet

Garnet from PNAC migmatites has an almandine composition, with a high content of spessartine (upto 8%). MnO is concentrated preferentially in the rim, being the core slightly depleted on this component (Fig. 5). Profiles are similar to those described by several authors in rocks from sillimanite-alcali feldspar metamorphic conditions. However, garnet from literature shows a large core without zoning, being all the zoning concentrated in the rim (*e.g.* Tracy *et al.*, 1976; Yardley, 1989). Garnets from the PNAC migmatites show a continuous zoning from core to rim, although some irregularities are found in the middle part of the largest crystals (Fig. 5).

The FeO profile is similar to the MnO one: concentration of these components increases towards to the rim. Regarding MgO, there exists depletion towards to the rim, although in some cases there are zones in between showing enrichment in comparison to the core. The depletion of MgO towards to the rim can be explained as cordierite has been generated at the expenses of garnet.

In general, a reverse zoning is observed on the garnets from the PNAC migmatites, related to its resorption and generation of cordierite and biotite at their expenses.

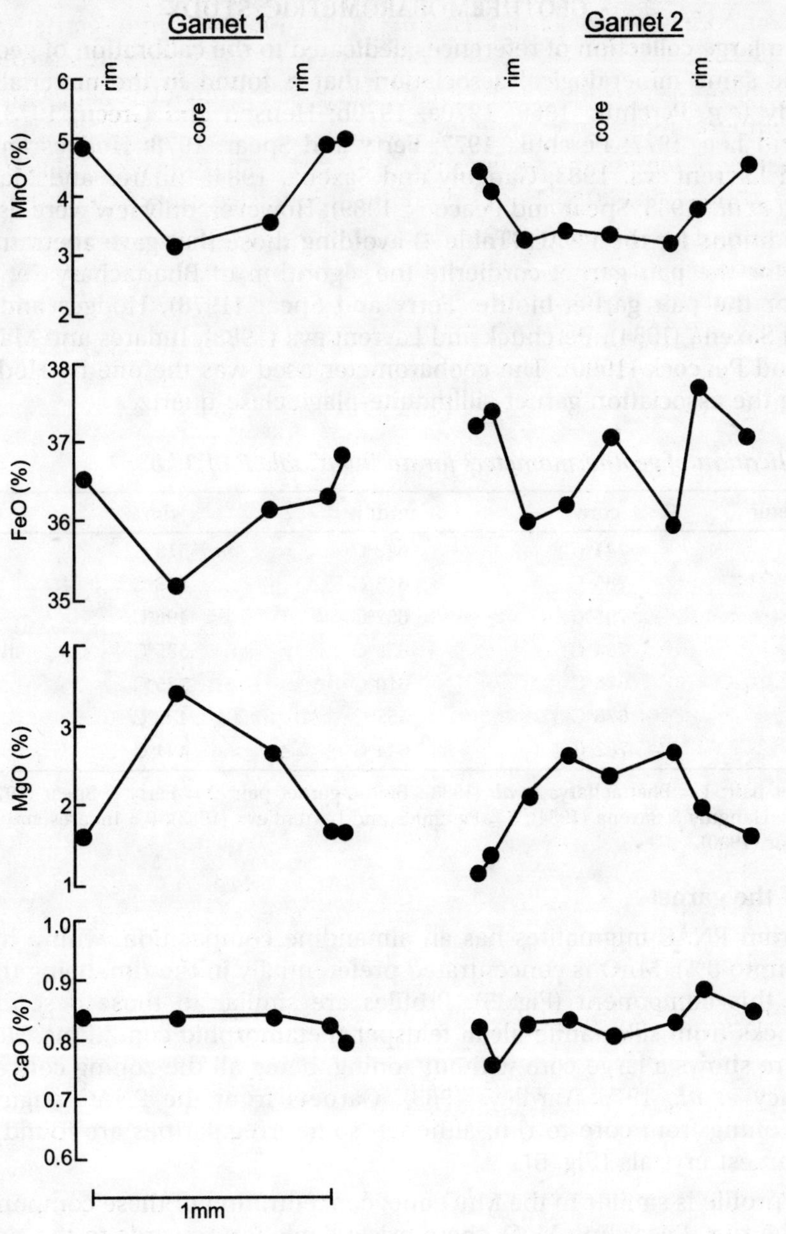

Fig. 5. Zoning profiles for MnO, FeO, MgO and CaO in garnets from migmatites in PNAC.

Geothermometry

Cordierite-garnet pair. Some of the calibrations considering the pair cordierite-garnet did not offer a reasonable result for the metamorphic climax. That is probably due to the lack of spessartine component of the garnet used for the calibration, but in garnet from PNAC can be very high (see above) with a high and constant enrichment towards to the crystal rim.

Only the thermometer calibrated by Bhattacharya *et al.* (1988) has being applied, obtaining reasonable temperature values (Table 4):

(1) The range of temperature values is 513 - 741°C. T has been calculated for a constant P of 3 kb, as this variable does not affect the temperature calculation.

(2) Depletion of MgO towards the rim has been explained as a consequence of cordierite generation (Grant and Weiblen, 1971; Tracy, *et al.*, 1976; Hollister, 1977; Tracy, 1982; Pereira and Bea, 1994).

(3) It can be observed that when using rim composition, the highest X_{Fe}/X_{Mg} values are obtained, together with the lowest temperatures. However, when using garnet core composition, where the lowest values for X_{Fe}/X_{Mg} are found, we get the maximum values of T, for a constant P = 3 kb. It is highly recommended to use core analysis when dealing with zoned minerals to get equilibrium temperatures, so we can conclude that the value around 740°C is the temperature for the equilibrium crystal-melt.

Biotite-garnet pair

Distribution of FeO and MgO between garnet core and biotite from the matrix is the more appropriated to estimate temperature conditions in prograde metamorphism (Tracy *et al.*, 1976; Chipera and Perkins, 1988). This is due to the high diffusion velocity of biotite in comparison to garnet, in the way that while zoning can be determined on the latter, biotite in contact with garnet do not show any zoning, as diffusion has deleted it (Robinson *et al.*, 1982; Spear, 1989).

Several thermometers have been applied, using the programme developed by Spear and Peacock (1990) on the biotite-garnet pair. Analyses from core, middle and rim have been used, attending to the garnet zoning, from leucosome and melanosome of migmatites of the PNAC. Although there is an agreement when using most of them, thermometers calibrated by Perchuck and Lavrent'eva (1983) and Indares and Martignole (1985) gave much lower values than the others. Temperatures obtained when using core or rim analysis are similar to the ones obtained with the calibrated method by Bhattacharya *et al.* (1988) on the garnet-cordierite pair.

Therefore, we can accept that the maximum T values obtained (750°C for constant P = 3 kb) represent the climax of metamorphism.

Geobarometry

Due to the mineralogical composition of the PNAC, only the barometer calibrated by Hodges and Spear (1982) has been used, on the association garnet-plagioclase-sillimanite-quartz. Composition of plagioclase in the migmatites is rather constant: core An_{20-26}, rim An_{11-18}.

Although in temperature calculation we used a constant P of 3 kb because the independence of the thermometers on this variable, a difference has been observed when calculating pressure for equilibrium conditions. For garnet core analysis, the maximum value of 4 kb is obtained, that drops down to 1-2 kb when using the rim analysis.

Using also the middle garnet analysis, we have represented all the values on a P-T diagram, obtaining a retrogressive path, from the equilibrium garnet melting ($P \approx 4$ kb and $T \approx 750$ºC) to the end of retrogression of the garnet ($P \approx 1$-2 kb and $T \approx 500$ºC) (Fig. 6).

On Fig. 6, the triple point is the one from Bohlen *et al.* (1991): $T \approx 530$ºC; $P \approx 4.2$ kb. H_2O concentration curves are from Johannes and Holtz (1990); muscovite (M) and biotite (B) melting curves are from Petö (1976) and Le Breton and Thompson (1988) respectively. Relevant mineral equilibrium regarding paragenesis in the PNAC materials is:

(a) musc + qtz = sil + Kfs + vapor as Chatterjee and Johannes (1974) modified by Holdaway and Lee (1977);

(b) bt + sil + qtz = crd (M=0.5) + Kfs + vapor,
bt + sil + qtz = crd + gr + Kfs + vapor, and
gr + sil + qtz + vapor = crd (M=0.5), M = Mg/(Mg+Fe) (Holdaway and Lee, 1977).

(c) bt + crd + qtz = Kfs + gr, from Thompson (1982).

P-T trajectory

Pre-anatectic paragenesis. We have not found pre-anatectic paragenesis within the PNAC, but in areas nearby there are very similar materials, where the anatexis isograde has not been reached yet (Beetsma, 1995) and the association biotite + sillimanite + quartz can be found stable. This point is taken as the first reference in the P-T trajectory of the metamorphism (point *A* in Fig. 7).

Origin of garnet. Garnet is a very rare component of migmatites from PNAC (see above). It only has been identified in a very restricted area, made of migmatite of schlieren facies. The first question to answer is whether there is a chemical difference between migmatites with and without garnet, specially regarding the aluminum index (ASI) and X_{Fe} value. Chemistry of all migmatites is virtually the same (Pereira, 1992) so we have to accept that generation of garnet was not due to a compositional peculiarity but to the change of P-T conditions that allowed this material to reach one of the equilibrium curves that represent the garnet generation reaction. From the geothermobarometric study, we can conclude that garnet was generated through the incongruent melting of biotite in the presence of sillimanite during the metamorphic paroxysm of the PNAC (point *B* in Fig. 7).

Retrogression of garnet. The scarce garnet one can find in the migmatites from the PNAC present always the same texture, being surrounded by cordierite and biotite. These biotite and cordierite have a different composition (less magnesium, more iron) than the other biotite and cordierite coexisting in the same samples (Tables 5, 6 and 7). Besides, garnet zoning profiles regarding FeO and MnO are always reverse, implying the effect of retrograde metamorphism. Retrogression of garnet was produced following a continuous reaction where cordierite and biotite were formed at the expenses of garnet (point *C* in Fig. 7) (Pereira, 1992; 1993; Pereira and Bea, 1994):

Alcali feldspar + garnet$_1$ + biotite$_1$ + cordierite$_1$ = garnet$_2$ + biotite$_2$ + cordierite$_2$ + quartz

From this reaction it is obtained:

(1) Garnet$_2$ is more ferrous and more manganesiferous than garnet$_1$. Garnet$_2$ is represented by the rim of the largest garnet crystals.

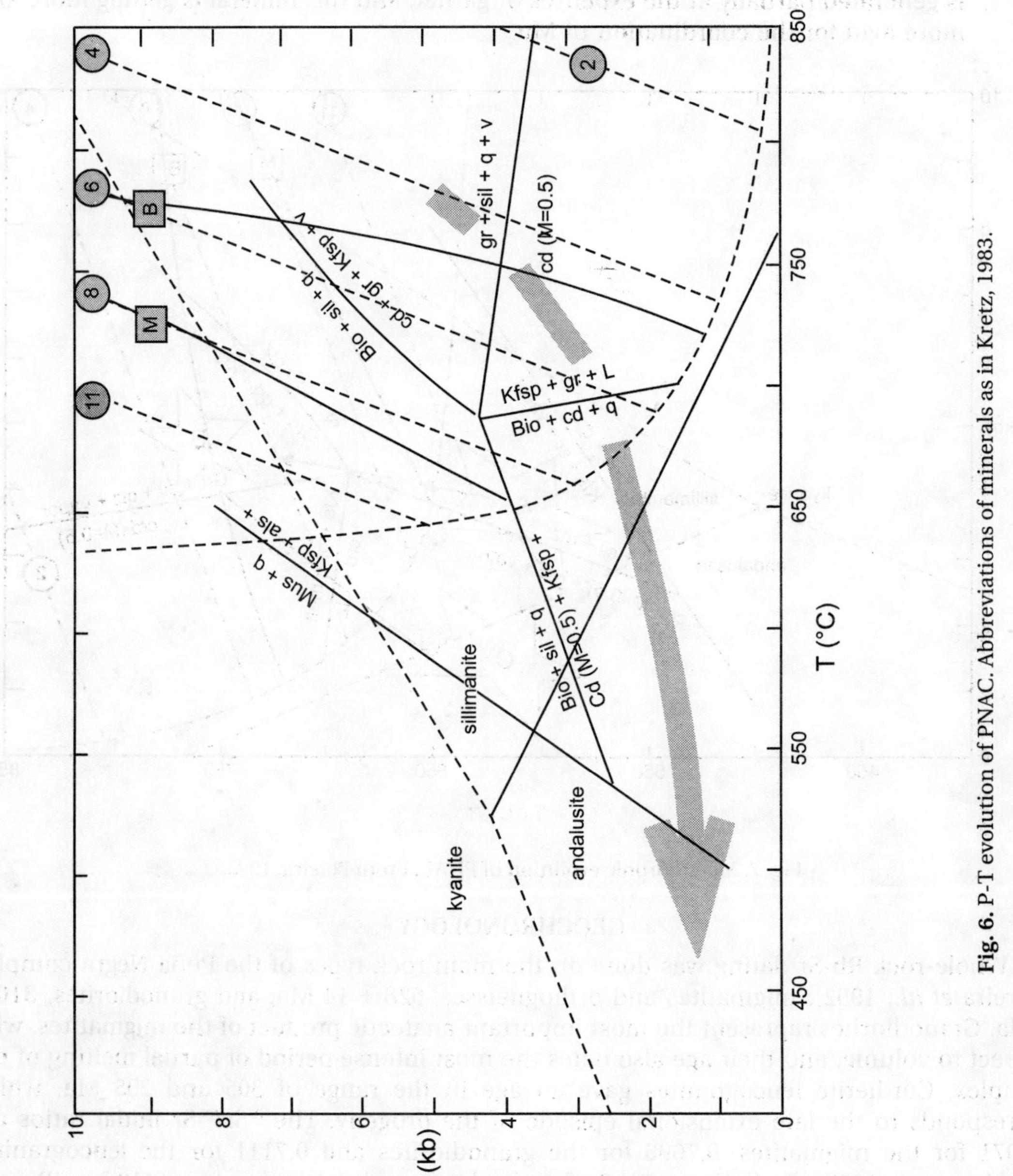

Fig. 6. P-T evolution of PNAC. Abbreviations of minerals as in Kretz, 1983.

(2) Biotite$_2$ is less magnesian and less titaniferous than biotite$_1$, because it is generated, at least partially, at the expenses of garnet.

(3) Cordierite$_2$ is less magnesian and less manganesiferous than cordierite$_1$, because it is generated partially at the expenses of garnet, and this mineral is getting more and more avid for the coordination of Mn.

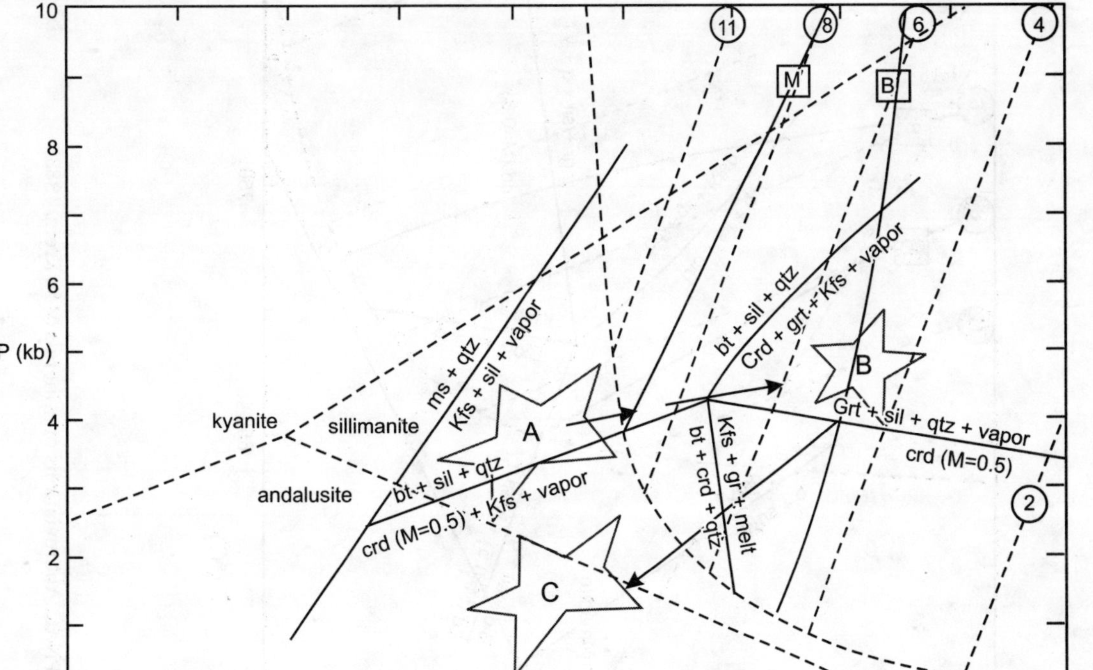

Fig. 7. Metamorphic evolution of PNAC. From Pereira, 1993.

GEOCHRONOLOGY

Whole-rock Rb-Sr dating was done on the main rock types of the Peña Negra complex (Pereira *et al.*, 1992): migmatites and orthogneisses, 528 ± 14 Ma, and granodiorites, 310 ± 6 Ma. Granodiorites represent the most important anatectic product of the migmatites, with respect to volume, and their age also dates the most intense period of partial melting of the complex. Cordierite leucogranites gave an age in the range of 305 and 295 Ma, which corresponds to the late extensional episode of the orogeny. The $^{87}Sr/^{86}Sr$ initial ratios are 0.7071 for the migmatites, 0.7096 for the granodiorites and 0.7111 for the leucogranites (Pereira *et al.*, 1992). Preliminary Rb-Sr data on basic rocks give ages from 416 Ma (Pereira *et al.*, 1992) to 340 Ma (Bea *et al.*, 1999). A more precise method is needed for these rocks, to be more accurate in their interpretation. Sm-Nd analyses are being carried out at these moments and results are expected to be available soon. Further geochronology studies

Table 5: *Analysis of garnet from Peña Negra migmatites.*

Crystal	1			2		3			4		
Zone	rim	mid	core	rim	mid	rim	mid	core	rim	mid	core
SiO_2	36.73	37.03	37.08	36.38	37.33	36.79	36.33	37.02	36.23	36.83	36.46
TiO_2	0.04	0.04	0.02	0.03	0.04	0.01	0.04	0.00	0.00	0.00	0.04
Al_2O_3	20.78	20.58	20.68	20.53	21.00	20.75	20.83	20.98	20.77	20.67	20.66
FeO	36.16	36.91	35.92	36.72	35.17	34.24	35.03	34.08	34.29	35.46	35.05
MgO	1.18	1.90	2.56	1.18	3.41	1.88	2.03	3.06	0.61	1.71	1.70
MnO	5.76	4.29	3.75	5.22	3.13	4.91	4.54	3.81	6.59	5.00	4.89
CaO	0.71	0.81	0.74	0.66	0.83	0.78	0.82	0.80	0.69	0.71	0.79
Na_2O	0.01	0.00	0.01	0.00	0.00	0.02	0.04	0.06	0.09	0.08	0.09
K_2O	0.00	0.00	0.00	0.00	0.00	0.00	0.00	0.01	0.02	0.01	0.00
Total	101.4	101.5	100.8	100.7	100.9	99.38	99.66	99.82	99.29	100.47	99.68
Structural formulae based on 12 oxygens											
Si	2.98	2.98	2.99	2.97	2.98	3.00	2.97	2.99	2.99	2.99	2.98
Al	1.98	1.95	1.96	1.97	1.98	2.00	2.00	2.00	2.02	1.98	1.99
Ti	0.00	0.00	0.00	0.00	0.00	0.00	0.00	0.00	0.00	0.00	0.00
Fe	2.45	2.49	2.42	2.51	2.35	2.34	2.39	2.30	2.37	2.41	2.40
Mg	0.14	0.22	0.30	0.14	0.40	0.22	0.24	0.36	0.07	0.20	0.20
Mn	0.39	0.29	0.25	0.36	0.21	0.34	0.31	0.26	0.46	0.34	0.33
Ca	0.06	0.07	0.06	0.05	0.07	0.06	0.07	0.06	0.06	0.06	0.06
Na	0.00	0.00	0.00	0.00	0.00	0.00	0.00	0.00	0.01	0.01	0.01
K	0.00	0.00	0.00	0.00	0.00	0.00	0.00	0.00	0.00	0.00	0.00
X_{fe}	0.80	0.80	0.79	0.81	0.77	0.78	0.79	0.76	0.79	0.79	0.79
X_{Mg}	0.04	0.07	0.10	0.04	0.13	0.07	0.08	0.12	0.02	0.06	0.06
X_{Mn}	0.13	0.09	0.08	0.11	0.07	0.11	0.10	0.08	0.15	0.11	0.11
X_{ca}	0.02	0.02	0.02	0.01	0.02	0.02	0.02	0.02	0.02	0.02	0.02

involving U-Pb in single crystal zircon and Sm-Nd in garnets could facilitate correlation of these rocks with other granitic rocks in the Spanish Central System (Barbero, 1995).

The second phase of the Hercynian orogeny had a very important role in the generation of melt. This phase was the response to a contractional stage, and was translated into sub-horizontal structures. In a general scheme of the PNAC, granodiorites define sub-horizontal sheets of 1 to 10 km length (Fig. 2 in Pereira and Bea, 1994) and can reach 200 m thickness. Sub-horizontal structures affect both the migmatites and the granodiorites. It is clear that these structures acted as conduits for fluids and volatiles that promoted the partial melting and controlled the geometry of the melt (Pereira and Shaw, 1999; Acosta *et al.*, 2000). A later extensional event produced the collapse of the migmatitic series. Locally, the planar fabric was affected by sub-vertical folding and sub-vertical shear zones (Díez Balda *et al.*, 1990).

GENERATION AND EVOLUTION OF THE COMPLEX

Mesocratic migmatites have a uniform low-pressure paragenesis: quartz + plagioclase + alkali feldspar + biotite + cordierite ± sillimanite. In a very specific area of the PNAC (see

Table 6: *Analysis of cordierite from Peña Negra.*

Crystal	1	2	3	4	5
SiO_2	47.64	48.25	48.17	48.53	47.83
TiO_2	0.00	0.02	0.01	0.00	0.00
Al_2O_3	32.04	32.85	31.64	32.27	32.86
FeO	11.95	10.41	13.08	11.58	10.77
MgO	6.10	6.50	5.34	6.09	6.08
MnO	0.23	0.22	0.33	0.27	0.26
CaO	0.00	0.01	0.00	0.00	0.01
Na_2O	0.18	0.39	0.15	0.08	0.25
K_2O	0.00	0.01	0.00	0.00	0.00
Total	98.14	98.66	98.72	98.82	98.06
Structural formulae based on 18 oxygens					
Si	5.004	5.004	5.053	5.045	4.998
Al	3.966	4.014	3.910	3.953	4.046
ΣT	8.970	9.019	8.963	8.997	9.043
Ti	0.000	0.002	0.001	0.000	0.000
Fe	1.050	0.903	1.147	1.006	0.941
Mg	0.954	1.004	0.835	0.943	0.946
Mn	0.021	0.019	0.030	0.024	0.023
ΣM	2.025	1.928	2.012	1.974	1.910
Ca	0.000	0.001	0.000	0.000	0.001
Na	0.037	0.078	0.031	0.016	0.051
K	0.000	0.001	0.001	0.000	0.000
ΣCh	0.037	0.081	0.032	0.016	0.052
Σcat.	11.032	11.027	11.007	10.987	11.005
M	0.471	0.521	0.415	0.478	0.495

Crystal 3 is a cordierite coexisting with garnet. In the structural formulae, ΣT, ΣM, ΣCh, represent the sum of tetrahedric, octahedric and channel cations respectively. M = Mg/(Fe+Mn+Mg).

above), garnet has been identified as a minor phase, invariably related to mesocratic migmatites of the schlieren facies (Pereira, 1992). Garnet-biotite and garnet-cordierite geothermobarometry reveals that the principal metamorphic episode occurred between T ≈ 750 and 800°C at a constant pressure of about 4 kbar (Pereira, 1998). Melting started under lower temperature conditions than the incongruent melting of biotite (see LeBreton and Thompson, 1988; Vielzeuf and Holloway, 1988). Cordierite is a dominant phase in all the rock types and the availability of this mineral at the beginning of anatexis increased the melt fraction due to its effect on lowering the solidus temperature in the haplogranitic systems at low pressure [see Pereira and Bea (1994) and references therein].

Textural observations are critical in the mass-balance study that evaluates the anatectic production of PNAC granites:

Table 7: *Analysis of biotite from Peña Negra migmatites.*

Crystal	1	2	3	4
SiO_2	34.40	34.60	35.03	34.72
TiO_2	2.34	2.30	2.43	2.39
Al_2O_3	19.80	19.46	18.29	19.13
FeO	22.67	23.13	21.31	20.62
MgO	5.82	5.90	8.06	8.10
MnO	0.21	0.06	0.20	0.22
CaO	0.01	0.04	0.00	0.00
Na_2O	0.23	0.22	0.08	0.20
K_2O	9.03	9.21	9.98	8.86
Total	95.51	94.92	95.38	94.24
Structural formulae based on 11 oxygens				
Si	2.681	2.692	2.705	2.686
^{IV}Al	1.319	1.308	1.295	1.314
^{VI}Al	0.500	0.477	0.369	0.430
Ti	0.137	0.135	0.141	0.139
Fe	0.676	0.684	0.927	0.933
Mg	1.477	1.505	1.376	1.334
Mn	0.014	0.004	0.013	0.014
ΣM	2.804	2.803	2.825	2.851
Ca	0.002	0.004	0.000	0.000
Na	0.036	0.033	0.012	0.030
K	0.898	0.914	0.983	0.874
ΣA	0.935	0.951	0.995	0.904
$\frac{Fe}{Fe+Mg}$	0.686	0.688	0.597	0.588

Crystals 1 and 2 are in contact with garnet.

1. Cordierite appears in different textural forms (Pereira and Bea, 1994): in the leucosome as euhedral crystals, suggesting a magmatic origin; in the melanosome as xenomorphic crystals with granoblastic textures; and surrounding garnet, as a consequence of retrograde metamorphism due to decompression.
2. In the melanosome, the association Biotite + Sillimanite + Quartz never shows textural equilibrium.
3. The most common paragenesis in the melanosome and in the restite enclaves is Biotite + Sillimanite + Cordierite + Ilmenite.

Complete analyses of the PNAC rock-forming minerals can be found in Bea *et al.* (1994) and in Pereira and Bea (1994).

The first stage in the metamorphic evolution is given by the pre-anatectic paragenesis that occurs in areas near the PNAC, and in which similar rocks have not reached the "melt-in" isograd and contain the association Bt + Sil + Qtz without textural evidence of reaction.

The reaction among biotite, sillimanite and quartz in the melanosome suggests the P-T conditions which permitted reaction (1) be totally displaced to the right:

$$Bt + Sil + Qtz = Crd + Kfs + vapor \quad (1)$$

Adjustment of the mass balance for reaction (1) to correspond to the mineralogy of the PNAC granitic products can be found in Pereira and Bea (1994). The metamorphic path which the PNAC followed went from melting of the protolith, through the retrograde conditions which generated cordierite + biotite from garnet (Pereira, 1998).

Petrogenetic duality for the granitic products

The two kinds of cordierite-bearing granitic rocks (leucogranite and granodiorite) result from the anatexis in migmatites in the PNAC. The most acceptable explanation of the field, petrographic and chemical observations, is that sub-autochthonous granodiorites were generated from a hybrid source composed of mesocratic migmatites and orthogneisses by partial melting. The small volume of residual rocks, which appear related to granodiorite, either as enclaves or sillimanite-rich rocks, suggests that restite-melt unmixing was inefficient. Sub-autochthonous granodiorites are directly related to sub-horizontal shear zones, where H_2O (derived from an external source and from dehydration of mica minerals) and volatiles (the role of boron in these rocks is explained in Pereira and Shaw, 1996; 1997; 1999) were concentrated. The importance of fault zones in concentrating H_2O and hydrous fluids has been considered by many workers (e.g., Strong and Hanmer, 1981) with regard to their significance in promoting extensive melting. Development of the granodiorites was conditioned by the source lithology and anatexis only occurred where the quartz-feldspar content exceeded a critical value. In this case, the melt fraction was higher than the Critical Melt Fraction (CMF = 30 vol. %, Wickham, 1987) and residual products were not separated. Applying the anatectic-mixing model of Bea (1989) a hybrid source is proposed, made up of migmatites and orthogneisses. This source would give a very low residual melt fraction. $^{87}Sr/^{86}Sr_i$ for the granodiorites agrees with the idea of their generation through partial melting of migmatites. However, the lower $d^{18}O$ values obtained for the granodiorite could be due to contamination with more basic rocks. The studies on mafic rocks described in areas nearby (Moreno-Ventas *et al.*, 1995, Castro *et al.*, 1999; Bea *et al.*, 1999) together with some tonalites from within the complex will help us to understand the genesis of these granodiorites and a possible interaction with basic rocks at a deeper level (Villaseca *et al.*, 1998).

The mechanism proposed for granodiorite generation contrasts with that for the cordierite leucogranites, whose mineralogy, textures and chemical composition are compatible with the segregation of a pure melt, largely devoid of restite material (Bea, 1991). In this case, CMF was not exceeded, perhaps because alkali feldspar and cordierite abundances played a limiting role, being the source of leucogranites locally depleted in these phases at this stage of anatexis. Segregation of leucogranites would yield a residue enriched in sillimanite and biotite, with some quartz and cordierite, which is represented by the restites. Melt extraction of the leucogranites represents a pure segregated fraction.

Therefore, the two different granitic rocks of anatectic origin found in the PNAC correspond to two different petrogenetic mechanisms of segregation of melt from restite, being low melt-fraction granite in the case of the leucogranites and high melt-fraction granites in the case of the granodiorites (both terms taken from Wickham, 1987). They never grade into each other and in both cases, the existence of a "fertile" source is the important factor controlling the segregation of one or the other kind of granitic fraction, although there is a close spatial and temporal association between the granitic rocks and shear structures, as described by Barbarin (1996, 1999). This situation is common within the Hercynian belt (Holtz, 1989; Holtz and Barbey, 1991) and is probably one of the causes of the characteristic granitic duality throughout the European Variscan belt.

Geodynamic evolution

On the basis of the geothermobarometric studies on the PNAC migmatites, a P-T trajectory is obtained indicating that the metamorphic paroxysm took place under maximum conditions of $T \approx 750^oC$ and a $P \approx 4$ kbar. Subsequently, a drop in temperature and pressure took place and retrograde metamorphism occurred under minimum conditions of $T \approx 500^oC$ and $P \approx 1\text{-}2$ kbar (Pereira, 1998). The same clockwise P-T path followed by the PNAC materials has been described for many others nearby Hercynian regions (e.g. Gil Ibarguchi and Martínez, 1982; Barbero, 1995). This path was responsible for the generation of important volumes of peraluminous granitic magma and has been interpreted to result from brief intervals of rapid tectonic thickening, followed by exhumation of materials (Hollister, 1994). The thickening was the result of thrusting during the contractional first phases of the Hercynian orogeny and, as a response to these events, granodiorite sheets were produced from partial melting of the migmatites. An extensional collapse at the end of the orogeny had an effect in decreasing pressure to produce the small bodies of leucogranite from a source already depleted in fertile components (Pereira and Rodríguez-Alonso, 2000).

DISCUSSION AND CONCLUSIONS

The partial melting of a highly heterogeneous protolith, the Schist Greywacke Complex, produced Migmatites of the Peña Negra anatectic complex only where a high volume of orthogneisses was available (Pereira and Rodríguez-Alonso, 2000). Owing to this heterogeneity, it is difficult to calculate an average composition for the SGC, but based on the relative abundance of lithologies, an attempt to estimate its average composition has been made:

SGC = [(15 * Black pelite) + (2 * Sandstone) + (3 * Conglomerate) + (0.5 * (volcanoclastic + amphibolic rocks)]/38.5

Rock types in the formulae represent the average composition of each one. Adjusting the mass balance to calculate the composition of the migmatites, following the method described by Bea (1989), we get the approximate expression:

Migmatites = 0.763 SGC + 0.049 * (Quartz + Orthoclase + Albite + Anorthite)

Gneisses have been dated by U-Pb zircon geochronology, giving Cadomian to Cambrian ages. More recently, Fernández-Suárez *et al.* (1999) have proposed that basement rocks range in age, from Palaeoproterozoic to Neoproterozoic, on the basis of zircon found in orthogneisses from northwestern Iberia. Homogenization of the protolith (SGC + orthogneisses) prior to the Hercynian anatexis gave rise to a compositional mixture, in which all the rocks (except for the most extreme lithologies, e.g. sandstones and limestones, that are found now as enclaves within the PNAC materials) progressively acquired essentially uniform characteristics.

Migmatites are the result of anatexis during the Hercynian orogeny, involving the breakdown of hydrous minerals. The existence of cordierite and the addition of boron and other volatile species led to lowering the solidus temperature and the minimum-melt composition was shifted to the left of the melt-production curve, producing peraluminous granodiorites and, at lower temperatures, leucogranites (Pereira and Bea, 1994; Guillot and Le Fort, 1995).

The granodiorite composition lies between that of migmatite and the orthogneiss precursor, suggesting a hybrid source in its production, with little change in bulk composition. This idea agrees with an episode of wholesale melting, with very little fractional crystallization. By contrast, the composition of the leucogranite supports an origin from a small degree of migmatite melting, leaving a residue impoverished in SiO_2 and Na_2O, and enriched in TiO_2, Al_2O_3 and CaO. The surprising feature for the leucogranites is the enrichment in K. The melanosome of the migmatites also is enriched in K, so it can be deduced that the leucogranites were produced from remelting migmatites where there was enrichment in melanocratic component, the breakdown of biotite providing the excess K for leucogranite generation.

The levels of concentration of the trace element and REE in the migmatites and granitic products provide evidence for the anatectic origin of the latter. The chemical differences between the granodiorites and leucogranites must be due to the composition of the source rocks. There is no place in the area where these two rock units grade into each other. Migmatites were formed from a very heterogeneous source, and the different layers defined by the proportions of feldspathic components, have different REE patterns in the anatectic products (Bea, 1991; Johannes *et al.*, 1995). The complementarity of some trace element anomalies of restites and leucogranite, *vis-à-vis* the migmatites, corroborates the anatectic hypothesis for the generation of the leucogranites. If there is a residue enriched in REE to compensate for the depletion in these elements for the leucogranites, then the leucogranites are likely to have formed from source-rocks with the characteristics of the migmatites. The enrichment in REE of this residue is mainly due to the high concentration of biotite, which is a conspicuous host for accessory phases, mainly zircon (Nabelek and Glascock, 1995). Such a residue might be represented by the sillimanite-rich rocks from the Peña Negra complex. The anatectic model proposed for the generation of the granites (Bea, 1989; Pereira, 1992) is only satisfied when chemical and mineralogical composition of granitic products is added to the composition of these sillimanite-rich rocks.

ACKNOWLEDGMENTS

I am indebted to Prof. Harel Thomas to give me the opportunity of participation in Prof. R.S. Sharma Felicitation volume. Data in this contribution were obtained in studies supported by the grants CICYT PB-95-1266 and PB-98-1345 from the Spanish government.

REFERENCES

Acosta, A., Pereira, M.D. & Shaw, D.M., 2000. Influence of volatiles in the generation of crustal anatectic melts. *J. Geochem. Exploration*, **69-70**, 339-42.

Barbarin, B., 1996. Genesis of the two main types of peraluminous granitoids. *Geology*, **24-4**, 295-98.

Barbarin, B., 1999. A review of the relationships between granitoid types, their origins and their geodynamic environments. *Lithos*, **46**, 605-26.

Barbero, L., 1995. Granulite-facies metamorphism in the Anatectic Complex of Toledo, Spain: late Hercynian tectonic evolution by crustal extension. *J. Geol. Soc. London*, **152**, 365-82.

Bea, F., 1989. A method for modelling mass-balance in partial melting and anatectic leucosome segregation. *J. Metamorphic Geol.*, 7, 619-28.

Bea, F., 1991. Geochemical modelling of low melt-fraction anatexis in a peraluminous system: The Peña Negra Complex (Central Spain). *Geochim. Cosmochim. Acta*, **55**, 1859-74.

Bea, F. Montero, P. & Molina, J.F., 1999. Mafic precursors, peraluminous granitoids and late lamprophyres in the avila Batholith: a model for the generation of Variscan Batholith in Iberia. *Journal of Geology*, **107**, 399-419.

Bea, F., Pereira, M.D. & Stroh, A., 1994. Mineral/leucosome trace-element partitioning in a peraluminous migmatite (a laser ablation-ICP-MS study). *Chem. Geol.*, **117**, 291-312.

Beetsma, J.J., 1995. *The Late Proterozoic/Paleozoic and Hercynian crustal evolution of the Iberian massif, N Portugal, as traced by geochemistry and Sr-Nd-Pb isotope systematics of pre-Hercynian terrigenous sediments and Hercynian granites*. PhD thesis, Vrije Universiteit, Amsterdam, The Netherlands.

Bhattacharya, A., Mazumdar, A.C. & Sen, S.K., 1988. Fe-Mg mixing in cordierite: Constraints from natural data and implications for cordierite-garnet geothermometry in granulites. *Am. Mineral.*, **vol. 73**, 338-44.

Bohlen, S.R., Montana, A. & Kerrick, D.M., 1991. Precise determination of the equilibria kyanite = sillimanite and kyanite = andalusite and a revised triple point for Al_2SiO_5 polymorphs. *Am. Mineral.*, **vol. 76**, 677-80.

Castro, A., Patiño Douce, A., Corretgé, L.G., de la Rosa, J., El-Biad, M. & El-Hmidi, H., 1999. Origin of peraluminous granites and granodiorites, Iberian massif, Spain: an experimental test of granite petrogenesis. *Contrib. Mineral. Petrol.*, **135**, 255-76.

Clayton, R.N. & Mayeda, T.K., 1963. The use of bromine pentafluoride in the extraction of oxygen from oxides and silicates for isotopic analysis. *Geochim. Cosmochim. Acta*, **27**, 43-52.

Currie, K.L., 1974. A note on the calibration of the garnet-cordierite geothermometer and geobarometer. *Contrib. Mineral. Petrol.*, **vol. 44**, 35-44.

Chatterjee, N.D. & Johannes, W., 1974. Thermal stability and standard thermodynamic properties of synthetic 2M1-muscovite $KAl_2[AlSi_3O_{10}(OH)_2]$. *Contr. Mineral. Petrol.*, **vol. 48**, 89-114.

Chipera, S.J. & Perkins, D., 1988. Evaluation of biotite-garnet geothermometers: application to the English River subprovince, Ontario. *Contrib. Mineral. Petrol.*, **vol. 98**, 40-48.

Díez Balda, M.A., Vegas, R. & González Lodeiro, F., 1990. Central Iberian Zone: Autochthonous sequences: structure. *In*: Dallmeyer R.D., Martínez García E. (eds) *Pre-Mesozoic geology of Iberia*. Springer Verlag, Berlin, Germany, 172-88.

Dostal, J. & Capedri, S., 1978. Uranium in Metamorphic rocks. *Contrib. Mineral. Petrol.* **66**, 409-514.

Fernández-Suárez, J., Gutiérrez-Alonso, G., Jenner, J.A. & Tubrett, M.N., 1999. Crustal sources in Lower Palaeozoic rocks from NW Iberia: insights from laser ablation U-Pb ages of detrital zircons. *J. Geol. Soc. London*, **156**, 1065-68.

Ferry, J.M. & Spear, F.S., 1978. Experimental calibration of partitioning of Fe and Mg between biotite and garnet. *Contrib. Mineral. Petrol.*, **vol. 66**, 113-17.

Ganguly, J. & Saxena, S.K., 1984. Mixing properties in aluminosilicate garnets: constraints from natural and experimental data and applications to geothermobarometry. *Am. Mineral.*, **vol. 69**, 88-97.

Gil Ibarguchi, I. & Martínez, F.J., 1982. Petrology of garnet-cordierite-sillimanite gneisses from the el Tormes thermal dome, Iberian Hercynian Foldbelt (W. Spain). *Contrib. Mineral. Petrol.*, **80**, 14-24.

Grant, J.A. & Weiblen, P.W., 1971. Retrograde zoning in garnet near the second sillimanite isograde. *Am. J. Science*, **vol. 270**, 281-96.

Guillot, S. & LeFort, P., 1995. Geochemical constraints on the bimodal origin of High Himalayan leucogranites. *Lithos* **35**, 221-34.

Hensen, B.J. & Green, D.H., 1973. Experimental study of the stability of cordierite and garnet in pelitic compositions at high pressures and temperatures. III. Synthesis of experimental data and geological applications. *Contrib. Mineral. Petrol.*, **vol. 38**, 151-66.

Hodges, K.V. & Spear, F.S., 1982. Geothermometry, geobarometry and the Al_2SiO_5 triple point at Mt. Moosilauke, New Hampshire. *Am. Mineral.*, **vol. 67**, 1118-34.

Holdaway, M.J. & Lee, S.M., 1977. Fe-Mg cordierite stability in high-grade pelitic rocks based on experimental, theoretical and natural observations. *Contrib. Mineral. Petrol.*, **vol. 63**, 175-98.

Hollister, L.S., 1994. Clockwise P-T-t paths: the View from the Coast Orogen, British Columbia. *EOS Transactions* **75**, 16, 186.

Hollister, L.S., 1977. The reaction forming cordierite from garnet, the Khtada Lake metamorphic complex, British Columbia. *Can. Mineral.*, **vol. 15**, 217-29.

Holtz, F., 1989. Importance of melt fraction and source rock composition in crustal genesis: The example of two granitic suites of northern Portugal. *Lithos*, **24**, 21-35.

Holtz, F. & Barbey, P., 1991. Genesis of peraluminous granites. II. Mineralogy and chemistry of the Tourem complex (North Portugal). Sequential melting vs restite unmixing. *J. Petrol.*, **32**, 959-78.

Ibarrola, E., Villaseca, C., Vialette, Y., Fuster, J.M., Navidad, M., Peinado, M. & Casquet, C., 1987. Dating of Hercynian Granites in the Sierra de Guadarrama (Spanish Central System). In: Bea F. *et al.* (eds) *Geología de los granitoides y rocas asociadas del Macizo Hespérico*. Rueda, Madrid, España, 377-83.

Indares, A. & Martignole, J., 1985. Biotite-garnet geothermometry in the granulite facies: the influence of Ti and Al in biotite. *Am. Mineral.*, **vol. 70**, 272-78.

Jamieson, R.A., 1991. P-T-t paths of collisional orogens. *Geologische Rundschau*, **vol. 80-2**, 321-32.

Johannes, W. & Holtz, F., 1990. Formation and composition of H_2O-undersaturated granitic melts. In: *High T metamorphism and crustal anatexis, Ashworth and Brown*, Eds., Unwin Hyman, 1990, 87-104.

Johannes, W., Holtz, F. & Möller, P., 1995. REE distribution in some layered migmatites: constraints on their petrogenesis. *Lithos*, **35**, 139-52.

Kempton, P.D. & Harmon, R.S., 1992. Oxygen isotope evidence for large-scale hybridization of the lower crust during magmatic underplating. *Geochim. Cosmochim. Acta*, **56**, 971-86.

Kretz, R., 1983. Symbols for rock-forming minerals. *Am. Mineral.*, **vol. 68**, 277-79.

Le Breton, N. & Thompson, A.B., 1988. Fluid-absent (dehydration) melting of biotite in metapelites in the early stages of crustal anatexis. *Contrib. Mineral. Petrol.*, **99**, 226-37.

McLellan, S.M., 1983. Contrasting textures in metamorphic and anatectic migmatites: an example from the Scottish Caledonides. *J. Metamorphic Geol.*, **1-3**, 241-62.

McLennan, S.M., 1989. Rare Earth Elements in sedimentary rocks: influence of provenance and sedimentary process. Geochemistry and Mineralogy of REE. *Rev. Mineral.*, **21**, 169-200.

Mehnert, K.R., 1987. The granitization problem – revisited. *Fortschr. Miner.*, **vol. 65**, 285-306.

Moreno-Ventas, I., Rogers, G. & Castro, A., 1995. The role of hybridization in the genesis of hercynian granitoids in the Gredos Massif, Spain: inferences from Sr-Nd isotopes. *Contrib. Mineral. Petrol.*, **120**, 137-49.

Nabelek, P.I. & Glascock, M.D., 1995. REE-depleted leucogranites, Black Hills, South Dakota: a consequence of disequilibrium melting of monazite-bearing schists. *J. Petrol.*, **36-40**, 1055-71.

Perchuk, L.L., 1969. The effect of temperature and pressure on the equilibrium of natural iron-magnesium minerals. *Intern. Geol.*, 875-901.

Perchuk, L.L., 1970a. Equilibria of rock-forming minerals. Nauka, Moscow, 1970.

Perchuk, L.L., 1970b. Equilibrium of biotite with garnet in metamorphic rocks. *Geoch. Intern.*, **vol. 7**, 157-79.

Perchuk, L.L., 1977. Thermodynamical control of metamorphic processes. *In: Energetics of geological processes*, Saxena, S.K. and Bhattacharji, S., Eds., New York: Springer-Verlag, 285-352.

Perchuk, L.L. & Lavrent'eva, I.V., 1983. Experimental investigation of exchange equilibria in the system cordierite-garnet-biotite. *In: Kinetics and Equilibrium in mineral reactions*, Saxena, S.K., Ed., New York: Springer-Verlag, 199-239.

Pereira, M.D., 1992. *El Complejo Anatéctico de la Peña Negra (Batolito de Avila): Un estudio de la anatexia cortical en condiciones de baja presión*. PhD thesis. Universidad de Salamanca, Salamanca, España.

Pereira, M.D., 1998. Pressure and Temperature conditions in the generation of an anatectic complex in central Spain: the Peña Negra complex. *Petrology,* **6-6**, 555-63.

Pereira, M.D. & Bea, F., 1994. Cordierite-producing reactions at the Peña Negra Complex, Avila batholith, central Spain: the key role of cordierite in low pressure anatexis. *Can. Mineral.*, **32-4**, 763-80.

Pereira, M.D., 1989. *Migmatización diatexítica y la génesis de las granodioritas subautóctonas del Complejo Anatéctico de la Peña Negra, Batolito de Avila.* MSc thesis, Universidad de Salamanca, Salamanca, España.

Pereira, M.D., 1993. Termometría de rocas con la asociación granate- cordierita-biotita; trayectorias P-T en el Complejo Anatéctico de la Peña Negra (Batolito de Avila). Implicaciones sobre el metamorfismo hercínico en la zona centro-ibérica. *Rev. Soc. Geol. España,* **vol. 6 (1-2)**, 131-40.

Pereira, M.D., Ronkin, Y. & Bea, F., 1992. Dataciones Rb/Sr en el Complejo Anatéctico de la Peña Negra (Batolito de Avila, España Central): Evidencias de magmatismo pre-hercínico. *Rev. Soc. Geol. España,* **5**, 129-34.

Pereira, M.D. & Shaw, D.M., 1996. Boron and lithium distribution in the Peña Negra complex: an alpha-track study. *Am. Mineral.*, **vol. 81**, 141-45.

Pereira, M.D. & Shaw, D.M., 1997. Behaviour of boron in the generation of an anatectic complex: the Peña Negra complex, central Spain. *Lithos,* **40**, 179-88.

Pereira, M.D. & Shaw, D.M., 1999. Relationship between geodynamics and generation of melt in central Spain. *Phys. Chem. Earth* A, **24**, 317-19.

Pereira, M.D. & M.D. Rodríguez Alonso, 2000. Duality of cordierite granites related to melt restite segregation in the Peña Negra anatectic complex, central Spain. *The Canadian Mineralogist,* **38**, 1329-46.

Petö, P., 1976. An experimental investigation of melting relations involving muscovite and paragonite in the silica saturated portion of the system K_2O-Na_2O-Al_2O_3- SiO_2-H_2O to 15 kb total pressure. *Prog. Exp. Petrol. NERC. Rept.*, **3**, 41-45.

Ramirez, J.A. & Grundvig, S., 2000. Causes of geochemical diversity in peraluminous granitic plutons: the Jálama pluton, Central-Iberian zone (Spain and Portugal). *Lithos,* **50**, 171-90.

Robinson, P., Hollocher, K.T., Tracy, R.J. & Dietsch, C.W., 1982. *High grade Acadian regional metamorphism in south-central Massachusetts,* Joesten, R. and Quarrier, S., Eds., NEIGC 74th Ann. Meet. Storrs. C.T.

Rodríguez Alonso, M.D. & Palacios, T., 1994. A Paleogeographical outline of Upper-Median-Lower Cambrian in the Iberian Peninsula. IGCP 319-376 Meeting. Nova Scotia, Canada, 1-2.

Spear, F.S., 1989. Relative thermobarometry and metamorphic P-T paths. In: Evolution of metamorphic belts, Daly, J.S., Cliff, R.A. and Yardley, B.W.D., Eds., *Geol. Soc. London Spec. Publ.*, 1989, **vol. 43**, 63-81.

Spear, F.S. & Peacock, S.M., 1989. Metamorphic Pressure-Temperature-Time Paths. In: Evolution of metamorphic belts, Daly, J.S., Cliff, R.A. and Yardley, B.W.D., Eds. Washington, D.C.: Am. Geoph. Union.

Spear, F.S. & Peacock, S.M., 1990. Metamorphic P-T-t Paths: Program manual and computer exercises for the calculation of metamorphic phase equilibria, pressure-temperature-time paths and thermal evolution of orogenic belts. Washington, D.C.: Am. Geoph. Union.

Strong, D.F. & Hanmer, S.K., 1981. The leucogranites of southern Brittany: origin by faulting, frictional heating, fluid flux and fractional melting. *Can. Mineral.*, **19**, 163-76.

Thompson, A.B., 1982. Dehydration melting of pelitic rocks and the generation of H_2O - undersaturated granitic liquids. *Am. J. Science,* **vol. 282**, 1567-95.

Tracy, R.J., 1982. Compositional zoning and inclusions in metamorphic minerals. *Rev. Mineral.*, **vol. 10**, 355-97.

Tracy, R.J., Robinson, P. & Thompson, A.B., 1976. Garnet composition and zoning in the determination of temperature and pressure metamorphism, central Massachusetts. *Am. Mineral,* **vol. 61**, 762-75.

Vidal, G., Palacios, T., Gámez-Vitaned, J.A., Díez Balda, M.A. & Grant, S.W.F., 1994. Neoproterozoic-early Cambrian geology and palaeontology of Iberia. *Geological Magazine,* **131** (6), 729-65.

Vielzeuf, D. & Holloway, J.R., 1988. Experimental determination of the fluid-absent melting relations in the pelitic systems. Consequences for crustal differentiation. *Contrib. Mineral. Petrol.*, **98**, 257-76.

Villaseca, C., Barbero, L. & Rogers, G., 1998. Crustal origin of Hercynian peraluminous granitic batholiths of central Spain; petrological, geochemical and isotopic (Sr, Nd) constraints. *Lithos*, **43-2**, 22-79.

Wickham, S.M., 1987. The segregation and emplacement of granitic magmas. *J. Geol. Soc. London*, **144**, 281-97.

Yardley, B.W.D., 1989. An Introduction to Metamorphic Petrology, Edinburgh: Longman.

5

POLYMETAMORPHISM IN THE ARCHAEAN GNEISS COMPLEX OF SHIVPURA GYANGARH DISTRICT BHILWARA, RAJASTHAN

H. THOMAS*

ABSTRACT

The Archaean rocks in North-Central Indian Shield consist mainly of tonalite gneiss, migmatite, aluminous schist/gneiss, amphibolite and granitoids, which collectively constitute the Banded Gneissic Complex (BGC). The BGC from north-central Rajasthan also contain isolated outcrops of granulites, which are mainly three type: (1) Garnet - sillimanite - kyanite± cordierite rock containing charnockite/enderbite as inclusions, is most dominant type and is interlayered with (2) Leptynite having quartz-feldspar-garnet assemblage (3) Garnet-bearing basic granulite is the next important rock which occurs as the structure base of the pelitic granulite. Metanorite dykes intrude the granulite and show a blastophitic texture. The granulite show a polymetamorphic history. The pelitic granulite documents the chronological development of Al-silicate phase in order: $kyanite_1 \rightarrow sillimanite \rightarrow kyanite_2$. The first generation kyanite occurs as prism with ragged terminations, whereas the second generation kyanite is mostly in needle form. The basic granulite exhibits coronitic textures wherein an early assemblage of hypersthene and plagioclase become incompatible in the second metamorphic event to form garnet-diopside-quartz corona. The norite dykes shows both igneous and metamorphic parageneses, with garnet at the interface of hypersthene and plagioclase, suggesting its development during the waning stage of granulite facies metamorphism.

The physical conditions of granulite-facies metamorphism gave pressure of 6.48-9.01 kbar and 6.65-10.26 kbar and temperature near 689°C and 785°C by appropriate geothermobarometric models applied to both pelitic and basic-granulite assemblages. On the basis of these P and T estimates and textural evidences on the crystallization sequence of the Al-silicate minerals in the order ($kyanite_1 \rightarrow sillimanite \rightarrow kyanite_2$) in the pelitic granulite, a movement path has been derived, which is reverse to the "normal" P-T path documented for collision belts placing considerable constraints on the geological evolution of the Precambrian rocks from this part of the Indian Shield.

KEY WORDS

Polymetamorphism, granulite, Thermobarometry, P-T path.

* Department of Geology, Mizoram University, Aizawl, Mizoram 796 001.
E-mail: harelthomas@yahoo.com

INTRODUCTION AND GEOLOGICAL SETTING

The Shivpura Gyangarh area is situated about 88 km NW of Bhilwara town (Fig. 1). It exposes indubitable metasediments of Pre-Aravalli age (Early Proterozoic, Gupta, 1934) and granulites of uncertain age (Fig. 2). The granulites occur at the structural base of the "basement" Banded Gneissic Complex (BGC).

The Banded Gneissic Complex, records two phases of Precambrian deformation which have been superimposed by the Proterozoic deformations (D_1, D_2 and D_3 Naha and Halyburtan, 1983; Roy, 1988) related to the Aravalli mountain orogeny (1700-900 Ma years). The tonalite gneiss and amphibolite components of the BGC are dated at 3.5 Ga and 2.85 Ga respectively (MacDougall *et al.*, 1983; Gopalan *et al.*, 1988). This rock-suite is intruded by granite of 2.9 Ga and the K-granite (Berach granite) of 2.6 Ga, as determined by Rb/Sr and U/Pb methods (Choudhary *et al.*, 1984; Sivaraman and Odom, 1982). There is no conclusive age data for the granulite facies metamorphism, although the intrusive charnockite/granodioritic comagmatic series within the BGC from Gyangarh has been dated at 1723 Ma (Sarkar *et al.*, 1989). It is from this terrain that the granulitic rocks have been studied for their metamorphism in relation to that of the surrounding BGC rocks.

Granulites Metamorphism

To demonstrate any possible petrological relationship between the granulites with the BGC rocks a detailed geological and mineralogical study was carried out for the Shivpura-Gyangarh areas (Fig. 2). The granulite complex forms a somewhat irregular, elliptical outcrop amidst the BGC rocks with which it shows a peripheral sheared contact. The granulite complex comprises the following main rock types:

(1) Garnet – sillimanite – kyanite – biotite with or without cordierite gneiss or pelitic granulite is the second dominant rock and contains discrete inclusions of charnockite and enderbite along its foliation. The pelitic granulites in investigated area are found to contain andalusite (Plate 1C, D & E) Thomas, (1995), which is attributed to the intrusion of granodiorite-charnockite comagmatic series (1723 Ma, Sarkar *et al.*, 1989).

(2) Garnet-leptynite or garnet-quartz-feldspar granulite is almost invariably interlayered with the pelitic granulite and their thickness varies from a few centimeters to over a meter.

(3) Garnet-bearing two-pyroxene granulite or basic granulite is an important rock that occurs at the structural base of the pelitic granulite. It is tholeiitic in composition.

(4) The metanorite dykes which crosscuts the granulite banding/foliation. It is not thoroughly recrystallized because of its emplacement in the waning stages of granulite metamorphism and hence shows both igneous and metamorphic paragenesis and ophitic to subophitic texture (Plate 1H).

(5) The garnet-bearing basic granulite is characterized by the development of garnet-clinopyroxene-quartz± hornblende coronas at the interface of hypersthene-plagioclase (Plate 1A). The garnet-clinopyroxene-quartz corona formed by the following equilibrium reaction (Plate 1A & B):

hypersthene + plagioclase = garnet ± clinopyroxene + quartz, which has a positive gentle dP/dT slope in P-T space. Obviously by this reaction the garnet-clinopyroxene-quartz coronas could have formed by isobaric cooling or by increased pressure with

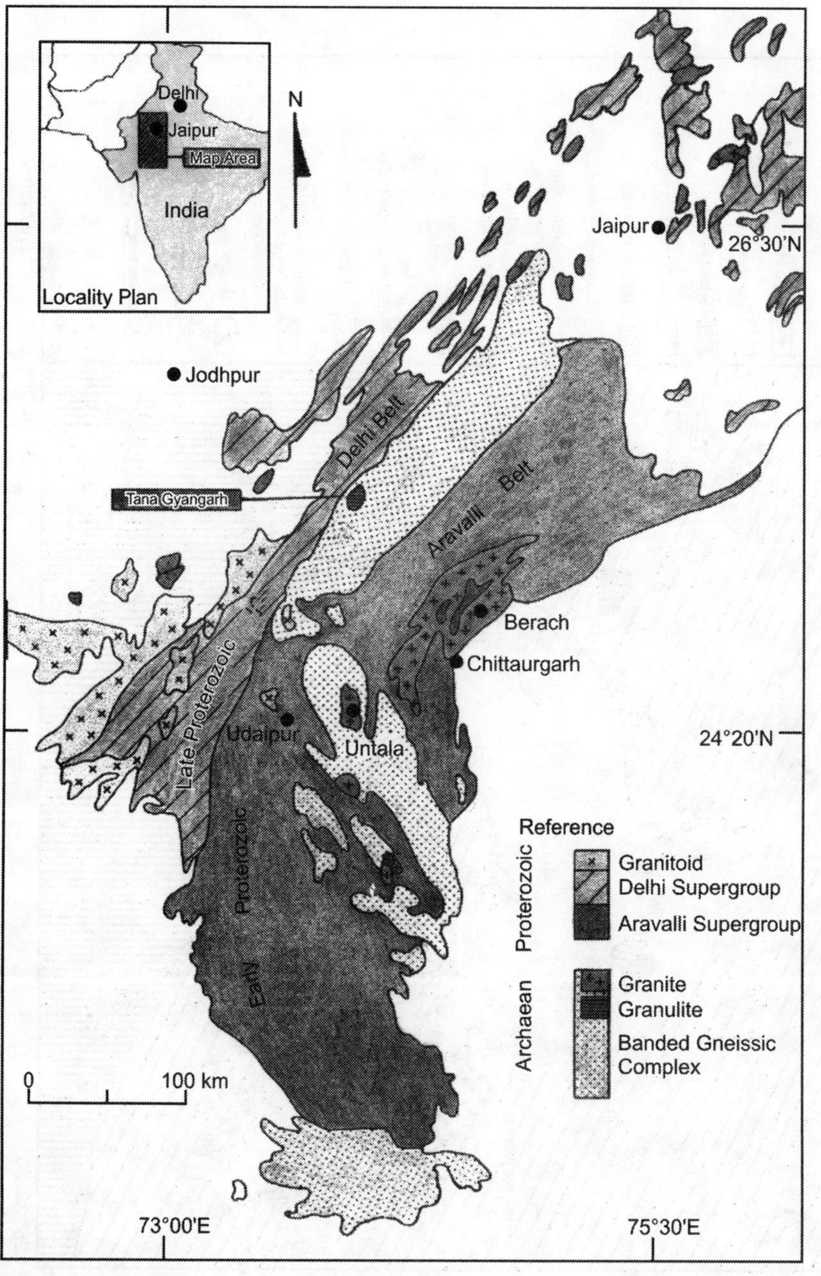

Fig. 1. Geological sketch map (simplified after G.S.I., 1969) showing the distribution of Precambrian rocks in Rajasthan, North-West India.

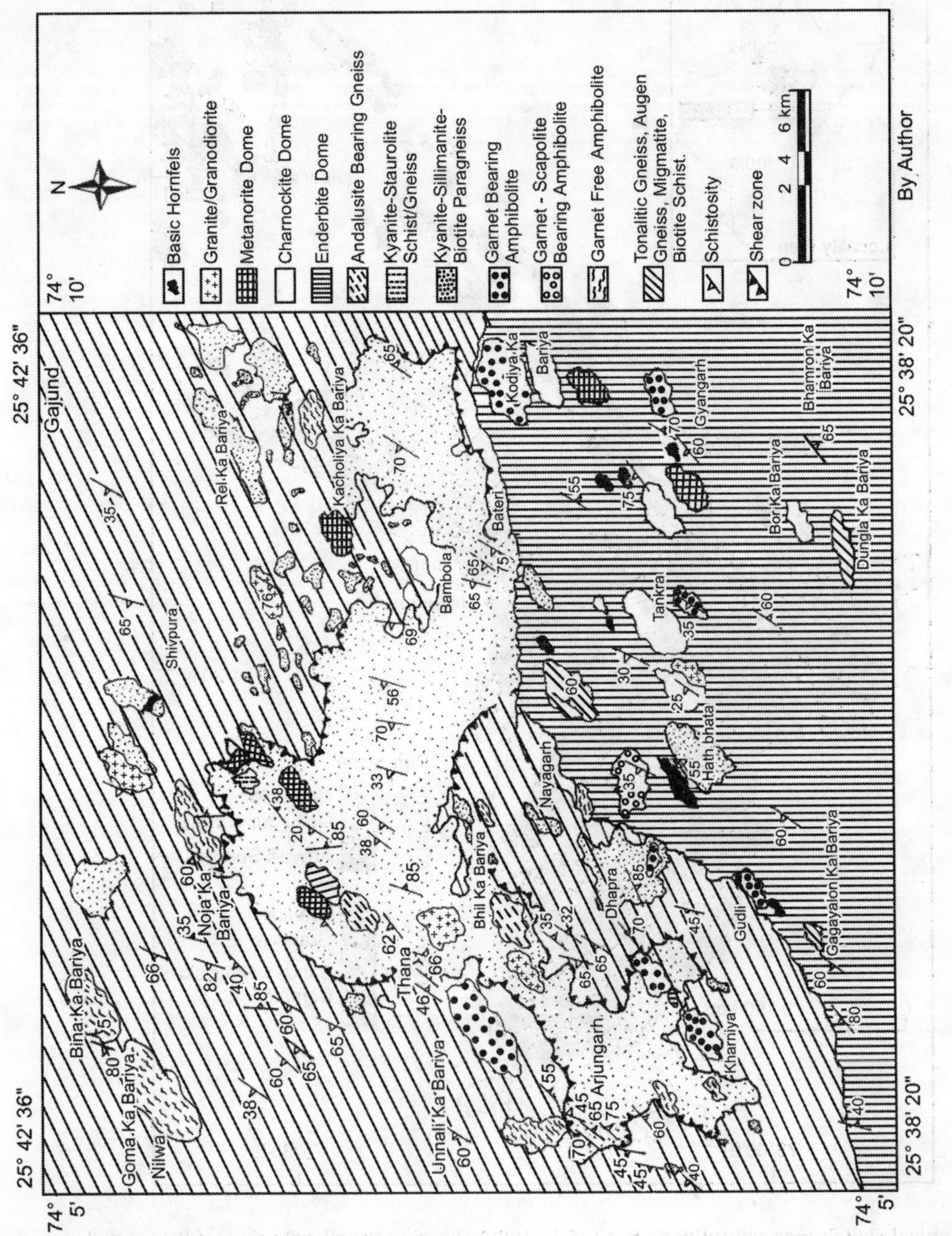

Fig. 2. A combined lithological and structural map of Shivapur-Gyangarh, District Bhilwara, Rajasthan.

or without cooling. It will be demonstrated that the corona, unlike symplectite was formed by compression during cooling; it is not a product of arrested reaction or isobaric cooling (Joshi *et al.*, 1993; Thomas, 1995).

The pelitic granulites are characterized by garnet–sillimanite–kyanite–biotite–quartz ± Ilmanite ± plagioclase (An_{16} to An_{45}) with or without cordierite. That the $P_{H_2O} > P_{load}$ condition prevailed during the metamorphism of these rocks is suggested by the presence of kyanite and cordierite (X_{Mg} = 0.86) (Hensen and Green, 1973). Two generation of kyanite, viz. $kyanite_1$ and $kyanite_2$ have been identified on the basis of textural relations (Plate 1F). The former kyanite ($kyanite_1$) occurs as large prism with ragged termination while the second-generation kyanite ($kyanite_2$) mostly occurs as randomly oriented short needles at the borders of slender sillimanite prism. This textural relationship suggests the development of aluminosilicates in the following chronological order: $kyanite_1 \rightarrow$ sillimanite $\rightarrow kyanite_2$ (Plate 1F & G).

Mineral Chemistry

A number of mineral phases have been analyzed by electron microprobe (Cambridge Instruments Microscan V EPMA at Leicester, U.K. and Chemeca SX 50 EPMA at Wurzburg, Germany) operated at 15kb accelerating voltage and 1-2 mm beam diameter. The ZAF corrected analyses are given in Tables 1-8.

The garnets in pelitic granulites of Shivpura-Gyangarh are dominantly almandine-pyrope, with a subordinate grossular content or X_{Ca} (0.0259 to 0.1127)(see Table 1).

The biotites in these rocks are homogeneous in composition and their titanium content ranges between X_{Ti} = (0.0344 to 0.0751) (Table 2). The plagioclase crystals in the pelitic granulites at places show minor normal zoning (never exceeding X_{Ca} = 0.4609) and the anorthite content varies between An_{16} to An_{46} (Table 3). Cordierite from this metapelite persistently shows high X_{Mg} (0.80) (Table 4).

The garnets developed in the corona in the basic granulites are commonly unzoned. However, they show a compositional variation depending upon whether they occur in contact with plagioclase or orthopyroxene. Garnet is mostly an almandine-pyrope-grossular solid solution (alm_{49-58} $pyrope_{15-29}$ $grossular_{20-29}$ mole% while spessartine less than 2 mole% (Table 5). The garnet in metanorite (sample no. R87/399) is zoned and pyrope content varies from 15 to 29 mole% (Table 5). The unzoned clinopyroxene in the coronas compositionally range from salite to ferrosalite, while clinopyroxene in the metanorite occurs as compositionally zoned large crystals. These clinopyroxenes distinctly show higher X_{Mg} at the rim (Table 6). Since the rims of orthopyroxenes in metanorite are commonly studded with iron oxide, only their core compositions were analyzed (Table 7).

The anorthite contents for plagioclase of basic granulite ranges between An_{47} to An_{76}. Individual plagioclase crystals, however, show variation of upto 5 mole% only and zoning is mild (Table 8).

Phase Equilibrium Relation

Orthopyroxene and plagioclase are recognized as the earliest phases in basic granulites. The absence of any relict hydrous phase (hornblende) in pyroxene indicated an igneous origin for basic granulites (Bohlen and Essene, 1978). However, hornblende occurring in the corona at the orthopyroxene-plagioclase interface is metamorphic, whereas this hornblende plagioclase along the cracks within pyroxene is of later origin. Such coronal textures yield

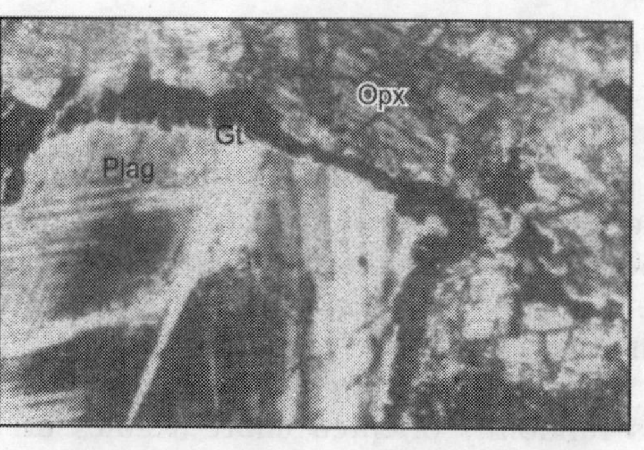

Hypersthene is rimmed by garnet and clinopyroxene against its contact with plagioclase suggesting the reaction Opx + plag = gt + cpx + qz

Plate 1A

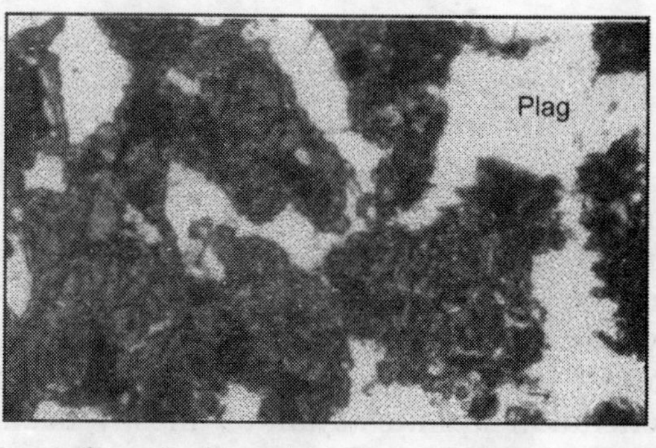

Hypersthene is rimmed by garnet against its contact with plagioclase.

Plate 1B

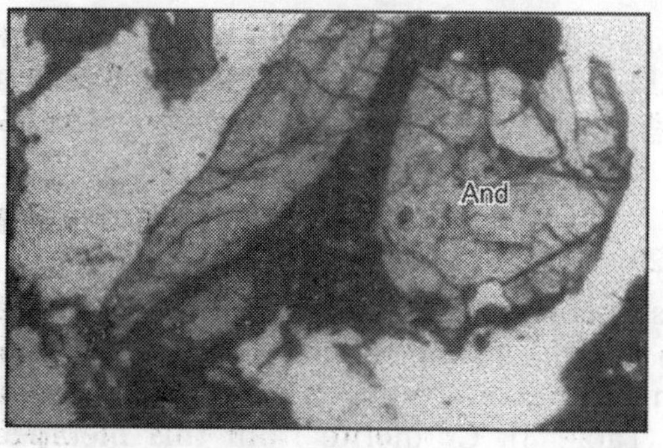

Andalusite porphyroblast in biotite-quartz feldspar-gneiss.

Plate 1C

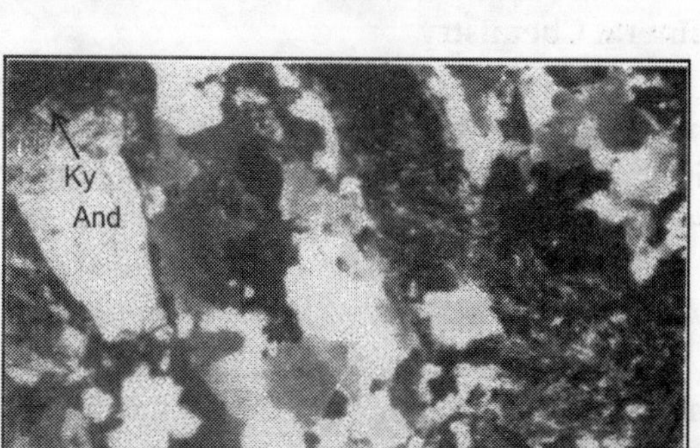

Andalusite having inclusion of kyanite (seen as needles in upper left part), suggesting andalusite development after kyanite.

Plate 1D

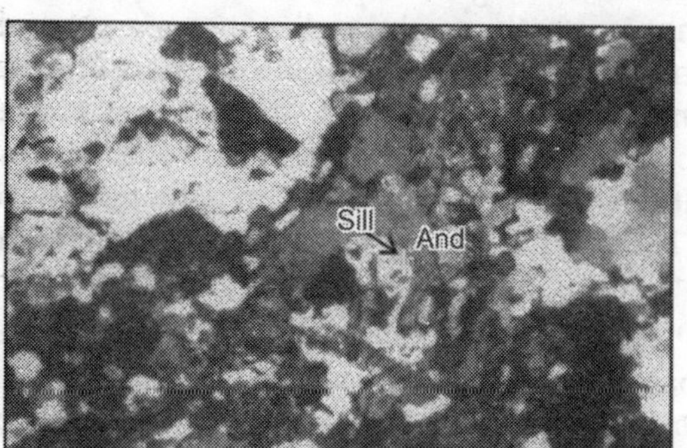

Andalusite having inclusions of sillimanite (Central part) while separate crystals of sillimanite also have inclusion of needle kyanite, confirming the chronological other Ky→Sill→And.

Plate 1E

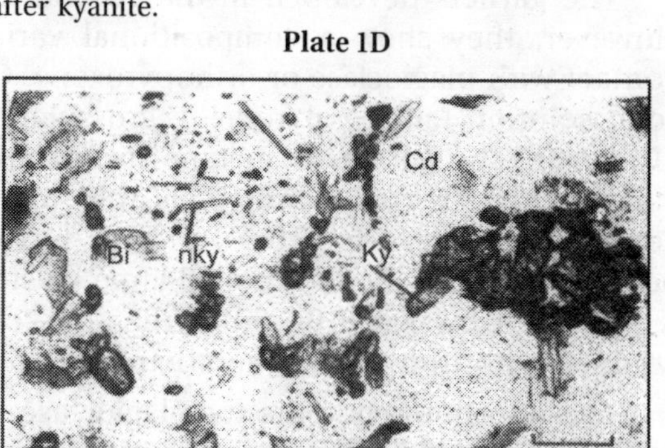

The stable association of cordierite (Cd) Kyanite (Ky)-biotite (Bi) in the paragneiss beside bladed kyanite (nKy) are also visible.

Plate 1F

(Plate *Contd...*)

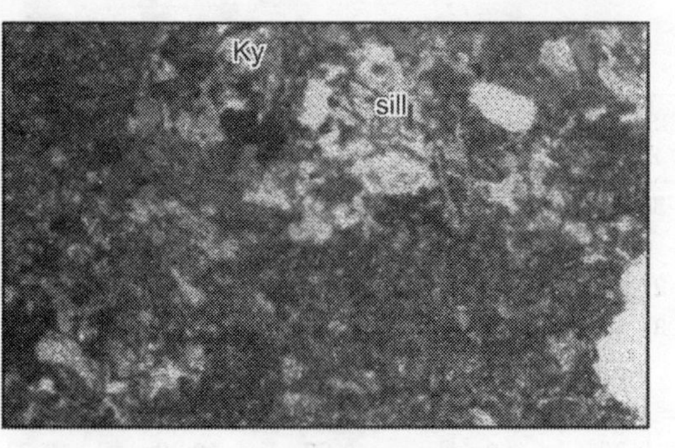

Relationship between sillimanite and kyanite which show (lower part) cross cutting relationship with each other.

Plate 1G

Hypersthene occurring as large crystals are seen intergrown with plagioclase to define subophitic to ophitic texture.

Plate 1H

Table 1: *Electron Microprobe Analyses of Garnet in Pelitic Granulite from Shivpura-Gyangarh, District Bhilwara, Rajasthan, India*

S.No.	R87/435 Core	R87/435 Core	R87/435 Pt. 23	R87/435 Pt. 1	R87/435 Pt. 3	R87/342 Pt. 26	R87/342 Pt. 31	R87/299 Pt. 2 C
SiO_2	37.38	37.52	37.52	37.97	37.64	37.53	37.55	37.81
TiO_2	0.00	0.02	0.00	0.04	0.00	0.01	0.05	0.01
Al_2O_3	21.28	21.23	21.52	21.22	21.10	21.79	21.43	21.06
FeO	33.00	33.59	33.95	33.15	35.38	30.96	31.68	33.97
MgO	5.95	5.88	5.61	5.97	4.76	7.51	6.97	4.21
CaO	0.92	1.06	1.16	1.06	1.12	2.07	1.12	2.32
MnO	0.60	0.70	0.56	0.67	0.60	0.26	0.28	0.67
Cr_2O_3	0.05	0.02	0.02	0.03	0.02	0.00	0.02	0.04
Fe_2O_3	0.01	0.42	0.00	0.00	0.00	0.14	0.00	0.44
Total	99.18	100.42	100.34	100.11	100.61	100.27	99.10	100.53
	24(O)	24(O)	24(O)	24 (O)	24(O)	24 (O)	24 (O)	24 (O)
Si	5.965	5.937	5.940	5.999	5.981	5.876	5.952	6.003
Ti	0.000	0.002	0.000	0.005	0.000	0.001	0.006	0.001
Al	4.002	3.960	4.017	3.952	3.953	4.022	4.005	3.942
Fe^{+2}	4.404	4.445	4.495	4.381	4.702	4.054	4.200	4.511
Mg	1.416	1.386	1.324	1.407	1.128	1.752	1.646	0.996
Ca	0.157	0.179	0.197	0.179	0.190	0.347	0.190	0.395
Mn	0.081	0.074	0.075	0.089	0.081	0.034	0.038	0.090
Cr	0.006	0.003	0.002	0.003	0.002	0.000	0.003	0.005
Fe^{+3}	0.001	0.050	0.000	0.000	0.000	0.016	0.000	0.053
Total	16.031	16.055	16.050	16.021	16.039	16.104	16.039	15.996
X_{Fe}	0.7269	0.7306	0.7379	0.7234	0.7707	0.6552	0.6914	0.7528
X_{Mg}	0.2337	0.2278	0.2173	0.2323	0.1848	0.2831	0.2709	0.1662
X_{Ca}	0.0259	0.0294	0.0323	0.0295	0.0311	0.0560	0.0312	0.0659
X_{Mn}	0.0133	0.0121	0.0123	0.0146	0.0133	0.0054	0.0062	0.0150

(Contd...)

Table 1: *Contd...*

S.No.	R87/299 Pt. 3 R	R87/299 Pt. 4 R	R87/291 Pt. 1	R87/291 Pt. 4	R87/291 Pt. 11	H90/16 Pt. 1 R	H90/16 Pt. 2 R	R87/452 Pt. 1 C	R87/452 Pt. 8 R
SiO_2	37.56	36.93	37.34	36.69	36.79	37.07	36.87	37.95	37.24
TiO_2	0.06	0.00	0.01	0.00	0.01	0.03	0.12	0.05	0.00
Al_2O_3	21.21	21.27	21.21	20.79	20.96	21.14	21.27	20.90	21.26
FeO	35.05	35.33	36.05	34.99	36.45	35.99	34.66	30.92	33.61
MgO	3.19	3.42	4.15	4.20	3.87	4.83	4.79	5.20	3.97
CaO	3.04	2.97	1.46	1.42	1.46	1.16	1.30	4.03	3.30
MnO	0.83	0.96	0.44	0.45	0.36	0.51	0.57	0.39	0.73
Cr_2O_3	0.03	0.09	0.06	0.03	0.01	0.10	0.09	0.05	0.05
Fe_2O_3	0.17	0.07	0.20	0.25	0.35	0.44	0.00	0.92	0.06
Total	101.14	101.04	100.92	98.82	100.25	101.27	99.67	100.41	100.22
	24 (O)	24 (O)	24 (O)	24 (O)	24 (O)	24(O)	24 (O)	24 (O)	24 (O)
Si	5.971	5.897	5.944	5.955	5.919	5.886	5.913	5.982	5.943
Ti	0.007	0.000	0.002	0.000	0.001	0.004	0.014	0.005	0.000
Al	3.975	4.004	3.980	3.977	3.975	3.957	4.022	3.882	3.999
Fe^{+2}	4.660	4.718	4.799	4.749	4.905	4.779	4.649	4.075	4.487
Mg	0.756	0.814	0.984	1.016	0.927	1.143	1.145	1.222	0.945
Ca	0.518	0.508	0.248	0.247	0.251	0.197	0.223	0.680	0.564
Mn	0.112	0.130	0.059	0.062	0.050	0.069	0.077	0.052	0.098
Cr	0.004	0.011	0.008	0.003	0.001	0.013	0.011	0.007	0.007
Fe^{+3}	0.020	0.008	0.024	0.031	0.042	0.053	0.000	0.109	0.007
Total	16.022	16.091	16.048	16.040	16.071	16.100	16.056	16.014	16.050
X_{Fe}	0.7707	0.7646	0.7880	0.7818	0.7997	0.7723	0.7628	0.6758	0.7362
X_{Mg}	0.1250	0.1319	0.1615	0.1672	0.1511	0.1847	0.1878	0.2026	0.1550
X_{Ca}	0.0856	0.0823	0.0407	0.0406	0.0409	0.0318	0.0365	0.1127	0.0925
X_{Mn}	0.0185	0.0210	0.0096	0.0102	0.0081	0.0111	0.0126	0.0086	0.0160

valuable information on fluctuation in P-T conditions during the course of tectonometamorphic evolution of the area; the garnet-clinopyroxene-quartz assemblage of the coronas in the basic granulites was thoroughly investigated. Equilibrium in this assemblage is indicated by the following criteria:

(a) The element partitioning data (not presented) exhibits excellent correlation for the mineral pairs (viz., garnet-biotite, orthopyroxene-garnet, orthopyroxene-plagioclase, garnet-clinopyroxene, clinopyroxene-plagioclase, orthopyroxene-clinopyroxene) with confidence level exceeding 90%. The only exception is the Mg-Fe distribution between garnet and orthopyroxene, but even in this case the confidence level is between 80-90%.

(b) The sharp grain contacts among the phases, suggest attainment of equilibrium.

(c) The K_D values of the participating phases in the corona-forming reaction (orthopyroxene + plagioclase = garnet) (clinopyroxene + quartz) (Plate 1A & B) strongly favours thermodynamic equilibrium. The evidence suggests that the corona-bearing granulite have maintained chemical equilibration during granulite facies metamorphism.

Table 2: *Electron Microprobe Analyses of Biotite in Pelitic Granulite from Shivpura-Gyangarh, District Bhilwara, Rajasthan, India*

S.No.	R87/435 In gt core	R87/435 Rim	R87/435 Pt. 3	R87/435 Pt. 2	R87/435 Pt. 4	R87/291 Pt. 2	R87/291 Pt. 3	R87/291 Pt. 12
SiO_2	36.20	36.96	35.34	33.91	36.04	36.33	35.48	35.77
TiO_2	3.79	4.06	2.55	3.61	3.12	2.37	2.36	2.52
Al_2O_3	16.69	16.86	17.46	19.15	16.73	18.23	17.65	17.76
FeO	12.22	11.65	13.93	12.82	16.52	14.69	15.04	16.09
MgO	15.17	15.09	13.39	13.07	12.26	13.24	13.33	12.58
CaO	0.02	0.01	0.00	0.00	0.00	0.00	0.00	0.03
MnO	0.00	0.00	0.03	0.03	0.06	0.01	0.00	0.02
Cr_2O_3	0.11	0.18	0.00	0.00	0.06	0.07	0.11	0.05
Na_2O	0.17	0.16	0.08	0.19	0.06	0.12	0.05	0.14
K_2O	9.65	9.64	9.79	9.32	9.95	9.67	9.76	9.57
H_2O	4.01	4.05	3.91	0.00	0.00	4.00	3.94	3.96
BaO	0.00	0.00	0.11	0.00	0.00	0.20	0.00	0.09
Total	98.02	98.67	98.49	92.10	94.80	98.93	97.72	98.58
	22 (O)	22 (O)	22 (O)	22 (O)	22 (O)	22 (O)	22 (O)	22 (O)
Si	5.422	5.471	5.361	5.200	5.469	5.440	5.399	5.415
Ti	0.426	0.452	0.290	0.417	0.357	0.267	0.270	0.287
Al	2.947	2.943	3.123	3.462	2.994	3.218	3.166	3.170
Fe	1.530	1.443	1.768	1.644	2.096	1.839	1.914	2.037
Mg	3.385	3.330	3.277	2.986	2.771	2.954	3.023	2.838
Ca	0.003	0.002	0.000	0.001	0.000	0.000	0.000	0.005
Mn	0.000	0.000	0.004	0.003	0.008	0.001	0.000	0.003
Cr	0.013	0.021	0.000	0.000	0.007	0.008	0.013	0.006
Na	0.048	0.046	0.022	0.057	0.019	0.035	0.015	0.041
K	1.843	1.821	1.894	1.823	1.927	1.847	1.895	1.848
Ba	0.000	0.000	0.006	0.000	0.000	0.012	0.000	0.005
Total	15.617	15.528	15.745	15.592	15.647	15.621	15.696	15.655
X_{Fe}	0.3113	0.3023	0.3504	0.3550	0.4306	0.3836	0.3877	0.4178
X_{Mg}	0.6887	0.6977	0.6495	0.6449	0.5693	0.6163	0.6123	0.5822
X_{Al}	0.3067	0.3151	0.3175	0.3451	0.3198	0.3439	0.3295	0.3333
X_{Ti}	0.0552	0.0590	0.0370	0.0540	0.0463	0.0345	0.0347	0.0370

S.No.	R87/299 Pt 1	R87/299 Pt. 5	R87/299 Pt. 6	R87/342 Pt. 25	R87/342 Pt. 30	H90/16 Pt. 4	H90/16 Pt. 5	R87/452 Pt. 4 C	R87/452 Pt. 6 R
SiO_2	37.60	35.40	34.77	36.80	36.55	37.89	37.96	36.37	35.40
TiO_2	2.43	3.76	3.82	5.11	5.14	2.53	2.63	3.27	2.91
Al_2O_3	19.73	18.37	18.53	16.45	16.45	17.97	17.76	18.60	17.76
FeO	15.08	18.86	18.87	12.71	12.95	12.40	12.27	16.05	18.09
MgO	12.79	9.11	9.18	13.94	13.35	14.91	15.05	11.73	11.23
CaO	0.00	0.03	0.00	0.04	0.04	0.00	0.01	0.01	0.00
MnO	0.00	0.00	0.06	0.05	0.00	0.06	0.03	0.06	0.03
Cr_2O_3	0.00	0.10	0.08	0.00	0.03	0.02	0.04	0.06	0.01
Na_2O	0.24	0.08	0.10	0.10	0.14	0.14	0.11	0.18	0.15
K_2O	8.93	9.32	9.34	9.42	9.51	9.45	9.56	9.05	9.50

(*Contd...*)

Table 2: *Contd...*

S.No.	R87/299 Pt 1	R87/299 Pt. 5	R87/299 Pt. 6	R87/342 Pt. 25	R87/342 Pt. 30	H90/16 Pt. 4	H90/16 Pt. 5	R87/452 Pt. 4 C	R87/452 Pt. 6 R
BaO	0.04	0.00	0.00	0.00	0.00	0.00	0.00	0.00	0.00
H_2O	4.13	3.94	3.92	4.03	4.01	4.09	4.10	0.00	0.00
Total	100.97	98.97	98.67	98.65	98.17	99.46	99.52	95.38	95.08
	22 (O)	22 (O)	22 (O)	22 (O)	22 (O)	22 (O)	22 (O)	22 (O)	22 (O)
Si	5.465	5.391	5.323	5.471	5.473	5.550	5.558	5.423	5.382
Ti	0.266	0.431	0.440	0.571	0.579	0.279	0.290	0.367	0.333
Al	3.381	3.298	3.344	2.883	2.904	3.103	3.066	3.270	3.184
Fe^{+2}	1.833	2.402	2.416	1.580	1.622	1.159	1.502	2.002	2.300
Mg	2.770	2.068	2.094	3.089	2.979	3.255	3.284	2.607	2.545
Ca	0.000	0.005	0.000	0.006	0.006	0.000	0.002	0.002	0.000
Mn	0.000	0.000	0.008	0.006	0.000	0.007	0.004	0.008	0.004
Cr	0.000	0.012	0.010	0.000	0.004	0.002	0.005	0.007	0.001
Ba	0.002	0.000	0.000	0.000	0.000	0.000	0.000	0.000	0.000
Na	0.068	0.024	0.030	0.029	0.041	0.040	0.031	0.052	0.044
K	1.656	1.811	1.824	1.787	1.817	1.766	1.786	1.722	1.843
Total	15.441	15.441	15.488	15.423	15.423	15.521	15.526	15.458	15.636
X_{FE}	0.3982	0.5373	0.5356	0.3384	0.3525	0.2625	0.3138	0.4343	0.4747
X_{Mg}	0.6017	0.4626	0.4643	0.6616	0.6474	0.7374	0.6861	0.5656	0.5252
X_{Al}	0.3688	0.3537	0.3493	0.3097	0.3143	0.3607	0.3403	0.3504	0.3311
X_{Ti}	0.0344	0.0566	0.0576	0.0751	0.0765	0.0379	0.0376	0.0477	0.0429

Table 3: *Electron Microprobe Analyses of Plagioclase in Pelitic Granulite from Shivpura-Gyangarh, District Bhilwara, Rajasthan, India*

Sample No.	R87/435	R87/452	R87/452
Spot No.	20	25	26
SiO_2	64.03	57.71	57.30
Al_2O_3	22.01	27.51	27.14
CaO	3.29	9.39	9.47
Na_2O	9.48	6.41	6.05
K_2O	0.10	0.06	0.11
Total	99.11	101.09	100.07
Cations	8 (O)	8 (O)	8 (O)
Si	2.847	2.559	2.566
Al	1.153	1.438	1.432
Ca	0.157	0.446	0.454
Na	0.817	0.551	0.525
K	0.006	0.004	0.006
Total	4.980	4.998	4.983
X_{Ca}	0.160	0.446	0.461
X_{Na}	0.834	0.550	0.533
X_K	0.006	0.004	0.006

Table 4: *Electron Microprobe Analyses of Cordierite and Ilmenite in Pelitic Granulite from Shivpura-Gyangarh, District Bhilwara, Rajasthan, India*

Sample No.	R87/452	R87/435	R87/452
Spot No.	Cd	Il	Il
SiO_2	48.91	0.04	0.19
TiO_2	0.00	55.51	44.48
Al_2O_3	33.23	0.00	0.01
FeO	5.96	39.42	40.73
MgO	10.10	0.17	0.03
Na_2O	0.20	0.00	0.00
Cr_2O_3	0.00	0.00	0.09
MnO	0.00	0.09	1.05
Fe_2O_3	0.00	0.00	10.31
Total	98.40	95.23	98.88
Cations	18 (O)	3 (O)	3 (O)
Si	4.978	0.002	0.010
Al	3.986	0.000	0.000
Ti	0.000	2.143	1.792
Fe^{+3}	0.000	0.000	0.398
Fe	0.508	1.692	1.746
Mg	1.533	0.013	0.002
Mn	0.000	0.004	0.046
Na	0.000	0.000	0.000
Total	6.027	3.854	3.994
X_{Fe}	0.249		
X_{Mg}	0.751		

(d) The pelitic assemblages from the granulite also show strong correlation for cationic distribution, particularly for the garnet-biotite pairs. Attainment of thermodynamic equilibrium is also supported by the estimated P-T values for similar assemblages at different locations that fall within the error ranges of the geothermobarometers used.

Geothermobarometry

Since basic granulite, metanorite and metapelitic granulites seem to attain equilibrium, the P-T values of suitable mineral assemblages were calculated on the basis of both core and rim compositions (Tables 9 to 12) through (Mpet programmes: Thomas, 1995b; 1998a, 1998b & 2003) developed by author himself on the basis of various published thermobarometric models given by different works.

Basic Granulite

Thermometry. Temperature calculated through orthopyroxene-clinopyroxene thermometry on the basis of Wood and Banno (1973) and Wells (1977) equations for different pairs in the investigated rocks are in the ranges of 802°C to 986°C. However, the temperature calculated by Wood and Banno (1973) equation is approximately average 60°C lower than

Table 5: *Electron Microprobe Analyses of Coronal Garnet in Granulite and Metanorite from Shivpura-Gyangarh, District Bhilwara, Rajasthan, India*

Sample No.	H90/9	H90/9	R87/399	R87/399	H87/399	H90/2
Spot No.	45	54	1 rim	30 core	21 core	In corona
SiO_2	37.67	38.15	37.97	38.82	37.44	37.15
TiO_2	0.05	0.10	0.21	0.06	0.31	0.06
Al_2O_3	21.61	21.25	21.47	22.04	20.66	21.44
FeO	25.32	24.69	25.46	23.25	27.06	25.25
MgO	5.77	5.36	4.91	7.56	4.11	4.05
CaO	8.21	9.42	7.31	7.23	8.23	10.78
MnO	0.82	0.51	2.64	0.87	1.16	0.97
Cr_2O_3	0.00	0.00	0.00	0.05	0.04	0.07
Fe_2O_3	0.17	0.87	0.12	0.05	0.95	0.39
Total	99.62	100.35	99.88	99.93	99.96	100.19
Cations	24 (O)	24 (O)	24 (O)	24 (O)	24 (O)	24 (O)
Si	5.906	5.939	5.955	5.975	5.929	5.862
Ti	0.006	0.012	0.025	0.007	0.037	0.008
Al	3.994	3.899	3.969	3.998	3.856	3.983
Fe	3.320	3.215	3.340	2.993	3.584	3.336
Mg	1.348	1.244	1.148	1.734	0.971	0.942
Ca	1.378	1.571	1.228	1.192	1.396	1.823
Mn	0.108	0.068	0.068	0.113	0.156	0.125
Cr	0.000	0.000	0.000	0.006	0.006	0.003
Fe^{+3}	0.020	0.102	0.014	0.000	0.113	0.036
Total	16.080	16.050	15.747	16.018	16.048	16.118
X_{Fe}	0.540	0.527	0.577	0.496	0.587	0.536
X_{Mg}	0.219	0.204	0.199	0.287	0.159	0.151
X_{Ca}	0.224	0.258	0.212	0.198	0.229	0.293
X_{Mn}	0.018	0.011	0.012	0.022	0.025	0.021

the temperature calculated by Wells (Table 9), this variation may be due to other parameters but not by differences in rate of cooling or re-equilibration.

Temperature calculated by (Lindsley, 1983; Lindsley and Anderson, 1983 and Davidson and Lindsley 1989; Berman, 1990 and Holland and Powell, 1990) has yielded inconsistent temperatures. Such inconsistency in T-estimates can possibly be attributed to excess non-quadrilateral components in clinopyroxenes.

Ellis and Green (1985) suggested, that the orthopyroxene solvus of Lindsley's (1983) method should not be used, because the independent orthopyroxene methods of Lindsley (1983) would indicate a temperature difference of 300°C, although orthopyroxene with either 1 or 4 mole% $CaSiO_3$ changes only by 20°C as compared to Wells (1977) methods. Therefore in the present study only coexisting clinopyroxene thermometers are considered.

According to Carswell and Gibbs (1987) and Adam *et al.* (1992) the Wells' calibration methods is currently the best ones as it yielded T-values nearly 50°C in excess to those calculated by Wood and Banno (1973).

Table 6: *Electron Microprobe Analyses of Clinopyroxene in Granulite and Metanorite from Shivpura-Gyangarh, District Bhilwara, Rajasthan, India*

Sample No.	H90/9	H90/9	H90/9	R87/399	R87/399	H90/2
Spot No.	41	42	49	25 rim	26 core	21
SiO_2	51.48	50.71	52.60	52.20	53.66	50.88
TiO_2	0.16	0.80	0.13	0.16	0.08	0.07
Al_2O_3	2.12	1.98	2.04	1.73	1.30	2.37
FeO	7.22	8.26	8.38	5.70	13.67	10.23
MgO	12.70	12.56	12.84	16.47	20.31	11.42
CaO	21.19	20.53	21.02	20.40	8.80	20.76
MnO	0.04	0.14	0.13	0.11	0.23	0.18
Cr_2O_3	0.02	0.00	0.01	0.19	0.03	0.00
Fe_2O_3	3.01	3.09	2.28	1.12	0.00	2.28
Na_2O	1.00	0.92	1.02	0.27	0.15	0.00
Total	98.94	98.99	100.45	98.35	98.23	98.19
Cations	6 (O)	6 (O)	6 (O)	6 (O)	6 (O)	6 (O)
Si	1.942	1.923	1.956	1.949	1.997	1.913
Ti	0.005	0.023	0.004	0.004	0.002	0.002
Al	0.094	0.088	0.089	0.076	0.057	0.105
Fe	0.228	0.262	0.261	0.178	0.425	0.289
Mg	0.714	0.710	0.712	0.916	1.126	0.640
Ca	0.857	0.834	0.837	0.816	0.351	0.836
Mn	0.001	0.004	0.004	0.004	0.007	0.006
Cr	0.001	0.000	0.000	0.005	0.001	0.000
Fe^{+3}	0.086	0.088	0.064	0.032	0.000	0.065
Na	0.074	0.067	0.073	0.019	0.011	0.000
Total	4.002	3.999	4.000	3.999	3.977	3.856

Table 7: *Electron Microprobe Analyses of Orthopyroxene in Granulite and Metanorite from Shivpura-Gyangarh, District Bhilwara, Rajasthan, India*

Sample No	H90/9	H90/9	H90/9	R87/399	H90/2
Spot No.	44	52	43	28 core	22 rim
SiO_2	51.86	52.76	51.34	53.71	50.55
TiO_2	0.04	0.04	0.00	0.00	0.02
Al_2O_3	1.01	1.09	1.02	2.13	0.93
FeO	24.70	24.39	24.00	21.33	25.95
MgO	20.55	20.50	20.52	22.87	18.57
CaO	0.36	0.50	0.37	0.23	0.54
MnO	0.26	0.16	0.25	0.43	0.48
Cr_2O_3	0.04	0.00	0.01	0.00	0.07
Fe_2O_3	1.61	0.00	2.28	0.00	2.51
Total	100.43	99.44	99.79	100.70	99.62
Cations	6 (O)	6 (O)	6 (O)	6 (O)	6 (O)
Si	1.953	1.989	1.945	1.967	1.943
Ti	0.001	0.001	0.000	0.000	0.001
Al	0.045	0.047	0.046	0.092	0.042
Fe	0.778	0.769	0.761	0.653	0.834
Mg	1.153	1.152	1.159	1.248	1.064
Ca	0.014	0.020	0.015	0.009	0.022
Mn	0.008	0.005	0.008	0.013	0.016
Cr	0.001	0.000	0.001	0.000	0.002
Fe^{+3}	0.046	0.000	0.065	0.000	0.073
Total	4.000	3.983	4.000	3.982	3.997
X_{Fe}	0.403	0.400	0.396	0.344	0.439
X_{Mg}	0.597	0.600	0.604	0.656	0.561

Table 8: *Electron Microprobe Analyses of Plagioclase in Granulite and Metanorite from Shivpura-Gyangarh, District Bhilwara, Rajasthan, India*

Sample No.	H90/9	H90/9	H90/9	R87/399	H90/2
Spot No.	47	55	24	27	19
SiO_2	56.34	56.77	54.52	48.49	54.45
Al_2O_3	27.28	27.76	28.84	31.34	28.53
CaO	9.80	9.96	11.49	15.33	11.17
Na_2O	5.99	5.75	5.18	2.62	5.03
K_2O	0.09	0.13	0.07	0.00	0.12
Total	99.50	100.37	100.10	97.78	99.30
Cations	8 (O)	8 (O)	8 (O)	8 (O)	8 (O)
Si	2.540	2.536	2.457	2.249	2.470
Al	1.450	1.462	1.532	1.713	1.525
Ca	0.473	0.477	0.555	0.762	0.543
Na	0.524	0.498	0.452	0.235	0.442
K	0.005	0.007	0.004	0.000	0.007
Total	4.000	4.980	5.000	4.959	4.987
X_{Ca}	0.471	0.486	0.549	0.764	0.547
X_{Na}	0.523	0.507	0.447	0.236	0.446
X_K	0.005	0.007	0.004	0.000	0.007

Table 9: *Temperature Estimates of Granulite Assemblage from Shivpura Gyangarh, District Bhilwara, Rajasthan, India*

Methods		Temperature in °C					
		H90/9 Basic Granulite		H90/2 Basic Granulite		R87/399 Metanorite	
Opx-Cpx pairs		C	R	C	R	C	R
Wood & Banno	1973	802	817	884	-	1143	957
Wells	1977	841	865	986	-	1385	1057
Gt-Cpx pairs							
Ellis & Green	1979	727	739	740	-	813	637
Ganguly & Saxena	1987	779	784	770	-	876	709
Berman	1990	785	828	753	-	-	478
Holland & Powell	1990	777	789	-	-	-	-
Gt-Opx pairs							
Bhattacharya *et al.*	1991	715	722	670	-	745	574
Lal	1993	728	737	694	-	744	592
Perchuk & Lavrant'eva	1990	683	691	654	-	697	529
Berman	1990	639	699	633	-	708	491
Lee & Ganguly	1988	790	799	751	-	832	637
Sen & Bhattacharya	1984	706	719	652	-	757	523
Harley's	1984	643	656	596	-	691	502
Gt-Bio pairs							
Ferry & Spear	1978	-	-	-	-	-	634
Thompson	1976	-	-	-	-	-	647
Perchuk & Lavrant'eva	1983	-	-	-	-	-	755
Hackler & Wood (Gt Mixing)	1989	-	-	-	-	-	659
Ganguly & Saxena (Gt Mixing)	1984	-	-	-	-	-	673
Berman	1990	-	-	-	-	-	692

The same observation has also supported by (Hormann *et al.*, 1980). Here the author also observed that the temperature calculated by opx-cpx are clearly near the peak of metamorphism (T_{max}) as pyroxene in Shivpura Gyangarh granulite are chemically homogeneous and temperature calculated from the rim composition, for which it may by accepted that the pyroxenes have most likely retained their peak metamorphic condition. This is also further supported from the comparable temperature yielded by garnet-clinopyroxene thermometer.

Application of the garnet-clinopyroxene on the basis of Ellis and Green (1979), Ganguly and Saxena (1987), Berman (1990) and Holland and Powell (1990) equations for different pairs yielded temperature in the range 727°C to 828°C (Table 9), which are comparable to T_{max} obtained by orthopyroxene-clinopyroxene thermometry. However, lower values in the vicinity of 700°C for the remaining samples can be regarded as indicating re-equilibration controlled by reaction kinetics. The temperature calculated by Ganguly and Saxena (1987), Berman (1990) and Holland and Powell (1990) equations are all nearly consistent, except the temperature calculated by Ellis and Green (1979) equation, which is approximately 50°C lower than other models (Table 9). The relatively small spread of the temperature calculated through individual calibration indicates near-isothermal Fe-Mg equilibration within garnet and clinopyroxene throughout the studied area.

Faulhaber and Raith (1991) suggest that temperature calculated through Ellis and Green (1979) equation is petrologically most reasonable temperature as compared with the values obtained from the other models. This seems logical because Berman's (1990) model for the garnet-clinopyroxene pairs gave a value of 828°C which is very close to the value obtained of the T_{max} (802-882°C) for the basic granulite.

Application of the garnet-orthopyroxene thermometry on the basis of Bhattacharya *et al.* (1991), Lal (1993), Perchuk and Lavrant'eva (1990), Berman (1990), Holland and Powell (1990), Lee and Ganguly (1988), Sen and Bhattacharya (1984) and Harley (1984) equations for different pairs yielded temperature in the range of 596°C to 799°C (Table 9).

The comparable temperatures of more than 700°C for core obtained from different geothermometers as applied to the assemblage of the granulite may be taken to testify the fact, that they record peak conditions of metamorphism, which probably lay close to or just below, the closure temperatures of the mineral assemblages (Frost and Chacko, 1990). The lower temperature of about 596°C obtained from rim by using Harley (1984) give overlapping temperatures and suggest equilibration subsequent to the uplift from deeper levels to shallower depth of the Shivpura Gyangarh rocks. Reaction kinetics may also have played a role in the overall attainment of equilibrium of the investigated assemblages.

The garnet-orthopyroxene Fe-Mg exchange thermometer of Bhattacharya *et al.* (1991) and Lal (1993), yield temperature estimates which agree well with independent petrologic temperature constraints. These models take into account non-ideal Fe-Mg mixing in orthopyroxene should be preferred over the other models since it show very little compositional dependence.

Metanorite

Thermometry. Temperature calculated through orthopyroxene-clinopyroxene thermometry on the basis of Wood and Banno (1973) and Wells (1977) equations for different pairs yielded in the range of 957 to 1385°C (Table 9).

Wells (1977) calibration, when used for metanorite gave 1385°C temperature for core composition of orthopyroxene-clinopyroxene pairs (sample No. R87/399; Table 9). This high T value suggests crystallization temperature of the norite from its melts.

At such high temperature the development of sillimanite in pelitic granulites in which the core of garnet with included biotite yielded temperature 726°C.

Application of the garnet-clinopyroxene thermometer on the basis of Ellis and Green (1979), Ganguly and Saxena (1987), Berman (1990) and Holland and Powell (1990) equations for different pairs yielded temperature in the range 637°C to 876°C from rim to core (Table 9). Ellis and Green (1985) garnet-clinopyroxene calibration is generally believed to yield reasonable temperatures, which are quite interestingly comparable to the value obtained from other models.

Application of the garnet-orthopyroxene thermometry on the basis of Bhattacharya *et al.* (1991) and Lal (1993), Perchuk and Lavrant'eva (1983), Berman (1990), Holland and Powell (1990), Lee and Ganguly (1988), Sen and Bhattacharya (1984) and Harley (1984) equations for different pairs from core to rim yielded in the range of 832°C to 447°C (Table 9).

The lower temperature of about 447°C obtained from rim by using Harley (1984) give overlapping temperatures and suggest equilibration subsequent to the uplift from deeper level to shallower depth of the Shivpura Gyangarh rocks. Reaction kinetics may also have played a role in the overall attainment of equilibrium of the investigated assemblages.

The temperature calculated through garnet-biotite thermometry on the basis of Ferry and Spear (1978), Thompson (1976), Perchuk and Lavrant'eva (1983), Bhattacharya *et al.*, A & B (1992) and Berman (1990) equation for different pairs yielded in the range of 634°C to 755°C (Table 9). The lower temperature of about 634°C in metanorite support the field and other criteria that the rocks has been uplifted along a shear one and emplaced at higher levels, where metamorphism ceased at about 600°C as recorded in both pelitic gneisses and granulites.

Pelitic Granulite

Thermometry. The temperature calculated through garnet-biotite thermometry by (Thomas, 1995b) on the basis of Ferry and Spears (1978), Thompson (1976), Perchuk and Lavrant'eva (1983), Bhattacharya *et al.*, A & B (1992) and Berman (1990) equations and garnet-cordierite thermometry on the basis of Thompson (1976) and Holdaway and Lee (1977) and Aranovich and Podlesskii (1989) equations for different pairs yielded in the range of 726 to 501°C. From core to rim (Table 10) with consideration of garnet activity composition.

The lower temperature of about 450°C obtained from rim by using Ferry and Spear (1978) give overlapping temperatures and suggest equilibration subsequent to the uplift from deeper levels to shallower depth of the Shivpura Gyangarh rocks. Reaction kinetics may also have played a role in the overall attainment of equilibrium of the investigated assemblages.

The significance of the temperature obtained near 600°C can be understood in the light of the garnet-orthopyroxene thermometer of the granulite besides the temperature overlap of nearly 800°C. The lower temperatures of about 500°C in pelitic gneisses support the field and other criteria that the rocks has been uplifted along a shear zone and emplaced at

Table 10: *Temperature Estimates of Granulite Assemblage from Shivpura-Gyangarh, District Bhilwara, Rajasthan, India*

Models		Temperature in °C							
		R87/435		R87/452		299	342	291	16
Gt-Bio pairs		C	R	C	R				
Ferry & Spear	1978	608	520	684	666	629	668	486	450
Thompson	1976	628	557	685	672	644	673	529	449
Perchuk & Lavrant'eva	1983	658	619	726	697	645	720	583	550
Bhattacharya *et al.* A & B Hackler & Wood (1989)	1992	604	555	665	653	625	637	544	483
Bhattacharya *et al.* A & B Ganguly & Saxena (1984)	1992	614	569	676	664	632	612	567	510
Berman	1990	576	516	702	653	–	–	–	–
Gt-Cd pairs									
Thompson	1976	–	–	611	523	–	–	–	–
Holdaway & Lee	1977	–	–	601	524	–	–	–	–
Aranovich & Podlesskii	1989	–	–	653	572	–	–	–	–

shallower level, where metamorphism ceased at about 500°C as recorded in both pelitic gneisses and metanorite.

Chipera and Perkins (1988) suggest that temperature calculated through Perchuk and Lavrant'eva (1983) thermometer yields the most precise and accurate result as compared with the values obtained from the other models. It is interesting to note that garnet cores with included biotite gave higher temperature (726°C) compared to garnet rim biotite composition (612°C), employing Perchuk and Lavrant'eva (1983) calibrations. This is taken to indicate that re-equilibration did not occur uniformly in all assemblages, probably due to the role of reaction kinetics.

Application of garnet-cordierite thermometry on the basis of Thompson (1976) and Holdaway and Lee (1977) and Aranovich and Podlesskii (1989) equations for different pairs yielded temperature 523°C to 653°C from rim to core (Table 10). Because of slight inhomogeneity in garnet crystal and sensitivity of the garnet-cordierite thermometer to X_{Fe}, the T value obtained is less than the T_{max}.

Basic Granulite

Barometry. The pressure calculated through clinopyroxene–garnet–plagioclase–quartz and orthopyroxene–garnet–plagioclase–quartz equilibria on the basis of Newton and Perkins (1982), Eckert *et al.* (1991) equation for different pairs yield in the range of 6.65 kbar to 10.26 kbar (Table 11).

According to Newton and Perkins (*ibid.*) this discrepancy in two barometers is within the combined uncertainties in -ΔG° for the reaction. But it could be really caused by the tendency for orthopyroxene to re-equilibrate with plagioclase during uplift to lower pressure.

Eckert *et al.* (1993) have provided an improved version of the geobarometric equations for both the orthopyroxene, clinopyroxene-garnet-plagioclase-quartz assemblage by utilizing the internally consistent thermodynamic dataset of Holland and Powell (1990) and Berman (1990), pressure estimated based on Eckert *et al.* (1993) model for different pairs yield a pressure range of 8.16 kbar to 10.01 kbar at average K_D temperature.

Table 11: *Pressure Estimates of Granulite Assemblage from Shivpura-Gyangarh, District Bhilwara, Rajasthan, India*

Sample No. P = Kbar; C = core; r = rim		H90/9 Basic Granulite		H90/2 Basic Granulite	R87/399 Metanorite	
Thermodynamic Models						
Gt-Cpx-Plag-Qz barometer						
Reactions		P_{Mg} at 724°C	P_{Mg} at 739°C	P_{Mg} at 700°C	P_{Mg} 707°C r	P_{Mg} at 876°C c
Newton & Perkins	1982	7.14	7.87	7.17	6.12	10.65
Eckert *et al.*	1991	9.50	10.26	9.54	8.45	13.16
Berman	1990	10.00	10.00	10.00	8.30	13.00
Holland & Powell	1990	9.24	10.01	9.24	8.19	13.02

Gt-Cpx-Plag-Qz barometer

Reactions		$P_{Fe} P_{Mg}$ at 799°C		$P_{Fe} P_{Mg}$ at 785°C		$P_{Fe} P_{Mg}$ at 700°C		$P_{Fe} P_{Mg}$ at 832°C c	$P_{Fe} P_{Mg}$ 642°C r
Newton & Perkins	1982	–	8.53	–	8.47	–	7.08	– ; 8.60	– ; 5.36
Bhattacharya *et al.*	1991	8.85	8.78	8.91	8.83	7.58	7.37	8.62; 8.72	6.86; 5.62
Lal	1993	8.59	8.49	8.64	8.54	7.61	7.55	8.32; 8.18	6.34; 6.24
Eckert *et al.*	1991	–	8.42	–	8.36	–	6.90	– ; 8.50	– ; 5.09
Perkins & Cipera	1985	10.18	9.10	10.28	9.15	8.88	7.93	10.35;8.56	8.16; 5.45
Berman	1990	–	8.26	–	8.16	–	6.65	10.00;8.70	8.60; 5.50
Holland & Powell	1990	–	8.38	–	8.33	–	6.97	– ; 8.43	– ; 5.24

Using the critically revised and internally consistent calibrations of the orthopyroxene–garnet–plagioclase–quartz equilibria of Perkins and Cipera (1985), Bhattacharya *et al.* (1991) and Lal (1993) equations for different pairs pressure of 9.15 kbar to 8.49 kbar is obtained based on P_{Mg} reaction and 10.28 kbar to 8.59 kbar is obtained based on P_{Fe} reaction for the composition of the coexisting phases in different pairs at average K_D temperature.

The P-values obtained for the basic granulite (Table 11) is characterized by two features:

The core of orthopyroxene – garnet – plagioclase yield lower P-estimates (P_{OPX} = 10.24 kbar/785°C based on P_{Fe} reaction and P_{OPX} = 9.15 kbar/785°C based on P_{Mg} reaction, than that obtained for the composition of the mineral phases (P_{CPX} = 10.26 kbar) which re-crystallized at 739°C.

The calculated P-values are the maximum values for this rock, since quartz is not present in this analyzed domain (O'Neil and Wood, 1979).

In the basic granulites, simultaneous solution of the garnet-clinopyroxene-plagioclase-quartz barometer (Eckert *et al.*, 1991) using the composition of coronal phases, yield P-T values of 9.50 kbar at 724°C and 10.26 kbar at 739°C.

The Shivpura Gyangarh granulite are clearly of deep crustal origin, metamorphosed at 6.65–10.26 kbar pressure, but subsequent uplift to lower pressure due to differential ingress of water in certain domain incited diffusion of cations and hence re-equilibration in certain domain only; dry assemblage were quenched at greater depth to give about 8-10 kbar pressures as the peak of metamorphism.

Metanorite

Barometry. The pressure calculated through clinopyroxene–garnet–plagioclase-quartz and orthopyroxene–garnet–plagioclase–quartz equilibria on the basis of Newton and Perkins (1982), Eckert *et al.* (1991) equations for different pairs yield in the range of 5.09 kbar to 13.16 kbar from rim to core (Table 11).

Pressures estimated based on Eckert *et al.* (1993) models for different pairs yield a pressure range of 8.19 kbar to 13.02 kbar from rim to core. The core of orthopyroxene-garnet-plagioclase yielded lower P-estimates (P_{OPX} = 10.35 kbar / 832°C based on P_{Fe} reaction and P_{OPX} = 8.56 kbar / 832°C based on P_{Mg} reaction, than that obtained for the core composition of the mineral phases (P_{CPX} = 13.16 kbar) which recrystallized at 876°C.

Pelitic Granulites

Barometry. The pressure calculated through garnet-sillimanite/kyanite-plagioclase-quartz equilibria by (Thomas, 1995; 1998) on the basis of Newton and Haselton (1981), Koziol and Newton (1988), Powell and Holland (1988) and Berman (1990) equations for different pairs yielded pressure in the range of 6.48 kbar to 9.01 kbar from rim to core (Table 12). Similar pressure value obtained on applying the garnet-cordierite-sillimanite-quartz equilibria. The pressure calculated through garnet-rutile-ilmenite-sillimanite-quartz equilibria on the basis of Bohlen *et al.* (1983) equation for different pairs yielded pressure in the range of 8.82 kbar to 11.53 kbar.

Table 12: *Pressure Estimates of Granulite Assemblage from Shivpura-Gyangarh, District Bhilwara, Rajasthan, India*

Sample No. P = Kbar; C = core; r = rim				R87/435					R87/452		
Thermodynamic Models											
Gt-Sill-Plag-Qz barometer											
		at 630°C (r)		at 630°C (c)		at 630°C (r)		at 630°C (c)		689°C (c)	
Newton & Haselton	1981	6.48	6.96	7.07	7.33	6.67	5.94	7.29	7.58	7.66	8.64
Koziol & Newton	1988	6.86	7.35	7.45	7.72	7.06	6.32	7.68	7.97	8.05	9.01
Holland & Powell	1985	6.76	7.27	7.38	7.66	6.96	6.19	7.62	7.92	8.01	8.89
Gt-Cd-Sill-Qz barometer											
							at 648°C (c)		at 572°C (r)		
Thompson	1976						6.40		6.68		
Wells	1979						6.29		6.47		
Gt-Sill-Ru-Ilm-Qz barometer											
		at 650°C (c)					at 876°C (c)		at 707°C (r)		
Bohlen *et al.*	1983b	8.82	8.85				11.53		9.75		

Two pairs of the pelitic granulites however, yielded very low pressures near 3 kbar (data not presented) and nearly 4.3 kbar which may be a record of recrystallization due to local shearing or a result of over printing by a later thermal event, related to the intrusive charnockite/granodiorite comagmatic series (1723 Ma).

All calculated pressures except two pairs in specimen No. R87/435 range from 6.48 to 11.53 kbar, thus confirming the re-equilibration pressures yielded by granulite assemblage. Pressure calibrations of Koziol and Newton (1988) and Bohlen *et al.* (1983) give the higher pressure values of about 9.01 to 11.53 kbar obtained from sample (R87/452) seems meaningful, since it conforms with the pressure value obtained from granulite. It seems most likely that these mineral composition and the resulting pressure values represent depth at which the rock had re-equilibrated after their exhumation from depth corresponding to 4.3 kbar.

Metamorphic History and P-T path

In the pelitic granulite a nearly complete homogenization at the peak of granulite facies metamorphism has evidence of earlier history, except that kyanite became metastable and sillimanite a stable phase in the assemblage. The randomly kyanite needles ($kyanite_2$) (Plate 1D & F) in this granulite indicate cooling in the kyanite field.

According to many workers granulites cool isobarically and corona texture (Plate 1A&B) in basic granulites is the result of this cooling (Bohlen, 1987). However, isobaric cooling is valid only if the recorded pressures do not represent maximum pressures, during metamorphism.

Garnet coronas in gabbroic rocks have been interpreted by some workers (Bohlen, 1987), as retrograde reaction (isobaric cooling) and by others (Harley, 1989) as arrested reaction products. However, the petrologically constrained P-T data indicate that the minerals in these coronas equilibrated at or near peak metamorphic conditions. In support of this conclusion is the fact that inhomogeneity in the coronal phases is very slight (within the uncertainty of electron microprobe analytical data), virtually insignificant chemical zoning in any of these phases, and the garnet-clinopyroxene and orthopyroxene-clinopyroxene thermometers are broadly consistent with inferred peak metamorphic temperatures. Furthermore, the metanorite intrusion may have supplied additional heat that prevented cooling of the country rocks during its emplacement in the late-stages of granulite facies metamorphism.

Interpretation of isobaric cooling in terms of the tectonics of granulite terranes has been a matter of considerable discussion. In Shivpura Gyangarh, this problem is discussed on the basis of petrography, mineral reactions and compositional zoning, and by the use of thermobarometry to deduce the P-T paths. It may be noted that the estimated P-T conditions for the mafic granulites are corroborated by the P-T estimates for the pelitic granulite.

The orthopyroxene–clinopyroxene composition give high temperatures of around 800-1300°C. At such high temperatures the development of sillimanite in pelitic granulites, in which the core of garnet with included biotite yielded comparable temperature (702°C). Such high temperatures are associated with high heat flow regimes (Sandiford and Powell, 1986). Which steepened the normal geothermal gradient in that segment of the crust. Metamorphic recrystallization of plagioclase with an increase in Ab content near the coronas and exsolution of opaque granulite at the borders and along the cleavage traces within the orthopyroxenes, the P-T estimates for garnet-pyroxene assemblages suggested that this recrystallization took place at about P_{max} (13.16kbar/876°C).

The K_D values of the participating phases in the corona forming [orthopyroxene + plagioclase = garnet + quartz (clinopyroxene)] strongly favour thermodynamic equilibrium. The evidence suggests that the corona-bearing granulites have maintained chemical equilibrium during granulite facies metamorphism. Furthermore, the pressure 13.16 kbar estimates for the metanorite intrusions, having higher-pressure values (and lower temperature) for garnet rim compositions than for the garnet core-pyroxene-plagioclase-quartz assemblage, suggests that corona development occurred during compression with cooling from T_{max}. A similar situation is reported from Broken Hill, Australia (Phillips and Wall, 1981), the Olary Block, Australia (Clarke *et al.*, 1987), MacRobertson Land East Antarctic (Clarke *et al.*, 1989) and Bunger Hill, East Antarctic (Stuwe and Powell, 1989).

From the evidence obtained by thermobarometry and textural relations, particularly in pelitic granulites were sillimanite grew between $kyanite_1$ and $kyanite_2$ events, the deduced P-T path is anticlockwise (ACW) for the investigated granulite terrane (Fig. 3). The CAW path is inconsistent with a continent-continent collision-underplating model of England and Thompson (1984).

It is suggested that underplating of the continental crust by asthenosphere-derived basaltic partial melts caused high temperature metamorphism. The substantial pressure

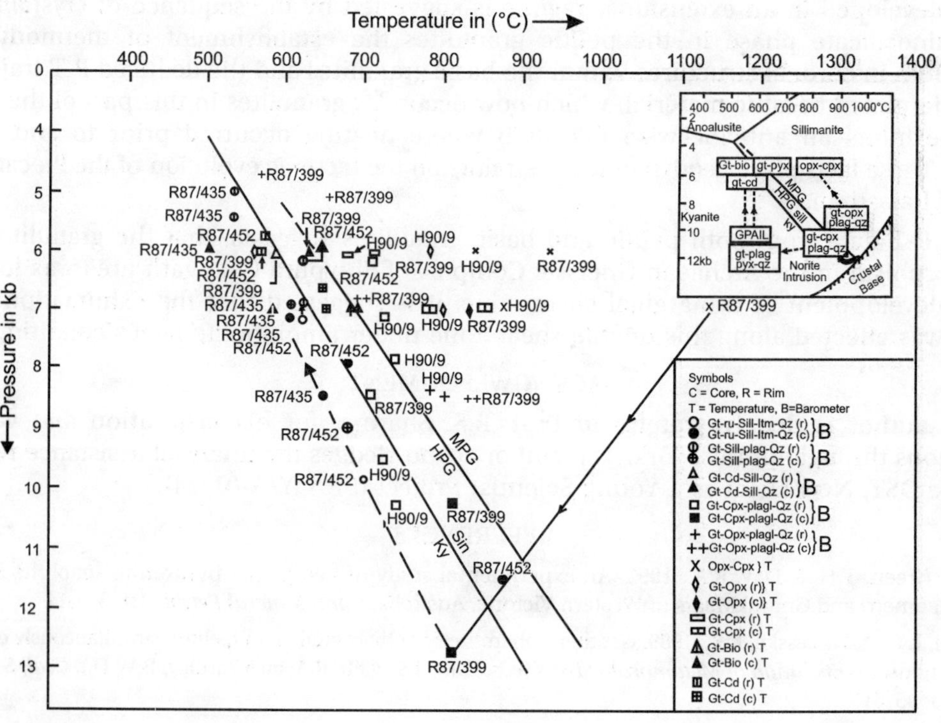

Fig. 3. Geothermobarometric estimates for granulite-facies from Shivapura Gyangarh, District Bhilwara, Rajasthan. The data for pelitic granulite and basic granulites are shown by different symbols indicated in the figure. The deduced P-T path is shown by arrows on the dashed lines. The medium pressure granulite/high-pressure granulite (MPG/HPG) boundary is after Green and Ringwood (1967) and the Al_2SiO_5 triple point and univariant curves are after Holdaway (1971). Inset shows the interpreted P-T path and tectonothermal evolution of the granulites. The boxes represent the different thermometers/barometers.

increase subsequent to this event was due to gravitational instability of locally thinned crust in the area. Gravitational sinking/collapse adjacent to the thicker crust resulted in increase pressure and this produced compression during cooling of the Shivpura Gyangarh granulites. As illustrated in (Fig. 3), the terrane was excavated from deeper crustal levels along marginal shear zones. The presence of marginal shear zone in Shivpura Gyangarh area (Sharma, 1988; 1999; 2003; Joshi *et al.*, 1993 and Thomas, 1995), together with concentration of temperatures at 7 Kbar constitute evidence in support of proposed mechanism. Stuwe and Powell (1989) suggested a similar mechanism for the Bunger Hills region and Jacobo *et al.* (2003) also suggest a similar mechanism for Ordenes Complex, nw Spain.

It appears that the terrane was exhumed rather quickly, near the kyanite/sillimanite mantle boundary, in view of the fact that some of the pressure plots between 12 kbar and 7 kbar are narrowly constrained around a temperature of 700°C (Fig. 3) and in view of the transformation of Sillimanite to random needles of kyanite.

CONCLUSIONS

The granulite facies metamorphism in the Archaean Gneissic Complex from Shivpura Gyangarh is characterized by a polymetamorphic history, resulting from the involvement of granulite complex in late Archaean and late Proterozoic orogenic events. That the granulite terrain developed in an extensional regime is suggested by the sequence of crystallization of aluminosilicate phase in the pelitic granulites the establishment of thermodynamic equilibrium in coronal structures within the basic granulites and the deduced P-T trajectory. The underplated basaltic material, which now occurs as granulites in this part of the Indian Shield, exhibits an anticlockwise P-T path where heating occurred prior to and during loading. These facts place geodynamic constraints on the tectonic evolution of the Precambrian rocks of Rajasthan.

The P-T data from both pelitic and basic granulites suggests, that the granulite facies metamorphism in the Archaean Gneissic Complex of Shivpura Gyangarh area was followed by the development of a marginal shear zone; it is suggested that the exhumation of the terrain was effected along this ductile shear zone under amphibolite facies conditions.

ACKNOWLEDGMENT

The author is deeply grateful to Prof. R.S. Sharma, for his inspiration and valuable suggestions throughout this work. The author acknowledges the financial assistance received from the DST, New Delhi as a Young Scientist Project (SR/OY/A-01/94).

REFERENCES

Adam, J., Green, T.H. & Day, R.A., 1992. An experimental study of two garnet pyroxenite xenoliths from the Bullenmerri and GnotukMaars of Western Victoria, Australia. *Cont. Mineral Petrol.*, III, 505-14.

Aranovich, L.Ya. & Podlesskii, K.K., 1989. Geothermobarometry of high-grade metapelites: simultaneously operating reactions. In: *Evolution of Metamorphic Belts* (eds. Daly, J.S., Cliff, R.A. and Yardley, B.W.D.), *Geol. Soc. Publ.*, v. 43, 45-61.

Bhattacharya, A., Krishnakumar, K.R., Raith, M., & Sen, S.K., 1991. An improved set of a-x parameters for Fe-Mg-Ca garnets and refinement of the orthopyroxene-garnet thermometer and orthopyroxene-garnet-plagioclase-quartz barometer. *Jour. Petrol.*, v. 32. 629-56.

Bhattacharya, A., Mohanty, L., Maji, A., Sen, S.K. & Raith, M., 1992. Non-ideal mixing in the phlogopite-annite binary constraints from experimental data on Fe-Mg partitioning and a reformulation of the biotite-garnet geothermometer. *Cont. Mineral Petrol.*, III, 87-93.

Berman, R.G., 1990. Mixing properties of Ca-Mg-Fe-Mn garnet. *Am Mineralogist,* **v. 75**, 328-44.

Bohlen, S.R., 1987. Pressure-temperature-time paths and a tectonic model for the evolution of granulites. *J. Geology,* **v. 95**, 617-23.

Bohlen, S.R. & Essene, E.J., 1978. Igneous pyroxenes metamorphosed anthrosite massifs. *Am. Mineral.,* **v. 65**, 433-42.

Bohlen, S.R., Wall, V.J. & Boettcher, A.L., 1983. Experimental investigation and geological application of equilibria in the system FeO-TiO_2-Al_2O3-SiO_2-H_2O., *Am. Mineral.,* **v. 68**, 1049-58.

Carswell, D.A. & Gibb, F.G.F., 1987. Evaluation of mineral thermometers and barometers applicable to garnet lherzolite assemblages. *Cont. Mineral. Petrol.,* **v. 95**, 499-511.

Chipera, S.J. & Perkins, D., 1988. Evaluation of biotite-garnet geothermometers: application to the English River Subprovince, Ontario., *Cont. Mineral Petrol.,* **v. 98**, 40-48.

Choudhary, A.L., Gopalan, K. & Sastry, C.A., 1984. Present status of geochronology of the Precambrian rocks of Rajasthan. *Tectonophysics.,* **v. 105**, 131-40.

Clarke, G.L., Guiraud, M., Powell, R. & Burg, J.P., 1987. Metamorphism in the Olary Block South Australia: Compression with cooling in a Proterozoic foldbelt., *Jour. Meta. Geology.,* **v. 5**, 291-306.

Clarke, G.L., Powell, R. & Guiraud, M., 1989. Low pressure granulite facies metamorphic assemblages and corona textures from MacRobertson Land, East Antarctica: the importance of Fe_2O_3 and TiO_2 in accounting for spinel bearing assemblages. *Jour. Meta. Geology,* **v. 7**, 323-35.

Davidson, P.M. & Lindsey, D.H., 1989. Thermodynamic analysis of pyroxene-olivine-quartz equilibria in the system CaO-MgO-FeO-SiO_2. *Am. Mineral.,* **v. 74**, 18-30.

Eckert, J.O. Jr., Newton R.C. & Kleppa, O.J., 1991. (H of reaction and recalibration of garnet-pyroxene-plagioclase-quartz geobarometers in CMAS system by solution calorimetry of stoichiometric mineral mixes., *Am. Mineral.,* **v. 79**, 148-60.

Eckert, J.O. Jr. & Newton, R.C., 1993. Palaeopressures of south Indian two-pyroxene garnet granulite from thermochemically calibrated barometers. *Jour. Meta. Geology,* **v. 11**, 845-54.

Ellis, D.J. & Green, D.H., 1985. Garnet forming reaction in mafic granulite from Enderby Land, Antarctica-implications for geothermometry and geobarometry. *Jour. Petrol.,* **v. 26**, 633-62.

England, P.C. & Thompson, A.B., 1984. Pressure-temperature-time paths of regional metamorphism. (I) Heat transfer during the evolution of regions of thickened continental crust. *Jour. Petrol.,* **v. 25**, 894-924.

Faulhaber, S. & Raith, M., 1991. Geothermometry and geobarometry of high-grade rocks: A case study of garnet-pyroxene granulites in southern Sri Lanka, *Mineralogical Mag.,* **v. 55**, 33-56.

Ferry, J.M. & Spear, F.S., 1978. Experimental calibration of the partitioning of Fe and Mg between biotite and garnet. *Cont. Mineral. Petrol.,* **v. 66**, 113-17.

Frost, B.R. & Chacko, T., 1989. The granulite uncertainty principle: Limitations on thermobarometry in granulites, *Jour. Geology.,* **v. 97**, 435-50.

Ganguly, J. & Saxena, S.K., 1984. Mixing properties of aluminosilicate garnet: Constraints from natural and experimental data and application to geothermobarometry. *Am. Mineral.,* **v. 69**, 88-97.

Ganguly, J. & Saxena, S.K., 1987. Mixtures and mineral reactions. In: A. El. Goresy and Von Engelhardt. W. (eds.) *Minerals and Rocks,* 19, Springer-Verlag, Berlin. Green, D.H. and Ringwood, A.E. (1967) An experimental investigation of the gabbro to eclogite transformation and its petrological applications. *Geochim. Cosmochim Acta.,* **v. 31**, 767-833.

Gopalan, K., Macdougall, J.D. & Roy, A.B., 1988. Sm/Nd systematics of the Banded Gneissic Complex, Northwest India. *Chem. Geol.,* **v. 70**, 140.

Green, D.H. & Ringwood, A.E., 1967. An experimental investigation of the gabbro to eclogite transformation and its petrological application. *Geochim. Cosmochim. Acta.,* **v. 31**, 767-833.

Gupta, B.C., 1934. The geology of Central Mewar. *Memoir Geol. Surv. India.,* **v. 65**(2), 107-69.

Hackler, R.T. & Wood, B.J., 1989. Experimental determination of Fe-Mg exchange between garnet and olivine and estimation of Fe-Mg garnet mixing properties. *Am. Mineral.*, **v. 74**, 994-99.

Harley, S.L., 1984. An experimental study of the partitioning of Fe and Mg between garnet and orthopyroxene. *Cont. Mineral Petrol.*, **v. 86**, 359-73.

Harley, S.L., 1989. The origin of granulites: A metamorphic perspective., *Geol. Mag.*, **v. 126**, 215-47.

Hensen, B.J. & Green, D.H., 1973. Experimental study of the stability of cordierite and garnet in pelitic compositions at high pressure and temperatures. III. Synthesis of experimental data and geological applications. *Cont. Mineral. Petrol.*, **v. 38**, 151-66.

Holdaway, M.J., 1971. Stability of andalusite and the aluminium silicate phase diagram. *Am. Jour. Sci.*, **v. 271**, 97-131.

Holdaway, M.J. & Lee, S.M., 1977. Fe-Mg cordierite stability in high grade pelitic rocks based on experimental, theoretical and natural observations. *Cont. Mineral Petrol.*, **v. 63**, 175-93.

Holland, T.J.B. & Powell, R., 1985. An enlarge and update internally consistent thermodynamic data set with uncertainties FeO-Fe_2O_3-TiO_2-SiO_2-H_2O-O_2. *Jour. Meta. Geol.*, **v. 8**, 9-124.

Holland, T.J.B. & Powell, R., 1990. An enlarged and updated internally consistent thermodynamic data set with uncertainties and correlations: The system K_2O-Na_2O-CaO-MgO-MnO-FeO-Fe_2O_3-Al_2O_3-TiO_2-SiO_2-C-H_2-O_2., *Jour. Meta. Geol.*, **v. 8**, 89-124.

Hormann, P.K., Raith, M., Raase, P., Ackermand, D. & Seifert, F., 1980. The granulite complex of Finnish Lapland: Petrology and metamorphic conditions in the Ivalojoki-Inanjarvi area., *Geol. Surv. Finl. Bull.*, **v. 308**, 1-95.

Jacobo, A., Ricardo, A., José, R.M.C. & Florentino, D.G., 2003. Anticlockwise P-T Path of Granulites from the Monte Castelo Gabbro (Órdenes Complex, NW Spain). *Jour. Petrol.*, **v. 44**, n. 2, 305-27.

Joshi, M., Thomas, H. & Sharma, R.S., 1993. Granulite facies metamorphism in the Archaean gneissic complex from North-Central Rajasthan. *Proc. Nat. Acad. Sci. India*, **v. 63** (A) I, 167-87.

Koziol, M.A. & Newton, R.C., 1988. Redetermination of the anorthite breakdown reaction and improvement of the plagioclase-garnet-Al_2SiO_5-quartz geobarometer. *Am. Mineral.*, **v. 73**, 216-33.

Lal, R.K., 1993. Internally consistent recalibrations of mineral equilibria for geothermobarometry involving garnet–orthopyroxene–plagioclase–quartz assemblages and their application to the South Indian granulites. *Jour. Meta. Geol.*, **v. 11**, 855-66.

Lee, H.Y. & Ganguly, J., 1988. Equilibrium composition of coexisting orthopyroxene and garnet: experimental determination in the system FeO-MgO-Al_2O_3-SiO_3 and application. *Jour. Petrol.*, **v. 29**, 93-113.

Lindsley, D.H., 1983. Pyroxene thermometry. *Am. Mineral.*, **v. 68**, 477-93.

Lindsley, D.H. & Anderson, D.J., 1983. A two-pyroxene thermometry. Proc. 13th earth and Plant Sci. Conf., *Jour. Geophy. Res.*, **v. 88**, 887-906.

Macdougall, T.D., Gopalan, K. & Roy, A.B., 1983. The Banded Gneissic Complex of Rajasthan, India: Early crust from depleted mantle at N 3.5 AE. *Trans Am. Geophys. Union*, **v. 64(18)**, 351.

Naha, K. & Halyburton, R.V., 1983. Late stress system deduced from conjugate folds and kinks bands in the Main Raialo Syncline, Udaipur district, Rajasthan, India. *Geol. Soc. Am. Bull.*, **v. 85**, 251-56.

Newton, R.C. & Haselton, H.T., 1981. Thermodynamics of garnet-plagioclase-Al_2SiO_5-quartz geobarometer. In: Newton, R.C., Navrotosky A and Wood, B.J. (eds) *Advances in Physical Geochemistry.*, Springer-Verlag, New York, I, 131-47.

Newton, R.C. & Perkins, D., 1982. Thermodynamic calibration of geobarometers for charnockites and basic granulites based on the assemblages garnet-plagioclase-orthoyroxene-(clinopyroxene)-quartz with application to high grade metamorphism. *Am. Mineral.*, **v. 67**, 203-22.

O'Neil H. St. C. & Wood, B.J., 1979. An experimental study of Fe-Mg partitioning between garnet and olivine and its calibration as a geothermometer. *Cont. Mineral Petrol.*, **v. 70**, 59-70.

Perchuk, L.L. & Lavrant'eva, I.V., 1983. Experimental investigation of exchange equilibria in the system cordierite-garnet-biotite In: Saxena, S.K. (ed.) *Kinetics and equilibrium in mineral reactions*. Springer. New York, 199-239.

Perchuk, L.L. and Lavrant'eva, I.V., 1990. Garnet-orthopyroxene and garnet-amphibole geothermobarometry: experiential data and thermodynamics. *International Geological Review*, **32**, 486-507.

Perkins, D. & Chipera, S.J., 1985. Garnet–orthopyroxene–quartz–plagioclase barometry refinement and application to English river subprovince and Minesota River Valley. *Cont. Mineral Petrol.*, **v. 89**, 69-80.

Phillips, G.N. & Wall, V.J., 1981. Evaluation of prograde regional metamorphic conditions: their implications for the heat source and water activity during metamorphism in the Willyama Complex, Broken Hill, *Australia Bull. Soc. Franc. Mineral Crystal.*, **v. 104**, 801-10.

Roy, A.B., 1988. Stratigraphic and tectonic framework of the Aravalli Mountain Range. *Geol. Soc. India Mem.*, **v. 7**, 3-31.

Sandiford, M.A. & Powell, R., 1986. Deep crustal metamorphism during continental extension: ancient and modern examples: *Earth Plant Sci. Lett.*, **v. 79**, 151-58.

Sarkar, G., Barman, T.R. & Corfu, F., 1989. Timing of continental arc-type magmatism in northwest India: evidence from U-Pb zircon geochron. *Jour. Geol.*, **v. 97**, 607-12.

Sen, S.K. & Bhattacharya, A., 1984. An orthopyroxene-garnet thermometer and its application to the Madras charnockites. *Cont. Mineral Petrol.*, **v. 88**, 64-71.

Sharma, R.S., 1988. Patterns of metamorphism in the Precambrian rocks of the Aravalli Mountain belt. *Geol. Soc. India. Mem.*, **v. 7**, 33-75.

Sharma, R.S., 1999. Crustal development in Rajasthan craton. *Indian J. Geology.*, **vol. 71**, 65-80.

Sharma, R.S., 2003. Evolution of Proterozoic fold belts of India: a case of the Aravalli Mountain belt of Rajasthan, NW India. *Geol. Soc. India, Mem.*, **52**, 145-62.

Sivaraman, T.V. & Odom, A.L., 1982. Zircon geochronology of Berach granite of Chittorgarh, Rajasthan. *J. Geol. Soc. India.*, **v. 23**, 575-77.

Stuwe, K. & Powell, R., 1989. Metamorphic evolution of the Bunger Hills, East Antarctica: evidence for substantial post-metamorphic peak compression with minimal cooling in a Proterozoic orogenic event. *J. Meta Geology*, **v. 7**, 449-64.

Thomas, H., 1995. Pressure Temperature considerations for granulite form Thana Gyangarh, District-Bhilwara, Rajasthan : Implications for Crustal Evolution. In : R.K. Srivastava & R. Chandra (ed.) *Magmatism in relation to Diverse Tectonic setting*. A.A. Balkema Pub. & Oxford & IBH Publishing Com Pvt. Ltd., 439-56.

Thomas, H., 1995b. MPET1: A computer program for coexisting garnet-biotite-aluminosilicate-plagioclase-quartz-geothermobarometry. *Journal of Himalayan Geology*, **v. 6(1)**, 69-74.

Thomas, H., 1998a. MPET2: A Quick Basic Program for coexisting Garnet-cordierite-sillimanite-quartz-geothermobarometry. *Himalayan Geology*, **v. 19(1)**, 49-54.

Thomas, H., 1998b. MPET4: A Computer program for coexisting garnet-clinopyroxene plagioclase-quartz-geothemobarometry. *Journal of Nepal Geological Society*, **v. 18**, 409-15.

Thomas, H., 2003. Garnet–Orthopyroxene–Plagioclase–Quartz Equilibria: *A Comparative study. Gond. Geol. Magz., Advances in Precambrian of Central India*, Special Volume **No. 7**, 327-37.

Thompson, A.B., 1976. Mineral reaction in pelitic rocks. II: Calculations of some P-T-X (Fe-Mg) phase relations. *Am. J. Sci.*, **v. 276**, 425-54.

Wells, P.R.A., 1977. Chemical and thermal evolution of Archean sialic crust, southwest Greenland., *J. Petrol.*, **v. 20**, 187-226.

Wood, B.J. & Banno, S., 1973. Garnet-orthopyroxene and orthopyroxene-clinopyroxene relationships in simple and complex system., *Cont. Mineral Petrol.*, **v. 42**, 109-24.

6

IBC GRANULITE IN CLOCKWISE PRESSURE-TEMPERATURE REGIME: A CASE FROM THE ORISSA SECTOR OF EASTERN GHATS MOBILE BELT

S.C. PATEL*

ABSTRACT

Bands of ferrodiorite and calc-silicate granulite are exposed in the northern part of Bolangir anorthosite complex in the Orissa sector of the Eastern Ghats Mobile Belt. Ferrodiorite commonly contains phenocrystic plagioclase set in a fine-grained schistose matrix of orthopyroxene and plagioclase. Phenocrystic plagioclase is ubiquitously surrounded by a corona of euhedral to subhedral grains of garnet, and is strongly zoned with composition $Ab_{54}An_{43}Or_3$ at the core to $Ab_{70}An_{25}Or_5$ at the rim in contact with garnet. Calc-silicate granulite is strongly gneissic in character with K-feldspar-dominated layers alternating with layers rich in scapolite, diopside, wollastonite and calcite. Garnet occurs in calc-silicate granulite as thin corona around both scapolite and wollastonite at the contact of each other or with calcite. The coronal garnet growth in both ferrodiorite and calc-silicate granulite is interpreted to have resulted from near isobaric cooling. Thermobarometric calculations in ferrodiorite indicate that the cooling occurred at a pressure of 7.5 kbar from a minimum temperature of 740°C. In view of the earlier reports of high-temperature decompression in Bolangir anorthositic rocks, and of clockwise P-T path from relict-kyanite-bearing Mg-Al granulite in the Hatimunda hill area of Orissa, the present near-isobaric cooling path is interpreted as part of a clockwise P-T trajectory.

KEY WORDS

Granulite, Corona, Clockwise Path, Bolangir, Eastern Ghats.

INTRODUCTION

Successful interpretation of the pressure (P)-temperature (T) history of granulite terranes is fundamental to the understanding of regional tectonic evolution. However, because granulite terranes are commonly ancient, complex and polymetamorphosed, the determination of P-T paths is rarely straightforward. During the 1980s, granulite terranes were grouped into two basic types according to the post-peak P-T evolution: those that record evidence of near-isothermal decompression (ITD) P-T paths and those that exhibit near-isobaric cooling (IBC) P-T paths (*see* review by Harley, 1989). This generated debate about clockwise vs

* Department of Earth Sciences, Indian Institute of Technology, Bombay, Powai, Mumbai - 400076. E-mail: scpatel@geos.iitb.ac.in

counterclockwise P-T paths and crustal thickening vs magmatic underplating for the origin of the extreme thermal perturbation required for granulite facies metamorphism.

An IBC path in lower- to mid-crustal granulites generally implies that cooling from the peak temperature of metamorphism occurred in crust of near-normal thickness, or at least a thickness in isostatic equilibrium with surrounding lithosphere. Therefore, IBC paths may result from a variety of tectono-magmatic processes which led to the production or retention of significant thermal anomalies in the crust some 25-40 km thick. Consequently, an IBC path may constitute part of either clockwise or counterclockwise P-T trajectory (Harley, 1989). In clockwise P-T paths IBC progresses at pressures somewhat less than the maximum recorded pressure, whereas in counterclockwise P-T paths IBC ensues from the peak pressure. The discrimination of the two types of paths depends on the documentation of the geometry of the prograde P-T path (e.g. Cartwright, 1990; Hiroi *et al.*, 1994; Kriegsman and Schumacher, 1999; Ouzegane *et al.*, 2003). However, prograde section of P-T paths is seldom well recorded in IBC granulites mainly because of the relatively fast reaction rates and approach to equilibrium maintained by fast diffusion at peak temperatures of 800-1000°C (Bohlen, 1991). Nevertheless, stepped P-T paths with an ITD segment preceding the IBC segment have been documented in numerous studies (e.g. Harley *et al.*, 1990; Carlswell and O'Brien, 1993; Hisada and Miyano, 1996; Owen and Dostal, 1996; Raith *et al.*, 1997; Fraser *et al.*, 2000; Zhao *et al.*, 2000; Guo *et al.*, 2002; Moraes *et al.*, 2002; Marschall *et al.*, 2003). Such a P-T trajectory is presumed to be the post-thermal peak segment of a clockwise P-T path, and generally this evolution is related to crustal thickening, possibly at convergent plate margins (Brown, 2001).

THE EASTERN GHATS BACKGROUND

The Eastern Ghats Mobile Belt (EGMB) of eastern India is a Proterozoic granulite facies terrane with opinions sharply divided on the question of clockwise or counterclockwise P-T path (Lal *et al.*, 1987; Sengupta *et al.*, 1990; Mukhopadhyay and Bhattacharya, 1997; Mohan *et al.*, 1997; Bhattacharya and Kar, 2002). All of these studies are on granulites in the Andhra Pradesh sector of the EGMB. However, studies on granulites in the Orissa sector of the EGMB mostly show evidences of clockwise P-T path. High temperature decompression has been reported from the Bolangir anorthositic rocks of Orissa (Patel *et al.*, 2001a). In the Hatimunda hill area of Orissa clockwise P-T path has been reported from Mg-Al granulites (Patel *et al.*, 2001b). Neogi and Das (2000) obtained high-temperature ITD path in Mg-Al granulites from the Deobhog area of Chhattisgarh, and interpreted it as a part of clockwise P-T path. Sen *et al.* (1995) reported high-temperature decompression followed by cooling in the Chilka Lake granulites of Orissa. It is against this background that this paper presents mineralogical data indicative of IBC path in ferrodiorite and calc-silicate granulite around Bolangir anorthosite complex of Orissa. This IBC is interpreted in the framework of clockwise P-T path reported from different parts of the Orissa sector of the EGMB.

GEOLOGICAL SETTING

The Bolangir anorthosite complex is a massif-type intrusion emplaced in the northwestern part of the EGMB. The complex is circular in shape with two tongue-like extensions in the north (Tak *et al.*, 1966; Bhattacharya *et al.*, 1998) (Fig. 1). The present study area falls in the northern part of the complex. Calc-silicate granulite, mafic granulite and khondalite occur as local patches in the quartzo-feldspathic gneiss country rock surrounding the complex. The complex is bordered in most part by granitoids, and ferrodiorite locally occurs as bands and lenses at the interface between the anorthositic rocks and the granitoids (Bhattacharya *et al.*, 1998). Bhattacharya *et al.* (1998) interpreted the ferrodiorite to represent late-stage residual melt that evolved through anorthosite crystallisation from high-Al basic magmas.

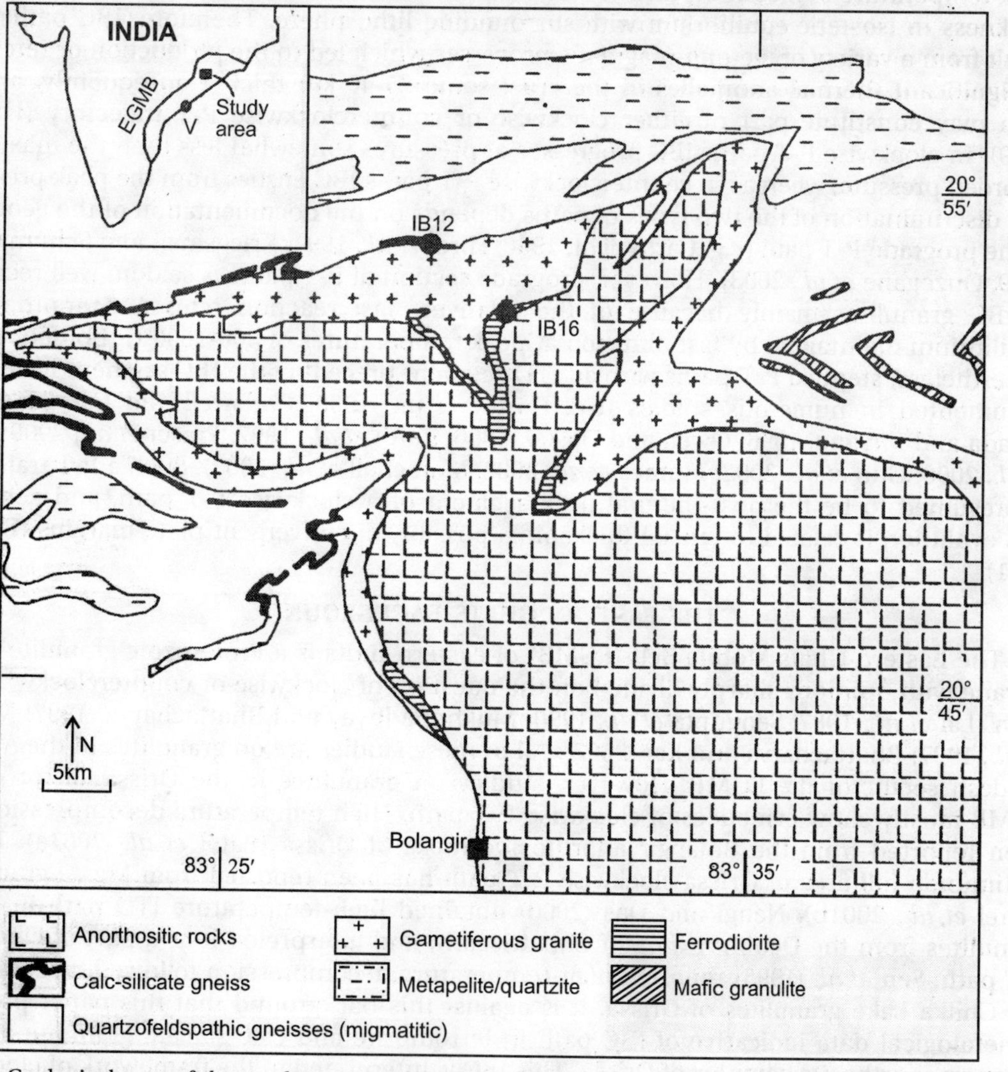

Fig. 1. Geological map of the northern part of Bolangir anorthosite complex (after Bhattacharya *et al.*, 1998). *Inset:* Location of Eastern Ghats Mobile Belt (EGMB) and study area in India. V - Vishakhapatnam.

The Bolangir area falls in the Western khondalite zone (Ramakrishnan *et al.*, 1998) and in the Crustal domain 4 (Rickers *et al.*, 2001) of the EGMB. The Crustal domain 4 is characterised by Nd model ages of 2.2-2.8 Ga in metasediments, and 2.9-3.9 Ga in orthogneisses. The Bolangir ferrodiorite has yielded U-Pb upper intercept age of 933 ± 32 Ma on zircons (Krause *et al.*, 2001). Available geochronological data in khondalite and mafic granulite from different areas of the EGMB north of Godavari graben indicate a granulite metamorphic event at 950-1000 Ma (Shaw *et al.*, 1997; Mezger and Cosca, 1999).

MINERAL TEXTURES AND CHEMISTRY

Electron microprobe analysis was carried out using the JEOL-JXA-8600M Superprobe at the University of Roorkee with an accelerating voltage of 15 kV, specimen current of 20 nA, and beam diameter of 3 μm. Mineral phases were used as standards. Representative analyses of minerals are given in Tables 1 and 2. The mineral abbreviations used in the paper are: *alm,* almandine; *an,* anorthite; *cc,* calcite; *cpx,* clinopyroxene; *en, enstatite*; *fs,* ferrosilite; *grt,* garnet; *grs,* grossularite; *ilm,* ilmenite; *kfs,* K-feldspar; *me,* meionite; *opx,* orthopyroxene; *pl,* plagioclase; *prp,* pyrope; *qtz,* quartz; *scp,* scapolite; *spn,* sphene; *sps,* spessartine; and *wo,* wollastonite.

Ferrodiorite

The detailed mineralogy and textural relations in the Bolangir ferrodiorite has been given by Bhattacharya *et al.* (1998). The ferrodiorite commonly contains plagioclase, orthopyroxene, and garnet with accessory ilmenite. It is constituted of a fine-grained (around 0.5 mm size) schistose matrix of plagioclase and orthopyroxene with occasional large crystals (upto 3 mm size) of plagioclase. Garnet occurs in different textural modes such as: (i) corona of euhedral to subhedral grains of < 1 mm size developed between phenocrystic plagioclase crystals and adjacent matrix grains of orthopyroxene (Fig. 2); (ii) aggregate of plagioclase and garnet grains; (iii) rounded to elongated clusters of grains; (iv) discrete crystals at the centres of plagioclase grains.

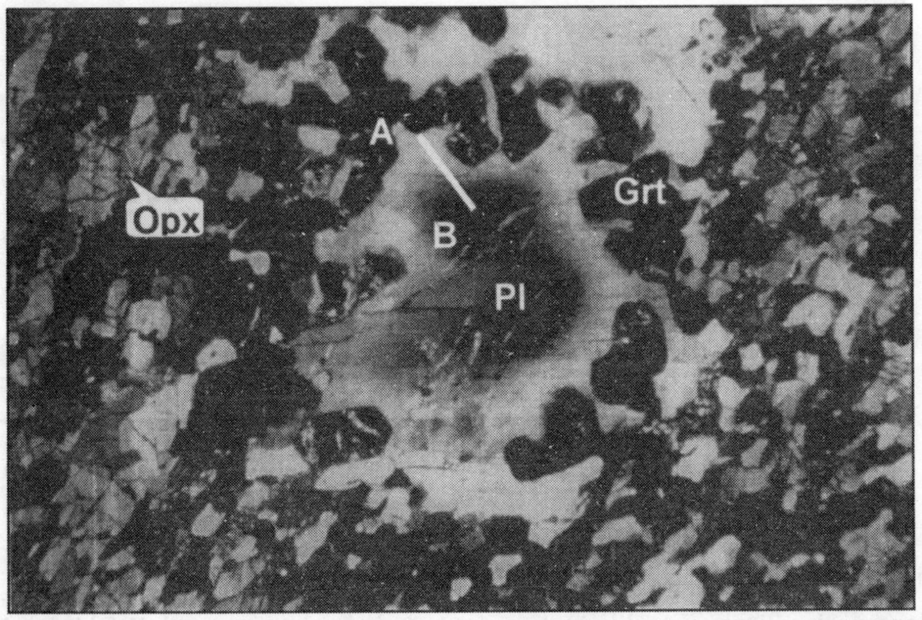

Fig. 2. Photomicrograph (crossed polars) of ferrodiorite showing garnet necklace around optically and chemically zoned phenocrystic plagioclase. EPMA scan was carried out along line AB. Sample IB12; field of view is 9 mm long.

Phenocrystic plagioclase is strongly zoned with compositions varying from $Ab_{54}An_{25}Or_3$ at the core to $Ab_{70}An_{25}Or_5$ at the rim in contact with garnet. The chemical zoning is reflected by the optical zoning seen under crossed polars (Fig. 2). Compositional zoning across the rim of a pheonocrystic plagioclase grain is shown in Figure 3. Continuous chemical variation rather than a step-function profile suggests that the rims are the product of diffusionally controlled re-equilibration rather than overgrowths precipitated from a fluid (e.g. St-Onge and Ijewliw, 1996). Garnet grains generally lack in chemical zoning, and have high X_{alm} (0.70 to 0.72) and low X_{sps} (0.03 to 0.04) with X_{grs} of 0.18, and X_{py} in the range of 0.06 to 0.08. Orthopyroxene shows strong pleochroism from bluish green to yellowish green, and is generally iron-rich with a uniform composition of X_{Mg} (= Mg/Mg+Fe) around 0.27. Ilmenite contains negligible MgO and MnO, and no Cr_2O_3.

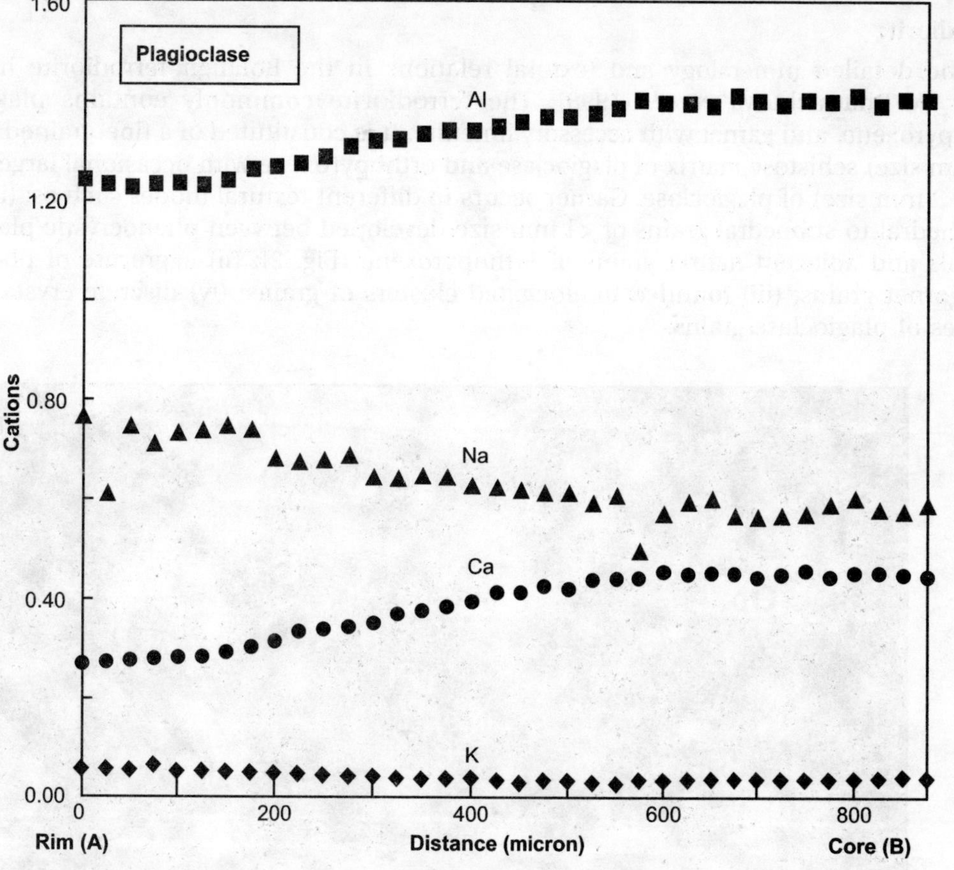

Fig. 3. Plagioclase zoning in ferrodiorite sample IB12 along line AB in Fig. 2.

Calc-silicate Granulite

Bolangir calc-silicate granulite is strongly gneissic in character with K-feldspar dominated layers alternating with layers rich in scapolite, diopside, wollastonite and calcite. Quartz and sphene occur in minor amounts. Garnet occurs as thin (<0.05 mm) corona around both scapolite and wollastonite at the contact of each other or with calcite. Quartz also occurs along with garnet as corona between scapolite and wollastonite. In addition, garnet locally occurs between scapolite and diopside.

Table 1: *Representative composition of minerals in ferrodiorite sample No. IB12*

	Matrix grains meeting at triple point						Garnet rim around plagioclase phenocryst				Matrix grains	
	Grt (rim)	Grt (core)	Pl (rim)	Pl (core)	Opx (rim)	Opx (core)	Grt (rim)	Grt (core)	Pl (rim)	Pl (core)	Opx	Ilm
SiO_2	35.39	35.35	61.60	63.14	45.95	46.37	35.65	35.97	60.61	56.99	46.35	0.44
TiO_2					0.10	0.13					0.10	49.70
Al_2O_3	21.54	21.24	23.48	23.31	0.56	0.54	21.15	21.95	23.46	26.19	0.42	0.32
FeO	32.85	32.85	0.23	0.27	42.04	41.68	32.17	32.21	0.04	0.07	41.92	48.86
MnO	1.71	1.31			0.51	0.60	1.30	1.40			0.53	0.09
MgO	1.67	1.89			8.88	8.42	2.10	2.14			8.65	0.31
CaO	6.59	6.50	5.76	5.79	0.91	0.96	6.50	6.60	5.61	9.29	0.97	
Na_2O			8.59	8.38	0.00	0.00			8.81	6.44	0.10	
K_2O			0.76	0.79	0.00	0.00			1.03	0.52	0.00	
Total	99.75	99.14	100.42	101.68	98.95	98.70	98.87	100.27	99.56	99.50	99.04	99.72
Oxygen	(12)	(12)	(8)	(8)	(6)	(6)	(12)	(12)	(8)	(8)	(6)	(6)
Si	2.882	2.893	2.740	2.767	1.930	1.948	2.914	2.894	2.725	2.578	1.943	0.023
Ti					0.003	0.004					0.003	1.900
Al	2.068	2.049	1.231	1.204	0.028	0.027	2.038	2.082	1.243	1.396	0.021	0.019
Fe	2.238	2.249	0.009	0.010	1.477	1.464	2.199	2.168	0.002	0.003	1.470	2.098
Mn	0.118	0.091			0.018	0.021	0.090	0.095			0.019	0.004
Mg	0.203	0.231			0.556	0.527	0.256	0.257			0.540	0.024
Ca	0.575	0.570	0.274	0.272	0.041	0.043	0.569	0.569	0.270	0.450	0.044	
Na			0.741	0.712	0.000	0.000			0.768	0.565	0.008	
K			0.043	0.044	0.000	0.000			0.059	0.030		
Total	8.084	8.083	5.038	5.009	4.053	4.034	8.066	8.065	5.067	5.022	4.048	4.068
X_{Mg}					0.27	0.26					0.27	
X_{An}			0.26	0.26					0.25	0.43		
X_{Alm}	0.71	0.72					0.71	0.70				
X_{Py}	0.06	0.07					0.08	0.08				
X_{Gr}	0.18	0.18					0.18	0.18				
X_{Sp}	0.04	0.03					0.03	0.03				

X_{Mg} = Mg/(Mg+Fe); X_{An} = Ca/(Ca+Na+K).
blank: not analysed

Table 2: *Representative compositions of minerals in calc-silicate granulite sample no. IB16*

	Scp (rim)	Scp (core)	Cpx	Wo	Grt	Kfs	Cc	Spn
SiO_2	45.62	45.66	49.61	51.27	38.89	64.40		30.17
TiO_2			0.07		0.10	0.13		33.09
Al_2O_3	27.91	27.46	0.98	0.00	19.86	17.86		4.46
Fe_2O_3								0.84
FeO	0.08	0.02	15.26	0.56	4.19	0.00	0.02	
MnO			0.33		0.02		0.00	0.06
MgO	0.00	0.00	8.74	0.01	0.01		0.00	0.00
CaO	19.38	19.35	23.69	47.54	35.80	0.01	56.45	28.60
Na_2O	2.50	3.01	0.10	0.19		1.11		
K_2O	0.13	0.14	0.02	0.02		15.88		
Total	95.62	95.64	98.80	99.59	98.65	99.26	56.47	97.22
Oxygen	*	*	(6)	(6)	(12)	(8)	(2)	Ψ
Si	6.972	7.022	1.950	1.996	3.020	3.003		0.968
Ti			0.002					0.814
Al	5.027	4.978	0.045	0.000	1.806	0.981		0.172
Fe^{3+}								0.021
Fe^{2+}	0.010	0.003	0.502	0.018	0.272	0.000	0.001	
Mn			0.011		0.001		0.000	0.002
Mg	0.000	0.000	0.512	0.001	0.001		0.000	0.000
Ca	3.173	3.189	0.998	1.983	2.978	0.000	1.999	1.003
Na	0.741	0.898	0.008	0.014		0.100		
K	0.025	0.027	0.001	0.001		0.944		
Total	15.948	16.117	4.049	4.013	8.078	5.028	2.000	3.000
Eq (An)	79.6	79.8						
Me	67.6	65.9						
X_{Mg}			0.50					
X_{Al}								0.17
X_{Gr}					0.92			
X_{Alm}					0.08			

Formula per Si + Al = 12; Ψ Formula based on Σcations = 3 and all iron as Fe_2O_3; Eq (An) =100 (Al-3)/3; Me = 100 (Σdivalent cations)/4; X_{Mg} = Mg/(Mg + Fe); X_{Al} = Al/ (Al+Ti+Fe)
blank: not analysed

Scapolite shows minor compositional variation with meionite contents of 65.9 to 67.6% and equivalent anorthite contents of 79.6 to 79.8%. Garnet is dominantly grossularite with a composition of $Gr_{0.92}Alm_{0.08}$. Clinopyroxene has an X_{Mg} value of 0.50. Microcline contains about 10 mole% albite. Calcite contains negligible FeO and no MgO or MnO. Wollastonite contains 0.6 wt.% FeO. Sphene is somewhat aluminous with an X_{Al} (= Al/Ti+Al+Fe) value of 0.17.

MINERAL REACTIONS

In ferrodiorite, the appearance of neoblastic garnet mantling plagioclase can be represented by the model end-member reactions (Perkins and Newton, 1981):

$$\underset{an}{3CaAl_2Si_2O_8} + \underset{en}{6MgSiO_3} = \underset{grs}{Ca_3Al_2Si_3O_{12}} + \underset{prp}{2Mg_3Al_2Si_3O_{12}} + \underset{qtz}{3SiO_2} \quad (1)$$

and

$$\underset{an}{3CaAl_2Si_2O_8} + \underset{fs}{6FeSiO_3} = \underset{grs}{Ca_3Al_2Si_3O_{12}} + \underset{alm}{2Fe_3Al_2Si_3O_{12}} + \underset{qtz}{3SiO_2} \quad (2)$$

Reactions (1) and (2) result in the loss of anorthite in plagioclase and concomitant gain of grossular content in garnet. The operation of these reactions is supported by zoning in phenocrystic plagioclase with anorthite-rich core, as lower anorthite contents are the product of cooling-driven readjustment. The Mg end-member reaction (1) has shallow dP/dT slope, whereas the Fe end-member reaction (2) has moderate positive dP/dT slope. Observed orthopyroxene compositions (Table 1) are closer to the Fe-end member reaction. The corona growth may have occurred during re-equilibration at lower T and higher P or both. Similar coronas have been described from IBC granulites in many parts of the world (Harley, 1989; Owen and Dostal, 1996; Druppel *et al.*, 2001), and provide evidence for IBC in Bolangir ferrodiorite. Furthermore, Shaw and Arima (1996b) interpreted an IBC path from coronal garnet-forming reaction from plagioclase in granulites of Rayagada in the Orissa sector of the EGMB.

In calc-silicate granulite, the formation of corona of grossular-rich garnet around wollastonite and scapolite suggests the model reaction (Harley *et al.*, 1994):

$$\underset{me}{Ca_4Al_6Si_6O_{24}CO_3} + \underset{wo}{3CaSiO_3} + \underset{cc}{2CaCO_3} = \underset{grs}{3Ca_3Al_2Si_3O_{12}} + CO_2 \quad (3)$$

Similarly, the formation of garnet + quartz corona between scapolite and wollastonite is consistent with the reaction:

$$\underset{me}{Ca_4Al_6Si_6O_{24}CO_3} + \underset{wo}{5CaSiO_3} = \underset{grs}{3Ca_3Al_2Si_3O_{12}} + \underset{qtz}{2SiO_2} + CO_2 \quad (4)$$

Grossular rims on scapolite coexisting with wollastonite, or with wollastonite + calcite, have been thought to reflect an influx of H_2O on the assumption that the dT/dX_{CO_2} slopes of all the reactions producing grossular + CO_2 are positive (Warren *et al.*, 1987). However, Harley *et al.* (1994) showed that these textures can form simply as a result of cooling below 850-780°C and at relatively low a_{co_2} in the range of 0.28-0.45, provided the pressures are greater than ca. 6 kbar. Shaw and Arima (1996a) described similar reaction textures in calc-silicate granulites from the Rayagada area in Orissa and interpreted it as indicative of IBC path. Thus, an IBC path is a reasonable model for the grossular corona texture in Bolangir calc-silicate granulite.

THERMOBAROMETRY

The TWQ programme of Berman (1991) was used for P-T calculation in ferrodiorite with the thermodynamic data base of Berman (1988) which was updated in 1997. Since plagioclase is the only phase preserving refractory compositions in ferrodiorite, core of zoned plagioclase grains cannot be used along with homogeneous garnet and orthopyroxene to constrain peak P-T conditions from equilibria (1) and (2). P-T calculation has been done using rim compositions of garnet, plagioclase and orthopyroxene meeting at triple point in the matrix. The P-T intersection of this assemblage along with quartz is at 7.5 kbar and 740°C (Fig. 4). Thus, the IBC path interpreted above on the basis of corona textures took place at a pressure of 7.5 kbar.

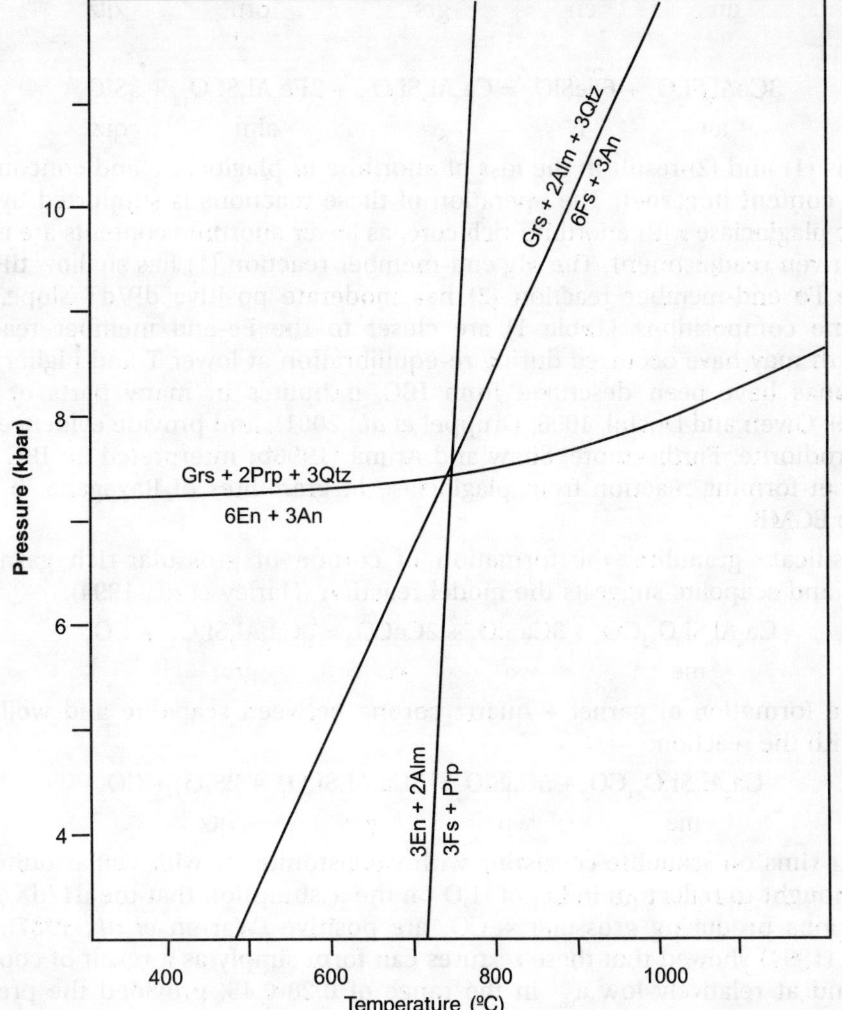

Fig. 4. P-T diagram for the assemblage grt-opx-pl-qtz in ferrodiorite sample IB12. Mineral compositions used are rim compositions of the triple point assemblage grt, pl and bt given in Table 1. Solution models used are garnet, orthopyroxene (Berman and Aranovich, 1996) and plagioclase (Fuhrman and Lindsley, 1988).

As with most granulite terranes, P-T trajectories can only be properly defined using a combination of techniques and textural observations obtained from a variety of rock types (Harley, 1989). Occurrence of kyanite in texturally early mode than sillimanite in any rock is a strong indication of clockwise P-T model. Patel *et al.* (2001b) reported relict kyanite in Mg-Al granulite of Hatimunda area of Orissa. The prograde metamorphic evolution in the Hatimunda area started in the kyanite stability field but evolved mainly within the sillimanite stability field. Several occurrences of minor amounts of kyanite in the nearshore sediments of coastal Andhra Pradesh (Reddy and Malathi, 2002, and references therein) suggest that EGMB is the likely provenance for this heavy mineral. Therefore, if detailed lithological mapping of the EGMB is carried out it may lead to discovery of many kyanite localities. In the Hatimunda rocks Patel *et al.* (2001b) constrained a clockwise P-T path with decompression from peak metamorphic conditions of 700-800°C and >8 kbar. In the study of Bolangir anorthositic rocks, Patel *et al.* (2001a) obtained high temperature decompression from minimum peak P-T condition of 8 kbar and 700°C. From these data it is evident that the IBC in Bolangir ferrodiorite took place at a pressure of 7.5 kbar following ITD from peak pressure of >8 kbar.

DISCUSSION

Available isotopic data on the EGMB point to several periods of granulite metamorphism (Shaw *et al.*, 1997; Mezger and Cosca, 1999). In such a case there is a possibility of getting apparent P-T paths, which actually link P-T vectors from isotopically distinct metamorphic cycles. In the absence of metamorphic dates on the rocks, it is difficult to ascertain whether the IBC path in the Bolangir ferrodiorite is a part of the same metamorphic cycle that produced ITD in the Bolangir anorthositic rocks and Hatimunda Mg-Al rich granulites. Nevertheless, the thermobarometric constraint of the occurrence of IBC following ITD makes clockwise P-T regime more likely than counterclockwise P-T regime in the Orissa sector of the EGMB.

Gupta *et al.* (2000) reported IBC path from the growth of coronal garnet at plagioclase-pyroxene interface in mafic granulites of Deobhog area in Chhattisgarh and interpreted it in terms of counterclockwise P-T path. However, in view of the ITD path at the peak temperature of metamorphism obtained by Neogi and Das (2000) from the same area, the counterclockwise P-T model of Gupta *et al.* (2000) is untenable, and the IBC path in Deobhog granulites is more likely to be part of a clockwise P-T regime. Gupta *et al.* (2000) also documented an ITD segment following IBC of Deobhog granulites. This later ITD is probably the result of final exhumation of the terrane through a separate event that is causally unrelated to prograde metamorphism (*see* Harley, 1989).

One problem still to be fully understood in granulite terranes showing clockwise P-T path is the cause of the thermal perturbation responsible for the metamorphism. In general, numerical models of metamorphism as a result of thickening and thermal relaxation do not generate granulite-facies conditions at appropriate depths (e.g. England and Thompson, 1984). This indicates that an additional heat source, most reasonably related to pre-, syn- or immediately post tectonic magmatic activities is necessary (Harley, 1989). Geophysical criteria and geobarometric information suggest that the EGMB represents an overthickened crust. However, detailed studies on the mechanism of thickening of this belt are lacking. Models proposed for the thickening include collisional overthrusting (Chetty and Murthy, 1998), and homogeneous shortening and folding (Bhattacharya, 1997). But available structural, petrological and geochronological data are rather meagre to constrain a hard tectonic model for the EGMB.

ACKNOWLEDGMENT

Sri Ibrat Faizan is thanked for help in mapping and sample collection.

REFERENCES

Berman, R.G., 1988. Internally-consistent thermodynamic data for stoichiometric minerals in the system Na_2O-K_2O-CaO-MgO-FeO-Fe_2O_3-Al_2O_3-SiO_2-TiO_2-H_2O-CO_2. *Jour. Petrol.*, v. **29**, 445-522.

Berman, R.G., 1991. Thermobarometry using multiequilibrium calculations: a new technique with petrological applications. *Can. Mineral.*, **v. 29**, 833-55.

Berman, R.G. & Aranovich, L.Ya., 1996. Optimised standard state and solution properties of minerals: I. Model calibration for olivine, orthopyroxene, cordierite, garnet and ilmenite in the system FeO-MgO-CaO-Al_2O_3-TiO_2-SiO_2. *Contrib. Mineral. Petrol.*, **v. 126**, 1-24.

Bhattacharya, A., Raith, M., Hoernes, S. & Banerjee, D., 1998. Geochemical evolution of the massif-type anorthosite complex at Bolangir in the Eastern Ghats belt of India. *Jour. Petrol.*, **v. 9**, 1169-95.

Bhattacharya, S., 1997. Evolution of the Eastern Ghats granulite belt of India in a compressional tectonic regime and juxtaposition against Iron Ore Craton of Singhbhum by oblique collision-transpression. *Proc. Ind. Acad. Sci.*, **v. 106**, 65-75.

Bhattacharya, S. & Kar, R., 2002. High-temperature dehydration melting and decompressive P-T path in a granulite complex from the Eastern ghats, India. *Contrib. Mineral. Petrol.*, **v. 143**, 175-91.

Bohlen, S.R., 1991. On the formation of granulites. *Jour. Metamorphic Geol.*, **v. 9**, 223-29.

Brown, M., 2001. From microscope to mountain belt: 150 years of petrology and its contribution to understanding geodynamics, particularly the tectonics of orogens. *Jour. Geodynamics*, **v. 32**, 115-64.

Carswell, D.A. & O'Brien, P.J., 1993. Thermobarometry and geotectonic significance of high-pressure granulites: Examples from the Moldanubian zone of the Bohemian massif in Lower Austria. *Jour. Petrol.*, **v. 34**, 447-59.

Cartwright, I., 1990. Prograde metamorphism, anatexis, and retrogression of the Scourian Complex, north-west Scotland. *In:* J.R. Ashworth and M. Brown (Eds.), *High-Temperature Metamorphism and Crustal Anatexis.* Kluwer Academic, London, 371-99.

Chetty, T.R.K. & Murthy, D.S.N., 1998. Regional tectonic framework of the Eastern Ghats Mobile Belt: A new interpretation. *Geol. Surv. India Spl. Publ.*, **no. 44**, 39-50.

Druppel, K., von Seckendorff, V. & Okrusch, M., 2001. Subsolidus reaction textures in the anorthositic rocks of the southern part of the Kunene Intrusive Complex, NW Namibia. *Eur. Jour. Mineral.*, **v. 13**, 289-309.

England, P.C. & Thompson, A.B., 1984. Pressure-temperature-time paths of regional metamorphism I. Heat transfer during the evolution of regions of thickened continental crust. *Jour. Petrol.*, **v. 25**, 894-928.

Fraser, G., McDougall, I., Ellis, D.J. & Williams, I.S., 2000. Timing and rate of isothermal decompression in Pan-African granulites from Rundvagshetta, East Antarctica. *Jour. Metamorphic Geol.*, **v. 18**, 441-54.

Fuhrman, M.L. & Lindsley, D.H., 1988. Ternary-feldspar modelling and thermometry. *Am. Mineral.*, **v. 73**, 201-16.

Guo, J.H., O'Brien, P.J. & Zhai, M., 2002. High-pressure granulites in the Sanggan area, North China craton: metamorphic evolution. P-T paths and geotectonic significance. *Jour. Metamorphic Geol.*, **v. 20**, 741-56.

Gupta, S., Bhattacharya, A., Raith, M. & Nanda, J.K., 2000. Contrasting pressure-temperature deformation history across a vestigal craton-mobile belt boundary: the western margin of the Eastern Ghats belt at Deobhog, India. *Jour. Metamorphic Geol.*, **v. 18**, 683-97.

Harley, S.L., 1989. The origin of granulites: a metamorphic perspective. *Geol. Mag.*, **v. 126**, 215-47.

Harley, S.L., Fitzsimons, I.C.W. & Buick, I.S., 1994. Reactions and textures in wollastonite-scapolite granulites and their significance for pressure-temperature-fluid histories of high-grade terranes. *Prec. Res.*, **v. 66**, 309-23.

Harley, S.L., Hensen, B.J. & Sheraton, J.W., 1990. Two-stage decompression in orthopyroxene-sillimanite granulites from Forefinger Point, Enderby Land, Antarctica: implications for the evolution of the Archaean Napier Complex. *Jour. Metamorphic Geol.*, **v. 8**, 591-613.

Hiroi, Y., Ogo, Y. & Namba, K., 1994. Evidence for prograde metamorphic evolution of Sri Lankan pelitic granulites, and implications for the development of continental crust. *Prec. Res.*, 245-63.

Hisada, K. & Miyano, T., 1996. Petrology and microthermometry of aluminous rocks in the Botswana Limpopo Central Zone: evidence for isothermal decompression and isobaric cooling. *Jour. Metamorphic Geol.*, **v. 14**, 183-97.

Krause, O., Dobmeier, C., Raith, M. & Mezger, K., 2001. Age of emplacement of massif-type anorthosites in the Eastern Ghats Belt, India: constraints from U-Pb zircon dating and structural studies. *Prec. Res.*, **v. 109**, 25-38.

Kriegsman, L.M. & Schumacher, J.C., 1999. Petrology of sapphirine-bearing and associated granulites from central Sri Lanka. *Jour. Petrol.*, **v. 40**, 1211-39.

Lal, R.K., Ackermand, D. & Upadhyay, H., 1987. P-T-X relationships deduced from corona textures in sapphirine-spinel-quartz assemblages from Paderu, southern India. *Jour. Petrol.*, **v. 28**, 1139-68.

Marschall, H.R., Kalt, A. & Hanel, M., 2003. P-T evolution of a Variscan lower-crustal segment: a study of granulites from the Schwarzwald. Germany. *Jour. Petrol.*, **v. 44**, 227-53.

Mezger, K. & Cosca, M.A., 1999. The thermal history of the Eastern Ghats Belt (India) as revealed by U-Pb and $^{40}Ar/^{39}Ar$ dating of metamorphic and magmatic minerals: implications for the SWEAT correlation. *Prec. Res.*, **v. 94**, 251-71.

Mohan, A., Tripathi, P. & Motoyoshi, M., 1997. Reaction History of sapphirine granulites and a decompressional P-T path in a granulite complex from the Eastern Ghats. *Proc. Ind. Acad. Sci.*, **v. 106**, 115-30.

Moraes, R., Brown, M., Fuck, R.A., Camargo, M.A. & Lima, T.M., 2002. Characterisation and P-T evolution of melt-bearing ultrahigh-temperature granulites: an example from the Anapolis-Itaucu Complex of the Brasilia Fold Belt, Brazil. *Jour. Petrol.*, **v. 43**, 1673-705.

Mukhopadhyay, A.K. & Bhattacharya, A., 1997. Tectonothermal evolution of the gneiss complex at Salur in the Eastern Ghats granulite belt of India. *Jour. Metamorphic Geol.*, **v. 15**, 719-34.

Neogi, S. & Das, N., 2000. Lithotectonic domains and petrotectonic evolution of the boundary zone between the Eastern Ghats mobile belt and the Bastar craton, Deobhog area, central India. *Geol. Surv. India Spl. Publ.*, **no. 57**, 180-204.

Ouzegane, K., Guiraud, M. & Kienast, J.R., 2003. Prograde and retrograde evolution in high-temperature corundum granulites (FMAS and KFMASH systems) from In Ouzzal terrane (NW Hoggar, Algeria). *Jour. Petrol.*, **v. 44**, 517-45.

Owen, J.V. & Dostal, J., 1996. Contrasting corona structures in mafic granulite from the Blansky Les Complex, Bohemian Massif, Czech Republic. *Can. Mineral.*, **v. 34**, 959-66.

Patel, S.C., Frost, B.R., Faizan, I. & Ghosh, T.K., 2001a. Metamorphism and contrasting corona structures in Bolangir anorthositic rocks. *Jour. Geol. Soc. India*, **v. 57**, 31-38.

Patel, S.C., Behera, S., Mohanty, A., Singh, A.K. & Patel, S.K., 2001b. Metamorphic history of sapphirine- and relict-kyanite-bearing Mg-Al rich rocks at Hatimunda hill near Junagarh, Kalahandi district, Orissa. *Jour. Geol. Soc. India*, **v. 57**, 417-27.

Perkins III, D. & Newton, R.C., 1981. Charnockite geobarometers based on coexisting garnet-pyroxene-plagioclase-quartz. *Nature*, **v. 292**, 144-46.

Raith, M., Karmakar, S. & Brown, M., 1997. Ultrahigh-temperature metamorphism and multi-stage decompressional evolution of sapphirine granulites from the Palni Hill Ranges, southern India. *Jour. Metamorphic Geol.*, **v. 15**, 379-99.

Ramakrishnan, M., Nanda, J.K. & Augustine, P.F., 1998. Geological evolution of the Proterozoic Eastern Ghats mobile belt. *Geol. Surv. India, Spec. Publ.*, **no. 44**, 1-2.

Reddy, D.R. & Malathi, V., 2002. Occurrence of kyanitic in the nearshore sediments of north coastal Andhra Pradesh and its implications. *Jour. Geol. Soc. India,* v. **60**, 329-31.

Rickers, K., Mezger, K. & Raith, M., 2001. Evolution of the continental crust in the Proterozoic Eastern Ghats belt, India and new constraints for Rodinia reconstruction: implications from Sm-Nd, Rb-Sr and Pb-Pb isotopes. *Proc. Res.*, v. **112**, 183-210.

Sen, S.K., Bhattacharya, S. & Acharyya, A., 1995. A multi-stage pressure-temperature record in the Chilka Lake granulites: the epitome of the metamorphic evolution of Eastern Ghats. *Jour. Metamorphic Geol.*, **v. 13**, 287-98.

Sengupta, P., Dasgupta, S., Bhattacharya, P.K., Fukuoka, M., Chakraborti, S. & Bhowmick, S., 1990. Petro-tectonic imprints in the sapphirine granulites from Anantagiri, Eastern Ghats Belt, India. *Jour. Petrol.*, **v. 31**, 971-96.

Shaw, R.K. & Arima, M., 1996a. High-temperature metamorphic imprint on calc-silicate granulites of Rayagada, Eastern Ghats. India: implication for the isobaric cooling path. *Contrib. Mineral. Petrol.*, v. **126**, 169-80.

Shaw, R.K. & Arima, M., 1996b. Mineral chemistry, reaction textures, thermobarometry and P-T path from orthopyroxene granulites of Rayagada, Eastern Ghats, India. *Jour. Southeast Asian Earth Sci.*, **v. 14**, 175-84.

Shaw, R.K., Arima, M., Kagamai, H., Fanning, C.M., Shiraishi, K. & Motoyoshi, Y., 1997. Proterozoic events in the Eastern Ghats granulite belt, India: Evidence from Rb-Sr, Sm-Nd systematics, and SHRIMP dating. *Jour. Geol.*, v. **105**, 645-56.

St-Onge, M.R. & Ijewliw, O.J., 1996. Mineral corona formation during high-P retrogression of granitic rocks. Ungava orogen, Canada. *Jour. Petrol.*, **v. 37**, 553-82.

Tak, M.W., Mitra, D. & Chatterjee, P.K., 1966. A note on the occurrence of anorthosite in the Bolangir-Patna district, Orissa. *Indian Minerals,* v. **20**, 339-42.

Warren, R.G., Hensen, B.J. & Rybum, R.J., 1987. Wollastonite and scapolite in Precambrian calc-silicate granulites from Australia and Antarctica. *Jour. Metamorphic Geol.*, v. **5**, 213-23.

Zhao, G., Wilde, S.A., Cawood, P.A. & Lu, L., 2000. Petrology and P-T path of the Fuping mafic granulites: implications for tectonic evolution of the central zone of the North China craton. *Jour. Metamorphic Geol.* **v. 18**, 375-91.

SECTION – II

FLUIDS

7

CARBONATES IN FELDSPATHIC GNEISSES FROM THE GRANULITE FACIES: IMPLICATIONS FOR THE FORMATION OF CO_2-RICH FLUID INCLUSIONS

WILLIAM LAMB*

ABSTRACT

Various processes have been proposed to explain the stabilization of those mineral assemblages that characterize the granulite facies, and to generate the relatively low values of H_2O activities (aH_2O) that are characteristic of granulite facies metamorphism. These granulite-forming processes include recrystallization of anhydrous lithologies, partial melting, and infiltration of fluids with low aH_2O. The presence of CO_2-rich fluid inclusions in rocks from granulite facies terranes has sometimes been taken as evidence for pervasive CO_2 infiltration. However, CO_2-rich fluids are, in some cases, inconsistent with values of CO_2 activities (aCO_2) determined from mineral equilibria. Unfortunately, the number of examples in which mineral equilibria have been used to constrain peak metamorphic aCO_2 are limited, in part because many quartzo-feldspathic granulite facies gneisses may not contain the appropriate mineralogy.

Small amounts of carbonate minerals are often present in granulite facies feldspathic gneisses, and the presence of these minerals may allow the determination of CO_2 activities using a variety of CO_2-buffering equilibria. However, calculations indicate that these carbonates are not in equilibrium with quartz ± orthopyroxene under granulite facies pressure-temperature conditions, and, therefore, these carbonates should not be used to determine peak metamorphic fluid compositions. In some cases the solid carbonate inclusions are closely associated with CO_2-rich fluid inclusions indicating they formed during the same event. Given that the solid carbonate inclusions should not be used to determine peak metamorphic fluid compositions, it is likely that the CO_2-rich fluid inclusions associated with the carbonates also are not reliable indicators of peak metamorphic fluid compositions.

KEY WORDS

Carbonates, Granulite facies and Fluid inclusions.

INTRODUCTION

Rocks metamorphosed to granulite facies conditions are thought to comprise significant portions of the lower continental crust (Heier, 1973; Hyndman *et al.*, 1968; Rudnick and

* Dept. of Geology and Geophysics, Texas A&M University, College Station, TX 77843-3115, U.S.A.
E-mail: lamb@geo.tamu.edu

Fountain, 1995; Sighinolfi, 1971). Consequently, understanding the processes involved in the stabilization of granulite facies mineral assemblages may provide some insight into the stabilization of continental crust. Rocks from granulite terranes share a variety of characteristics. They were metamorphosed at relatively high temperatures (T) and pressures (P), with peak metamorphic T approximately 700 to 1000°C, and peak metamorphic P generally between 6 and 12 kbar (Bohlen, 1987; Harley, 1989). Also, granulite facies metamorphism is generally characterized by H_2O activities (aH_2O) that are significantly less than unity (where $aH_2O = 1$ for pure H_2O at the P & T in question, Bhattacharya and Sen, 1986; Lamb and Valley, 1988; Phillips, 1980; Valley *et al.*, 1990; Valley *et al.*, 1983).

Various ideas have been proposed to explain the stabilization of the minerals that characterize the granulite facies, and account for reduced aH_2O. In some cases granulite facies metamorphism may involve recrystallization of relatively anhydrous lithologies, such as certain meta-igneous rocks or rocks that had previously undergone high-grade metamorphism. In this case, because rocks were "dry" prior to granulite facies metamorphism, no additional mechanism, operating at peak granulite conditions, is required to account for reduced aH_2O (Lamb and Valley, 1988; Valley *et al.*, 1990).

Partial melting may also stabilize granulite facies mineral assemblages. During melting, granulite facies assemblages are produced as H_2O is dissolved into the melt phase. Loss of the melt by movement to higher crustal levels will, therefore, remove the H_2O. The rock that remains may contain no free fluid (i.e., vapor) phase, or, if a free fluid phase is present, it may become enriched in components with low solubilities in melts relative to H_2O. Thus, melting reactions may generate a CO_2-rich fluid phase if H_2O is partitioned preferentially into a melt relative to CO_2 (Clemens and Vielzeuf, 1987; Eggler and Kadik, 1979; Fyfe, 1973; Lamb and Valley, 1988; LeBreton and Thompson, 1988; Waters, 1988).

Infiltration of anhydrous fluids, such as CO_2, may dilute H_2O and stabilize granulite facies mineral assemblages (Newton, 1992; Newton *et al.*, 1980; Touret, 1992). In this case, CO_2-infiltration must act to dilute H_2O driven off by dehydration reactions, and, therefore, the CO_2 must infiltrate in a pervasive manner and in large amounts. Touret (1992) has argued that the deep crust is a storehouse for CO_2 and that quantities are sufficiently large such that release of this CO_2 to the atmosphere will affect surface temperatures of the earth on a global scale. More recently it has been suggested that hypersaline brines, and not CO_2, are low aH_2O fluids that play a critical role in high-grade metamorphism (Newton *et al.*, 1998).

The debate surrounding the role that fluid play in the formation of granulites can probably be best characterized as a debate between those who argue in favour of the pervasive actions of large amounts of externally derived fluids versus those who argue that fluids act in a more heterogeneous and localized manner. In the first case, pervasive fluids, it has been argued that granulite facies mineral assemblages have been stabilized by the infiltration of large amounts of low aH_2O fluids. The quantities of fluids must be sufficient to drive dehydration reactions (e.g., reactions that form orthopyroxene at the expense of biotite or amphibole), and dilute the H_2O produced by these reactions to maintain low aH_2O. Alternatively, fluid infiltration may be more restricted, affecting rocks in a localized manner, while in the same terrane, and at approximately the same time, other processes (e.g., partial melting and recrystallization of dry rock) also operate to stabilize granulite facies mineral assemblages. For example, evidence for fluid-absent peak metamorphic conditions has been found in samples from the Adirondack Mountains (Lamb and Valley, 1984; Lamb and Valley, 1985; Valley *et al.*, 1990). However, it is possible that peak metamorphic fluids existed in other samples from the same terrane. Thus, CO_2-rich fluids may have occurred locally, however, a pervasive CO_2-rich fluid was not present (Valley *et al.*, 1990). The evidence for heterogeneous fluid conditions during Adirondack granulite facies metamorphism is in contrast to models that call upon the pervasive infiltration of low aH_2O fluids (Newton, 1992; Newton *et al.*, 1980), and ideas that the lower crust is a storehouse for large amounts of CO_2 (Touret, 1992). However, it is possible that differences exist between various granulite facies terranes, and these differences may correlate with age or tectonic setting (Lamb and Valley, 1984; Lamb and Valley, 1985).

The composition of granulite facies fluids has been inferred from investigations of fluid inclusions (e.g., Hansen *et al.*, 1984a; Lamb, 1990; Touret, 1985; Touret, 1989; Vry and Brown, 1991; Wilmart *et al.*, 1991). Alternatively, granulite facies fluid compositions may be inferred through the use of equilibria between minerals and a volatile such as H_2O, CO_2, or O_2 (e.g., Bhattacharya and Sen, 1986; Lamb and Valley, 1984; Lamb and Valley, 1985; Lamb and Valley, 1988; Moecher, 1993; Phillips, 1980; Valley *et al.*, 1990). Only rarely have both techniques, fluid inclusions and phase equilibria, been applied to the same sample (e.g., Lamb *et al.*, 1987). This apparent lack of comparisons between fluid inclusions and fluid-buffering equilibria may arise, in part, from a lack of CO_2-buffering equilibria that may be applied to typical granulite facies quartzo-feldspathic gneisses. In other words, while dehydration equilibria may be used to constrain values of aH_2O, and typically indicate that $aH_2O << 1$, it is much more difficult to constrain values of aCO_2. Determining values of aCO_2 is important in order to determine if granulite facies fluids are CO_2-rich as suggested by fluid inclusion studies. Small amounts of carbonates have been reported in various quartzo-feldspathic gneisses from granulite terranes. A variety of CO_2-buffering equilibria may be written between the carbonates and the various silicates that occur in these same rocks. Thus, the presence of these carbonates offers the potential of providing a method to determine the activities of CO_2 in granulites. This paper explores the possibility of using equilibria involving carbonates and other minerals common to many quartzo-feldspathic granulite facies gneisses to determine peak metamorphic CO_2 activities.

A BRIEF REVIEW OF THE COMPOSITIONS OF GRANULITE FACIES FLUIDS AS INFERRED FROM PHASE EQUILIBRIA AND FLUID INCLUSIONS

Before investigating the usefulness of carbonate equilibria to determine values of aCO_2 in granulite facies quartzo-feldspathic gneisses, it is important to briefly review some of the evidence regarding the compositions of granulite facies fluids. For example, the presence of CO_2-rich fluid inclusions in granulites has sometimes been taken as evidence of pervasive CO_2-infiltration. However, these fluid inclusions might also contain the fluid that remains after partial melting and removal of the melt phase. In either case, use of fluid inclusions to make inferences about the compositions of peak metamorphic fluids assumes that these inclusions were formed during the peak of metamorphism, and that there was no change in inclusion composition during retrograde cooling and depressurization. However, evidence from synthetic and natural fluid inclusions demonstrates that peak metamorphic fluid inclusions may, in some cases, undergo compositional changes during cooling and depressurization (Audétat and Günther, 1999; Bakker and Jansen, 1991; Bakker and Jansen, 1994; Vityk *et al.*, 2000). Furthermore, the formation of CO_2-rich fluid inclusions in granulite facies gneisses occurs, in at least some cases, after the peak of metamorphism (Lamb, 1990; Lamb and Morrison, 1997; Lamb *et al.*, 1987; Morrison and Valley, 1988; Morrison and Valley, 1991).

Various methods have been employed in an effort to determine if fluid inclusions contain samples of the peak metamorphic fluid. For example, the density of the fluid in any given fluid inclusion can be compared with the density that a fluid of equivalent composition would have at peak metamorphic P-T. Agreement between these two densities has sometimes been used as an argument in favour of peak metamorphic fluid inclusion formation (Hansen *et al.*, 1984a; Hansen *et al.*, 1984b). However, any given density value is consistent with a wide range of P-T conditions (i.e., all P-T values along the appropriate isochore). For example, Lamb and Morrison (1997) document the presence of CO_2-rich fluid inclusions in retrograde veins that crosscut the Archean Shawmere anorthosite. These CO_2-rich fluid inclusions are clearly retrograde, yet some of these fluid inclusions have densities consistent with peak metamorphic pressures and temperatures. Furthermore, CO_2-rich fluid inclusions from a number of granulite facies terranes have highly variable fluid densities suggesting that, in many instances, the density of the fluid in the inclusion was modified, or the inclusions were formed, during retrograde metamorphism (Lamb, 1990).

Textural evidence has also been used to differentiate between inclusions formed during crystal growth (primary inclusions) and inclusions formed by healing of small fractures that occur after crystal growth (secondary inclusions). However, many CO_2-rich fluid inclusions in granulite facies samples clearly formed along healed fractures, and, therefore, textural evidence for formation of fluid inclusions during granulite facies metamorphism is often ambiguous (Lamb, 1990).

Another approach that may be used to determine the timing of fluid inclusion formation relative to the peak of granulite facies metamorphism is to compare the compositions of the fluids trapped in inclusions with peak metamorphic fluid compositions as determined using mineral equilibria (Lamb *et al.*, 1987). Devolatilization equilibria, when too many rocks from granulites facies terranes, indicate that values of aH_2O are typically less than 1. These aH_2O determinations are generally based on dehydration reactions, often involving mica or amphibole (e.g., Lamb and Valley, 1988), and these equilibria provide no direct information concerning CO_2 activities. Carbonate-bearing rocks, such as marbles or calc-silicates, may contain all the phases necessary to determine aCO_2 (i.e., CO_2-buffering equilibria), and so it is sometimes possible to use samples from these rock types to constrain activities of CO_2 (Lamb *et al.*, 1987; Valley and Essene, 1980). Scapolite equilibria have been used to evaluate the activities of fluids other than H_2O in granulite facies feldspathic gneisses. For example, Moecher (1993; 1991) applied scapolite equilibria to determine peak metamorphic aCO_2, and found values of CO_2 activities ranging from 0.1 to >1.0. Thus, some values of aCO_2 are consistent with the presence of a CO_2-rich fluid phase, while other values are not. Furthermore, when these results are combined with determinations of the carbon and oxygen isotopic compositions of scapolite, they indicate that CO_2 in the lower crust occurs only locally, and, therefore, did not act in a pervasive manner (Moecher, 1993; Moecher and Essene, 1991; Moecher *et al.*, 1994). The isotopic compositions of carbon in cordierite also indicates that many granulite facies terranes did not experience pervasive infiltration of large amounts of CO_2 (Vry *et al.*, 1990; Vry *et al.*, 1988).

Lamb and Valley (1984; 1985) also applied mineral equilibria to evaluate the activities of fluids other than H_2O in granulite facies quartzo-feldspathic gneisses. They used calculations of fluid speciation in the C-O-H system (*see* also French, 1966; Ohmoto and Kerrick, 1977) to demonstrate that values of oxygen fugacity (fO_2) approximately 0.3 log bars more reducing than the C + CO_2 oxygen-buffer are not consistent with a CO_2-rich fluid. Lamb and Valley (1984; 1985) documented that a wide range of fO_2 values have been determined for samples from various granulite terranes, and some values of fO_2 are sufficiently reducing such that they are not consistent with a CO_2-rich peak metamorphic fluid (also see Valley *et al.*, 1990). In more general terms, it may be tempting to use calculations of fluid speciation in the C-O-H system to indicate that relatively oxidizing values of fO_2 (e.g., with approximate values $\geq$ the C + CO_2 oxygen-buffer) are consistent with CO_2-rich fluids (e.g., Harlov, 2000). Similarly, it might be argued that relatively reducing values of fO_2 (e.g., with approximate values > 1 log unit more reducing than the C + CO_2 oxygen-buffer) are consistent with CH_4-rich fluids. However, such statements ignore the possibility of fluid-absent peak metamorphic conditions, or the possibility of non-C-O-H species in the peak metamorphic fluid. Consequently, calculations of fluid speciation in the C-O-H system can better be used to demonstrate that relatively reducing conditions are inconsistent with CO_2-rich fluids and relatively oxidizing conditions are inconsistent with CH_4-rich fluids. Furthermore, intermediate values of fO_2 that are approximately within ± 0.6 log units of the "H_2O maximum" described by Ohmoto and Kerrick (1977), when taken in conjunction with low values of aH_2O (e.g., aH_2O < about 0.2) characteristic of many granulites, provide evidence for fluid-absent metamorphism (Lamb and Valley, 1984; Lamb and Valley, 1985).

Occurrence of Carbonates in Granulite Facies Gneisses

It is difficult to quantify the percentage of granulite facies feldspathic gneisses that contain carbonate minerals. However, a review of 40 studies that describe the presence of CO_2-rich fluid inclusions in granulite facies feldspathic gneisses revealed that 15 of these studies also report the presence of carbonate minerals, suggesting carbonates may be relatively common. Table 1 lists these 15 studies, including the location and the peak metamorphic P-T estimates. The peak metamorphic mineralogy is also listed, however, for a more detailed discussion of the mineralogy in each sample, particularly for those studies in which a suite of samples are described, the original reference should be consulted. Table 1 also notes which carbonate phase is present, and also if the carbonate was interpreted to be part of the peak metamorphic mineralogy (primary), or if the author(s) of the study in question indicated that the carbonate was formed after the peak of granulite facies metamorphism (secondary).

In general, the amounts of carbonates reported in the rocks from the studies listed in Table 1 are rather small, and they often occur as small inclusions within other phases. Morrison and Valley (1991), found that cathodoluminesence was useful in locating small amounts of carbonate minerals in samples from the Adirondack granulite terrane. This work suggests that small amounts of carbonates may be more prevalent in granulite facies gneisses than generally recognized, and Touret (1985) argues they are ubiquitous. In many cases the carbonates are intimately associated with CO_2-rich fluid inclusions, such that it is difficult to decouple the formation of the fluid inclusions from the formation of the carbonate minerals (e.g., Hansen *et al.*, 1984a; Touret and Hansteen, 1988). In some cases carbonates may also occur as daughter minerals in CO_2-rich fluid inclusions (Table 1).

Table 1: *Examples of granulite facies gneisses containing carbonates.*

Author(s), Year, & Location	Peak P-T		Peak Metamorphic Mineralogy[1]								Carbonate Mineralogy		
	T(°C)	P-kb	Qtz	Pl	Opx	Cpx	Grt	Bt	Amp	Other	Min[1]	Type[1,2]	P/S[3]
Bolder-Schrijver *et al.* (2000) Sri Lanka	830	9			X		X	X		Si, Sa, Crd	Mgs	D	P
DeMaesschalck *et al.* (1991) Weddagala, Sri Lanka	800	8	X	X	X		X	X	X	Ap, Gr, Kfs	U	I (Ap)	P
Elvevold, & Andersen (1993) Øksfjord, Norway	725	5.5	X	X	X		X	X		Kfs	Cal?	D	P
Hansen *et al.* (1984a) S. Karnataka, India	700-800	5-8	X	X							U	I (Pl)	P
Herms & Schenk (1992) Calabria, Italy	700	6.5	X	X			X	X		Gr, Crd	Cal, Mgs	I (Crd, Grt) D	S?
Hoefs & Touret (1975) Bamble, S. Norway	750	7	X	X			±X	X			U		S
Lamb *et al.* (1991) Adirondacks, N.Y., U.S.A.	700-775	7-8	X			X	X			Cal, Wo	U	D	S
Morrison & Valley (1991) Adirondacks, N.Y., U.S.A.	750	7.5	X	X	X	X	X	X	X	Il, Ap	Cal	I (Qtz, Pl)	S
Munyanyiwa *et al.* (1993) Magondi, Zimbabwe	700-800	5-7	X	X	X	±X	X	X	±X	±Kfs	U	I	P?
Rudnick *et al.* (1984) Kapuskasing, Ont., Canada	700-800	6-8	X	±X			±X	±X	X	±Ap	U	D, I (Qtz)	P?
Srikantappa *et al.* (1992) Nilgiri, India	750	7-9.5	X	X	X		X	X	±X	Il, ±Kfs	Cal, Mgs	D	P
Touret (1985) Bamble, S. Norway	750	7	X	X	X		±X		X	Il	Ank, Cal?	I (Pl)	P
Touret & Hansteen (1988) Doddabetta, SW India	800	8	X	X	X		X	X			Ank, Cal, U	I (Grt), D	P
Vry & Brown (1991) Pikwitonei, Manit., Canada	750-830	7-8	X	X	X	±X	X			Mag, Il, Kfs	Ank, Cal	I (Grt), D	P, S
Wilmart *et al.* (1991) Rogaland, Norway	800-850	6-7.5	X	X	X	X				Mag, Il	Cal	D	P?, S

1. Amp = amphibole, Ank = ankerite, Ap = apatite, Bt = Biotite, Cal = calcite, Cpx = clinopyroxene, Crd = cordierite, Gr = graphite, Gt = garnet, Il = ilmenite, Kfs = K-feldspar, Mag = magnetite, Mgs = magnesite, Opx = orthopyroxene, Pl = plagioclase, Qtz = quartz, Sa = Sapphirine, Si = sillimanite, Wo = wollastonite. U = unknown.
2. I = Inclusion; the carbonate occurs as an inclusion in another mineral, the identity of the host phase(s) are listed within the parentheses. D = Daughter; the carbonate occurs as a daughter mineral within fluid inclusions.
3. P = primary; S = secondary.

Carbonate Equilibria in Granulites

It is possible to derive a variety of chemical reactions involving CO_2, carbonates and various minerals common to many granulite facies rocks. Thus, the presence of carbonates in rocks from the granulite-facies may allow direct determination of CO_2 activities. Accurate fluid activity calculations require chemical analyses of the phases involved in any given fluid-buffering reaction. However, in many of the studies that describe the occurrence of carbonates in feldspathic gneisses from granulite facies terranes, the carbonate mineral has not been identified, or, if identification was possible, detailed compositional information is not available. If information on the composition of the carbonate phase is lacking, a range of carbonate compositions may be employed to evaluate the stability of the carbonate relative to inferred peak pressures and temperatures. Furthermore, in cases where chemistries of the silicate phases are not known, examination of end-member reactions may still place limits on carbonate stability or CO_2 activities. The following discussion focuses on equilibria involving quartz and pyroxenes as these are some of the most common silicates in rock samples described in the 15 studies listed in Table 1.

Anovitz and Essene (1987) and Davidson (1994) have reviewed existing experimentally derived data on the stability of carbonate phases in the $CaCO_3$-$MgCO_3$-$FeCO_3$ system to formulate thermodynamic models and examine phase relations in this ternary system. Although some differences exist between these two thermodynamic models, some general conclusions can be drawn about the extent of solid solution in the ternary Ca-Mg-Fe carbonate system. At temperatures of approximately 700 to 750°C miscibility gaps exist between calcite and dolomite, dolomite and magnesite, and calcite and siderite. There is complete Fe-Mg solid solution between siderite and magnesite, and the maximum $CaCO_3$ content is less than approximately 20 mole %. There is nearly complete solid solution between dolomite and ankerite, although end-member ankerite may not be stable at a temperature of approximately 750°C as the maximum Fe-content may be approximately 90 mole % $FeCO_3$ (Davidson, 1994). At $T \approx 700$ to 800°C calcite may contain as much as approximately 20 mole % $MgCO_3$ and approximately 40 mole % $FeCO_3$ (Anovitz and Essene, 1987; Davidson, 1994). Given these constraints on the compositions of phases in the ternary $CaCO_3$-$MgCO_3$-$FeCO_3$ system at relatively high temperatures, evaluation of the stability of carbonates in granulites can largely be accomplished by examining the stability of calcite, dolomite-ankerite solid solutions, and magnesite-siderite solid solutions.

In the following discussion, the stability of carbonates at granulite facies pressures and temperatures will be evaluated primarily by using end-member compositions of carbonates and co-existing silicate phases. However, this discussion also includes an examination of the effects of solid-solution in the carbonates given the limits on carbonate composition outlined above. Determination of the P-T stability of various equilibria involving carbonates was performed using the computer program THERMOCALC that employs the internally consistent database of Holland and Powell (1990).

Calcite

Many of the carbonate-bearing granulite facies samples described in the literature contain quartz and orthopyroxene. Thus, the following end-member equilibria are potentially important:

$$\text{Calcite + Enstatite + Quartz = Diopside} + CO_2 \quad (1)$$

$$\text{Calcite + Ferrosilite + Quartz = Hedenbergite} + CO_2 \quad (2)$$

The P-T location of these two end-member equilibria are shown in Fig. 1 at $XCO_2 = 1.0$. Many granulites contain hydrous phases such as amphibole or mica, and, therefore, if a free fluid phase were present during metamorphism it must contain some H_2O. For equilibria (1) and (2), addition of H_2O to the fluid phase would stabilize clinopyroxene relative to calcite + orthopyroxene + quartz.

The results of these calculations show that calcite should not co-exist with quartz and end-member orthopyroxenes at granulite facies P-T. Evaluation of calcite stability should also include the effects of solid solution in various phases on the stability of equilibria (1) and (2). For example, most pyroxenes in granulites are not Fe or Mg end-members, but rather have some intermediate Fe/Mg ratio and so the stability of carbonates in equilibrium with two pyroxenes and quartz depends, in part, upon the partitioning of Fe and Mg between co-existing ortho- and clinopyroxene. This partitioning is a function of temperature and provides the basis for various formulations of the two-pyroxene geothermometer. However, natural pyroxenes metamorphosed under the same, or very similar, P-T conditions exhibit significant variability in values of the distribution coefficient $\left(K_{DMg\text{-}Fe}^{opx\text{-}cpx}\right)$, presumably due to the presence of non-quadrilateral components. Fig. 1 shows the stability of the equilibria

$$\text{calcite + orthopyroxene + quartz = diopside + } CO_2 \quad (3)$$

assuming an intermediate orthopyroxene composition (En_{50}) and a $K_{DMg\text{-}Fe}^{opx\text{-}cpx}$ of 1.8, a value consistent with compositions of granulite facies pyroxenes (Bohlen and Essene, 1980; Kretz, 1961). The results of this calculation suggest that Fe-Mg solid solution in pyroxenes will not operate to stabilize calcite + orthopyroxene + quartz at granulite facies P-T.

Solid solution in calcite should increase the stability of calcite + orthopyroxene + quartz relative to clinopyroxene + CO_2. In order to determine the highest temperature at which calcite + orthopyroxene + quartz may be stable the Fe end-member reaction was employed as a starting point as this equilibria lies at a higher temperature than the analogous Mg end-member equilibria (Fig. 1). At granulite facies temperatures calcite can accommodate more Fe than Mg (*see* previous discussion). Thus, Fe end-member pyroxenes and a calcite activity of 0.6 (based on a maximum $FeCO_3$ content of 40 mole % and assuming ideal mixing) should yield the highest temperature at which calcite + orthopyroxene + quartz is stable, and even in this extreme case there is very little overlap between the stability of calcite + orthopyroxene + quartz and granulite facies P-T (Fig. 1).

Although calcite + orthopyroxene + quartz is apparently not a stable assemblage at granulite facies P-T, many feldspathic granulites do not contain orthopyroxene. The assemblage calcite + quartz, which reacts to form wollastonite + CO_2 at relatively high temperatures, may be stable over a wide range of P-T conditions common to many granulite facies terranes (Greenwood, 1967; Valley *et al.*, 1990; Zeigenbein and Johannes, 1974). Thus, calcite could occur in many feldspathic granulites (e.g., those with no orthopyroxene). Furthermore, wollastonite is common in calcareous granulites, and, when it occurs with calcite + quartz, it is possible to determine values of aCO_2 (e.g., Valley *et al.*, 1990).

Dolomite-Ankerite

A variety of equilibria involving dolomite or ankerite and minerals common to many of the carbonate-bearing granulite facies rocks may be written. Quartz is common to most of these studies (Table 1), and so the equilibria

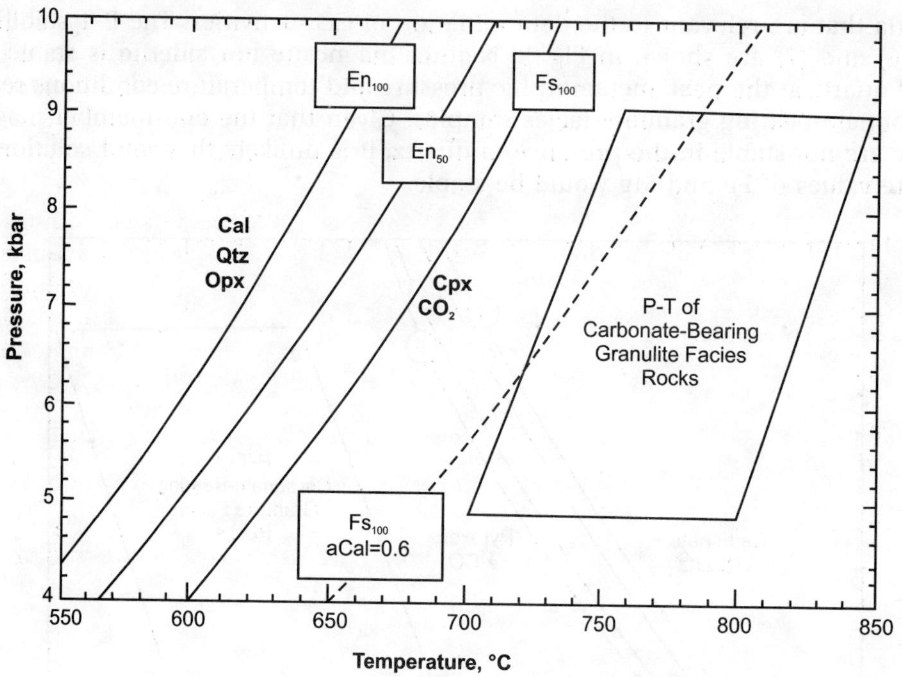

Fig. 1. Calculated stability of the equilibria calcite + quartz+ orthopyroxene = clinopyroxene + CO_2. Two end-member equilibria are shown, Calcite + Enstatite + Quartz = Diopside + CO_2 (En_{100}), and Calcite + Ferrosilite + Quartz = Hedenbergite + CO_2 (Fs_{100}), as is the position of this equilibria for one set of intermediate pyroxene compositions (En_{50}). The upper limit of the assemblage calcite + quartz + orthopyroxene was determined assuming Fe end-member pyroxenes and a calcite activity of 0.6 (dashed curve, see text), and, even in this extreme case, there is little overlap between the stability of calcite + quartz + orthopyroxene and granulite facies P-T. All calculations were performed using THERMOCALC (*see* text).

$$\text{Dolomite + Quartz = Diopside + } CO_2 \quad (4)$$

$$\text{Ankerite + Quartz = Hedenbergite + } CO_2 \quad (5)$$

are applicable to many of the rocks in question. The P-T stability of equilibria of equilibria (4) and (5) are shown in Fig. 2, assuming $XCO_2 = 1$. Fluid compositions in which $XCO_2 < 1$ will stabilize the products of equilibria (4) and (5) and, therefore, Fig. 2 shows the maximum stability of dolomite + quartz or of ankerite + quartz. Thus, these end-member carbonates are not stable at granulite facies pressures and temperatures in the presence of quartz, indicating that intermediate carbonates compositions would also react with quartz to produce clinopyroxene at relatively low temperatures.

Magnesite-Siderite

If the carbonate phase contained in a rock from the granulite facies is magnesite or siderite, and the rock contains quartz, then

$$\text{Magnesite + Quartz = Enstatite + } CO_2 \quad (6)$$

and

$$\text{Siderite + Quartz = Ferrosilite + } CO_2 \quad (7)$$

are equilibria that are relevant to the determination of CO_2 activities. The P-T stabilities of equilibria (6) and (7) are shown in Fig. 2. Neither magnesite nor siderite is stable in the presence of quartz at the peak metamorphic pressure and temperature conditions reported for the carbonate-bearing granulite-facies samples. Given that the end-member magnesite and siderite are not stable in the presence of quartz, it is unlikely that solid solutions with intermediate values of Fe and Mg would be stable.

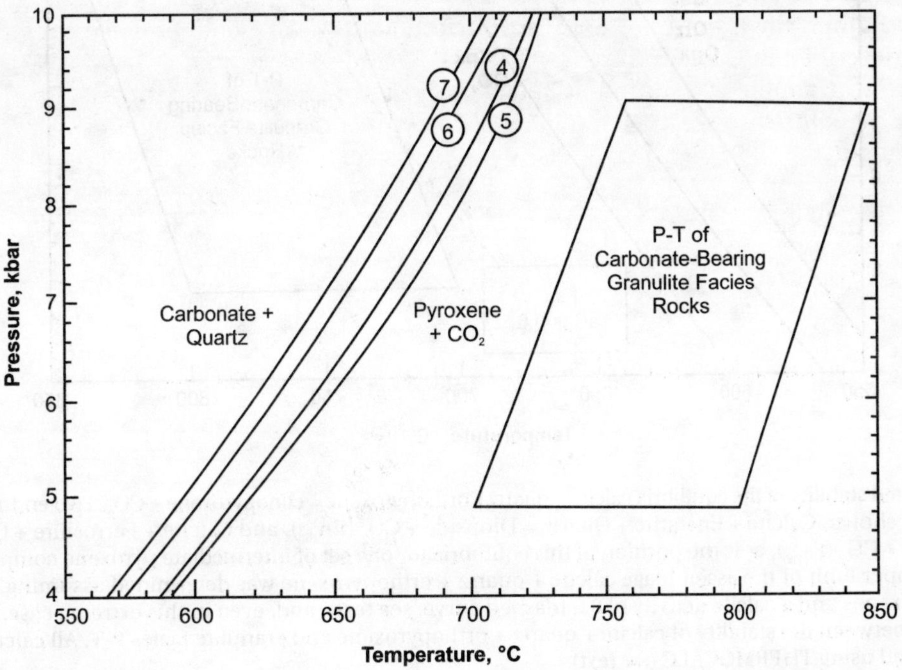

Fig. 2. Calculated stability of the four end-member carbonates, dolomite, ankerite, magnesite, and siderite, in the presence of quartz. The products of all four equilibria are either clinopyroxene or orthopyroxene and CO_2. The numbers on each curve corresponds to numbers assigned to each equilibria in the text; e.g., Dolomite + Quartz = Diopside + CO_2 is equilibria (4).

CARBONATE STABILITY: GENERAL CONSIDERATIONS

Equilibria (1) through (7) were chosen primarily to examine CO_2-buffering equilibria between carbonates and other minerals common to those feldspathic granulites that contain small amounts of carbonates (e.g., *see* Table 1). These reactions [equilibria (1) through (7)] were not chosen to provide a definitive examination of the stability of carbonates in feldspathic granulites. For example, application of equilibria (1) through (3) is not meant to imply that calcite + orthopyroxene has a stability field at metamorphic grades that are below the granulite facies. Instead equilibria (1) through (3) is probably metastable relative to equilibria (4) through (6) in the crust because the stability of calcite + orthopyroxene is limited by the reaction

calcite + orthopyroxene = dolomite + clinopyroxene (8)

(personal communication with Eric Essene). The relations between equilibrium (8) and other equilibria are shown in Fig. 3, which shows that, for the Mg end-member system, equilibrium (1) is metastable relative to equilibrium (8). However, the phase relations illustrated in Fig. 3, provide additional evidence that calcite is not stable in orthopyroxene-bearing feldspathic granulites under typical granulite facies P-T conditions.

It is important to emphasize that the preceding discussion is not meant to imply that carbonates are never stable in feldspathic gneisses under granulite facies pressure-temperature conditions. Rather, the important point is that carbonate-bearing mineral assemblages that have been described in the literature (*see* Table 1) include phases that are apparently not in equilibrium at granulite P-T. In addition to the examples already provided, the reported occurrence of co-existing calcite and magnesite (Table 1) is a further indication of the

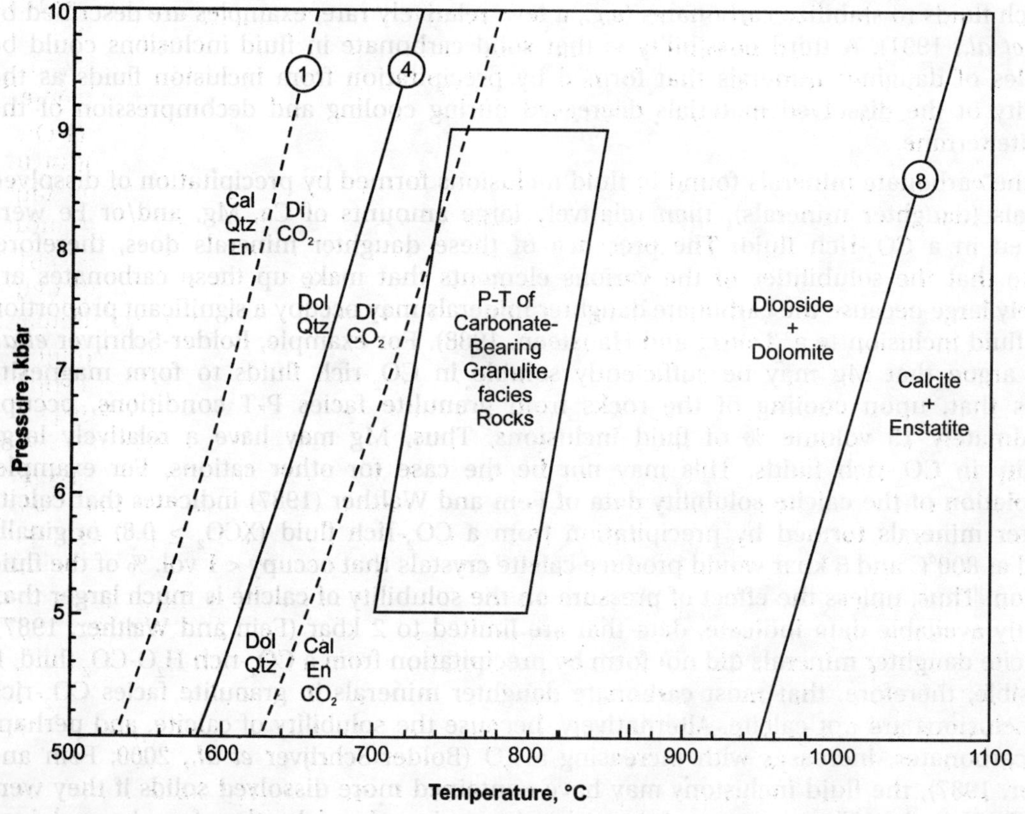

Fig. 3. Calculated stability of various equilibria involving calcite + dolomite + enstatite + diopside + CO_2. These are Mg end-member phases and, for reactions involving a fluid, this fluid is pure CO_2. Solid curves are stable reactions, whereas the dashed lines indicate metastable equilibria. Equilibrium (8), diopside + dolomite = calcite + enstatite, lies at relatively high temperatures and indicates that calcite + enstatite is not stable at pressures and temperatures typical of carbonate-bearing granulite facies gneisses (Table 1). Although equilibrium (1), calcite + quartz + enstatite = diopside + CO_2, is metastable, the assemblage calcite + quartz + enstatite is not in equilibrium at the P-T of interest (*see* text).

presence of minerals that are not in equilibrium, as calcite and magnesite should react to form dolomite at granulite P-T.

APPLICATION OF CARBONATE EQUILIBRIA TO GRANULITES: CARBONATES IN FLUID INCLUSIONS

The carbonates in granulite facies feldspathic gneisses may occur as individual grains that were (apparently) solid during the peak of granulite facies metamorphism, however, in some cases carbonates occur within fluid inclusions (*see* Table 1). Carbonates, or other solid phases, located within fluid inclusions may form in a variety of ways. For example, the carbonates may have been solid at peak metamorphic temperatures and pressure, and have formed via accidental trapping in fluid inclusions. A second mechanism to explain the presence of solid carbonates in fluid inclusions is through mixing of CO_2-rich fluids with H_2O + salt-rich fluids. In this case the CO_2 may react with cations originally dissolved in the H_2O-rich fluids to stabilize carbonates (*e.g.*, a few, relatively rare, examples are described by Lamb *et al.*, 1991). A third possibility is that solid carbonate in fluid inclusions could be examples of daughter minerals that formed by precipitation from inclusion fluids as the solubility of the dissolved materials decreased during cooling and decompression of the granulite terrane.

If the carbonate minerals found in fluid inclusions formed by precipitation of dissolved materials (daughter minerals), then relatively large amounts of Ca, Mg, and/or Fe were dissolved in a CO_2-rich fluid. The presence of these daughter minerals does, therefore, indicate that the solubilities of the various elements that make up these carbonates are relatively large because the carbonate daughter minerals may occupy a significant proportion of the fluid inclusion (e.g. Touret and Hansteen, 1988). For example, Bolder-Schrijver *et al.* (2000) argue that Mg may be sufficiently soluble in CO_2-rich fluids to form magnesite crystals that, upon cooling of the rocks from granulite facies P-T conditions, occupy approximately 15 volume % of fluid inclusions. Thus, Mg may have a relatively large solubility in CO_2-rich fluids. This may not be the case for other cations. For example, extrapolation of the calcite solubility data of Fein and Walther (1987) indicates that calcite daughter minerals formed by precipitation from a CO_2-rich fluid ($XCO_2 \geq 0.8$) originally formed at 800°C and 8 kbar would produce calcite crystals that occupy < 1 vol. % of the fluid inclusion. Thus, unless the effect of pressure on the solubility of calcite is much larger than currently available data indicate, data that are limited to 2 kbar (Fein and Walther, 1987), the calcite daughter minerals did not form by precipitation from a CO_2-rich H_2O-CO_2 fluid. It is possible, therefore, that most carbonate daughter minerals in granulite facies CO_2-rich fluid inclusions are not calcite. Alternatively, because the solubility of calcite, and perhaps other carbonates, increases with increasing aH_2O (Bolder-Schrijver *et al.*, 2000; Fein and Walther, 1987), the fluid inclusions may have contained more dissolved solids if they were originally H_2O-rich. The presence of these daughter minerals might, therefore, be explained by preferential loss of H_2O (e.g., Bakker and Jansen, 1991; Bakker and Jansen, 1994). It is likely that a complete understanding of carbonate daughter minerals found in CO_2-rich granulite facies fluid inclusions will require more data on the solubility of carbonates in CO_2-rich fluids at elevated temperatures and pressures. Furthermore, information of the equilibrium partitioning of Fe and Mg between CO_2-rich fluids and minerals that host fluid inclusions (e.g., garnet) may provide further information that may restrict the conditions of carbonate formation.

Bolder-Schrijver applied equilibria (6) by determining the activity of $MgCO_3$ based on the amount of Mg dissolved in a CO_2-rich fluid inclusion. In other words, by assuming ideal mixing of cations that were apparently dissolved in a CO_2-rich fluid to calculate the activity of magnesite, they were able to apply mineral equilibria to constrain peak metamorphic aCO_2. However, application of mineral equilibria, including carbonate equilibria (1) through (7) described previously, is probably most readily applied to solid phases, as described in the following section.

APPLICATION OF CARBONATE EQUILIBRIA TO GRANULITES: SOLID CARBONATE GRAINS

Carbonate minerals are a common constituent of many granulite facies feldspathic gneisses (Table 1 and see the previous discussion). In many cases the carbonates in granulites are thought to be secondary, however, there are cases in which solid carbonate inclusions are thought to have been present during the peak of granulite facies metamorphism (Table 1). However, application of equilibria (1) through (7) show that, for a variety of carbonate species and carbonate compositions, these carbonates are not stable at peak P-T (i.e., not in equilibrium with other minerals in the same rock). This discussion of carbonate stability, however, has focused largely on end-member equilibria. A more definitive determination of carbonate stability is possible for any given sample from the granulite facies if the compositions of all co-existing phases are known. Unfortunately, in most studies that report the presence of carbonates in granulite facies feldspathic gneisses, the compositions of all pertinent phases have not been reported. However, in a careful study of samples from Doddabetta, S. India, Touret and Hansteen (1988) describe inclusions of Fe-rich Dolomite in garnets which also contain fluid inclusion, and their published chemical analyses for Garnet, Orthopyroxene, and Plagioclase provide the basis to calculate the stability of the ankerite. The stability of the reaction

$$\text{Almandine} + \text{Ankerite} + \text{Quartz} = \text{Ferrosilite} + \text{Anorthite} + CO_2 \quad (9)$$

was calculated using their analyses and various activity models for the silicate phases. The composition of the Fe-rich Dolomite is not reported, and, therefore, calculations were performed using ankerite activities that ranged from 0.2 to 1.0 (Fig. 4). Equilibrium (9) lies approximately 125 to 180°C below peak T at peak P. Thus, the carbonate is most likely not in equilibrium with the garnet and quartz under peak P-T conditions.

Examination of the stability of ankerite in samples from Doddabetta, S. India supports the general conclusion, based on equilibria (1) through (8), that solid carbonate grains are not in equilibrium with quartz and orthopyroxene under typical granulite facies pressure-temperature conditions. Thus, in many cases, the carbonates are likely not to be part of the peak metamorphic mineral assemblage.

SUMMARY AND DISCUSSION

Calculation of the stability of co-existing Ca-Fe-Mg carbonate, orthopyroxene and quartz indicates that these minerals are not in equilibrium with one another under granulite facies pressure-temperature conditions. These calculations focus largely on end-member equilibria, although calculations performed to account for variations in major element contents (i.e., Ca-Fe-Mg) of pyroxenes and carbonates indicate that solid solution generally does not enhance the stability of carbonates sufficiently to permit equilibrium between carbonates and quartz + orthopyroxene at granulite P-T. Thus, it is unlikely that solid carbonates are

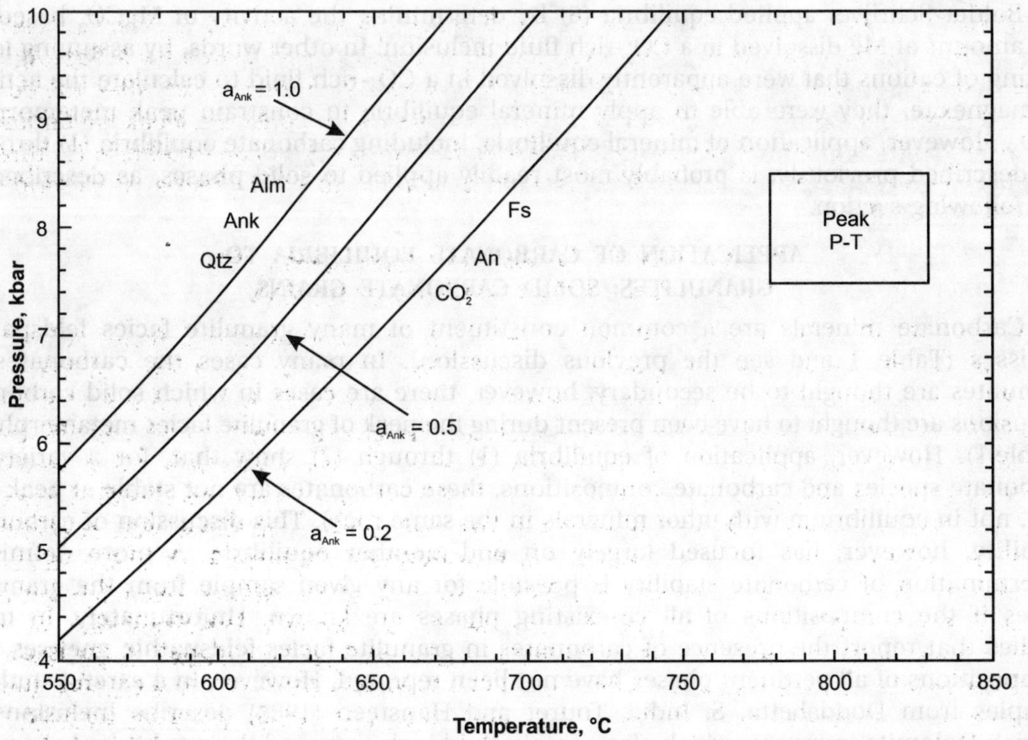

Fig. 4. The stability of Almandine + Ankerite + Quartz = Ferrosilite + Anorthite + CO_2 is shown for three activities of $CaFe(CO_3)_2$ (1.0, 0.5 and 0.2). The activities of the silicate phases were calculated at peak metamorphic P-T using the activity models of Berman (1990) for garnet, Furhman and Lindsley (1988) for plagioclase, and an ideal model for orthopyroxene ($a = X_{Fe}^{M1} \bullet X_{Fe}^{M2}$ using compositions published by Touret and Hansteen (1988). Because the information on the composition of the Fe-bearing dolomite is semi-quantitative a variety of $CaFe(CO_3)_2$ activities have been employed for these calculations, however, even for this wide range in activities, the carbonate is not in equilibrium with garnet and quartz at peak metamorphic P-T.

part of the peak metamorphic equilibrium assemblage in most of the rocks where carbonates are reported (Table 1), although more analyses of carbonates and co-existing phases in quartzo-feldspathic granulite facies gneisses are needed. It is also important to emphasize that these conclusions are applicable to carbonates that were solid at the peak of granulite facies metamorphism. The presence of carbonates within CO_2-rich fluid inclusions (daughter minerals?) has sometimes been taken to indicate that CO_2-rich fluid inclusions contain appreciable dissolved material that precipitates during retrograde metamorphism. However, a greater understanding of the significance of these inclusions may require more experimental information on the solubilities of various elements (e.g., Ca, Mg, and Fe) in CO_2-rich fluids.

Solid carbonate inclusions are common in granulite facies feldspathic gneisses, and these carbonates are sometimes interpreted to have been present at the peak of granulite facies metamorphism. However, if solid carbonate is not stable at typical granulite facies P-T conditions, how is it possible to account for the presence of carbonates in granulite facies gneisses containing quartz and orthopyroxene? In some cases carbonates may have been present prior to the peak of metamorphism and, given that they often occur as solid inclusions in other phases (e.g., garnet), they may have been armored against reaction with other phases during the peak of metamorphism. This assumes that the phase(s) which host the carbonates grew prior to attainment of granulite facies conditions. Alternatively, the samples interacted with retrograde fluids, perhaps very small amount of retrograde fluids, and these fluids contained some CO_2 that reacted with the rocks to stabilize the small amounts of carbonates present in these rocks. This explanation is consistent with textural evidence described in a few of the 15 studies tabulated in Table 1, as some carbonates are clearly secondary.

Whether the carbonates are prograde or retrograde phases, these minerals do not provide information regarding peak metamorphic conditions. Thus, the equilibria between carbonates and silicates described in this paper do not support the presence of a CO_2-rich peak metamorphic fluid, nor do they preclude the presence of such a fluid. It might be argued that many of the carbonates found in feldspathic granulite facies gneisses formed by retrograde reaction between a CO_2-rich peak metamorphic fluid and the silicate phases. This mechanism would readily explain carbonates present along grain boundaries; however, micro-cracking would be required to explain the presence of carbonate inclusions in other phases such as garnet and plagioclase. Also, it is difficult to determine if the fluid that reacted with the rock was a pre-existing metamorphic fluid, or if it were a fluid that was introduced into the rock during retrograde metamorphism.

While the results of the calculations presented in this paper do not constrain the compositions of peak metamorphic fluids, they do have important implications for the presence of CO_2-rich fluid inclusions that are associated with the carbonate inclusions. Some authors report a close association between fluid inclusions and carbonate minerals. For example Hansen *et al.* (1984a) show that CO_2-rich fluid inclusions and carbonate inclusions occur along the same trail, indicating that the fluid inclusions and carbonate inclusions formed simultaneously (*see* their Fig. 3b). Touret and Hansteen (1988) also report a close association between carbonate minerals and CO_2-rich fluid inclusions, as garnets that contain CO_2-rich fluid inclusions also contain carbonate inclusions. The carbonate inclusions, and the associated fluid inclusions, both have "primary" textures consistent with formation of the inclusions during the peak of granulite facies metamorphism. However, because the carbonate is not in equilibrium with the peak metamorphic mineral assemblage it is inadvisable to use fluid inclusions that are intimately associated with these carbonates to determine peak metamorphic fluid compositions.

In summary, solid carbonate is probably not in equilibrium with quartz and orthopyroxene, an assemblage reported in many studies (Table 1), at peak granulite facies pressures and temperatures. Consequently, it is often not possible to use the presence of small amounts of carbonates to determine peak metamorphic fluid compositions, and it is also inadvisable to use fluid inclusions that are intimately associated with these carbonates to determine peak metamorphic fluid compositions. It is interesting to note that some of the CO_2-rich fluid inclusions that are associated with carbonates have other characteristics, such as textures and fluid densities, that have sometimes been taken as indicative of primary, or peak metamorphic, fluid inclusions. Thus, calculation of the stability of carbonates in quartzo-feldspathic granulite facies gneisses demonstrates that the identification of those fluid inclusions that actually contain peak metamorphic fluids is difficult. Furthermore, some inclusions that were previously thought to contain peak metamorphic fluids may have been trapped or modified after the peak metamorphic event.

ACKNOWLEDGMENTS

An early version of this manuscript benefited from reviews by Eric Essene and Virginia Sisson. Part of this work was conducted while the author was on Faculty Development Leave at the University of Utrecht, Netherlands, and special thanks is extended to Dr. Martyn Drury and others in the Aardwetenschappen at Utrecht for their support. The financial support of the National Science Foundation (EAR-9405629), the Nederlandse Organisatie voor Wetenschappelijk Onderzoek, and Texas A&M University is acknowledged and appreciated.

REFERENCES

Anovitz, L.M. & Essene, E.J., 1987. Phase equilibria in the system $CaCO_3$-$MgCO_3$-$FeCO_3$. *Jour. Petrology,* **28**, 389-414.

Audétat, A. & Günther, D., 1999. Mobility and H_2O loss from fluid inclusions in natural quartz crystals. *Contrib. Mineral. Petrol.*, **137**, 1-14.

Bakker, R.J. & Jansen, J.B.H., 1991. Experimental post-entrapment water loss from synthetic CO_2-H_2O inclusions in natural quartz. *Geochimica et Cosmochimica Acta,* **55**, 2215-30.

Bakker, R.J. & Jansen, J.B.H., 1994. A mechanism for preferential H_2O leakage from fluid inclusions in quartz, based on TEM observations. *Contrib. Mineral. Petrol.*, **116**, 7-20.

Berman, R.G., 1990. Mixing properties of Ca-Mg-Fe-Mn garnet. *American Mineralogist,* **75**, 328-44.

Bhattacharya, A. & Sen, S.K., 1986. Granulite metamorphism, fluid buffering, and dehydration melting in the Madras Charnockites and Metapelites. *Jour. Petrology,* 27, 1119-41.

Bohlen, S., 1987. Pressure-temperature-time paths and a tectonic model for the evolution of granulites. *Jour. Geology,* **95**, 617-32.

Bohlen, S.R. & Essene, E.J., 1980. Evaluation of coexisting garnet-biotite, garnet-clinopyroxene, and other Mg-Fe exchange thermometers in Adirondack granulites. *Geological Society of America Bulletin,* **91**, 685-719.

Bolder-Schrijver, L.J.A., Kriegsman, L.M. & Touret, J.L.R., 2000. Primary carbonate/CO_2 inclusions in sapphirine-bearing granulites from central Sri Lanka. *Jour. Metamorphic Geol.*, **18**, 259-69.

Clemens, J.D. & Vielzeuf, D., 1987. Constraints on melting and magma production in the crust. *Earth Planet. Science Letters,* **86**, 287-306.

Davidson, P.M., 1994. Ternary iron, magnesium, calcium carbonates: A thermodynamic model for dolomite as an ordered derivative of calcite-structure solutions. *Am. Mineralogist,* **79**, 332-39.

DeMaesschalck, A.A., Touret, J.L.R., Maaskant, P. & K., D., 1991. Petrology and fluid inclusions in garnetiferous gneisses and charnockites from Weddagala (Ratnapua District, Sri Lanka). *Jour. Geology,* **99**, 443-56.

Eggler, D.H. & Kadik, A.A., 1979. The system $NaAlSi_3O_8$-H_2O-CO_2 to 20 kbar pressure: I. Compositional and thermodynamic relations of liquids and vapors coexisting with albite. *Am. Mineralogist,* **64**, 1036-48.

Elvevold, S. & Andersen, T., 1993. Fluid evolution during metamorphism at increasing pressure: carbonic- and nitrogen-bearing fluid inclusions in granulites from Øksfjord, north Norwegian Caledonides. *Contrib. Mineral. and Petrol.*, **114**, 236-46.

Fein, J. & Walther, J., 1987. Calcite solubility in supercritical CO_2-H_2O fluids. *Geochemica et Cosmochimica Acta,* **51**, 1665-73.

French, B.M., 1966. Some geological implications of equilibrium between graphite and a C-O-H gas phase at high temperatures and pressures. *Rev. Geophys,* **4**, 223-53.

Furhman, M.L. & Lindsley, D.H., 1988. Ternary-feldspar modeling and thermometry. *Am. Mineralogist,* **73**, 201-15.

Fyfe, W.S., 1973. The granulite facies, partial melting and the Archean crust. *Phil. Trans. R. Soc. London,* A **273**, 457-61.

Greenwood, H.J., 1967. Wollastonite: stability in H_2O-CO_2 mixtures and occurrence in a contact-metamorphic aureole near Salmo, British Columbia, Canada. *Am. Mineralogist,* **52**, 1669-80.

Hansen, E.C., Newton, R.C. & Janardhan, A.S., 1984a. Fluid inclusions in rocks from the amphibolite-facies gneiss to charnockite progression in southern Karnataka, India: direct evidence concerning the fluids of granulite metamorphism. *Jour. Metamorphic Geology*, **2**, 249-64.

Hansen, E.C., Newton, R.C. & Janardhan, A.S., 1984b. Pressures, temperatures and Metamorphic Fluids Across an Unbroken Amphibolite Facies to Granulite Facies Transition in Southern Karnataka, India. In Kroner, Goodwin, and Hanson, Eds. *Archean Geochemistry*, 161-81. Springer-Verlag.

Harley, S.L., 1989. The origins of granulites: a metamorphic perspective. *Geological Magazine*, **126**, 215-47.

Harlov, D.E., 2000. Titaniferous magnetite-ilmenite thermometry and titaniferous magnetite-ilmenite-orthopyroxene-quartz oxygen barometry in granulite facies gneisses, Bamble Sector, SE Norway: implications for the role of high-grade CO_2-rich fluids during granulite genesis. *Contrib. Mineral. Petrol.*, **1139**, 180-97.

Heier, K.S., 1973. Geochemistry of granulite facies rocks and problems of their origin. *Philosophical Transactions of the Royal Society of London*, A **273**, 429-42.

Herms, P. & Schenk, V., 1992. Fluid inclusions in granulite-facies metapelites of the Hercynian ancient lower crust of the Serre, Calabria, Southern Italy. *Contrib. Mineral. Petrol.*, **112**, 393-404.

Hoefs, J. & Touret, J.L.R., 1975. Fluid inclusion and carbon isotope study from the Bamble Granulites (South Norway). *Contrib. Mineral. Petrol.*, **52**, 165-74.

Holland, T.J.B. & Powell, R., 1990. An enlarged and updated internally consistent thermodynamic dataset with uncertainties and correlations: the system K_2O-Na_2O-CaO-MgO-FeO-Fe_2O_3-Al_2O_3-TiO_2-SiO_2-C-H_2-O_2. *Jour. Metamorphic Geol.*, **8**, 89-124.

Hyndman, R.D., Lambert, I.B., Heier, K.S., Jaeger, J.C. & Ringwood, A.E., 1968. Heat flow and surface radioactivity measurements in the Precambrian shield of Western Australia. *Phys. Earth Pl. Int.* **1**, 129-35.

Kretz, R., 1961. Some applications of thermodynamics to coexisting minerals of variable composition. Examples: orthopyroxene-clinopyroxene and orthopyroxene garnet. *Jour. Geology*, 361-87.

Lamb, W.M., 1990. Fluid inclusions in granulites; Peak vs. retrograde formation. In D. Vielzeuf, and P. Vidal, Eds. *Granulites and Crustal Evolution*, **311**, 419-33.

Lamb, W.M., Brown, P.E. & Valley, J.W., 1991. Fluid inclusions in Adirondack Granulites: Implications for the Retrograde P-T path. *Contrib. Mineral. Petrol.*, **107**, 472-83.

Lamb, W.M. & Morrison, J., 1997. Retrograde fluids in the Archean Shawmere anorthosite, Kapuskasing structural zone, Ontario, Canada. *Contrib. Mineral. Petrol.*, **129**, 105-19.

Lamb, W.M. & Valley, J.W., 1984. Metamorphism of reduced granulites in low-CO_2, vapor-free environment. *Nature*, **312**, 56-58.

Lamb, W.M. & Valley, J.W., 1985. C-O-H fluid calculations and granulite genessis. In A. Tobi, and J. Touret, Eds. *The deep Proterozoic crust in the North Atlantic Provinces*, 119-31. Reidel Pub.

Lamb, W.M. & Valley, J.W., 1988. Granulite facies amphibole and biotite equilibria, and calculated peak-metamorphic water activities. *Contrib. Mineral. Petrol.*, **100**, 349-60.

Lamb, W.M., Valley, J.W. & Brown, P.E., 1987. Post-metamorphic CO_2-rich fluid inclusions in granulites. *Contrib. Mineral. Petrol.*, **96**, 485-95.

LeBreton, N. & Thompson, A.B., 1988. Fluid-absent (dehydration) melting of biotite in metapelites in the early stages of anatexis. *Contrib. Mineral. Petrol*, **99**, 226-37.

Moecher, D.P., 1993. Scapolite phase equilibria and carbon isotopes: constraints on the nature and distribution of CO_2 in the lower continental crust. *Chemical Geology*, **108**, 163-74.

Moecher, D.P. & Essene, E.J., 1991. Calculations of CO_2 activities using scapolite equilibria: constraints on the presence and composition of a fluid phase during high grade metamorphism. *Contrib. Mineral. Petrol.*, **108**, 219-40.

Moecher, D.P., Valley, J.W. & Essene, E.J., 1994. Extraction and carbon isotope analysis of CO_2 from scapolite in deep crustal granulites and xenoliths. *Geochimica et Cosmochimica Acta*, **58**, 959-67.

Morrison, J. & Valley, J.W., 1988. Post-granulite facies fluid infiltration in the Adirondack Mountains. *Geology*, **16**, 513-616.

Morrison, J. & Valley, J.W., 1991. Retrograde fluids in granulites: Stable isotope evidence of fluid migration. *Jour. Geology*, **99**, 559-70.

Munyanyiwa, H., Touret, J.L.R. & Jelsma, H.A., 1993. Thermobarometry and fluid evolution of enderbites within the Magondi Mobile Belt, northern Zimbabwe. *Lithos*, **29**, 163-76.

Newton, R.C., 1992. Charnockitic alteration: evidence for CO_2 infiltration in granulite facies metamorphism. *Jour. Metamorphic Geol.*, **10**, 383-400.

Newton, R.C., Aranovich, L.Y., Hansen, E.C. & Vandenheuvel, B.A., 1998. Hypersaline fluids in Precambrian deep-crustal metamorphism. *Precambrian Research*, **91**, 41-63.

Newton, R.C., Smith, J.V. & Windley, B.F., 1980. Carbonic metamorphism, granulites and crustal growth. *Nature*, **288**, 45-50.

Ohmoto, H. & Kerrick, D., 1977. Devolatilization equilibria in graphitic systems. *Am. Jour. Sci.*, **277**, 1013-44.

Phillips, G.N., 1980. Water activity changes across an amphibolite-granulites facies transition, Broken Hill Australia. *Contrib. Mineral. Petrol.*, **75**, 377-86.

Rudnick, R.L., Ashwal, L.D. & Henry, D.J., 1984. Fluid inclusions in high-grade gneisses of the Kapuskasing structural zone, Ontario: metamorphic fluids and uplift/erosion path. *Contrib. Mineral. Petrol.*, **87**, 399-406.

Rudnick, R.L. & Fountain, D.M., 1995. Nature and composition of the continental-crust – A lower crustal perspective. *Reviews of Geophysics*, **33**, 267-309.

Sighinolfi, G.P., 1971. Investigations into deep crustal levels: Fractionating effects and geochemical trends related to high-grade metamorphism. *Geochimica et Cosmochimica Acta*, **35**, 1005-21.

Srikantappa, C., Raith, M. & Touret, J.L.R., 1992. Synmetamorphic high-density carbonic fluids in the lower crust: Evidence from the Nilgiri Granulites, Southern India. *Jour. Petrology*, **33**, 733-60.

Touret, J.L.R., 1985. Fluid regime in southern Norway: the record of fluid inclusions. In A.C. Tobi, and J.L.R. Touret, Eds. *The Deep Proterozoic Crust in the North Atlantic Provinces*, 517-49. Reidel, Dordrecht.

Touret, J.L.R., 1989. Fluid control in the lower crust. *North German Academy of Sciences (Hannover) Publications*, **1**, 57-68.

Touret, J.L.R., 1992. CO_2 transfer between the upper mantle and the atmosphere: temporary storage in the lower continental crust. *Terra Nova*, **4**, 87-98.

Touret, J.L.R. & Hansteen, T.H., 1988. Geothermometry and fluid inclusions in a rock from the Doddabetta charnockite complex, Southwest India. *Rendicotta della Italiana Di Mineralogie E Petrologie*, **43**, 65-82.

Valley, J.W., Bohlen, S., Essene, E.J. & Lamb, W.M., 1990. Metamorphism in the Adirondacks: II. The Role of Fluids. *Jour. Petrology*, **31**, 555-96.

Valley, J.W. & Essene, E.J., 1980. Calc-silicate reactions in Adirondack marbles: The role of fluids and solid solutions. *Geological Society of America Bulletin*, **91**, 720-815.

Valley, J.W., McLelland, J., Essene, E.J. & Lamb, W.M., 1983. Metamorphic fluids in the deep crust: Evidence from the Adirondack Mountains, N.Y. *Nature*, **301**, 226-28.

Vityk, M.O., Bodnar, R.J. & Doukan, J.C., 2000. Synthetic fluid inclusions. XV. TEM investigation of plastic flow associated with reequilibration of fluid inclusions in natural quartz. *Contrib. Mineral. Petrol.*, **139**, 285-97.

Vry, J. & Brown, P.E., 1991. Texturally-early fluid inclusions in garnets: evidence of the prograde metamorphic path. *Contrib. Mineral. Petrol.*, **108**, 271-82.

Vry, J.K., Brown, P.E. & Valley, J.W., 1990. Cordierite volatile content and the role of CO_2 in high-grade metamorphism. *Amer. Mineralogist*, **75**, 71-88.

Vry, J.K., Brown, P.E., Valley, J.W. & Morrison, J., 1988. Constraints on granulite genesis from carbon isotope compositions of cordierite and graphite. *Nature*, **332**, 66-68.

Waters, D.J., 1988. Partial melting and the formation of granulite facies assemblages in Namaqualand, South Africa. *Jour. Metamorphic Geol.*, **6**, 387-404.

Wilmart, E., Clocchiatti, R., Duchesne, J.C. & Touret, J.L.R., 1991. Fluid inclusions in charnockites from the Bjerkreim-Sokndal massif (Rogaland, southwestern Norway): fluid origin and *in situ* evolution. *Contrib. Mineral. Petrol.*, **108**, 453-62.

Zeigenbein, P. & Johannes, W., 1974. Wollastonitbildung aus Quarz und Calcit bei Pf = 2, 4, and 6 kb. *Forschr. Mineral.*, **52**, 77-79.

SECTION—III

HIMALAYAN METAMORPHISM

8

GROWTH AND EXHUMATION OF THE LOWER CRUST OF THE KOHISTAN ARC, NW HIMALAYAS

T. YOSHINO* AND T. OKUDAIRA**

ABSTRACT

The Kamila Amphibolite Belt, which is representative of the intermediate to lower crust of the Cretaceous Kohistan island arc (NW Himalayas), recorded polyphase deformation structures in relation to different tectonic events. These deformation structures could be classified into four phases (D1–D4). Based on structural analysis in combination with metamorphic *P–T* estimation for each phase, we elaborate a tectonic model from growth to exhumation for the lower crust of the Kohistan Arc. The model is classified into three tectonic stages: crustal thickening, crustal thinning and collision. Pervasive gneissose foliation (D1) develops in relatively undeformed pods or lenses consisting of granulite facies mineral assemblage. Anastomosed mylonitic shear zones of D2 with SW thrust sense distinctly cut the D1 fabric. D2 lasted after formation of the granulites developed in the lowermost portion of the Kohistan Arc. Based on the metamorphic conditions of D1 and D2, cooling path from D1 to D2 was nearly isobaric. These structures facilitated the crustal thickening of arc during subduction of Tethys oceanic lithosphere. D3 structures showing the movement of top-to-the-NW normal faults cut or deflect the foliation of the D2 shear zones. Mineralogical difference between early stage of D3 and late stage of D3 suggests an isothermal decompression path under lower amphibolite facies condition. The isothermal decompression associated with formation of D3 suggests significant crustal thinning and exhumation of high-pressure rocks of the Kohistan Arc. Based on the age considerations, since both exhumation of the lowermost crust and intra-arc rifting of the Kohistan Arc occurred at the same time (~80 Ma), the normal faulting probably produced a tectonic configuration of frontal convergence/intra-arc extension. Localized shear zones (D4) formed under greenschist facies after the Indian collision. The D4 shear zones did not contribute to the further exhumation of the Kohistan Arc because there is little evidence for significant decompression during D4.

KEY WORDS

Growth; Exhumation; Kohistan Arc; polyphase deformation; ectonic events and Kamila amphibolite belt.

* Institute for Study of the Earth's Interior, Okayama University, Tottori 682-0193, Japan.

** Department of Geosciences, Osaka City University, Osaka 558-8585, Japan.

INTRODUCTION

The Cretaceous Kohistan paleo-island arc (NW Himalayas) has been considered to be squeezed between the Asian and Indian continents (Fig. 1: Tahirkheli *et al.*, 1979; Bard, 1983; Coward *et al.*, 1982, 1986). It comprises an exposed lower crustal section of the Cretaceous island arc, and provides significant clue to understand its evolution. Yoshino *et al.* (1998) summarized the evolution of the Kamila Amphibolite Belt, the main constituent of the intermediate to the lower crust of the Kohistan Arc, as of three major successive events: (1) intrusion of basaltic arc-derived magma; (2) formation of pyroxene and garnet granulites under medium to high pressure granulite facies metamorphism; and (3) hydration of the granulites under amphibolite facies condition. The major deformation phases were classified by Treloar *et al.* (1990) into: (1) high-temperature ductile deformation during the main crustal thickening and imbrication of the island arc, and (2) low-temperature cataclastic deformation during exhumation due to collision between the Asian plate and Kohistan Arc. Recently, based on detail structural analysis, Arbaret *et al.* (2000) concluded that high temperature ductile shear zones represents arc-related deformation during subduction of the oceanic lithosphere below the Kohistan Arc rather than deformation during the collision. However, these metamorphic and structural data have not been combined well to understand the overall tectonic history of the lower crust of the Kohistan Arc.

In the present paper, we investigate polyphase deformation structures of the metabasic rocks in the Kamila Amphibolite Belt combined with estimation of metamorphic conditions. It is revealed that deformation structures were developed under different metamorphic conditions and were related to different tectonic events during formation and subsequent exhumation of the deeper portion of the island arc. Finally we propose a tectonic model for the evolution of the lower crust of the Kohistan Arc.

OUTLINE OF GEOLOGY

In the northwestern Himalayas, the Kohistan and Ladakh paleo-arcs are located between the Asian and the Indian plates (Fig. 1). The Kohistan Arc is separated by the Northern Suture (Main Karakoram Thrust) from the Asian plate in the north (Coward *et al.*, 1986; Pudsey, 1986), and by the Main Mantle Thrust from the Indian plate in the south (Tahirkheli *et al.*, 1979; Bard, 1983). The Kohistan Arc underwent two collision events. Based on Rb–Sr and K–Ar ages of syn-collisional granites and post collisional basic dykes (Petterson & Windley, 1985, 1992; Treloar *et al.*, 1989), the first collision between the Kohistan Arc and the Asian plate along the Northern Suture was inferred to be during 75–102 Ma. The second collision between the Kohistan arc and the Indian plate along the Main Mantle Thrust was at about 50 Ma (Patriat & Achache, 1985).

The northern part of the arc consists of the Cretaceous sediments and volcanic rocks, called the Yasin Group, the Chalt Volcanics and the Jaglot Group, intruded by the Kohistan Batholith. The southern part of the Kohistan Arc comprises three main geological units with basaltic composition: the Chilas Complex, the Kamila Amphibolite Belt and the Jijal Complex from north to south (Fig. 1). The Chilas Complex occupies a 300 km long and 40 km wide mafic-ultramafic body characterized by a calk-alkaline affinity and has been considered as a thick (>10 km) stratiform body that represents a sub-arc chamber (Coward *et al.*, 1986; Khan *et al.*, 1989). The Chilas Complex forms a large, tight antiform (Coward *et al.*, 1987). More than 85% of the Chilas Complex are made up of gabbronorites, which include a string of lenses consisting of sequence of peridotites, pyroxenites, anorthosites, and two pyroxene

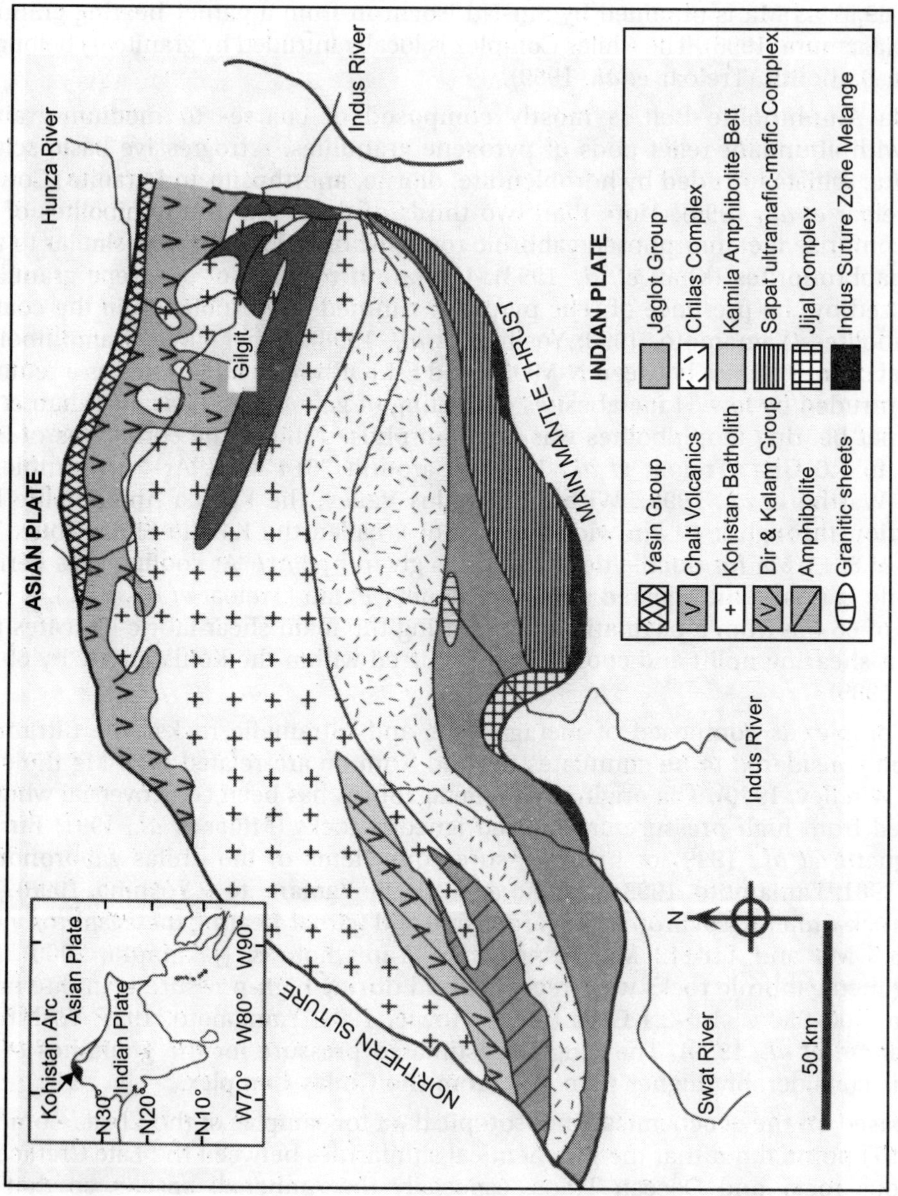

Fig. 1. Regional geological map of the Kohistan complex, modified after Searle *et al.* (1998) and Yoshino *et al.* (1998). Inset: Locality of the Kohistan arc.

diorites. The rock sequence is referred as the ultramafic-mafic-anorthosite (UMA) association (Khan *et al.*, 1993) from the apices of intra-arc diapirs along the axial plane of the antiform (Burg *et al.*, 1998). After magmatic crystallization, the Chilas gabbronorites underwent granulite facies metamorphism at 750–850°C and at 500 to 650 MPa (Jan & Howie, 1980). Minimum cooling age of 69.5±9.3 Ma is obtained by Sm-Nd isochron from a garnet-bearing granulite (Yamamoto & Nakamura, 1996). The Chilas Complex is locally intruded by granitoids belonging to the Kohistan Batholith (Treloar *et al.*, 1989).

The Kamila Amphibolite Belt is mostly composed of coarse- to medium-grained amphibolites with abundant relict pods of pyroxene granulites, retrogressive basic schists and minor pelitic schists intruded by hornblendite, diorite, anorthosite and granite (Coward *et al.*, 1987; Treloar *et al.*, 1990). More than two-thirds of these low-Ti amphibolites of the Indus Valleys comprise metamorphosed gabbroic rocks, whose REE pattern is similar to that of the Chilas gabbronorites (Khan *et al.*, 1993). Hydration reaction of pyroxene granulites can be recognized by the presence of relic pyroxene rimmed by hornblende in the coarse-grained amphibolites (Yamamoto, 1993; Yoshino *et al.*, 1998). Other high-Ti amphibolites with transitional characteristics between N-MORB and E-MORB were interpreted as a remnant oceanic crust intruded by low-Ti metabasites consisting of gabbros and basalts (Khan *et al.*, 1993). The garnet-bearing amphibolites was metamorphosed under the conditions of 550–650°C and 0.9 to 1.0 GPa (Treloar *et al.*, 1990) at around 76±4 Ma (^{40}Ar–^{39}Ar hornblende cooling age of Wartho *et al.*, 1996). Within the Indus Valley, the Kamila Amphibolite Belt provides a section through a 40 km wide shear zones, called the Kamila Shear Zone. The minimum age of 83±1 Ma for ductile deformation is given by $^{40}Ar/^{39}Ar$ cooling ages derived from hornblende grains in the sheared amphibolite near Patan (Treloar *et al.*, 1990). A K–Ar muscovite age of 66 Ma from a pegmatite cross-cutting the main shear fabric indicates that substantial post-shearing uplift and cooling had occurred within the Kohistan Arc by 60 Ma (Treloar *et al.*, 1989).

The Jijal Complex is composed of metagabbros and ultramafic rocks. The ultramafic rocks have been considered to be cumulates derived from an arc-related high-Mg tholeiitic magma (Jan & Windley, 1990). The origin of the metagabbros has been controversial whether they are derived from high-pressure crystallized igneous rocks (Miller *et al.*, 1991; Burg *et al.*, 1998; Ringuette *et al.*, 1999) or high-pressure equivalents of the Chilas gabbronorites (Jan & Howie, 1981; Yamamoto, 1993; Yoshino *et al.*, 1998; Yamamoto & Yoshino, 1998). The Sm–Nd whole rock-mineral isochron from garnet-free and garnet-bearing metagabbros yields ages of 91.0±6.3 Ma and 118±12 Ma, respectively (Yamamoto & Nakamura, 2000). The garnet-bearing metagabbroic rocks were recrystallized during high-pressure granulite facies metamorphism (700–950°C, 1.0–2.1 GPa) (Jan & Howie, 1981; Yamamoto, 1993; Yoshino *et al.*, 1998; Ringuette *et al.*, 1999). The range of estimated pressure for the granulites of the Jijal Complex is considerably higher than that from the Chilas Complex.

Recently, based on the geochemical and isotopic data for samples of the Chilas Complex, Khan *et al.* (1997) pointed out that the geochemical similarities between the Late Cretaceous Kohistan igneous rocks and Deccan Traps, especially the Ambenali basalts, so that the Kohistan Arc formed at or south of the present equator. Although these ideas are interesting for information about source mantle of the Kohistan igneous rocks, we do not discuss about their source mantle. We focus on pressure–temperature–time–deformation history of the Kamila Amphibolite Belt, to clarify the tectono-metamorphic evolution of the Kohistan Arc.

DEFORMATION STRUCTURES AND PETROGRAPHY

The Kamila Amphibolite Belt can be classified into the southern and northern sub-belts based on their lithological features, as Dasu and Patan Units (Fig. 2). The Dasu Unit in the northern-part of the Kamila Amphibolite Belt includes fine-grained banded amphibolites interlayered with minor calc-silicates and sediments (Jan, 1988; Treloar *et al.*, 1990), intruded by numbers of granitic sheets or pegmatites near Dasu. The presence of pelagic sediments and MORB type basic schists (Khan *et al.*, 1993) suggests that the Dasu Unit is derived mainly from a remnant oceanic crust on which the mature arc was built. Dominant lithology throughout the Patan Unit in the southern-part of the Kamila Amphibolite Belt is composed of amphibole-bearing metagabbros and hornblende-diorites, intruded by hornblendites (Treloar *et al.*, 1990; Khan *et al.*, 1993). In the Patan Unit, the pyroxene granulite pods are recognized within the amphibolites. The pyroxene granulites are petrologically quite similar to the Chilas gabbronorites. The abundance of the relict pyroxene granulite pods decreases with increasing intensity of deformation and occurrence is restricted between Keru and Patan along the Indus Valley (Fig. 3). Igneous layering (S0) within the undeformed coarse-grained metagabbros and relict pyroxene granulites can be recognized as the oldest structure. Such igneous structure is defined as a compositional banding structure composed of the plagioclase-rich and pyroxene or amphibole-rich layers. These primary structures are dominated by a NE–SW to an EW-striking layering (Fig. 4a). Tectonic planar structures can be subdivided into four types as follows: (1) steeply dipping pyroxene-bearing penetrative gneissose foliation (S1; Fig. 4b), (2) hornblende-bearing anastomosed shear zone with sub-horizontal to moderate N- to NW-dipping foliation (S2; Fig. 4c), (3) epidote-rich subvertical to moderate N-dipping schistose foliation (S3; Fig. 4d), and (4) chlorite-bearing narrow shear zones with distinct foliation planes (S4) (Fig. 4e). Based on the sequence of these foliations (S1–S4), we can then classify deformation in the Kamila Amphibolite Belt into four different phases, D1–D4.

Tables 1 and 2 summarize the deformational features and the estimated *P–T* conditions of each event. Structural data and shear sense determined by asymmetric structures for each deformation phase are illustrated in Figs. 5 and 6, respectively. To understand a tectonic history in the lower crust of the Kohistan Arc, interrelationship of the deformation structures, mineral chemistry and metamorphic conditions for each deformation phase are described in detail. Mineral compositions were determined using JEOL 733 MKII and JXA 8900L electron microprobes at the Geological Institute, University of Tokyo. Standard analytical conditions were 15 kV acceleration voltage, 12 nA beam current, tight beam diameter and 10 sec counting times. Representative mineral compositions are shown in Tables 3, 4 and 5. To estimate the *P–T* metamorphic condition, microprobe analyses were carried out at the contacts of hornblende, plagioclase and garnet, and amphibole formula were calculated by using the average Fe^{3+} option of Spear & Kimball (1984).

The first phase deformation (D1)

The S1 foliation defined by tabular alignment of hornblende and pyroxene grains is preserved well in the southern part of the Patan Unit composed of coarse-grained metagabbros. Mineral assemblage forming S1 is of clinopyroxene, orthopyroxene, brown-colored hornblende and plagioclase with minor amounts of garnet, biotite, quartz and Fe–Ti oxides. Mineral and stretching lineations were not recognized well. Absence of low-grade minerals such as clinozoisite, actinolite and chlorite, indicates that the S1 foliation developed under upper amphibolite facies metamorphic condition.

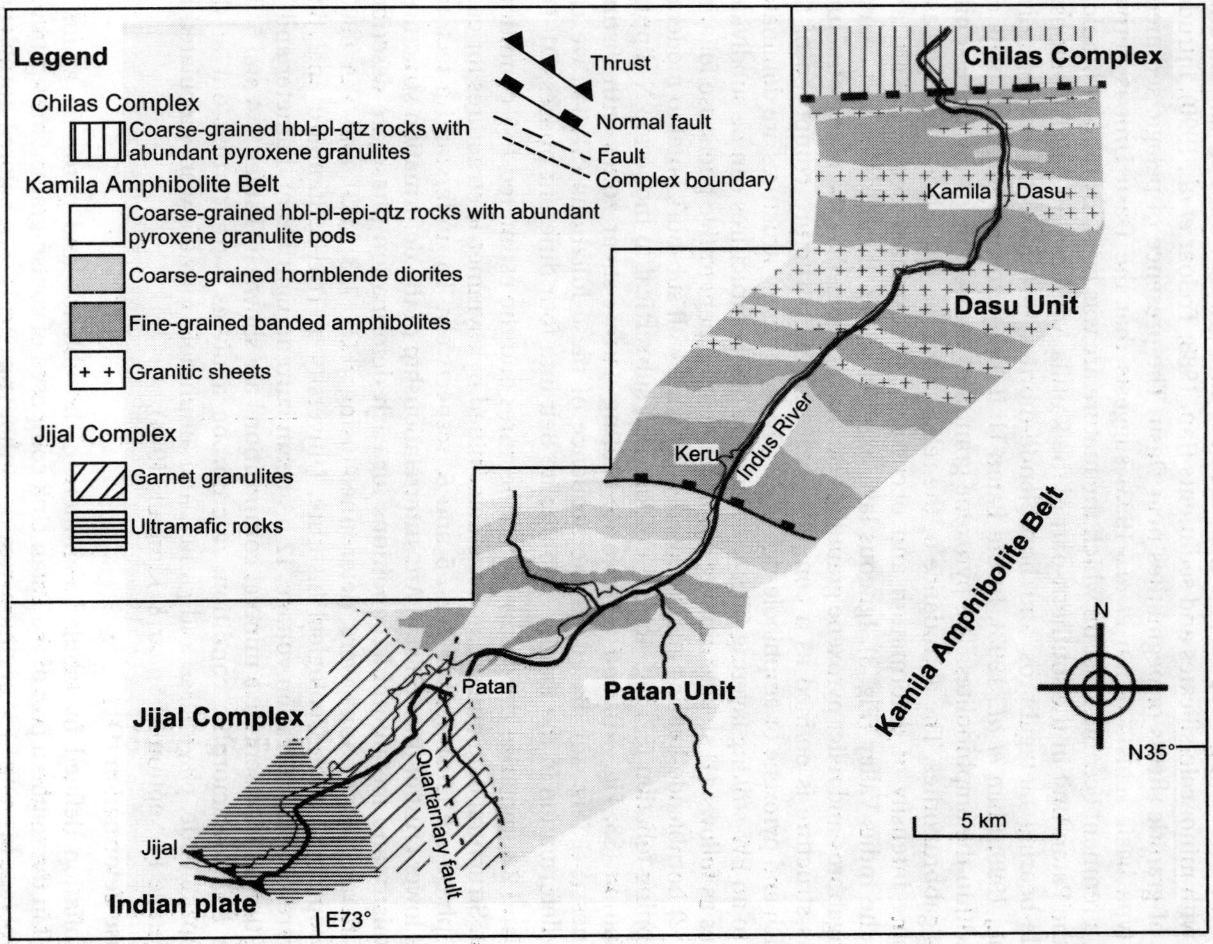

Fig. 2. Lithology map along the Indus Valley traverse through the Jijal Complex and the Kamila Amphibolite Belt, modified after Yoshino *et al.* (1998).

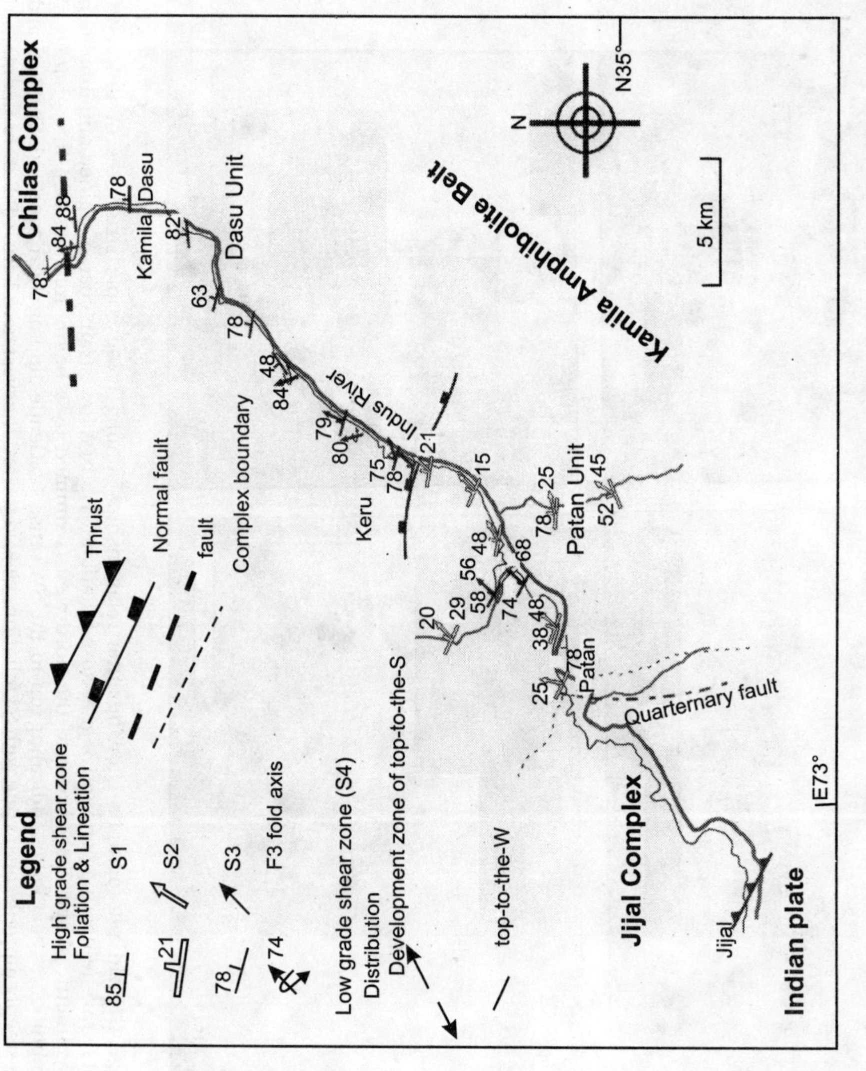

Fig. 3. A structural map along the Indus Valley route. A map to show structural elements observed.

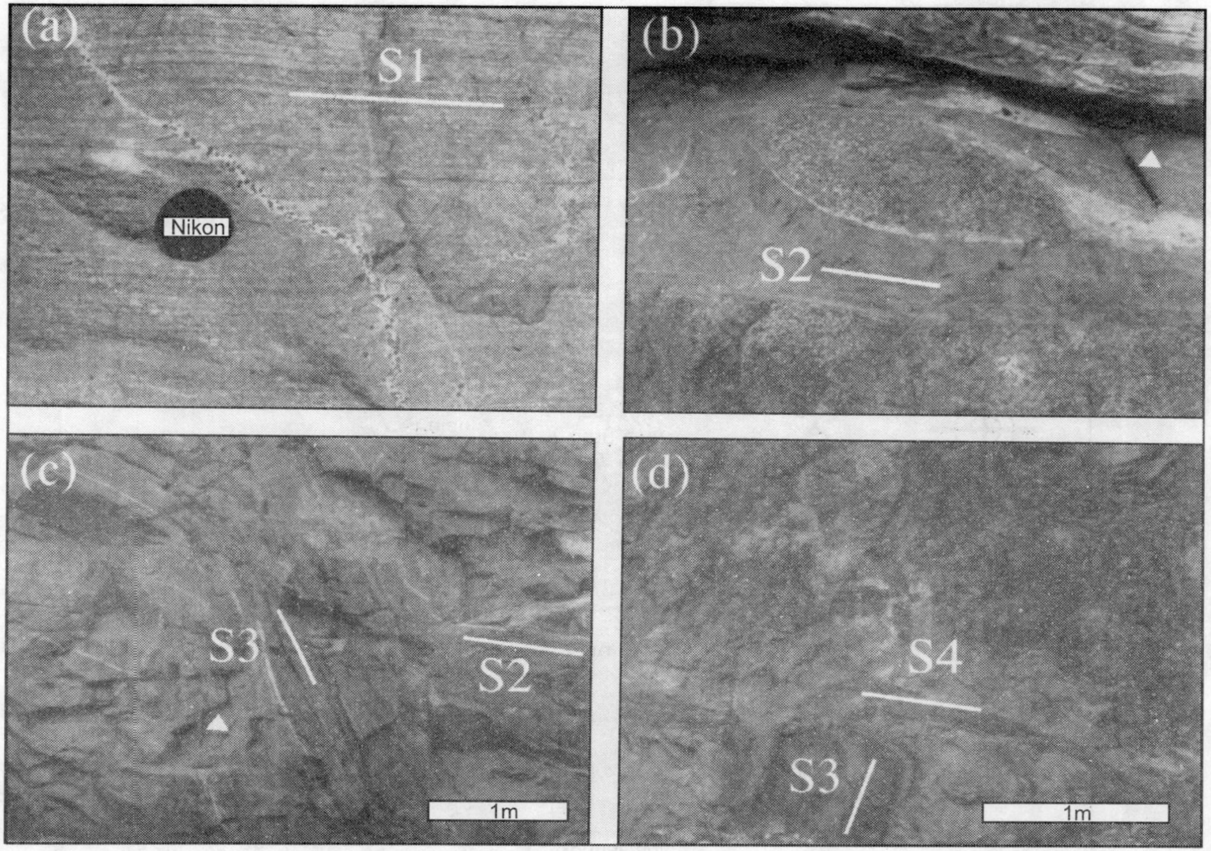

Fig. 4. Field photographs showing inter-relationships among mesoscopic structures of different phases. (a) Pyroxene-bearing penetrative gneissose foliation (S1). (b) A low-strain coarse-grained amphibolite pod with asymmetric shape viewed toward the northwest is rimmed by a highly foliated (S2), fine-grained mylonitic amphibolites near Patan, indicating top-to-the-SW sense. (c) Gentle dipping shear zone with S2 foliation deflected by steeply dipping shear zone with S3 foliation, viewed toward the west near Keru. (d) Sub-horizontal shear zone with S4 foliation indicative of top-to-the-S sense cut the steeply dipping gneissose foliation (S3) viewed toward the west near Mandraza.

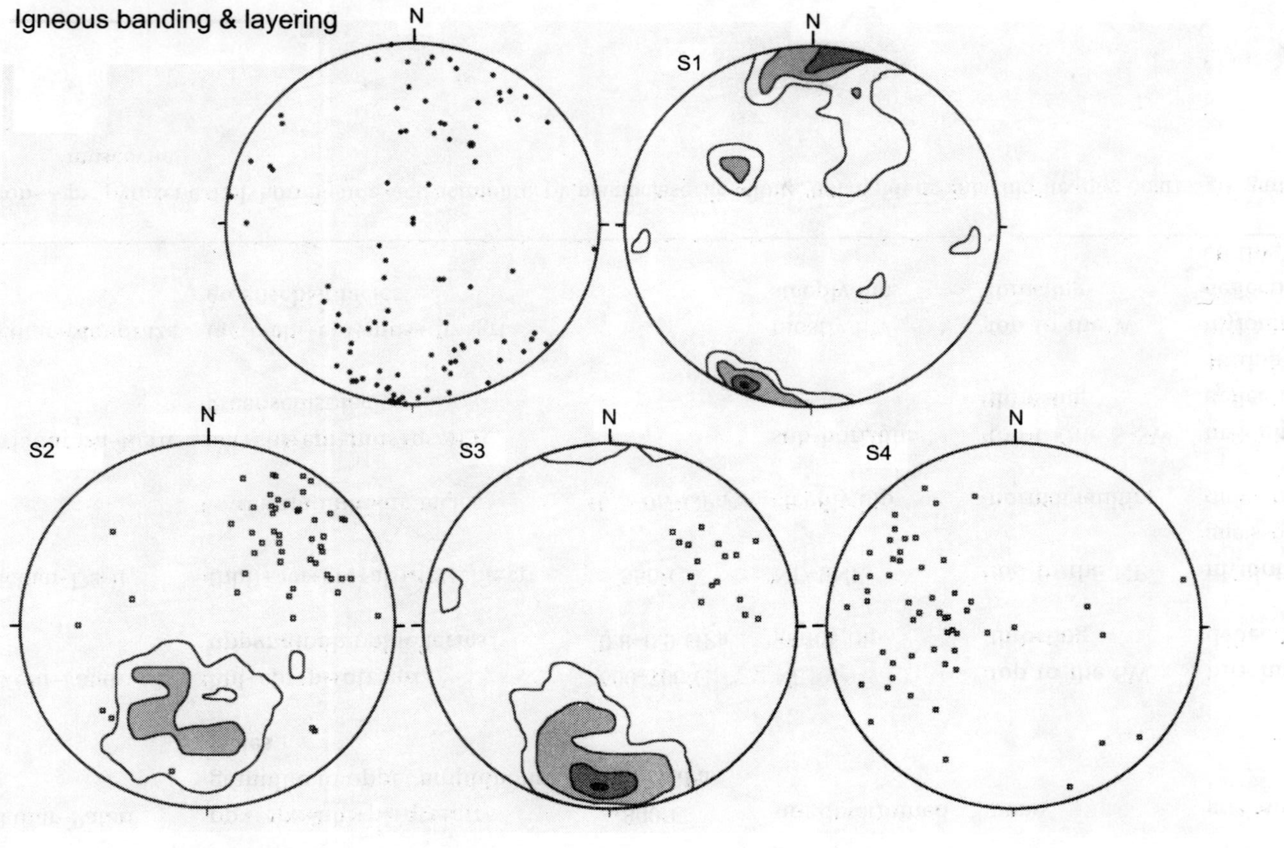

Fig. 5. Equal-area lower hemisphere stereographic projections of pole to the shear plane orientation, mineral lineation and fold axis in the Kamila Amphibolite Belt from Patan to Lutar: (a) S0, closed circles: Lutar to Keru; open circles: Keru to Patan, (b) S1, contour map: pole to S1 foliation, (c) S2, closed squares: mineral and stretching lineation; contour map: pole to S2 foliation, (d) S3, closed squares: mineral and stretching lineation; open squares: F3 fold axis; contour map: pole to S3 foliation, and (e) S4, closed circles: pole to foliation of top-to-the-S shear zone; open squares: pole to foliation of top-to-the-NW shear zone.

Table 1: *A summary of characteristics of each deformational phase in the Kamila Amphibolite Belt*

Phase	Distribution	Mineral assemblage	*P–T*	Attitude of L	Sense of shear	Occurrence
D1	Lutar–Patan	opx+cpx+hbl+pl+qtz±grt granulite to upper amphibolite facies	~ 800°C ~ 1.1 GPa	not determined	none	gneissose
D2	Keru–Patan	hbl+epi+pl+qtz±grt upper amphibolite facies	600–700°C 0.8–0.9 GPa	NE gently dip	top to the SW thrusting	mylonitic deflecting S1
D3	Patan–Dasu	(hbl)+act+epi+ab+qtz±chl±grt lower amphibolite facies	> 550°C 0.5–0.7 GPa	NE–ENE steeply dip	top to the NE normal faulting	mylonitic and gneissose deflecting S2
D4	Mandraza–Keru	act+chl+epi+mus+qtz±grt greenschist facies		sub-horizontal	top to the S–SW thrusting	mylonitic deflecting and cutting S3
	Lutar–Mandraza	(act)+chl+epi+mus+qtz±grt greenschist facies		mostly EW steeply dip	top to the W thrusting	mylonitic deflecting and cutting S3

Abbreviation — px: pyroxene, hbl: hornblende, act: actinolite, pl: plagioclase, ab: albite, epi: epidote, chl: chlorite, qtz: quartz, grt: garnet, mus: muscovite

Table 2: *Growth periods of the main metamorphic minerals with respect to the deformation phases*

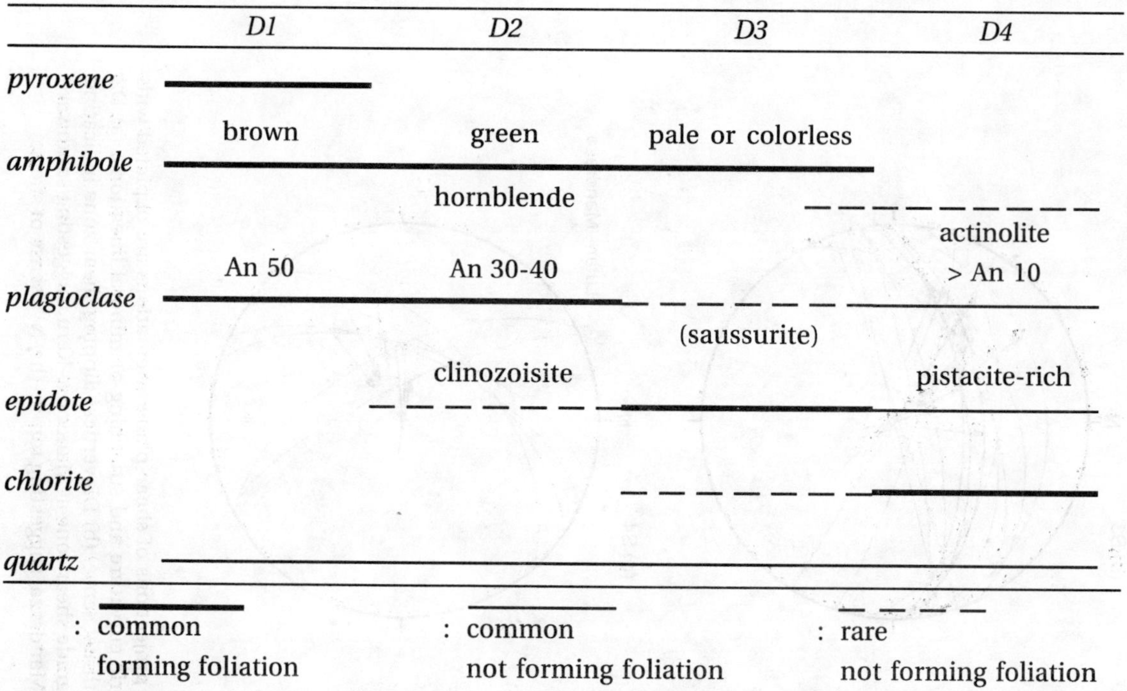

	D1	*D2*	*D3*	*D4*
pyroxene				
amphibole	brown	green hornblende	pale or colorless	actinolite
plagioclase	An 50	An 30-40	(saussurite)	> An 10
epidote		clinozoisite		pistacite-rich
chlorite				
quartz				

: common forming foliation   : common not forming foliation   : rare not forming foliation

In the pyroxene granulites with the weakly developed S1 foliation, large pyroxene with exsolution lamella is rimmed by fine-grained polygonal pyroxene aggregate. The fine polygonal pyroxene grains do not have the exsolution lamella, indicating that the pyroxene grains were not crystallized from magma but were recrystallized under granulite facies condition. It suggests that S1 formed under granulite facies metamorphic conditions. In fact, aluminum content in pyroxene is different between igneous pyroxenes with exsolution lamella and metamorphic polygonal grains (Fig. 8). Core of the metamorphic grains has high Al_2O_3 content (3–5 wt.%) similar to that of the highest Al_2O_3 in the undeformed rocks (Table 5). It suggests that recrystallized pyroxene grains formed near the peak *P–T* metamorphic conditions. Yoshino *et al.* (1998) estimated that metamorphic condition from a pyroxene granulite (PU5) with S1 foliation, by using Al-zoning in clinopyroxene and plagioclase, is of 810°C and 1080 MPa.

The dynamically recrystallized plagioclase aggregate represents tabular shape, and most of recrystallized plagioclase grains have straighter grain boundaries and isotropic, polygonal optically strain-free with 120° triple junctions (Fig. 7a). The texture suggests post-deformation annealing (Ji & Mainprice, 1987; Rutter & Brodie, 1992). Although strain features have been destroyed by recovery, the compositional difference between porphyroclasts and neoblasts is clear; relatively large plagioclase porphyroclasts show higher anorthite contents (An_{60-70}) than recrystallized grains (An_{45-55}) (Fig. 9). Anorthite content generally decreases with increasing modal hornblende. Chemical compositions of hornblende characterized by high Ti and K contents are similar to that in the undeformed pyroxene granulites (Fig. 10c). Garnet occurs

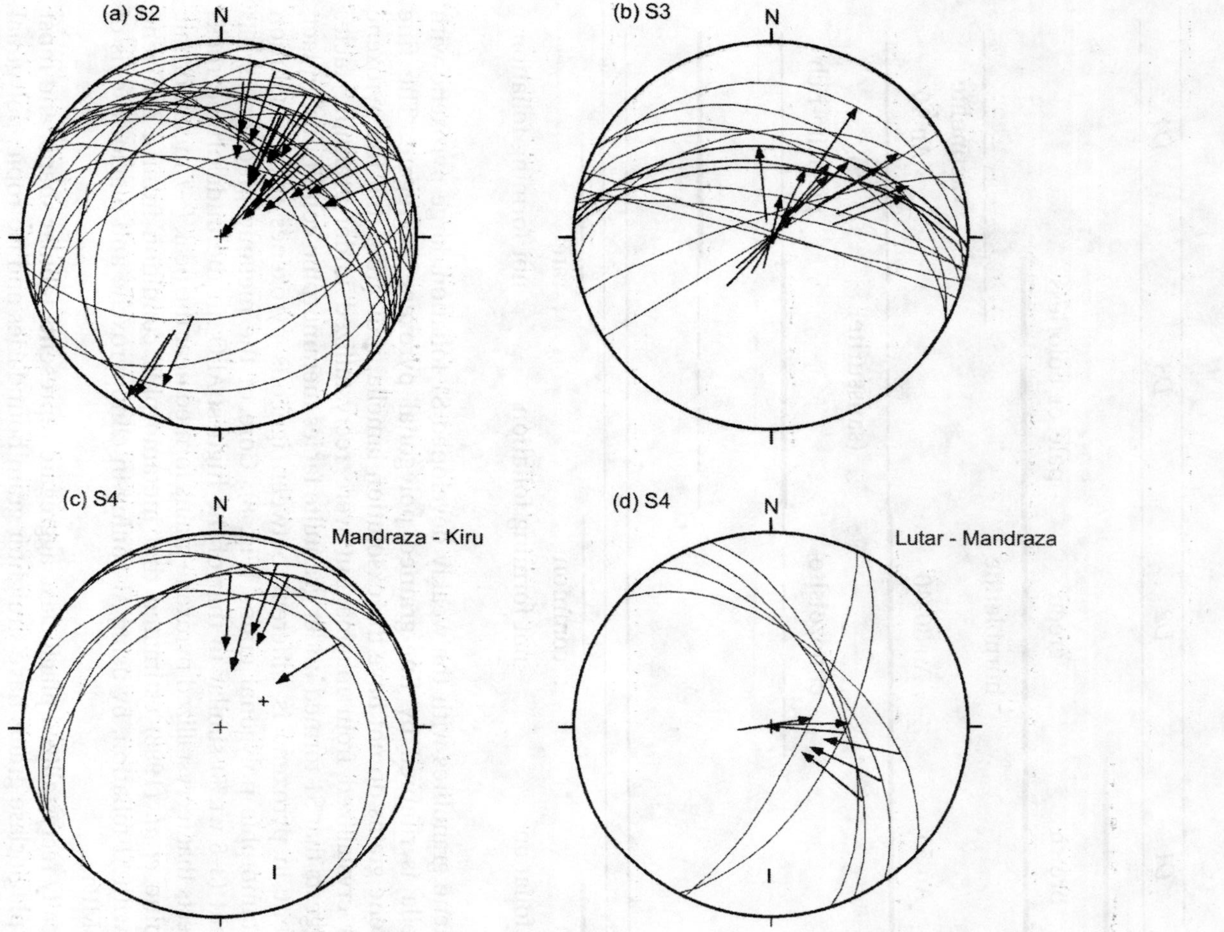

Fig. 6. Equal-area lower hemisphere stereographic projections of shear-plane orientations are depicted with movement directions (arrow) interpreted from fabric curvature and stretching or mineral lineations. (a) D2 anastomosed shear zone indicating top-to-the-SW thrust sense. (b) D3 vertical dipping shear zone indicating north side down NE normal fault sense. (c) D4 retrograde shear zone at Mandraza–Keru, suggesting southward thrust sense. (d) D4 retrograde shear zone at Lutar–Mandraza, suggesting top-to-the-NW sense of shear.

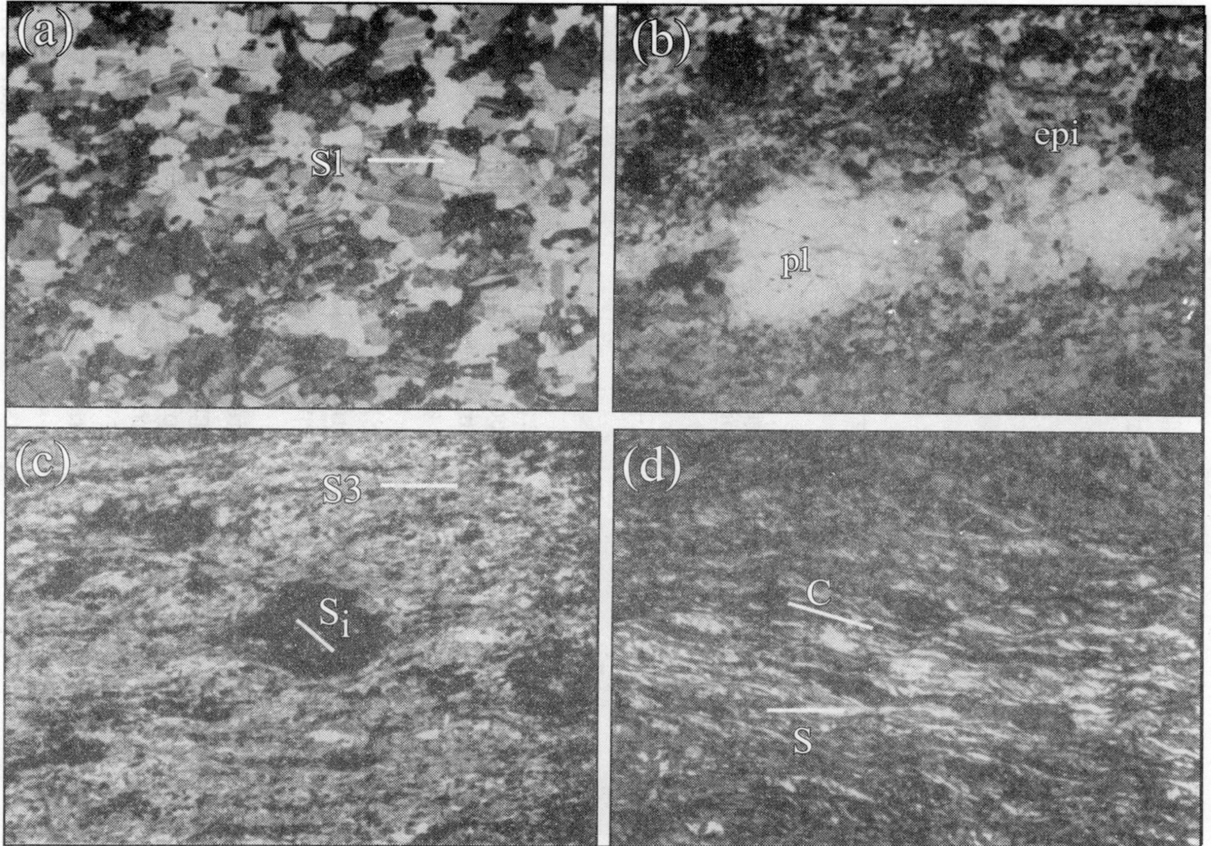

Fig. 7. Photomicrographs showing typical microstructures of these deformation phases. Section parallel to the stretching lineation and normal to the foliation. Width of view is of 18 mm. Crossed polarized light. (a) A high-temperature structure (D1) showing clinopyroxene, orthopyroxene and hornblende in recrystallized plagioclase aggregates. (b) A structure of D2 shear zone indicating plagioclase porphyroclasts with subgrain and optically strain-free clinozoisite porphyroblasts in fine-grained plagioclase-hornblende matrix. (c) The deformed amphibolites (D3) showing that saussuritized plagioclase with inclusion trails (S_i) behaves as a porphyroclast during D3, since the S3 foliation defined by alignment of acicular and platy clinozoisite and hornblende porphyroblasts is deflected. Dextral shear sense. (d) A low temperature basic mylonite (D4) suggesting *S–C* fabric defined by quartz ribbon and alignment of fine-grained chlorite. Sinistral shear sense. Dark round patches are due to thin section preparation.

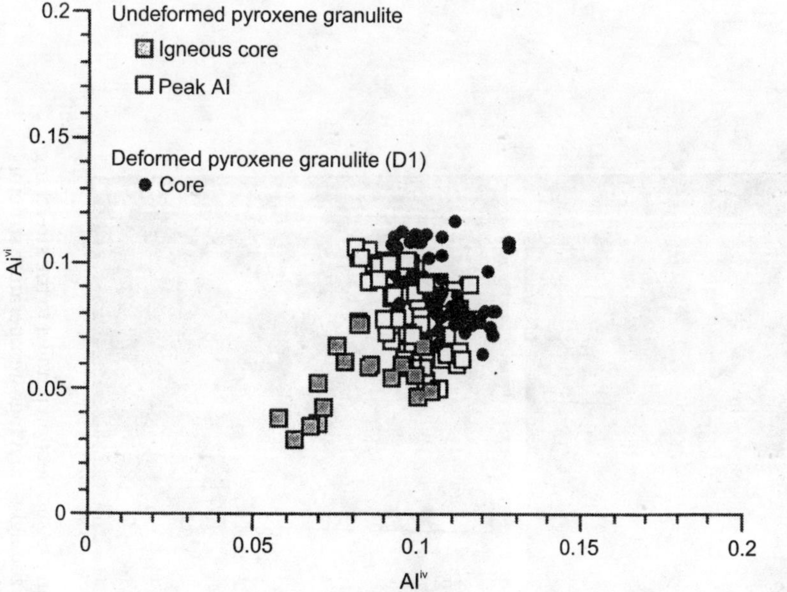

Fig. 8. Al^{iv} vs. Al^{vi} in clinopyroxene forming the S1 foliation and igneous clinopyroxene in the undeformed pyroxene granulites. Note that plotted area of clinopyroxene forming S1 is similar to the peak Al content in clinopyroxene with exsolution lamellae.

occasionally along the quartzo-feldspathic vein and has pyrope-rich (X_{pyr}~0.3) and low spessartine component (X_{sps}<0.02).

The second phase deformation (D2)

In the Patan Unit, distinct anastomosed shear zones are well recognized as hydrated zone, that is Amphibolite. The S2 foliation in the localized shear zones clearly cut both the garnet-free and garnet-bearing granulites at the boundary between the Jijal Complex and the Kamila amphibolite Belt. On the S2 plane, mineral and stretching lineations (L2) are well developed (Fig. 3). The S0 or S1 foliation in the undeformed pods consisting of pyroxene granulites are deflected or truncated by the S2 foliation (Fig. 5b). The S2 dips toward the NNE–NE, and the L2 stretching lineation concentrates to the NE. Kinematic directions are indicative of top-to-the-SW thrust movement (Fig. 6a). This deformation phase has also been described by Arbaret *et al.* (2000).

The D2 amphibolites consist of plagioclase, blue-green hornblende, epidote, quartz, Fe–Ti oxides, and occasionally garnet. L2 is defined by alignment of hornblende and ribbon-like aggregates of plagioclase and quartz. Pyroxene completely disappears within the fine-grained matrix in the D2 amphibolites. Green-colored hornblende grains occur as porphyroclasts and as fine-grained matrix grains (Fig. 7b). It indicates that some hornblende crystals were presented before D2. The chemical composition of hornblende is characterized by lower K and Ti contents, and higher Al^{iv} content and A-site occupancy than that of the S1 foliation (Fig. 10). These chemical changes caused by Ti-tschermak substitution result from retrograde metamorphism, because of a temperature dependence of Ti content of amphibole coexisting with ilmenite (e.g., Spear, 1981; Raase *et al.*, 1986). Plagioclase porphyroclasts embedded in a plagioclase matrix generally have recrystallized at their margin (Fig. 7b).

Table 3: *Representative chemical compositions of amphibole in each deformation phase*

	S1		S2			S3							
Sample No.	P5	PL2	KY15	KY09	PA3-1	KM8		1409		1425		1427	
						PB	NB	PB	NB	PB	NB	PB	NB
SiO_2	43.07	40.28	41.55	42.41	42.25	48.38	56.32	47.71	54.70	43.78	48.35	45.90	52.61
Al_2O_3	14.89	14.1	16.84	16.49	16.89	10.77	2.11	11.92	1.71	17.60	9.17	14.14	6.16
TiO_2	0.41	2.42	0.41	0.63	0.61	0.45	0.03	0.97	0.04	0.06	0.27	0.15	0.27
FeO*	15.58	14.07	15.74	13.92	13.34	8.36	7.36	9.39	8.61	8.43	7.06	8.01	6.16
MnO	0.17	0.07	0.25	0.08	0.21	0.16	0.18	0.18	0.20	0.07	0.08	0.13	0.12
MgO	8.62	10.33	8.84	9.43	9.77	15.32	18.65	15.07	17.92	14.24	16.78	14.64	18.36
CaO	11.61	11.48	10.86	11.25	11.27	12.33	13.06	10.03	12.60	11.73	12.48	12.30	12.65
Na_2O	0.82	1.21	2.00	1.68	1.61	1.33	0.25	2.05	0.23	1.88	1.25	1.62	0.58
K_2O	1.50	2.20	0.50	0.51	0.61	0.65	0.06	0.28	0.07	0.10	0.04	0.10	0.02
Cr_2O_3	0.03	0.08	0.01	0.00	0.00	0.03	0.00	0.01	0.13	0.05	0.55	0.04	0.28
Total	96.70	96.24	97.00	96.40	96.55	97.76	98.01	97.61	96.22	97.94	96.04	97.02	97.22
O	23	23	23	23	23	23	23	23	23	23	23	23	23
Si	6.44	6.09	6.12	6.26	6.21	6.88	7.86	6.67	7.82	6.12	6.92	6.53	7.36
Al_{total}	2.62	2.51	2.92	2.87	2.93	1.71	0.35	1.96	0.29	2.90	1.52	2.37	1.02
Al^{iv}	1.56	1.91	1.88	1.74	1.80	1.12	0.14	1.33	0.18	1.88	1.08	1.47	0.64
Al^{vi}	1.06	0.60	1.04	1.13	1.13	0.69	0.21	0.63	0.11	1.03	0.47	0.91	0.38
Ti	0.05	0.28	0.05	0.07	0.07	0.05	0.00	0.10	0.01	0.01	0.03	0.02	0.03
Cr	0.00	0.01	0.00	0.00	0.00	0.00	0.00	0.01	0.02	0.00	0.06	0.00	0.03
Fe^{3+}**	0.16	0.26	0.65	0.34	0.41	0.09	0.00	0.88	0.11	0.51	0.30	0.31	0.22
Fe^{2+}	1.78	1.52	1.29	1.38	1.22	0.90	0.86	0.22	0.92	0.45	0.55	0.64	0.50
Mn	0.02	0.01	0.03	0.01	0.03	0.02	0.02	0.02	0.02	0.03	0.01	0.02	0.01
Mg	1.92	2.33	1.94	2.08	2.14	3.25	3.88	3.14	3.82	2.85	3.58	3.11	3.83
Ca	1.86	1.86	1.71	1.78	1.77	1.88	1.95	1.50	1.93	1.75	1.91	1.88	1.90
Na_{M4}	0.14	0.14	0.29	0.22	0.23	0.12	0.05	0.50	0.06	0.26	0.09	0.12	0.10
Na_A	0.10	0.21	0.29	0.26	0.23	0.24	0.02	0.06	0.00	0.24	0.26	0.32	0.06
K	0.29	0.42	0.10	0.10	0.11	0.12	0.01	0.05	0.01	0.05	0.08	0.01	0.00
Mg#	0.52	0.61	0.60	0.60	0.64	0.78	0.82	0.94	0.81	0.86	0.87	0.83	0.88

* FeO* was calculated as total iron.

** Fe^{3+} in amphibole were calculated based on Robinson *et al.* (1982).

Abbreviations: PB = porphyroblast; NB = neoblast forming the S3 foliation.

Table 4: *Representative chemical compositions of plagioclase in each deformation phase*

	S1		S2						S3				
Sample No.	P5	PL2	KY15		PA3-1		PA3-2	PU4	KM8		1409	1425	1427
			PC	NB	PC	NB	NB	NB	PC	SG	SG	PC	PC
SiO_2	55.11	55.25	57.43	61.00	54.01	60.21	61.37	59.12	61.52	68.43	69.54	49.16	49.16
Al_2O_3	29.13	28.25	26.52	24.52	29.27	25.79	24.38	26.12	23.99	19.92	19.17	31.39	31.39
FeO*	0.04	0.25	0.07	0.13	0.05	0.06	0.03	0.17	0.04	0.05	0.01	0.52	0.52
CaO	11.36	10.73	8.92	7.17	11.58	7.18	6.52	8.01	6.73	0.72	0.22	17.08	17.08
Na_2O	4.64	4.75	5.97	7.23	4.52	7.27	7.25	7.18	7.49	10.67	11.01	2.16	2.16
K_2O	0.35	0.35	0.57	0.10	0.02	0.02	0.09	0.15	0.07	0.06	0.15	0.01	0.01
Total	100.63	99.58	99.48	100.15	99.45	100.53	99.64	100.75	99.84	99.85	100.10	100.32	100.32
O	8	8	8	8	8	8	8	8	8	8	8	8	8
Si	2.48	2.51	2.60	2.73	2.44	2.66	2.76	2.62	2.75	3.02	3.05	2.26	2.26
Al_{total}	1.55	1.52	1.41	1.29	1.56	1.34	1.29	1.37	1.27	1.03	0.99	1.70	1.70
Fe^{3+}**	0.00	0.00	0.00	0.00	0.00	0.00	0.00	0.00	0.00	0.00	0.00	0.01	0.01
Ca	0.55	0.52	0.43	0.34	0.56	0.34	0.31	0.38	0.32	0.03	0.01	0.84	0.84
Na	0.40	0.42	0.52	0.63	0.40	0.62	0.63	0.61	0.65	0.91	0.94	0.19	0.19
K	0.02	0.02	0.03	0.01	0.00	0.00	0.01	0.01	0.00	0.00	0.01	0.00	0.00
An	0.57	0.54	0.44	0.35	0.59	0.35	0.33	0.38	0.33	0.03	0.01	0.82	0.82
Ab	0.41	0.44	0.53	0.64	0.41	0.65	0.66	0.61	0.67	0.97	0.98	0.18	0.18
Or	0.02	0.02	0.03	0.01	0.00	0.00	0.01	0.01	0.00	0.00	0.01	0.00	0.00

* FeO* was calculated as total iron.

** All Fe in plagioclase were calculated as ferric iron.

Abbreviations: PC = porphyroclast; NB = neoblast forming the S2 foliation; SG = secondary grain after formation of the S3 foliation.

Anorthite contents of plagioclase porphyroclast (An_{50}) are higher than that of matrix grains (An_{30-40}), and are consistent with that of matrix grain in the D1 granulite (Fig. 9). Clinozoisite within the S2 foliation displays no optical intracrystalline strain feature. Garnet frequently shows an elongate shape parallel to the S2 foliation. Compositional zoning profiles of garnet show a discontinuous pattern between the core and rim (Fig. 11). Grossular and spessartine components abruptly increase at the boundary between core and rim. Core compositions of the garnet are nearly consistent with those of the D1 granulite, whereas rim compositions are quite similar to those of the D2 amphibolites (Table 5). From these observations, it may be acceptable for the following speculation that the garnet core formed under granulite facies conditions during the D1, and then garnet rims overgrew or were re-equilibrated during the D2.

The third phase deformation (D3)

The flat-lying S2 foliation is asymmetrically bent on a centimetre to metre scale by tight shear bands in which generally new axial planar cleavage is developed (Fig. 5c). These shear bands define a top-to-the-NE normal fault sense (Fig. 6b). S3 foliation developed in the Dasu Unit, especially the south of Keru (Fig. 2) is defined by alignment of idiomorphic

Table 5: *Representative chemical compositions of clinopyroxene and garnet in each deformed phase*

	clinopyroxene				garnet						
Sample No.	S0		S1		S1		S2				S3
	PL2		PU5		PL2	P5	KY15	KY09	PA3-1		MA2
	core	rim	core	rim					core	rim	
SiO_2	52.19	50.20	49.90	50.05	38.58	39.18	39.49	38.31	38.18	38.11	37.93
Al_2O_3	2.48	5.02	4.61	4.90	21.41	21.77	22.04	20.88	21.58	21.24	21.06
TiO_2	0.23	0.37	0.37	0.45	0.10	0.08	0.02	0.04	0.12	0.10	0.00
FeO*	7.91	10.11	12.34	10.10	24.17	24.58	26.02	25.66	27.12	25.78	29.39
MnO	0.04	0.08	0.30	0.16	0.48	0.64	1.79	2.44	0.95	2.49	1.20
MgO	13.73	11.22	10.27	11.21	7.15	8.02	4.79	4.68	6.36	4.61	4.06
CaO	22.39	21.22	21.15	21.79	7.70	6.73	7.67	8.02	5.55	7.87	6.62
Na_2O	0.57	0.91	0.80	0.81	0.00	0.00	0.02	0.01	0.00	0.02	0.00
K_2O	0.02	0.00	0.01	0.01	0.00	0.00	0.00	0.01	0.00	0.00	0.00
Cr_2O_3	0.00	0.00	0.00	0.04	0.00	0.00	0.02	0.00	0.00	0.02	0.00
Total	99.56	99.13	99.75	99.52	99.59	101.00	101.86	100.05	99.86	100.24	100.26
O	6	6	6	6	12	12	12	12	12	12	12
Si	1.94	1.89	1.89	1.88	2.98	2.99	3.03	3.00	2.98	2.98	2.99
Al	0.11	0.22	0.21	0.22	1.90	1.96	1.99	1.93	1.98	1.96	1.96
Ti	0.01	0.01	0.01	0.01	0.00	0.01	0.00	0.00	0.01	0.01	0.00
Cr	0.00	0.00	0.00	0.00	0.00	0.00	0.00	0.00	0.00	0.00	0.00
Fe^{3+}**	0.04	0.04	0.06	0.05	0.00	0.00	0.00	0.07	0.05	0.07	0.06
Fe^{2+}	0.21	0.28	0.33	0.26	1.56	1.57	1.67	1.62	1.72	1.62	1.87
Mn	0.00	0.00	0.01	0.01	0.03	0.04	0.12	0.16	0.06	0.17	0.08
Mg	0.76	0.63	0.58	0.63	0.83	0.91	0.55	0.55	0.74	0.54	0.48
Ca	0.89	0.86	0.86	0.88	0.64	0.55	0.63	0.67	0.46	0.66	0.56
Na	0.04	0.07	0.06	0.06	0.00	0.00	0.00	0.00	0.00	0.00	0.00
K	0.00	0.00	0.00	0.00	0.00	0.00	0.00	0.00	0.00	0.00	0.00

* FeO* was calculated as total iron.

** Fe^{3+} were calculated from stoichiometry.

acicular or platy green to pale green or colorless calcic amphibole and pleochroic pink or colorless epidote. The orientation of S3, predominantly dipping steeply to the north, is shown in Fig. 3. Mineral lineation (L3) is defined by alignment of hornblende or actinolite and epidote on the S3 plane. Well-developed mineral lineations trend steeply or moderately towards around N50°E near Keru (Fig. 5b). Asymmetric inclusion trails around plagioclase porphyroclasts (Fig. 7c) are indicative of a top-to-the-NE normal fault sense. The S3 foliation commonly accompanies leucocratic veins and granitic sheets. The reclined folds (F3) widely develop from Dasu to Keru. The fold axes dip steeply toward the W (Fig. 4d). The S3

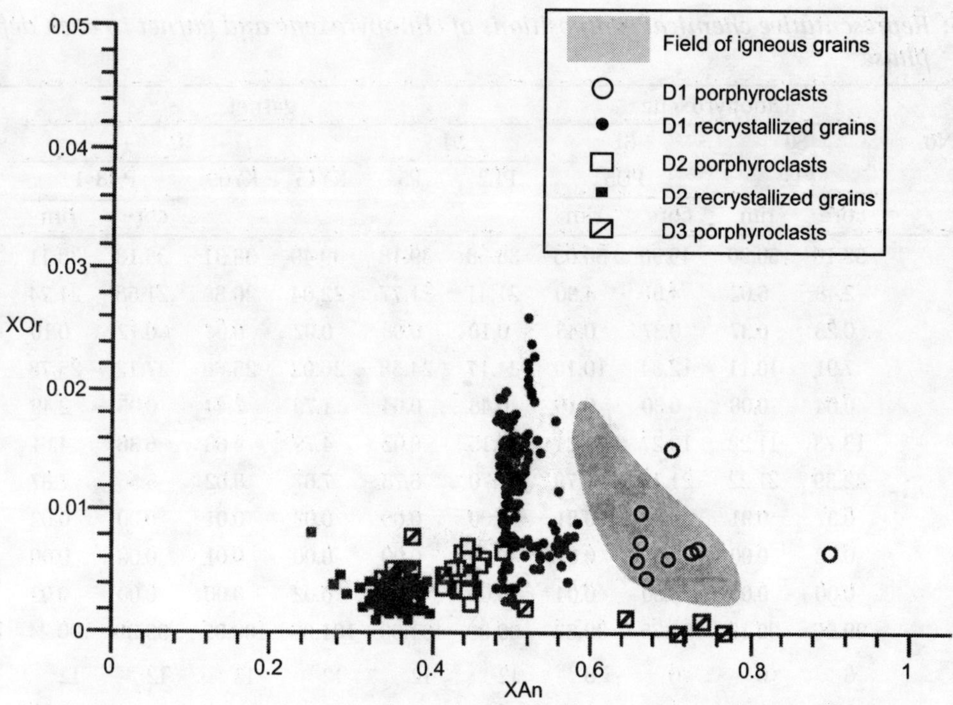

Fig. 9. Compositional variations of plagioclase in each deformation phase. Plot of anorthite (An) vs. orthoclase (Or) mole fraction.

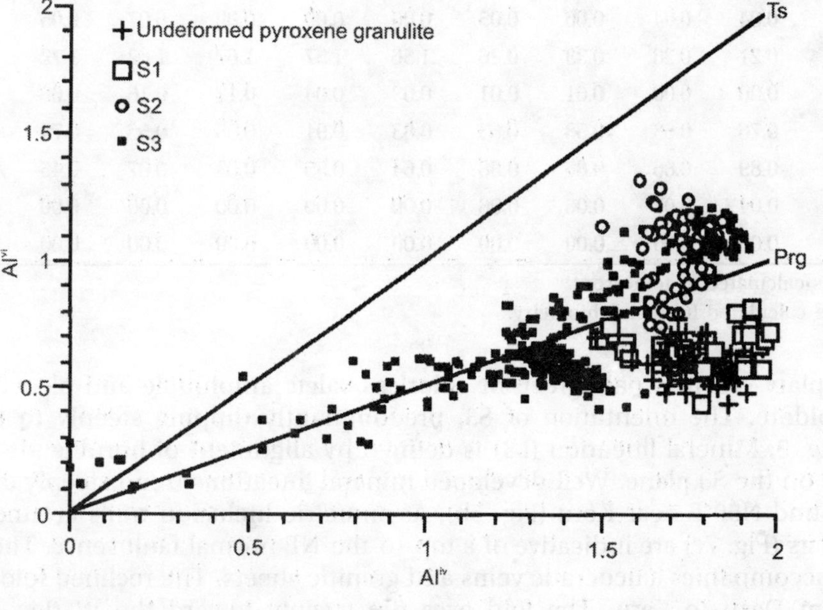

Fig. 10 (a)

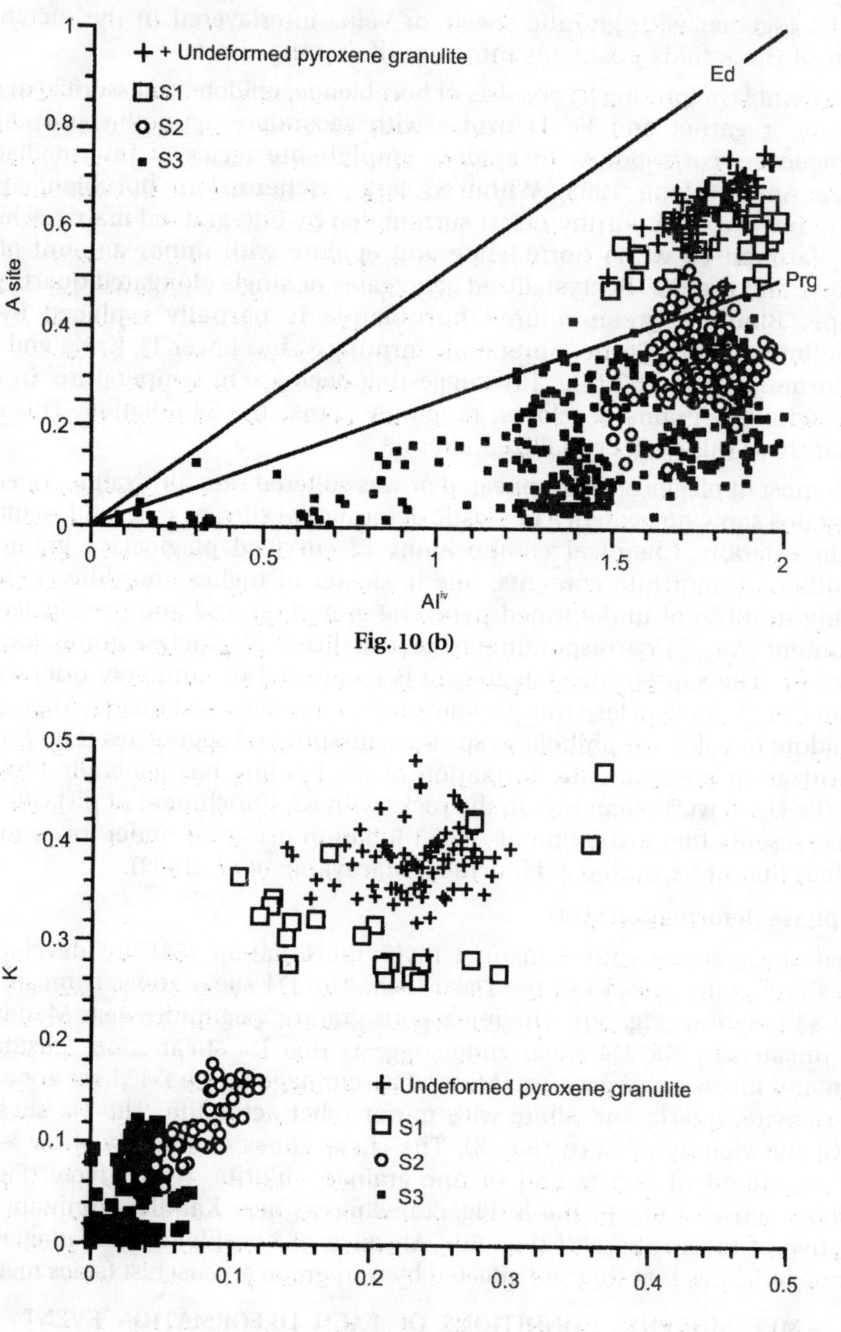

Fig. 10. Variations in compositions of amphibole in each deformation phase. Amphibole formulae were calculated on the basis of 13 CNK (Robinson *et al.*, 1982). (a) Plot of Al^{iv} vs. Al^{vi}. (b) Plot of Al^{iv} vs. A site occupancy. (c) Plot of Ti vs. K. Abbreviations: Ts = tschermakite; Ed = edenite; Prg = pargasite.

deflects by F3 together with granitic sheets or veins interlayered in the metabasites. The development of these folds postdates intrusion of granitic sheets.

Mineral assemblage forming S3 consists of hornblende, epidote, (saussuritized) plagioclase, quartz, sphene, ± garnet and Fe–Ti oxides with secondary actinolite and chlorite. This mineral paragenesis corresponds to epidote amphibolite facies at intermediate pressure (500–700 MPa: Apted & Liou, 1983). Within S3, larger tschermakitic hornblende grain (up to 15 wt.% Al_2O_3) behaves as a porphyroclast surrounded by fine-grained matrix, which consists of low Al_2O_3 (around 10 wt.%) hornblende and epidote with minor amount of actinolite, chlorite, quartz and titanite. Recrystallized aggregates or single elongated quartz grains form ribbon shape. Blue- to green-colored hornblende is partially replaced by low-Al_2O_3 (2–3 wt.%) actinolite and chlorite. Amphibole forming S3 has lower Ti, K, Na and Al contents than those forming S1 and S2 (Fig. 10), suggesting decrease in temperature. In most cases, chlorite and actinolite grains developed randomly across the S3 foliation. This observation indicates that these minerals crystallized after S3.

Although most of plagioclase disappeared or was suffered saussuritization, rarely preserved plagioclase grains show no evidence of plastic deformation such as core-and-mantle structure and subgrain rotation. Chemical compositions of survived plagioclase grains show two clusters of different anorthite contents; one is cluster of higher anorthite content (An_{50-60}) corresponding to those of undeformed pyroxene granulites and another cluster is of lower anorthite content (An_{30-40}) corresponding to recrystallized plagioclase grains forming the S2 foliation (Fig. 9). The saussuritized aggregate is composed of randomly oriented very fine-grained albite (An_{0-3}) or oligoclase (An_{20}), clinozoisite, muscovite and quartz. Mineral alignment of matrix epidote or calcic amphibole wraps the saussuritized aggregates (Fig. 7c). It suggests that saussuritization postdates the formation of S3. Epidote has generally higher pistasite component (Fe_2O_3> 5 wt.%) than that in the rocks with S2. Enrichment of pistasite component in epidote represents that formation of the S3 foliation occurred under lower metamorphic condition than that of S2 (Apted & Liou 1983; Maruyama *et al.*, 1983).

The fourth phase deformation (D4)

Localized shear zones with a distinct mylonitic foliation (S4) are developed in both amphibolites and granitic rocks in the Dasu Unit. The D4 shear zones truncate sharply the pre-existing S3 foliation (Fig. 5d). The micaceous granitic pegmatite near Mandraza cut the S3 and is truncated by the D4 shear zone suggests that D4 shear zones postdated the D3 deformation and intrusion of granites. Mineral assemblage in the D4 shear zones is chlorite, epidote, muscovite, quartz and albite with minor relict actinolite. The D4 shear zones are developed in the vicinity of Keru (Fig. 3). The shear zones commonly show *S–C* structure defined by alignment of quartz and of fine-grained chlorite, respectively (Fig. 7d). They represent shear sense of top-to-the-S (Fig. 6c), whereas near Kamila, dominant shear sense is an indicative of top-to-the-NW (Fig. 6d). Absence of hornblende and plagioclase within the shear zone indicates that they are affected by retrograde greenschist facies metamorphism.

METAMORPHIC CONDITIONS OF EACH DEFORMATION EVENT

The metamorphic mineral assemblages of each deformation phase are summarized in Table 2, which indicates that the later phase deformation took place under lower metamorphic grade. To compare the *P–T* conditions for each phase, the metamorphic conditions were estimated using the most common mineral assemblage (garnet–hornblende–plagioclase–

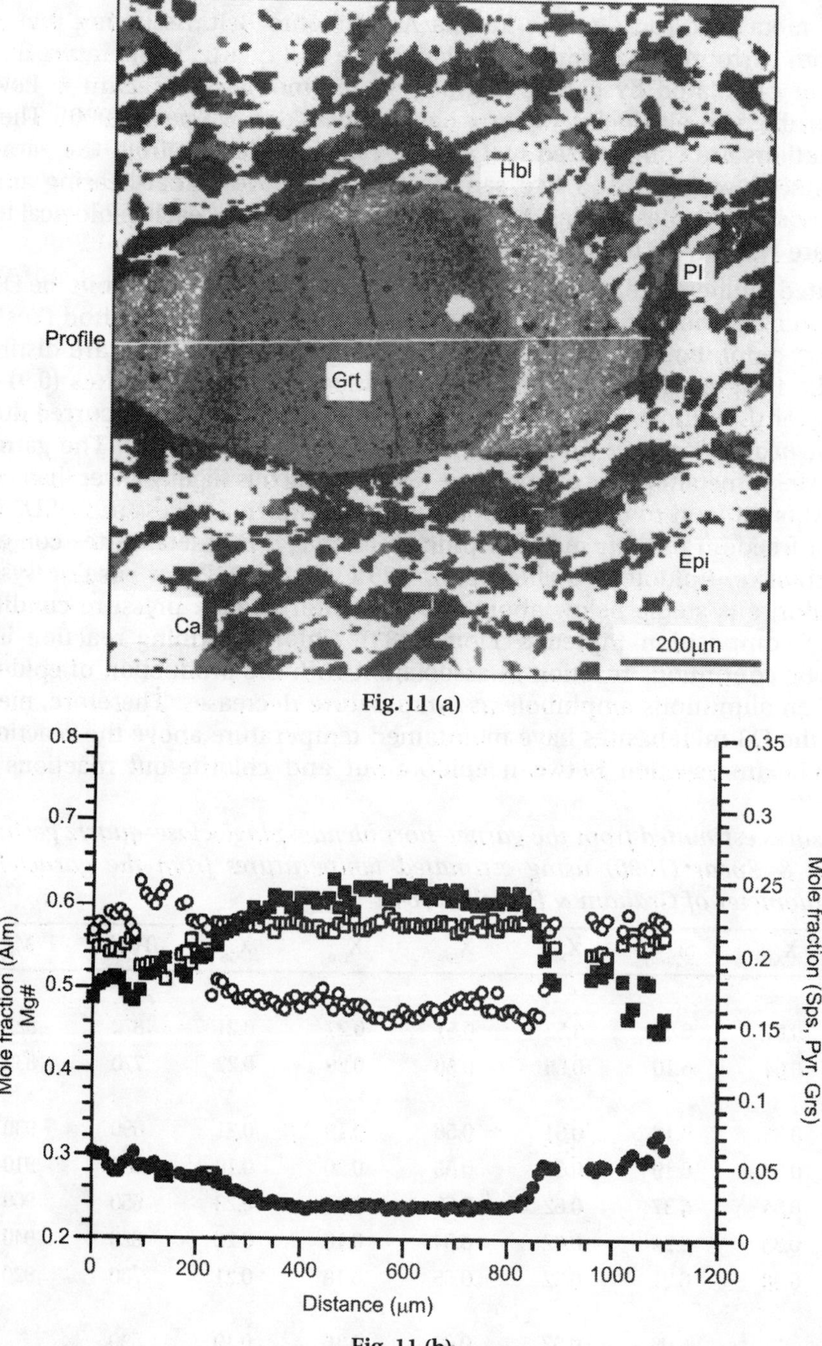

Fig. 11 (a)

Fig. 11 (b)

Fig. 11. Zoned garnet porphyroblasts within S2 foliation in sample PA3 from the SKA. (a) A Ca element map of garnet. Spot analyses at 5 μm intervals. Lighter pattern represents an enrichment in Ca content. Note discontinuous zoning in garnet between core and rim. (b) A compositional zoning profile in the garnet grain of (a). Mole fractions of almandine, pyrope, spessartine and grossular are denoted by closed square, open square, closed circle and open circle, respectively.

quartz) in the metabasic rocks of the Kamila Amphibolite Belt. Assuming that garnet rim equilibrated with surrounding hornblende, plagioclase and quartz, temperature and pressure conditions were calculated by garnet-hornblende thermometry (Graham & Powell, 1984) and garnet–hornblende–plagioclase–quartz barometry (Kohn & Spear, 1990). The results of the *P–T* calculations are summarized in Table 6. Two samples are from the garnet-bearing pyroxene granulites with S1, and five samples are from the garnet-bearing amphibolites within ansatomosing D2 shear zones. Metamorphic *P–T* paths in each geological units of the Kohistan Arc are tentatively illustrated in Fig. 12.

The estimated temperature (770–870°C) and pressure (0.83–0.93 GPa) from the D1 pyroxene granulites are in agreement with those obtained from the Al-zoning method (Yoshino *et al.*, 1998). As for D2 amphibolites, the estimated temperatures (600–750°C) are distinctly lower than that of the D1 granulites. The pressure condition of D2 amphibolites (0.91–0.99 GPa) are slightly higher than those of the D1 granulites, indicating that D2 occurred during lower temperature at nearly constant pressure, i.e. isobaric cooling (Fig. 12). The garnet-bearing D3 metabasite yields metamorphic temperature (590°C), which is slightly lower than temperature ranges of D2. It is difficult to obtain quantitatively metamorphic pressures of D3 because of large degree of freedom in their metamorphic assemblage. D3 metabasites contain epidote and exclude chlorite. Epidote is stable below 625°C at 700 MPa in basaltic system (QFM buffer) and chlorite is stable below about 550°C at intermediate pressure condition in the common basic composition (Apted & Liou, 1983). Chlorite-forming reaction is generally considered to be continuous reaction in association with the production of epidote and the breakdown of an aluminous amphibole as temperature decreases. Therefore, metamorphic conditions of the D3 metabasites have maintained temperature above the reaction curve of chlorite-out. The intersection between epidote-out and chlorite-out reactions locates at

Table 6: *Pressures estimated from the garnet–hornblende–plagioclase–quartz geobarometer of Kohn & Spear (1990) using estimated temperatures from the garnet–hornblende thermometer of Graham & Powell (1984).*

Sample No.	X_{An}	Si_{Amp}	X_{Mg}	X_{Alm}	X_{Prp}	X_{Grs}	T(°C)	P(Mg)†	P(Fe)†
S1									
P5	0.57	6.44	0.52	0.51	0.27	0.21	870	820	910
PL2	0.54	6.10	0.59	0.56	0.20	0.22	770	830	920
S2									
KY15	0.35	6.17	0.51	0.56	0.19	0.21	690	930	1000
KY09	0.35	6.19	0.53	0.55	0.20	0.19	680	910	980
PA3-1	0.35	6.37	0.62	0.52	0.20	0.23	650	900	970
PA3-2	0.33	6.29	0.62	0.54	0.18	0.22	620	940	950
PU4	0.38	6.21	0.52	0.55	0.16	0.21	700	920	920
S3									
MA2*		6.06	0.57	0.63	0.16	0.19	590		

* plagioclase-free assemblage (D3).

† P(Mg) and P(Fe) represent the pressures in MPa estimated from the calibrations of Mg- and Fe-endmember reactions, respectively.

Fe^{3+} contents in hornblende were estimated by average option of Spear & Kimball (1982).

T~550°C and P~330 MPa (Spear, 1993). The formation of D3 structure underwent under slightly higher P conditions than the intersection (Fig. 12). Retrogressive overprint developed in matrix grain includes chlorite and actinolite, which are characteristic assemblage of metabasite in the greenschist facies. Since the actinolite grains within matrix of the D3 metabasites have little glaucophane component, the metamorphic pressure can be restricted less than 500 MPa at the retrogression period (Otsuki & Banno, 1990). It suggests that the Kamila Amphibolite Belt underwent the decompression at least ΔP~400 MPa during D3. The metamorphic temperature between 550°C and 625°C deduced from the mineral assemblage leads to a conclusion that D3 structure formed under isothermal decompression around 600°C.

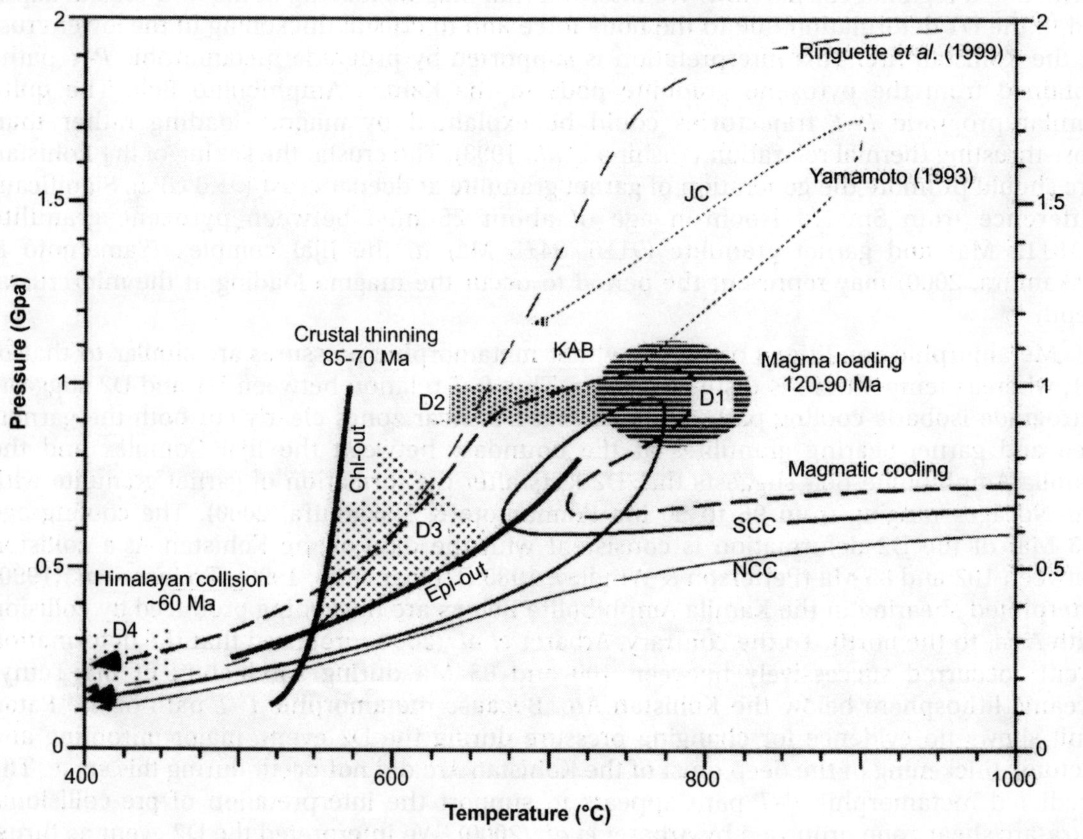

Fig. 12. Predicted P–T paths of each geological unit belonging to the lower crustal sequence of the Kohistan arc along the Indus Valley route. *P–T* path estimations of the Jijal Complex after Yamamoto (1993) and Ringuette *et al.* (1999), and *P–T* path of the NKA and the Chilas Complex modified after Yoshino *et al.* (1998) and Yamamoto (1993). Abbreviations: JC, Jijal Complex; KAB, Kamila Amphibolite Belt; SCC, the southern part of Chilas Complex; NCC, the northern part of Chilas Complex.

Localized D4 shear zones involve chlorite, epidote, muscovite, quartz and albite with minor relict actinolite are developed in both metabasites and granitic rocks. The assemblage indicates greenschist facies metamorphic temperatures of approximately 300°C at low-pressure condition (Frey *et al.*, 1991). The D4 shear zones did not accompany a substantial pressure change and the lower part of the Kohistan Arc underwent further cooling.

TECTONO-METAMORPHIC EVOLUTION OF THE LOWER CRUST OF THE KOHISTAN ARC

Integrating our structural and metamorphic data we distinguish three main events of fabric generation in the tectono-metamorphic evolution of the lower crust of the Kohistan Arc (Fig. 13): (1) crustal thickening, (2) crustal thinning and (3) collision.

Crustal thickening event (D1 and D2)

The general absence of asymmetric kinematic indicators of D1 suggests that D1 fabrics formed in a regional coaxial flow. We interpret that magma loading at the mid-crustal depth led to the D1 deformation due to the body force and to crustal thickening in the lower crust of the Kohistan Arc. This interpretation is supported by prograde metamorphic *P–T* paths obtained from the pyroxene granulite pods in the Kamila Amphibolite Belt. The quite similar prograde *P–T* trajectories could be explained by magma loading rather than post-thrusting thermal relaxation (Yoshino *et al.*, 1998). The crustal thickening of the Kohistan Arc should promote the generation of garnet granulite at deeper crust (>1.0 GPa). Significant difference from Sm–Nd isochron age of about 25 m.y. between pyroxene granulite (118±12 Ma) and garnet granulite (91±6, 94±5 Ma) in the Jijal complex (Yamamoto & Nakamura, 2000) may represent the period to occur the magma loading at the mid-crustal depth.

Metamorphic conditions of D2 show that metamorphic pressures are similar to that of D1, whereas temperature is distinctly lower. This *P–T* relation between D1 and D2 suggests retrograde isobaric cooling path. The fact that D2 shear zones clearly cut both the garnet-free and garnet-bearing granulites at the boundary between the Jijal Complex and the Kamila Amphibolite Belt suggests that D2 lasts after the formation of garnet granulite with Sm–Nd ages ranging from 96 to 90 Ma (Yamamoto & Nakamura, 2000). The cooling age (83 Ma) of the D2 deformation is consistent with the date of the Kohistan–Asia collision between 102 and 85 Ma (Petterson & Windley, 1985; Treloar *et al.*, 1990). Treloar *et al.* (1990) interpreted shearing in the Kamila Amphibolite Belt as arc thickening produced by collision with Asia, to the north. To the contrary, Arbaret *et al.* (2000) proposed that the deformation events occurred successively between 100 and 83 Ma during subduction of the Tethys oceanic lithosphere below the Kohistan Arc. Because metamorphic *P–T* path of the Patan Unit shows no evidence for changing pressure during the D2 event, major unroofing and tectonic thickening of the deep crust of the Kohistan Arc did not occur during this stage. The predicted metamorphic *P–T* path appears to support the interpretation of pre-collisional intra-arc shear zone proposed by Arbaret *et al.* (2000). We interpreted the D2 event as thrust imbrication on the subduction zone-hanging wall due to the northward subduction of the Tethys oceanic slab after the high-pressure granulite facies metamorphism caused by magma loading. Formation of abundant garnet-bearing rocks at the deep crust would lead to the development of the D2 shear zones.

Crustal thinning event (D3)

Retrograde *P–T* paths in the Kamila Amphibolite Belt inferred from mineral assemblages and mineral chemistries for D2 and D3 events represent isothermal decompression in temperature range between 550 and 625°C (Fig. 12). These retrograde *P–T* paths show quite

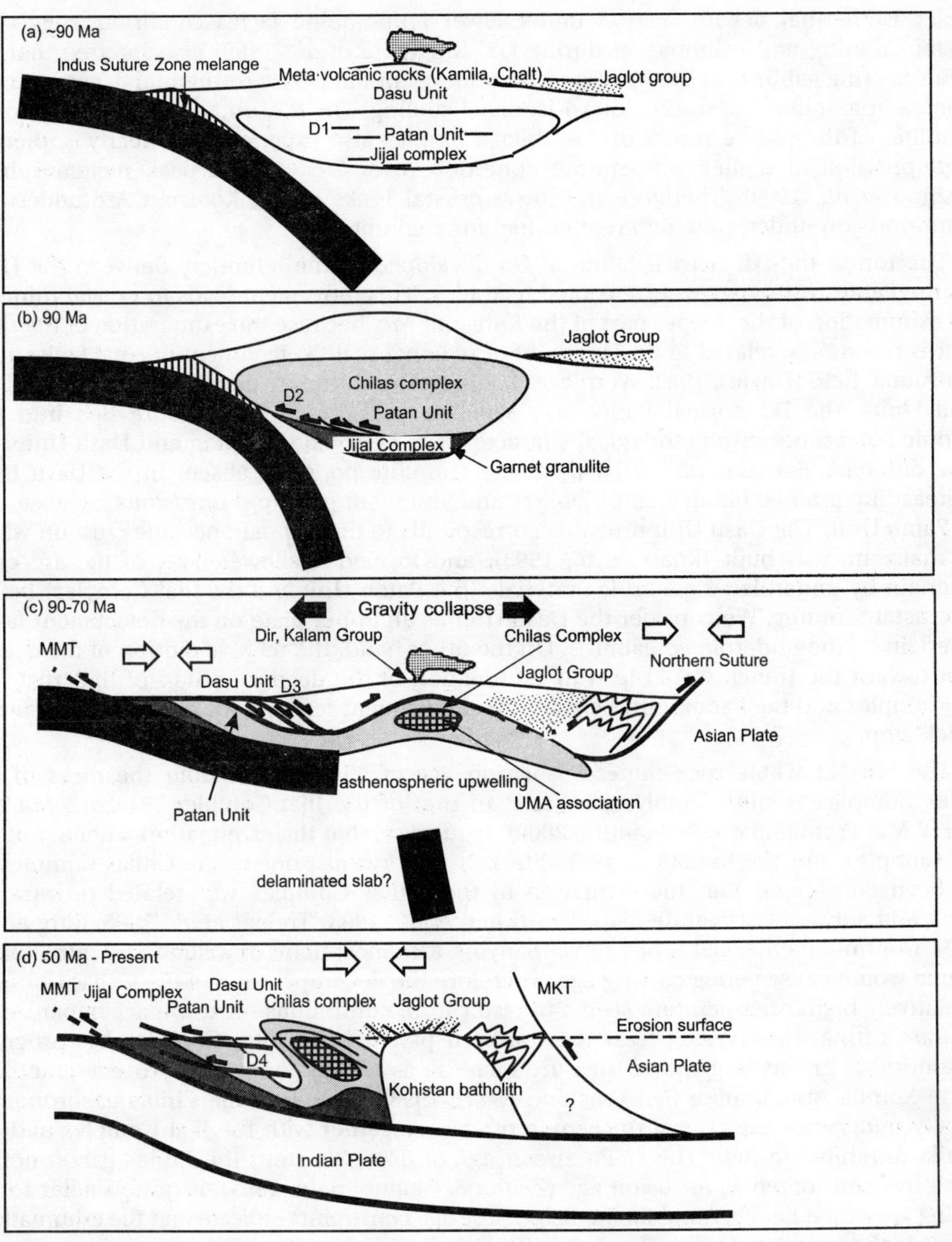

Fig. 13. Schematic model of the tectonic evolution in the lower crust of the Kohistan arc. (a) D1 stage is characterized by horizontal pure shear in the lower crust due to the magma loading at the mid-crustal depth. (b) D2 stage shows that localized shearing occurred at the boundary between the Jijal garnet granulites and the garnet-free rocks above the subduction thrust. (c) Crustal thinning resulted from gravitically collapsed. (d) D4 stage was triggered by convergence due to the main Himalayan collision and then antiform of the Chilas Complex and tilting of the lower crustal sequence occurred.

similar isothermal decompression under lower amphibolite facies condition, suggesting crustal thinning and exhumation during D3. Ringuette *et al.* (1999) also inferred that the garnet-bearing gabbros of the Jijal Complex underwent the nearly isothermal decompression under amphibolite facies after quasi-isobaric cooling. On the other hand, the pyroxene granulites of the southern part of the Chilas Complex also experienced a nearly isothermal decompression at higher temperature condition (over 700°C) after peak metamorphism (Yoshino *et al.*, 1998). Therefore the lower crustal rocks of the Kohistan Arc underwent decompression under quite different temperature conditions.

The top-to-the-NE normal faults of D3 developed at the boundary between the Dasu and the Patan Units suggest extensional tectonics. This movement leads to crustal thinning and exhumation of the deeper part of the Kohistan Arc, because the exhumation of the deep crust is commonly related to development of regional scale detachment normal faults at an extensional field (Davies, 1983; Wernicke, 1985). At the boundary between the Patan and the Dasu Units, the D3 normal faults may separate the Kamila Amphibolite Belt into two tectonic blocks, because petrological characteristics between the Patan and Dasu Units are quite different. For example, relict pyroxene granulite pods are absent in the Dasu Unit, whereas fine-grained banded amphibolites and abundant granitoid intrusions are absent in the Patan Unit. The Dasu Unit probably corresponds to the remnant oceanic crust on which the main arc was built (Khan *et al.*, 1993), and formed shallower level of the arc crust underlain by underplated gabbroic materials (the Patan Unit and the Jijal Complex) before the crustal thinning. We consider the Dasu Unit as an upper plate on the detachment faults underlain by the underplated gabbros. On the other hand, the relative motion of the deeper crust toward the trench would lead to exhumation of the deeper portion of the crust (the Jijal Complex and the Patan unit) beneath the detachment faults, if the oceanic slab acted as a back-stop.

The Sm–Nd whole rock–mineral isochron age of 69.5±9.3 Ma from the rocks of the Chilas Complex is considerably younger than that of the Jijal Complex (91.0±6.3 Ma and 118±12 Ma) (Yamamoto & Nakamura, 2000). It suggests that the exhumation (cooling) of the Jijal Complex and the Kamila amphibolite Belt underwent prior to the Chilas Complex. It has been considered that the formation of the Chilas Complex was related to intra-arc rifting and subsequent mantle diapirism (Khan *et al.*, 1993; Treloar *et al.*, 1996; Burg *et al.*, 1998). Continuous thermal input accompanying asthenospheric upwelling and intrusion of magma would cause young cooling age. Therefore the decompression of the Chilas Complex at relatively higher temperature seems to result from continuous intrusion accompanied by intra-arc rifting. However, at least the southern part of the Chilas Complex has prograde metamorphic history suggesting pressure increase as well as the relict pyroxene granulites in the Kamila Amphibolite Belt (Yoshino *et al.*, 1998). Therefore the Chilas gabbronorites initially underwent the crustal thickening process, together with the Jijal Complex and the Kamila Amphibolite Belt. The U–Pb zircon age of 84±5 Ma from the Chilas gabbronorite, which has considered as intrusion age (Zeitler & Chamberlain, 1991), is quite similar to the cooling age of the Kamila Amphibolite Belt. These age constraints indicate that the exhumations of the deeper crust at fore-arc and the rifting at intra-arc occurred in the Kohistan at the short time interval.

Two cases can be considered as driving force for crustal thinning. First Treloar *et al.* (1996) considered emplacement of the Chilas Complex as extension by subduction rollback after suturing of the arc to Asia. However, the slab retreating makes it difficult to occur emplacement onto the foreland lithosphere of dense materials such as ultramafic bodies (Royden, 1993; Waschbusch & Beaumont, 1996). In addition, there is no clear trace of wide accretion prism between the Kohistan Arc and the Indian plate, although subduction retreat should facilitate frontal accretion to refill the space (Waschbusch & Beaumont, 1996). Therefore, the exhumation of dense lower crust requires other process rather than slab retreating.

Alternative interpretation is gravity collapse of abnormal thickened crust after lithospheric delamination. Tectonic denudation including extensional faults and detachments is a well-documented process for syn- to post-collisional exhumation of deep crustal rocks in many orogenic belts due to catastrophic removal of a thickened thermal boundary layer in the lithospheric mantle of the deep root of the orogen (Dewey, 1988; Platt & Vissers, 1989; Chemenda *et al.*, 2000). After lithospheric delamination, abnormal thickened crust with high topography becomes gravitationally unstable and should collapse by body force, even if convergence continued. The predicted isobaric cooling paths from the lower part of the Kohistan Arc represent significant cooling after peak metamorphism. Then lithospheric mantle below the Kohistan Arc would increase the thickness. The delamination of lithospheric mantle would occur when the lithosphere below the Kohistan Arc was cooler and denser than the surrounding athenospheric mantle.

Main Himalayan collision event (D4)

The D4 developed under low *P–T* conditions (Fig. 12). Because the D4 shear zones develop within micaceous pegmatitic rocks as the last magmatic activity (66 Ma: Treloar *et al.*, 1989), this shear deformation presumably occurred after 66 Ma. The corresponding tectonic event may be the Main Himalayan collision between the Indian and Asian continents initiated at Eocene (Patriat & Achache, 1985). The movement of the D4 shear zones indicating top-to-the-S shear sense is consistent with northward subduction of the Tethys oceanic plate. On the other hand, the D4 shear zones with top-to-the-NW sense may be related to doming of the Nanga Parbat-Haramosh massif, showing northwestward thrusting of Indian crustal rocks over the Kohistan Arc (Zeitler, 1985; Butler & Prior, 1988; Zeitler *et al.*, 1989; George *et al.*, 1995).

CONCLUSIONS

The growth of arc's crust occurred during D1 and D2. D1 is characterized by a penetrative foliation with high-grade mineral paragenesis. This structure occurred as an intra-arc deformation during crustal thickening by magma accretion at middle crustal depth. D2 is characterized by development of anastomosed shear zones with SW thrust sense under upper amphibolite facies. This shear event lasted after formation of garnet granulites in the Jijal complex during nearly isobaric cooling. The shear zones may represent pre-collisional deformation during subduction of the Tethys oceanic lithosphere below the Kohistan Arc.

The exhumation of deep arc rust occurred during D3, showing the top-to-the-NW normal faulting under lower amphibolite facies. High-pressure rocks forming the lower arc could exhume to the surface due to the back-stop of oceanic lithosphere. We interpret that the crustal thickening could trigger lithospheric delamination and gravity collapse of thickened crust. These relationships suggest gravitational collapse of the thickened crust as possible cause for exhumation of deep-seated rocks of the Kohistan Arc. Sub-horizontal top-to-the-S and top-to-the-NW D4 shear zones formed under greenschist facies after the Kohistan–Asia collision. This event locally occurred after onset of the Main Himalayan collision.

ACKNOWLEDGMENTS

The manuscript has benefited from discussions with M. Toriumi and G. Kimura. Early version of the manuscript has improved by critical comments of J.P. Burg, J. Miller and P.J. Treloar. Electron-probe work was carried out in the Geological Institute, University of Tokyo with a helpful assistance of H. Yoshida. This work is partially supported by the Grant-in-Aids of the Fukada Geological Institute. We thank H. Thomas who gave us the opportunity to present this work in this special volume.

REFERENCES

Apted, M.J. & Liou, J.G., 1983. Phase relations among greenschist, epidote-amphibolite, and amphibolite in a basaltic system. *American Journal of Science,* **283A,** 328-54.

Arbaret, L., Burg, J.-P., Zeillinger, G., Chaudhry, N., Hussain, S. & Dawood, H., 2000. Pre-collisional anastomosing shear zones in the Kohistan arc, NW Pakistan. *Geological Society of London, Special Publications,* **170**, 295-311.

Bard, J.-P., 1983. Metamorphism of an obducted island arc: example of the Kohistan sequence (Pakistan) in the Himalayan collided range. *Earth and Planetary Science Letters,* **65**, 133-44.

Burg, J.-P., Bodinier, J.-L., Chaudhry, S., Hussain, S. & Dawood, H., 1998. Infra-arc-mantle transition and intra-arc mantle diapirs in the Kohistan Complex (Pakistani Himalaya): petro-structural evidence. *Terra Nova,* **10,** 74-80.

Butler, R.W.H. & Prior, D.J., 1988. Tectonic controls on the uplift of the Nanga Parbat Massif, Pakistan Himalayas. *Nature,* **333**, 247-50.

Chemenda, A.I., Burg, J.-P. & Mattauer, M., 2000. Evolutionary model of the Himalaya-Tibet system: geopoem based on new modelling, geological and geophysical data. *Earth and Planetary Science Letters,* **174**, 397-409.

Coward, M.P., Butler, R.W.H., Khan, M.A. & Knipe, R.J., 1987. The tectonic history of Kohistan and its implications for Himalayan structure. *Journal of the Geological Society of London,* **144**, 377-91.

Coward, M.P., Jan, M.Q., Rex, D.C., Tarney, J., Thirlwall, M. & Windley, B.F., 1982. Geo-tectonic framework of the Himalaya of N. Pakistan. *Journal of the Geological Society of London,* **139**, 299-308.

Coward, M.P., Broughton, R.D., Luff, I.W., Petterson, M.G., Pudsey, C.J., Rex, D.C. & Khan, M.A., 1986. Collision tectonics in the NW Himalayas. *Geological Society of London, Special Publications,* **19**, 203-19.

Davies, G.H., 1983. Shear-zone model for the origin of metamorphic core complexes. *Geology,* **11**, 342-47.

Dewey, J.F., 1988. Extensional collapse of orogens. *Tectonics,* **7,** 1123-39.

Frey, M., de Capitani, C. & Liou, J.G., 1991. A new petrogenetic grid for low-grade metabasites. *Journal of Metamorphic Geology,* **9**, 497-509.

George, M., Reddy, S. & Harris, N.B.W., 1995. Isotopic constraints on the cooling history of the Nanga Parbat–Haramosh Massif and Kohistan arc, western Himalaya. *Tectonics,* **14**, 237-52.

Graham, C.M. & Powell, R., 1984. A garnet-hornblende geothermometer: calibration, testing, and application to the Perona schist, Southern California. *Journal of Metamorphic Geology,* **2**, 13-21.

Jan, M.Q., 1988. Geochemistry of amphibolites from the southern part of Kohistan arc, N. Pakistan. *Mineralogical Magazine,* **52**, 147-59.

Jan, M.Q. & Howie, R.A., 1980. Ortho- and clinopyroxenes from pyroxene granulites of Swat Kohistan, northern Pakistan. *Mineralogical Magazine,* **43**, 715-28.

Jan, M.Q. & Howie, R.A., 1981. The mineralogy and geochemistry of the metamorphosed basic and ultrabasic rocks of the Jijal Complex, Kohistan, NW Pakistan. *Journal of Petrology,* **22**, 85-126.

Jan, M.Q. & Windley, B.F., 1990. Chromian spinel-silicate chemistry in ultramafic rocks of the Jijal Complex, northwest Pakistan. *Journal of Petrology,* **31**, 667-715.

Ji, S. & Mainprice, D., 1987. Natural deformation fabrics of plagioclase: implications for slip systems and seismic anisotropy. *Tectonophysics,* **147**, 145-63.

Khan, M.A., Jan, M.Q. & Weaver, B.L., 1993. Evolution of the lower arc crust in Kohistan: temporal arc magmatism through early, mature and intra-arc rift stages. *Geological Society of London, Special Publications,* **74**, 123-38.

Khan, M.A., Jan, M.Q., Windley, B.F., Tarney, J. & Thirlwall, M.F., 1989. The Chilas mafic-ultramafic igneous complex; the root of the Kohistan island arc in the Himalaya of northern Pakistan. *Geological Society of America, Special Papers,* **232**, 75-94.

Khan, M.A., Stern, R.J., Gribble, R.F. & Windley, B.F., 1997. Geochemical and isotopic constraints on subduction polarity, magma sources, and paleogeography of the Kohistan intra-oceanic arc, northern Pakistan Himalaya. *Journal of the Geological Society of London,* **154**, 935-46.

Kohn, M.J. & Spear, F.S., 1990. Two new geobarometers for garnet amphibolites with applications to southeastern Vermont. *American Mineralogists,* **75**, 89-96.

Maruyama, S., Suzuki, K. & Liou, J.G., 1983. Greenschist–amphibolite transition equilibria at low pressures. *Journal of Petrology,* **24**, 583-604.

Miller, D.J., Loucks, R.R. & Ashraf, M., 1991. Platinum-group elements mineralization in the Jijal layered ultramafic–mafic complex, Pakistani Himalayas. *Economic Geology,* **86**, 1093-1102.

Otsuki, M. & Banno, S., 1990. Prograde and retrograde metamorphism of hematite-bearing basic schists in the Sanbagawa belt, central Shikoku. *Journal of Metamorphic Geology,* **8**, 425-39.

Patriat, P. & Achache, J., 1985. India–Eurasia collision chronology and its implications for crustal shortening and driving mechanism of plates. *Nature,* **311**, 615-21.

Petterson, M.G. & Windley, B.F., 1985. Rb–Sr dating of the Kohistan arc-batholith in the Trans-Himalaya of N. Pakistan and tectonic implications. *Earth and Planetary Science Letters,* **74**, 54-75.

Petterson, M.G. & Windley, B.F., 1992. Field relations, geochemical and petrogenesis of the Cretaceous basaltic Jutal dykes, Kohistan, northern Pakistan. *Journal of the Geological Society of London,* **149**, 107-14.

Platt, J.P. & Vissers, R.L.M., 1989. Extensional collapse of thickened continental lithosphere: A working hypothesis for the Alboran Sea and Gibraltar arc. *Geology,* **17**, 540-43.

Pudsey, C.J., 1986. The Northern Suture, Pakistan: margin of a Cretaceous island arc. *Geological Magazine,* **123**, 405-23.

Raase, P., Raith, M., Ackermand, D. & Lal, R.K., 1986. Progressive metamorphism of mafic rocks from greenschist to granulite facies in the Dhawar craton of southern India. *Journal of Geology,* **94**, 261-82.

Ringuette, L., Martignole, J. & Windley, B.F., 1999. Magmatic crystallization, isobaric cooling, and decompression of the garnet-bearing assemblages of the Jijal sequence (Kohistan terrane, western Himalayas). *Geology,* **27**, 139-42.

Robinson, P., Spear, F.S., Schmacher, J.C., Laird, J., Klein, C., Evans, B.W. & Doolan, B.L., 1982. Phase relations of metamorphic amphiboles: natural occurrence and theory. *Reviews in Mineralogy,* **9B**, 1-227.

Royden, L.H., 1993. The tectonic expression slab pull at continental convergent boundaries. *Tectonics,* **12**, 303-25.

Rutter, E.H. & Brodie, K.H., 1992. Rheology of the lower crust. In: *Continental Lower crust* (eds. Fountain, D.M., Arculus, R. & Amsterdam, R.W.) Elsevier, 201-58.

Searle, M.P., Khan, M.A., Jan, M.Q., DiPietro, J.A., Pogue, K.R., Pivnik, D.A., Sercombe, W.J., Izatt, C.N., Blisniuk, P.M., Treloar, P.J., Gaetani, M. & Zanchi, A., 1998. Geological map of northern Pakistan and adjacent areas of northern Ladakh and western Tibet. scale 1:650,000.

Spear, F.S., 1993. Metamorphic phase equilibria and pressure–temperature–time path. Washington, D.C., Mineralogical Society of America, 799.

Spear, F.S., 1981. An experimental study of hornblende stability and compositional variability in amphibolite. *American Journal of Science,* **281**, 697-734.

Spear, F.S. & Kimball, K.L., 1984. RECAMP–A FORTRAN IV program for estimating Fe^{3+} contents in amphiboles. *Computers in Geology,* **10**, 317–25.

Tahirkheli, R.A.K., Mattauer, M., Proust, F. & Tapponier, P., 1979. The India Eurasia suture zone in northern Pakistan: synthesis and interpretation of recent data at plate scale. In: *Geodynamics of Pakistan* (eds. Farah, A. & De Jong, K.A.), Geological Survey of Pakistan, Quetta, 125-30.

Treloar, P.J., Petterson, M.G., Jan, M.Q. & Sullivan, M.A., 1996. A re-evaluation of the stratigraphy and evolution of the Kohistan arc sequence, Pakistan Himalaya: implications for magmatic and tectonic arc-building processes. *Journal of the Geological Society of London,* **153**, 681–93.

Treloar, P.J., Brodie, K.H., Coward, M.P., Jan, M.Q., Khan, M.A., Knipe, R.J., Rex, D.C. & Williams, M.P., 1990. The evolution of the Kamila shear zone, Kohistan, Pakistan. In: *Exposed cross-sections of the continental crust* (eds. Salisbury, M.H. & Fountain, D.M.), Kluwer Academic Publishers, Netherlands, 175-214.

Treloar, P.J., Rex, D.C., Guise, P.G., Coward, M.P., Searle, M.P., Windley, B.F., Petterson, M.G., Jan, M.Q. & Luff, I.W., 1989. K–Ar and Ar–Ar geochronology of the Himalayan collision in NW Pakistan: Constraints on the timing of suturing, deformation, metamorphism and uplift. *Tectonics*, **8**, 881-909.

Wartho, J.-A., Rex, D.C. & Guise, P.G., 1996. Excess argon in amphiboles linked to greenschist facies alteration in the Kamila Amphibolite Belt, Kohistan island arc system, northern Pakistan: insights from $^{40}Ar/^{39}Ar$ step-heating and acid leaching experiments. *Geological Magazine*, **133**, 595-609.

Waschbusch, P. & Beaumont, C., 1996. Effect of a retreating subduction zone on deformation in simple regions of plate convergence. *Journal of Geophysical Research*, **101**, 28133-48.

Wernicke, B., 1985. Uniform-sense normal simple shear of the continental lithosphere. *Canadian Journal of Earth Science*, **22**, 108-25.

Yamomoto, H., 1993. Contrasting metamorphic *P–T*–time paths of the Kohistan granulites and tectonics of the western Himalayas. *Journal of the Geological Society of London*, **150**, 843-56.

Yamomoto, H. & Nakamura, E., 1996. Sm–Nd dating of garnet granulites from the Kohistan complex, northern Pakistan. *Journal of the Geological Society of London*, **153**, 965-69.

Yamomoto, H. & Nakamura, E., 2000. Timing of magmatic and metamorphic events in the Jijal Complex of the Kohistan arc deduced from Sm–Nd dating of mafic granulites. *Geological Society of London, Special Publications*, **170**, 313-19.

Yamomoto, H. & Yoshino, T., 1998. Superposition of replacements in the mafic granulite of the Jijal Complex of the Kohistan arc, northern Pakistan: Dehydration and rehydration within deep arc crust. *Lithos*, **43**, 219-34.

Yoshino, T., Yamomoto, H., Okudaira, T. & Toriumi, M., 1998. Crustal thickening of the lower crust of the Kohistan arc (N. Pakistan) deduced from Al zoning in clinopyroxene and plagioclase. *Journal of Metamorphic Geology*, **16**, 729-48.

Zeitler, P.K., 1985. Cooling history of the N.W. Himalaya, Pakistan. *Tectonics*, **4**, 127-51.

Zeitler, P.K. & Chamberlain, C.P., 1991. petrogenetic and tectonic significance of young leucogranites from the northwestern Himalaya, Pakistan. *Tectonics*, **10**, 729-41.

Zeitler, P.K., Sutter, J.F., Williams, I.S., Zartman, R. & Tahirkheli, R.A.K., 1989. Geochronology and temperature history of the Nanga–Parbat–Haramosh massif, Pakistan. *Geological Society of London, Special Publications*, **74**, 1-22.

9

EVIDENCE OF UPPER AMPHIBOLITE FACIES METAMORPHISM FROM ALMORA NAPPE, KUMAUN HIMALAYA

MALLICKARJUN JOSHI* AND A.N. TIWARI*

ABSTRACT

Pelitic schists and gneisses of Saryu Formation comprising the higher structural levels of Almora Nappe have escaped the thrust related Tertiary mylonitization. These gneisses and associated diatexites are products of upper amphibolite facies metamorphism that crossed the first sillimanite isograd. The peak metamorphic conditions were $\geq$ 700°C at 7.5 ± 0.4 kbar.

These schists and gneisses have been folded by tight to isoclinal post metamorphic folds resulting in the repetition of reaction isograds on map scale. The folded sequence is intruded by Early Palaeozoic granitoids and the attendant contact metamorphic textures overprint the regional foliation. The upper amphibolite facies metamorphism is highly likely to be of Pre-Cambrian age.

KEY WORDS

Almora Nappe; amphibolite facies; Almora gneisses; diatexites; mylonites; *P–T* conditions; sillimanite isograd.

INTRODUCTION

Tectonometamorphic reworking of leading edge of the Indian plate during the Cenozoic continent-continent collision has largely modified the pre-Himalayan structural and metamorphic signatures of these rocks comprising Himalayan Metamorphic Belt (HMB) in the Higher Himalaya. One of the major tectonic consequence of the collision was the Eocene-Oligocene large scale southward transport of the HMB over the Main Central Thrust (MCT) (De Celles *et al.*, 2001). Thus the HMB in the Lesser Himalaya is represented by the relicts of a once continuous basal shear zone with MCT comprising the sole thrust delimiting the overlying metamorphics from the underlying sedimentaries of the Lesser Himalaya. This basal thrust is known by various local names from the far western to the far eastern Lesser Himalaya. Almora Nappe is one of the largest nappes of Lesser Himalaya and attracted the attention of Middlemiss way back in 1887. Auden (1937) believed the Central Crystallines to be the root zone of Almora Nappe. Heim and Gansser (1939) and Gansser (1964) identified the Almora Nappe with the North Almora Thrust and the South Almora Thrust as its boundaries. Valdiya's (1980) stratigraphy of the Almora Nappe was modified by Joshi (1999)

* Department of Geology, Banaras Hindu University, Varanasi-221 005, India.

and the latter has been followed here. Lithotectonically speaking the Almora Nappe comprises two units, viz. the Ramgarh Nappe and the tectonically overlying Almora Nappe (sensu stricto). However, the term Almora Nappe in the geological literature of Kumaun Himalaya is entrenched to mean the composite nappe which includes both the Ramgarh and Almora nappes. The Ramgarh Group of rocks comprising the basal part of Almora Nappe have undergone intense mylonitization close to the sole of the nappe. The degree of mylonitization decreases upsection (Joshi, 1999). The schists and gneisses of Almora Group comprising the central parts of nappe have largely escaped mylonitization (Fig. 1).

This paper presents the metamorphic evolution of Almora gneisses and describes a detailed petrology of these metapelitic gneisses with particular emphasis on the relationship between deformation and metamorphism. These gneisses dated at 1865 ± 60 Ma (Rb - Sr, Trivedi *et al.*, 1984) are likely to be the relicts of Pre-Cambrian metamorphics which have escaped the Tertiary metamorphism. The Almora gneisses were metamorphosed under high - T, intermediate - P conditions and represent the culmination of regional metamorphism in Almora Nappe. The paper presents a $P - T$ array for the Almora gneisses which underwent isobaric cooling following the peak metamorphism.

REGIONAL GEOLOGICAL SETTING

Along the Chhara–Someshwar transect (Fig. 1), the Almora Nappe comprises two distinct lithostratigraphic units, viz. the Ramgarh Group and the Almora Group. Ramgarh Group is the lower tectonic subdivision of the Almora Nappe and is separated from the underlying sedimentaries by the North Ramgarh Thrust (NRT) and South Ramgarh Thrust (SRT) in the northern and southern flanks respectively (Joshi, 1999). Similarly, the Ramgarh Group is tectonically separated from the overlying Almora Group by North Almora Thrust (NAT) and South Almora Thrust (SAT) respectively in northern and southern flanks of the nappe.

The Ramgarh Group of rocks are exposed both in the northern and southern extremes of the nappe, and have been divided into two formations, viz. the Debguru Porphyroids at the base followed higher up in the succession by the Nathuakhan Formation. The Debguru Porphyroids comprise hard, black and greenish black protomylonites, mylonites, ultramylonites and foliated ultramylonites derived by reworking of augen granitic gneisses. The Nathuakhan Formation comprises brown, gray and olive green mylonites and ultramylonites derived from phyllites, schists, fine grained quartzites and metasiltstones. All the rock types of Ramgarh Group are largely mylonitized on account of their involvement in large scale thrusting during emplacement of the nappe and have undergone low grade metamorphism (phyllonites) under greenschist facies conditions.

The Almora Group comprises medium- to high-grade metapelites of the Saryu Formation and the low-grade metapelites of the Gumalikhet Formation. The Saryu Formation is constituted of mica schists, succeeded by alternating sequence of garnetiferous mica schists, kyanite schists and micaceous quartzites. Upper part of the Formation is characterized by lenses and bands of K-feldspar-sillimanite bearing gneisses and migmatites (Fig. 2a). The average strike of gneisses of Saryu Formation varies from E-W to WNW-ESE. These sillimanite bearing gneisses are products of metamorphism reaching the upper amphibolite facies conditions and grade into the diatexites (Fig. 2b).

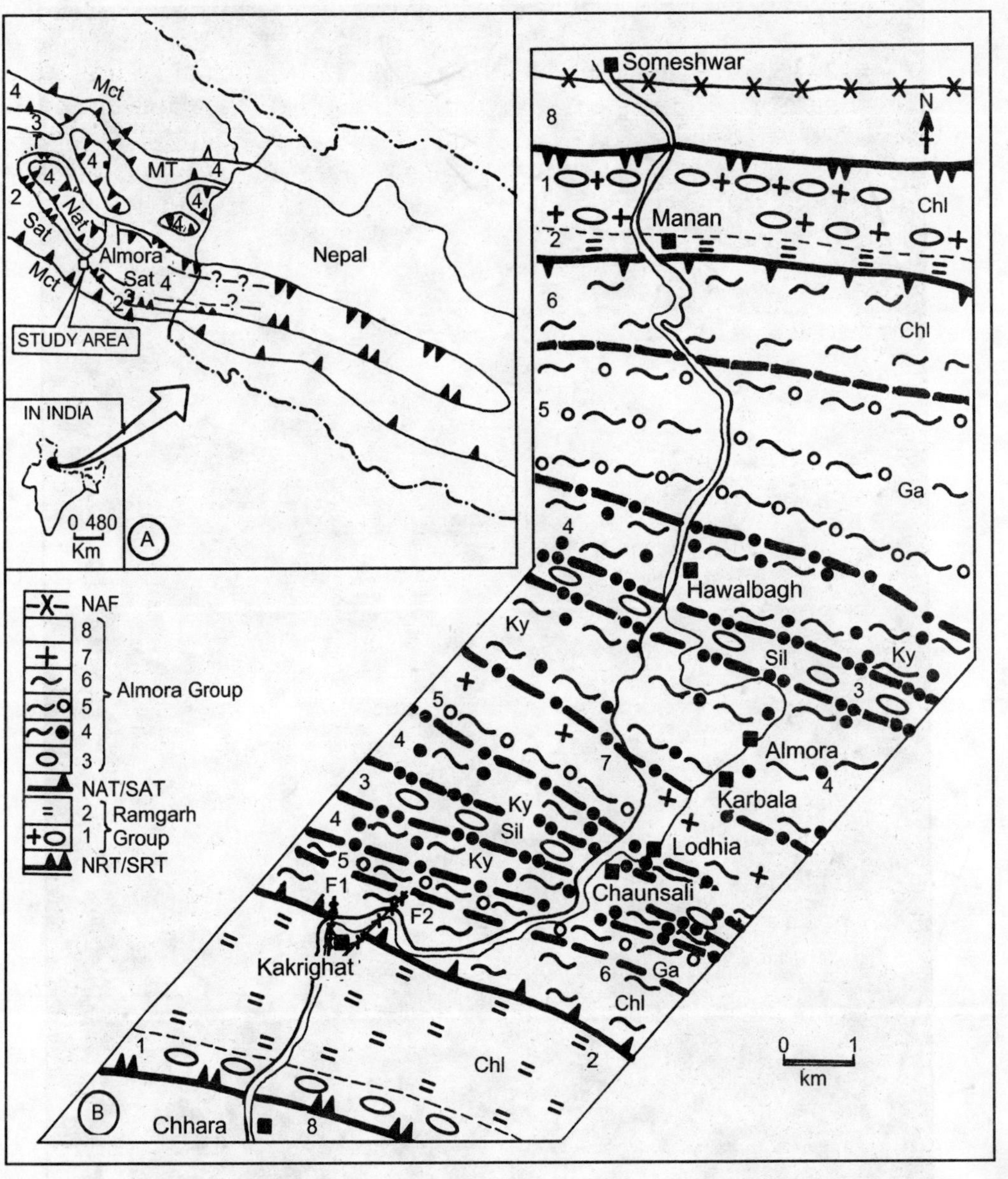

Fig. 1. A : Location map of the study area 1. Damtha, Tejam and Jaunsar Groups, 2. Jaunsar Group, 3. Ramgarh Group, 4. Almora Group. B : Lithotectonic units and metamorphic isograds in Almora Nappe along Chhara – Someshwar transect. NRT/SRT – North/South Ramgarh Thrust, NAT/SAT – North/South Almora Thrust, NAF – North Almora Fault; 1 - Debguru Porphyroids, 2 - phyllites, 3 - Almora gneisses, 4 - kyanite mica schists, 5 - garnet mica schists, 6 - mica schists, 7 – intrusive granite (= Cambro-Ordovician Champawat Granitoids Singh *et al.*, 1993; Champawat Granodiorite, Valdiya, 1962), 8 - autochthonous sedimentaries. Chl – chlotite - biotite zone, Ga – garnet - biotite zone, Ky – kyanite - biotite zone, Sil – sillimanite - K-feldspar zone. Thick dash lines indicate garnet - biotite isograd, thick dash lines with single dot indicate kyanite - biotite isograd, thick dash lines with double dots indicate sillimanite - K-feldspar isograd boundaries.

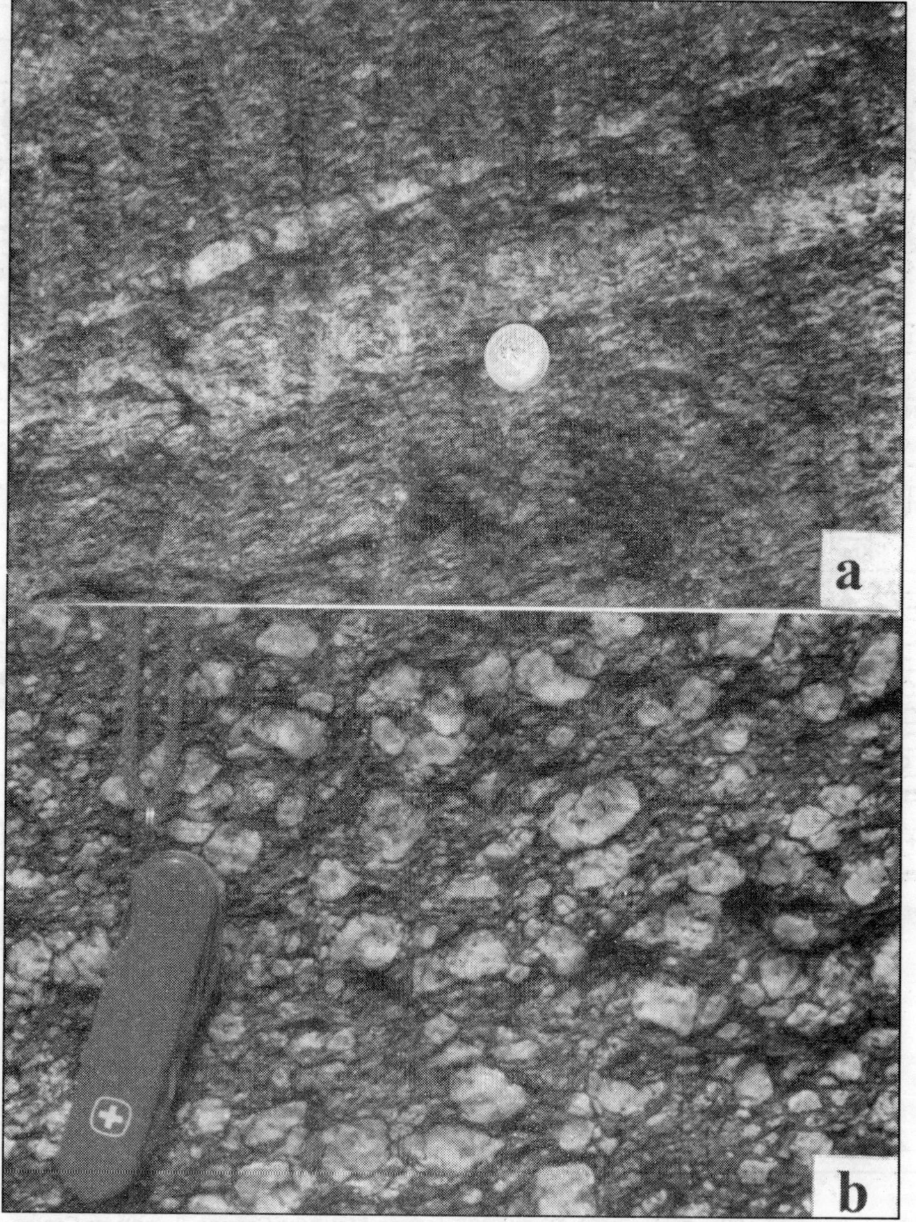

Fig. 2. Field photograph (a) showing the bands of leucosomes and melanosomes in migmatite, loc. near Chaunsali, (b) diatexites showing the development of large grains of K-feldspar.

The studied area comprises three clearly distinguishable structural regimes, viz. the pre-shear zone structures, the syn-shear zone structures and the post-shear zone structures. The pre-shear zone structures include the lithological banding (S_0) striking E-W to WNW-ESE and dipping due south and due north respectively in the northern and southern halves of the central part of the nappe. The F_1- and F_2- tight to isoclinal folds plunging sub-horizontal to about 25° in NNE direction and the F_3- open folds with chevron type hinges with plunges varying between 10-35° in NNW direction are preserved in the unmylonitized quartzites of the Saryu Formation included under the pre-shear zone structures. The syn-shear zone structures are preserved in the mylonites of Ramgarh Group and Almora Group and are represented by a range of shear sense indicators, viz. shear bands, asymmetric porphyroclast systems, (σ_a- and σ_b- type), mica fish, pressure shadows, fractured and displaced grains, folded mylonitic layering and the characteristic Lattice Preferred Orientation (LPO) of quartz. The steep brittle faults affecting the mylonites of the shear zone represent the post-shear zone structures. These brittle faults are responsible for the neotectonic rejuvenation of the Almora Nappe due to a still continuing northward movement of the Indian plate. The north-south transect across the Almora Nappe clearly documents the repetition of metamorphic reaction isograds due to tight to isoclinal WNW-ESE, F_2-folding (Fig. 1B). As the mineral assemblages of the repeated reaction isograds are identical their petrography, mineralogy and metamorphism have been discussed in an integrated manner.

Cambro-Ordovician Champawat Granitoids (dated at 560 ± 20 Ma, Rb/Sr. Trivedi *et al.*, 1984) are intruded between garnet-mica schists and kyanite schists of the Saryu Formation. These granitoids are emplaced roughly parallel to the F_2-fold axes. The contact metamorphic imprints of the intrusion are (1) well preserved euhedral andalusite crystals and (2) the decussate texture defined by micas, overprinting the foliation of garnet mica schists.

Table 1(a): *Representative probe analyses of minerals in sample 94/39.*

Mineral	garnet	biotite	plagioclase
SiO_2	37.78	37.83	63.61
Al_2O_3	21.61	18.83	22.98
FeO	33.44	17.05	–
MgO	3.63	11.12	–
MnO	1.33	0.04	–
CaO	2.18	0.03	3.87
K_2O	–	8.46	0.09
Na_2O	–	0.40	9.63
TiO_2	–	1.79	–
Total	99.97	95.55	100.18
Si	3.01	5.61	2.81
Al	2.03	3.30	1.19
Fe	2.23	2.11	–
Mg	0.43	2.46	–
Mn	0.09	0.01	–
Ca	0.19	0.05	0.18
K	–	1.60	0.01
Na	–	0.11	0.82
Ti	–	0.20	–
Total	7.98	15.45	5.01
X_{Fe}	0.84	0.46	–

Table 1(b): *Representative probe analyses of minerals in sample 94/34.*

Mineral	garnet	biotite	plagioclase
SiO_2	37.26	37.62	62.12
Al_2O_3	21.74	18.07	23.74
FeO	23.07	17.85	–
MgO	1.77	10.48	–
MnO	6.41	0.18	–
CaO	8.61	0.04	4.93
K_2O	–	7.81	0.08
Na_2O	–	0.11	8.77
TiO_2	–	1.77	–
Total	98.86	93.93	99.64
Si	2.99	5.68	2.76
Al	2.06	3.21	1.24
Fe	1.55	2.25	–
Mg	0.21	2.35	–
Mn	0.44	0.02	–
Ca	0.74	0.01	0.24
K	–	1.50	0.01
Na	–	0.03	0.76
Ti	–	0.20	–
Total	7.99	15.25	5.01
X_{Fe}	0.88	0.49	–

PETROGRAPHY

Petrographically the area has been divided into four metamorphic zones, characterized by the mineral assemblages : (1) muscovite - chlorite - biotite - plagioclase - quartz, (2) muscovite - chlorite - biotite - garnet - plagioclase - quartz, (3) muscovite - biotite - garnet - kyanite - plagioclase - quartz, (4) muscovite - biotite - garnet - kyanite - sillimanite - plagioclase - K-feldspar - quartz.

The Almora gneisses are mainly constituted of quartz, plagioclase, K-feldspar, biotite, muscovite, garnet, kyanite and sillimanite. The gneisses are largely psammitic in composition with a predominance of quartz, plagioclase and K-feldspar while muscovite, biotite, garnet, kyanite and sillimanite are present in small modal amounts (<10%). The characteristic feldspar augens in the gneisses lie with their longest axes aligned parallel to the foliation. Preferred alignment of muscovite defines the dominant foliation while boundaries of muscovite lepidoblasts are corroded and show a reaction relationship with quartz and plagioclase (Fig. 3a). Most of the quartz grains show a reaction boundary with muscovite suggesting the reaction : muscovite + quartz = K-feldspar + sillimanite + H_2O. Few fresh muscovite flakes with sharp boundaries are observed in association with quartz which suggest they developed during the regressive phase of the regional metamorphism. Garnet is anhedral in shape and occurs as highly fragmented grains. Kyanite occurs as tabular crystals which are at times bent. Sillimanite occurs both as needles and as a fibrolite. The needles are enclosed in K-feldspar (Fig. 3b) and associated with muscovite, plagioclase and quartz. Fibrolite sillimanite occurs as radiating bunch of fibers in aggregates and is associated with muscovite. The intimate association of K-feldspar with sillimanite needles substantiates the reaction : muscovite + plagioclase + quartz = K-feldspar + sillimanite + melt; (Thompson and Tracy, 1979) responsible for the formation of high grade Almora gneisses and occasional diatexites.

The K-feldspar–sillimanite reaction isograds can be defined by two important muscovite dehydration melting reactions viz. muscovite + quartz + plagioclase = K-feldspar + sillimanite + melt (Thompson & Tracy, 1979) and at a higher temperature biotite may be produced by this reaction as a peritectic phase (Vielzeuf and Holloway, 1988). Patiño-Douce and Harris (1998) suggested that the second reaction has the form : 22 muscovite + 7 plagioclase + 8 quartz = 25 melt + 3K-feldspar + 5 sillimanite + 2 biotite. The absence of fresh biotites in association with K-feldspar and sillimanites in the area clearly indicates that only the first sillimanite isograd was crossed during the metamorphism in the Almora Group of rocks.

MINERAL CHEMISTRY

The Electron Probe Micro Analyses (EPMA) were carried out at University Science Instrumentation Centre, Roorkee on JEOL JXA - 8600M Electron Probe Micro Analyzer (EPMA). Acceleration voltage of 15 kV and specimen current of 20 nano ampere were used. The beam diameter was 2 micron for all the minerals excluding mica and plagioclase for which the beam diameter was kept at 0.5 micron. The oxide ZAF correction procedures were used.

The garnets from Almora gneisses are almandine-rich, spessartine- and grossular-poor. The X_{Fe} values range between 0.84 - 0.88 and suggest that the pyrope content of garnets is maximum in sillimanite - K-feldspar bearing gneisses representing the highest grade of metamorphism in the area.

In biotite X_{Fe} ranges between 0.46 and 0.49. Values of $TiO_2/(TiO_2+Al_2O_3)$ vary between 0.087 and 0.089. The Al^{iv} content in biotite is in the range of 2.32 to 2.39. The anorthite content of plagioclase varies from 17.82 to 23.76 mole per cent.

COMPOSITE PETROGENETIC GRID FOR THE MINERAL ASSEMBLAGES AND P – T PATHS FOR ALMORA GNEISSES IN THE ALMORA NAPPE

Four metamorphic zones have been delineated in the area on the basis of reaction isograds. They have been subjected to post metamorphic (regional) tight folding (F_2). The zone - I (chlorite-biotite zone) is characterized by the abundance of chlorite, biotite and muscovite. The first reaction isograd, chlorite + muscovite + quartz = garnet + biotite + H_2O ...(1) (Thompson and Norton, 1968) encountered in the area is characterized by the development of intimately associated garnet and biotite at the expense of chlorite and muscovite. The presence of almandine garnet in zone - II suggests that the rocks were on higher pressure (>4kb) side of the reaction, chlorite + muscovite + quartz = almandine + biotite + Al_2SiO_5 + H_2O ...(2) (Hirschbergh and Winkler, 1968). The close association of kyanite and Mg-rich biotite (X_{Mg} = 0.65) in sample no. A.50 and the common presence of garnet with the removal of Mg-rich chlorite in the rocks suggest the reaction Mg-chlorite + muscovite = kyanite + Mg-biotite (phlogopite) + H_2O ...(3) (Spear, 1993) and represents the kyanite - biotite isograd. The presence of plagioclase along with kyanite and absence of zoisite in these rocks suggests that the metamorphic conditions were below the reaction

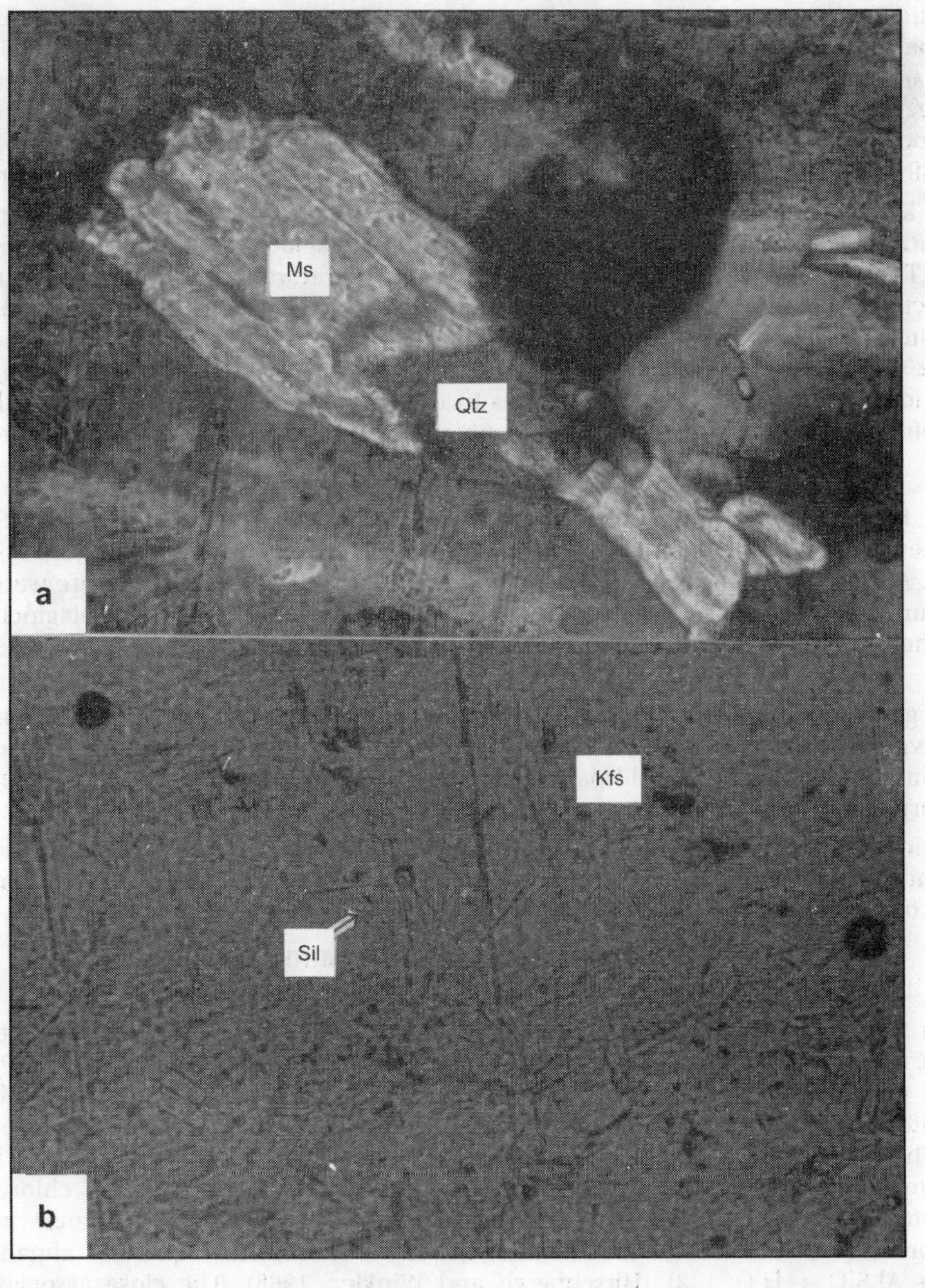

Fig. 3. Photomicrograph (a) showing the reaction boundary between muscovite and quartz suggesting the reaction : muscovite + quartz = K-feldspar + sillimanite + melt (the dark spot is an air bubble!), (b) showing the sillimanite needles enclosed in K-feldspar suggesting the reaction : muscovite + plagioclase + quartz = K-feldspar + sillimanite + melt.

curve of kyanite + plagioclase + water = paragonite + zoisite ...(4) (Spear, 1993) in the stability field of kyanite - biotite zone. The probable $P-T$ array is shown by the arrow A in Fig. 4. The close association of K-feldspar and sillimanite needles with the presence of corroded muscovite and plagioclase correspond to the vapour absent melting reaction (5), viz. muscovite + plagioclase + quartz = K-feldspar + sillimanite + melt in the petrogenetic grid (Fig. 4). The presence of this mineral assemblage in Almora gneisses suggests that the prograde $P-T$ array crossed the reaction (5) in petrogenetic grid and this reaction represents the culmination of regional metamorphism at the $P-T$ conditions of $\geq$ 700°C at ~ 7.5 kbar. Fresh muscovites show sharp contacts with plagioclase in the same thin section, suggesting that this association developed at the expense of K-feldspar and it can be inferred that the regressive phase of the prograde metamorphism begins through this reaction. The presence of kyanite with sillimanite and absence of low pressure minerals, viz. andalusite and cordierite in the regional metamorphic assemblages suggests that the path for the regressive phase (Fig. 4) would be passing through kyanite/sillimanite stability field, but on the higher pressure side of reaction (6). The presence of fresh sillimanite needles and absence of cordierite in the Almora group gneisses suggests that the peak metamorphic conditions did not cross the biotite dehydration melting (dm) reaction : biotite + sillimanite + plagioclase + quartz = garnet + cordierite + K-feldspar + melt (Le Breton and Thompson, 1988) to the higher temperature side as shown in Fig. 4.

THERMOBAROMETRY

To deduce the $P-T$ conditions of metamorphism, the characteristic coexisting mineral assemblages, viz. garnet-biotite for temperature and garnet-biotite - muscovite - plagioclase, garnet - Al_2SiO_5 - quartz - plagioclase for pressure have been used. For temperature the garnet - biotite thermometry models viz. Thompson (1976); Holdaway and Lee (1977); Ferry and Spear (1978); Pigage and Greenwood (1982); Hoinkes (1986); Bhattacharya *et al.* (1992) using garnet mixing model of Ganguly and Saxena (1984) and Berman (1991). For pressure, garnet - biotite - muscovite - plagioclase barometry model of Ghent and Stout (1981) and garnet - Al_2SiO_5 - quartz - plagioclase barometry model using the TWEEQU program (Berman, 1991), based on internally consistent thermodynamic dataset, and Mpet programs (Thomas, 1995) were used.

The temperatures calculated for the garnet-biotite zone after Thompson (1976) range between 453°-551°C; Holdaway and Lee (1977), 449°-536°C; Ferry and Spear (1978), 400°-517°C; Pigage and Greenwood (1982), 504°-577°C; Hoinkes (1986), 485°-567°C; Bhattacharya *et al.* (1992) using Ganguly and Saxena (1984) garnet mixing model, 402°-573°C. For kyanite - biotite zone the temperatures after Thompson (1976) range between 500°-561°C; Holdaway and Lee (1977), 491°-548°C; Ferry and Spear (1978), 455°-546°C; Pigage and Greenwood (1982), 547°-693°C; Hoinkes (1986), 537°-630°C; Bhattacharya *et al.* (1992) using Ganguly and Saxena (1984) garnet mixing model, 480°-576°C. For sillimanite - K-feldspar zone the temperatures calculated after Thompson (1976) range between 584°-593°C; Holdaway and Lee (1977), 565°-573°C; Ferry and Spear (1978), 558°-570°C; Pigage and Greenwood (1982), 612°-709°C; Hoinkes (1986), 614°-633°C; Bhattacharya *et al.* (1992) using Ganguly and Saxena (1984) garnet mixing model, 598°-604°C. In addition temperatures were calculated using the TWEEQU program, based on internally consistent thermodynamic datasets of Berman (1991), which yielded temperature range of 504°-535°C for garnet-biotite zone, 511°-600°C for kyanite - biotite zone and 597°-611°C for sillimanite - K-feldspar zone.

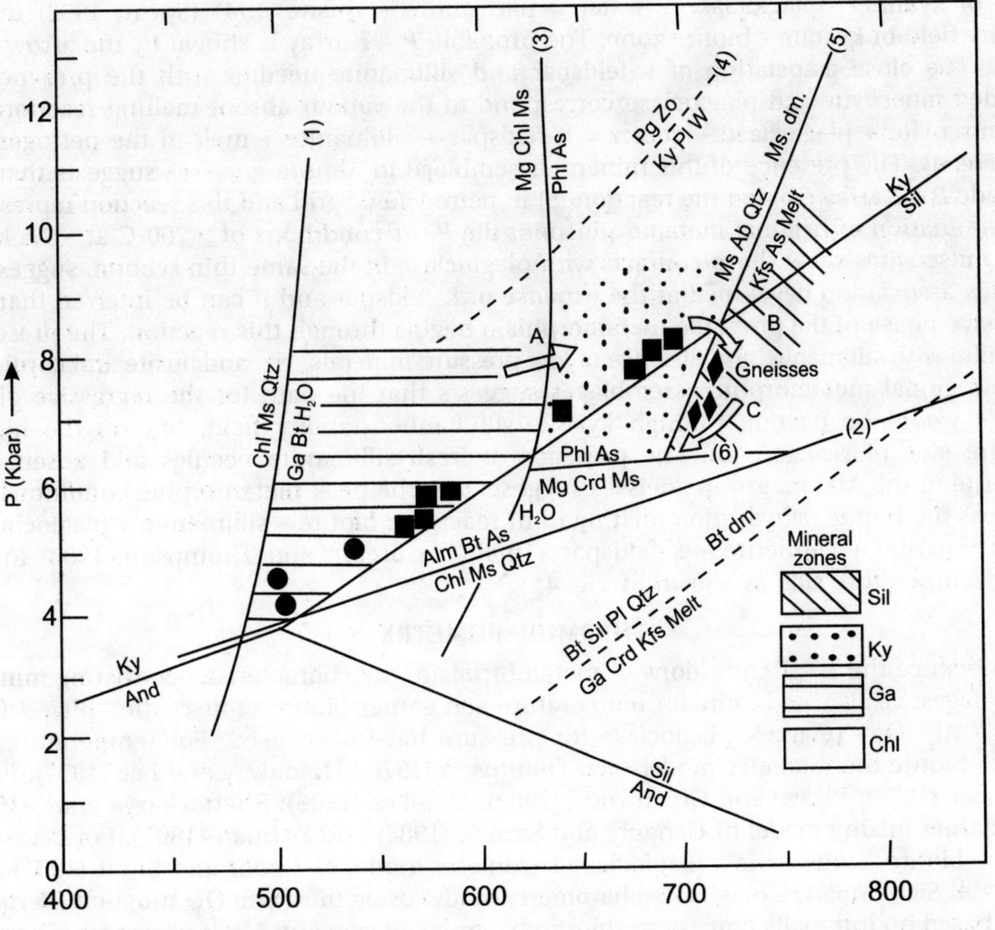

Fig. 4. Composite petrogenetic grid showing the relative peak metamorphic conditions in Almora Nappe for the mineral assemblages developed in Almora gneisses.
Abbreviations used: Ab = Albite, Alm = Almandine, And = Andalusite, As = Aluminosilicate, Bt = Biotite, Chl = Chlorite, Crd = Cordierite, dm = dehydration melting, Fib = Fibrolite, Ga = Garnet, Kfs = K-feldspar, Ky = Kyanite, Mg = Magnesium, Ms = Muscovite, Pg = Paragonite, Phl = Phlogopite, Pl = Plagioclase, Qtz = Quartz, Sil = Sillimanite, W = Water, Zo = Zoisite
● calculated P-T value for garnet-biotite zone
■ calculated P-T value for kyanite-biotite zone
♦ calculated P-T value for sillimanite-K-feldspar zone
Al_2SiO_5 triple point and emerging univariant curves after Holdaway (1971)
Numbers refer to reactions discussed in text.

The pressure range obtained employing garnet - biotite - muscovite - plagioclase model after Ghent and Stout (1981) yielded 4.48 - 4.96 kbar, for garnet - biotite zone. Using garnet - Al_2SiO_5 - quartz - plagioclase barometry model after Berman (1991), the estimated pressure for the rocks of kyanite - biotite zone are 5.13-7.79 kbar and for the rocks of sillimanite - K-feldspar zone are 6.9-7.9 kbar.

DISCUSSION AND CONCLUSIONS

Majority of earlier workers believe that the interbanded metapelites and metapsammites of Almora Nappe were subjected to low grade metamorphism with only rare occurrence of kyanite in Dudhatoli area (e.g. Valdiya, 1980). Most of these workers are noncommittal about the age of regional metamorphism.

Regionally speaking the grade of metamorphism for the Kulu - Almora - Baijnath Group (KAB) is greenschist facies with basal parts showing chlorite grade and the upper parts chlorite - biotite - garnet grade (Bhargava and Bassi, 1998). Most of the workers favour an exclusively Tertiary age for the Himalayan metamorphics (e.g. Powell and Conaghan, 1973; Le Forte *et al.*, 1983; Roy and Valdiya, 1988; Verma, 1989). However, there is a gradually increasing group of workers who believe that the first regional metamorphism in Himalaya was of Pre-Cambrian age (e.g. Sinha Roy, 1974; Acharya, 1979; Gansser, 1981, 1990; Sharma, 1983; Chakraborty, 1988; Joshi and Singh, 1990; Joshi *et al.*, 1994; Bhargava and Bassi, 1994, 1998; Mukhopadhyay *et al.*, 1997; Joshi and Rai, 2003).

In the absence of reliable mineral dates the age of regional metamorphism in the Almora Nappe should not be uncritically accepted as Tertiary just because the age of metamorphism is Tertiary for its root zone, viz., the HMB. The degree of both the structural and metamorphic reworking for the Himalayan metamorphics is different for the LHC *vis-a-vis* the HHC. The ductile shearing is pervasive for the HHC (Jain and Manickavasagam, 1993) while it is generally restricted to the basal parts of the Lesser Himalayan nappes (Joshi, 1999; Tiwari, 2000; Devi, 2001). Moreover, the HHC are intruded by Tertiary granites (Harrison *et al.*, 1997; Searle *et al.*, 1999 and Harrison & McKeegan, 1994) which are conspicuous by their absence in the Lesser Himalayan nappes suggesting that they were intruded after the thrust sheets were transported to south during Eocene-Oligocene (De Celles *et al.*, 2001). Thus the post-Oligocene tectonometamorphic histories for the LHC and the HHC must be different.

Long back Sharma (1970) using only plagioclase compositions concluded that "the oligoclase bearing granites and granite gneisses are of anatectic origin".

We have demonstrated above that gneisses of Saryu Formation of the Almora Group crossed the first sillimanite isograd and were subjected to temperatures around 700°C at about 7.5 kbar pressures. Similar metamorphic conditions were also suggested from the Dhunaghat (Singh *et al.*, 1993), Champawat (Mamgai, 1992) and Dwarahat (Devi, 2001) areas of the Almora Nappe. Thus it is amply clear that the regional metamorphism reached upper amphibolite facies conditions in many parts of the nappe.

Admittedly, there is evidence that the dominant metamorphism in the root zone of the LHC, viz., the HMB is related to the Tertiary Himalayan Orogeny. However, the age of dominant regional metamorphisms in the Lesser Himalayan nappes need not be necessarily Tertiary. The most significant difference between the nature of metamorphism in the HMB and the LHC is that "the signatures of inverted metamorphism are definitely weaker in the LHC than the HHC" (Joshi and Rai, 2003). The inverted metamorphism in the Lesser Himalayan nappes is apparent only in the vicinity of the basal thrusts of the nappes while structurally upsection pre-Himalayan metamorphic slices are preserved (Joshi *et al.*, 1994; Singh *et al.*, 1995; Mukhopadhyay *et al.*, 1997; Joshi and Rai, 2003).

The evidence favouring pre-Himalayan (Pre-Cambrian) metamorphism from various sources is summarized below:

1. E - W trending F_1 and F_2 folds oriented transverse to the NW - SE regional strike with attendant metamorphism were identified long back in Jutogh rocks of Himachal Himalaya (Ray and Naha, 1971). This is ratified by amphibolites folded by F_2 folds which are exposed west of Karbala near Almora town.
2. Tight to isoclinally folded (F_2) regional metamorphic isograds with the Pre-Cambrian gneisses (1865±60 Ma, Trivedi *et al.*, 1984) formed as a product of culmination of regional metamorphism in the Almora Nappe (Singh, 1990; Singh *et al.*, 1995; Tiwari, 2000; Devi, 2002 and the present work).
3. The older metamorphic foliation abuts against the intrusive Early Palaeozoic granitoids (Mamgai,1992; Bhargava and Bassi, 1998).
4. Randomly oriented schistose enclaves are preserved in these Early Palaeozoic granitoids (560±20 Ma, Trivedi *et al.*, 1984) intruding the LHC (Singh, 1990; Bhargava, 2000).
5. Well preserved contact metamorphic aureoles with randomly oriented fresh chlorites, chloritoids and andalusites occur around the Early Palaeozoic granitoid. The contact metamorphic assemblages overprint the regional schistosity (Joshi *et al.*, 1994 and the present work).

On the basis of the foregoing discussion it is concluded that the upper amphibolite facies metamorphism in the Almora Nappe which culminated in anatexis is highly likely to be of Pre-Cambrian age.

ACKNOWLEDGMENTS

We gratefully acknowledge the financial assistance by the DST for the project No. ESS/23/132/91. We are also grateful to the DST for using the computer facilities provided under Project No. ESS/16/091/97. We thank the Head, Department of Geology for providing necessary facilities. ANT is thankful to the UGC for financial assistance as a Senior Research Fellowship. Help rendered by Dr. T.K. Ghosh, USIC, Roorkee, during Electron Probe Micro Analyses is gratefully acknowledged.

REFERENCES

Acharya, S.K., 1979. Pre-Tertiary fabric and metamorphism in the Eastern Himalaya. In Verma, P.K. (Ed). *Metamorphic rock sequences of the Eastern Himalaya.* Today's and Tomorrow's Publishers, New Delhi, 67-82.

Auden, J.B., 1937. Structure of the Himalaya in Garhwal. *Record Geological Survey of India,* **71**, 347-54.

Berman, R.G., 1991. Thermobarometry using multi–equilibrium calculations : A new technique, with petrological applications. *Canadian Mineralogist,* **29**, 833-55.

Bhargava, O.N., 2000. The Pre-Cambrian sequences in the western Himalaya. *Geological Survey of India Special Publication,* **55**, 69-84.

Bhargava, O.N. & Bassi, U.K., 1994. The crystalline thrust sheet in the Himachal Himalaya and the age of amphibolite facies metamorphism. *Journal of Geological Society of India,* **43(4)**, 343-52.

Bhargava, O.N. & Bassi, U.K., 1998. Geology of Spiti-Kinnaur, Himachal Himalaya. *Memoir Geological Survey of India,* 124, 210.

Bhattacharya, A., Mohanty, L., Maji, A., Sen, S.K. & Raith, M., 1992. Non-ideal mixing in the phlogopite-annite binary constraints from experimental data on Mg - Fe partitioning and a reformulation of the biotite-garnet geothermometer. *Contributions to Mineralogy and Petrology,* **111**, 87-93.

Chakraborty, K., 1988. Deformation and metamorphism of the Pre-Cambrian sequence of the Simla Himalaya and some observations on energy transformation and recrystallization. *Indian Mineralogist,* **42**, 266-80.

Devi, P.M., 2001. Structural and Petrological study in Chaukhutia area of Almora Nappe Kumaun Himalaya. *Unpublished Ph.D. thesis,* Banaras Hindu University, 170.

DeCelles, P.G., Robinson, D.M., Quade, J., Ojha, T.P., Garzione, C.N., Copeland, P. & Upreti, B.N., 2001. Stratigraphy, structure and tectonic evolution of the Himalayan fold–thrust belt in western Nepal. *Tectonics,* **20(4)**, 487-509.

Ferry, J.M. & Spear, F.S., 1978. Experimental calibration of the partitioning of Fe and Mg between biotite and garnet. *Contributions to Mineralogy and Petrology,* **66**, 113-17.

Ganguly, J. & Saxena, S.K., 1984. Mixing properties of aluminosilicate garnets: constraints from natural and experimental data and application to geothermobarometry. *American Mineralogist,* **69**, 88-97.

Gansser, A., 1964. Geology of the Himalayas. Interscience Publishers, London, 273.

Gansser, A., 1981. The geodynamic history of the Himalaya - In: H.K. Gupta and I.M. Delany (Eds) Zagros-Hindkush-Himalaya Geodynamic Evolution, *Amer. Geophy. Union,* Washington, D.C., 111-21.

Gansser, A., 1990. Facts and theories on the Himalayas. *Himalaya - Tibet - Karakoram Workshop Milan.* March 1-2, 1990: 16-17.

Ghent, E.D. & Stout, M.Z., 1981. Geobarometry and geothermometry of plagioclase - biotite - garnet - muscovite assemblages. *Contributions to Mineralogy and Petrology,* **76**, 92-97.

Harrison, T.M. & McKeegan, K.D., 1994. $^{208}Pb/^{232}Th$ ion microprobe dating of monazite and Himalayan tectonics. *Geological Society of America* Abstract Prog., **26(7)**, A-367.

Harrison, T.M., Lovera, O.M. & Groove, M., 1997. New insights into the origin of two contrasting Himalayan granite belts. *Geology,* **25**, 899-902.

Heim, A. & Gansser, A., 1939. Central Himalaya, Geological observations of the Swiss expedition in 1936. *Mem. Soc. Helv. Sci. Nat.,* **73**, 1-245.

Hirschberg, A. & Winkler, H.G.F., 1968. Stabilitätbeziehungen Zwischen Chlorit, Cordierit und Almandin bei der Metamorphose. *Contributions to Mineralogy and Petrology,* **18**, 17-42.

Hoinkes, G., 1986. Effect of grossular-content in garnet on the partitioning of Fe and Mg between garnet and biotite: An empirical investigation on staurolite zone samples from the Austroalpine Schneeberg Complex. *Contributions to Mineralogy and Petrology,* **92**, 393-99.

Holdaway, M.J., 1971. Stability of andalusite and the aluminum silicate phase diagram. *American Journal of Science,* **271**, 97-131.

Holdaway, M.J. & Lee, S.M., 1977. Fe-Mg cordierite stability in high-grade pelitic rocks based on experimental, theoretical and natural observations. *Contributions to Mineralogy and Petrology,* **63**, 175-98.

Jain, A.K. & Manickavasagam, R.M., 1993. Inverted metamorphism in the intracontinental ductile shear zone during Himalayan collision tectonics. *Geology,* **21**, 407-10.

Joshi, M. & Singh, B.N., 1990. Metamorphism and related deformation around Dhunaghat area of Almora Crystallines, Eastern Kumaon, Lesser Himalaya U.P. *Proc. Seminar Himalayan Geology.* Wadia Institute of Himalayan Geology Dehra Dun, 6-7 April, 42-43 *(Abs).*

Joshi, M., Singh, B.N. & Goel, O.P., 1994. Metamorphic conditions of the aureole rock from Dhunaghat area, Kumaun Lesser Himalaya. *Current Science,* **67(3)**, 185-88.

Joshi, M., 1999. Evolution of the Basal Shear Zone of the Almora Nappe, Kumaun Himalaya. *Gondwana Research Group Memoir,* **6**, 69-80.

Joshi, M. & Rai, J.K., 2003. An Appraisal of Models for Inverted Himalayan Metamorphism. *Memoir52 Geological Society of India* (in press).

Le Breton, N. & Thompson, A.B., 1988. Fluid-absent (dehydration) melting of biotite in metapelites in the early stages of crustal anatexis. *Contributions to Mineralogy and Petrology,* **99**, 226-37.

Le Forte, P., Debon, F. & Sonet, J., 1983. The Lower Palaeozoic Lesser Himalayan Granite: Emphasis on Simachar Pluton of Central Nepal. In F.A. Shams (Ed) *Granites of Himalaya, Karakoram and Hindkush,* Punjab Univ., Lahore, 235-55.

Mamgai, D.C., 1992. Tectonics and Petrological studies of rocks around Champawat-Lohaghat area, District Pithoragarh (U.P). *Unpublished Ph.D thesis,* Kumaun University, 176.

Middlemiss, C.S., 1887. Crystalline and metamorphic rocks of the Lower Himalaya, Garhwal and Kumaun (section 1). *Record Geological Survey of India,* **XX**, 134-43.

Mukhopadhyay, D.K., Ghose, T.K., Bhadra, B.K. & Srivastava, D.C., 1997. Structural and metamorphic evolution of the rocks of the Jutogh Group, Chur half-klippe, Himachal Himalayas: A summary and comparison with the Shimla area. *Earth Planetary Sciences,* **106**(4), 197-207.

Patiño-Douce, A.E. & Harris, N.B.W., 1998. Experimental constraints on Himalayan anatexis. *Journal of Petrology,* **39**, 689-710.

Pigage, L.C. & Greenwood, H.G., 1982. Internally consistent estimates of pressure and temperature; the staurolite problem. *American Journal of Science,* **282**, 943-69.

Powell, C. Mc. & Conaghan, P.J., 1973. Polyphase deformation in metamorphic rocks of Central Himalayan Gneiss, Northwest India, *Journal of Geeology,* **8**, 127-43.

Ray, S.K. & Naha, K., 1971. Structural and metamorphic history of Simla Klippe, A Summary. *Himalayan Geology,* **1**, 1-24.

Roy, A.B. & Valdiya, K.S., 1988. Tectonometamorphic evolution of the Great Himalayan Thrust Sheets in Garhwal region, Kumaon Himalaya. *Journal of Geological Society of India,* **32**(2), 106-24.

Searle, M.P., Noble, S.R., Hurford, A.J. & Rex, D.C. (1999). Age of crustal melting, emplacement and exhumation history of the Shivling leucogranite, Garhwal Himalaya. *Geological Magazine,* **136**(5), 513-25.

Sharma, K.K., 1983. Granitoid Belts of the Himalaya. In F.A. Shams (Ed) *Granites of Himalayas, Karakoram and Hindukush.* Institute of Science, Punjab University, Lahore, 11-37.

Sharma, R.S., 1970. Plagioclase composition and some aspects of the gneissic rocks of Almora–Dudatoli region, Lower Kumaun Himalaya. *Publication of the Centre of Advanced Study in Geology, Punjab University,* Chandigarh, 1-13.

Singh, B.N., 1990. Geochemical Evolution of the Granitoids and associated rocks of Kumaun Lesser Himalaya occurring around Dhunaghat, District Pithoragarh, U.P. *Unpublished Ph.D. Thesis,* Kumaun University, 261.

Singh, B.N., Goel, O.P., Joshi, M. & Shereton, J.W., 1993. Geochemistry and petrogenesis of the Champawat Granitoids occurring around Dhunaghat, Distt. Pithoragarh, U.P., *Journal of Geological Society of India,* **42**, 289-302.

Singh, B.N., Goel, O.P., Joshi, M. & Shereton, J.W., 1995. Geochemistry and origin of the granite gneisses of the Almora Group from Dhunaghat area, Dist. Pithoragarh, U.P. *Journal of Geological Society of India,* **46**(7), 27-36.

Sinha-Roy, S., 1974. Polymetamorphism in the Daling rocks from a part of Eastern Himalayas and some problems of the Himalayan metamorphism. *Himalayan Geology,* **4**, 74-101.

Spear, F.S., 1993. Metamorphic phase equilibria and Pressure – Temperature – time paths. *Mineralogical Society of America, Washington D.C.,* 799.

Thomas, H., 1996b. MPET1: A computer program for coexisting garnet - biotite - aluminosilicate - plagioclase Quartz - geothermobarometry. *Journal of Himalayan Geology,* **v. 19**(1), 49-54.

Thompson, A.B., 1976. Mineral reactions in pelitic rocks. I. Prediction of P-T-X (Fe - Mg) phase relations. II. Calculations of some P-T-X (Fe-Mg) phase relations. *American Journal of Science,* **276**, 401-54.

Thompson, A.B. & Tracy, R.J., 1979. Model systems for anatexis of pelitic rocks ; facies series melting and reactions in the system CaO-KalO_2-NaAlO_2-Al_2O_3-SiO_2-H_2O. *Contributions to Mineralogy and Petrology,* **70**, 429-38.

Thompson, J.B. Jr. & Norton, S.A., 1968. Palaeozoic regional metamorphism in New England and adjacent areas. In : *"Studies of Applachian Geology": "Northern and Maritime"*, E–an–Zen, W.S. White, J.B. Hadley and J.B. Thompson Jr. (eds.), Interscience Wiley, New York, 319-27.

Tiwari, A.N., 2000. Structural and Metamorphic evolution of Almora Nappe along Chhara-Someshwar Transect Kumaun Himalaya. *Unpublished Ph.D. thesis*, Banaras Hindu University, 142.

Trivedi, J.R., Gopalan, K. & Valdiya, K.S., 1984. Rb-Sr ages of granites within the Lesser Himalayan Nappes, Kumaun Himalaya. *Journal of Geological Society of India*, **25**, 641-54.

Valdiya, K.S., 1980. Geology of Kumaun Lesser Himalaya.WIHG, Dehradun, 291.

Verma, P.K., 1989. The Himalayan metamorphism. In J.S. Daly, R.A. Cliff. and B.W.D. Yardly (Eds). *Evolution of Metamorphic Belts.* Blackwell Scientific Publications, 377-83.

Vielzeuf, D. & Holloway, J.R., 1988. Experimental determination of the fluid-absent melting relations in the pelitic system. Consequences for crustal differentiation. *Contributions to Mineralogy and Petrology*, **98**, 257-76.

10

IS MUSCOVITE IN THE MANDI GRANITE PRIMARY? A GUIDE TO DISTINCTION BETWEEN THE LOWER PALEOZOIC AND TERTIARY GRANITES OF THE HIMALAYAS

S. NAG,* S. SENGUPTA* AND P.K. VERMA*

ABSTRACT

The present work embodies results of mineral chemical and crystallographic observations on the Mandi Granite, in the Lesser Himalayas of Himachal Pradesh. The Mandi Granite belongs to the well-known sequence of "500 Ma" granites that intrude low-grade metasediments of the Precambrian age in the Himalayas. Previous workers have concentrated on the petrography, geochronology, and genetic aspects of the Mandi Granite, as well as on the metamorphic evolution of the metasediments. The Mandi Granite is S-type granite and the Himalayan metamorphism is the major metamorphic event in the region. The present study explores the hitherto untouched aspect, i.e. metamorphism of the Mandi Granite. The metamorphic influence involves the compositional adjustment of the perthitic lamellae, the alteration of alkali feldspars to sericite and its subsequent crystallization as muscovite, generation of garnet, and other effects. The implications of this study are that the primary muscovite in the Mandi Granite is either absent or is negligible making it two-feldspar biotite granite. Because of the presence of muscovite, the melt generation occurred at a deeper level than what was earlier thought. Since the Mandi Granite is a typical example of 500 Ma granites in the Himalayas, the results may be extrapolated to all such studies. In the end it is pointed out that these granites differ from the Tertiary post-tectonic granites in having muscovite of metamorphic origin.

KEY WORDS

Muscovite; Primary; Metamorphism; Mineral Chemical; Crystallographic and Mandi Granite.

INTRODUCTION

There are numerous occurrences of granites in the Himalayas and these, obviously, have been subjected to intense and diverse studies (Gansser, 1964; Shams, 1983; Verma, 2003). Broadly speaking, there is a belt of 1800 Ma old granites that forms a part of the Vaikrita Group, then a 500 Ma granitic suite that intrudes this group. Finally, we have a series of plutons that are related to the Upper Tertiary Himalayan metamorphism. Texturally, the oldest suite exhibits the severest deformation, and it is practically absent from the

* Department of Geology, University of Delhi.

youngest suite. The youngest suite is interpreted as the end product of the Himalayan Metamorphism; therefore, the older suites would show the imprints of the metamorphism. However, the oldest suite is deformed in a complicated way, so the middle suite, i.e. 500 Ma granites, appears to be the best candidate for evaluating the effects of the Himalayan metamorphism on a granite body. It is believed that the overall mineralogy of all the three suites is that of two-feldspar-two-mica granite; this mineralogy hardly changes during a metamorphic episode. To critically evaluate metamorphic changes, if any, it was decided to give a close petrologic examination to one of the 500 Ma granites. The Mandi Granite was chosen because of its easy accessibility, and also because the region has been studied in detail by a large number of workers during past 50 years or so (Gupta, 1974; Chatterjee, 1976; Frank *et al.*, 1977; Thoni, 1977; Purkayastha, 1997; Srikantia and Bhargava, 1998; Jain *et al.*, 1999).

The problem of metamorphism of granites shall be tackled by presenting pertinent textures in detail. Then a thermobarometric analysis of relevant assemblages and individual minerals would be presented. The results of study reveal that the Mandi Granite is now a metamorphic unit. The original mineralogy consisting of quartz-plagioclase-K-feldspar-biotite ± zircon ± tourmaline ± apatite was transformed to the present one consisting of quartz - plagioclase - K-feldspar-biotite-muscovite ± zircon ± tourmaline ± apatite.

The previous authors covered petrographic, geochronologic, genesis of the Mandi Granite and also the metamorphic evolution of the host pelitic rocks (Purkayastha, 1997). The metamorphic evolution of the Mandi Granite remained untouched. Hence the present paper focuses on this aspect of the Mandi Granite.

GEOLOGY

Himachal Pradesh is characterized by the Tethyan sequence, which is demarcated by the Precambrian rocks (Srikantia and Bhargava, 1998) which are present as a number of thrust sheets (like the Jutogh thrust, the Vaikrita thrust, the Kulu thrust and the Krol thrust) and windows (like Rampur Window and Lajri Window), as well as a root zone. The present area forms a part of the Mandi district, and belongs to the Vaikrita Thrust sheet. The region is characterized by the metasediments belonging to the Vaikrita Group, along with the Mandi Granite.

The present area is characterized by an interbedded sequence of fine-grained metasediments of the Vaikrita Group and intrusive the Mandi Granites; within the intrusive, there is an occurrence of a gabbro body (Mandi Gabbro) near the northern end (Fig. 1). The Mandi Granite, the focus of the present investigation, makes an oval shaped outcrop in the central part of the area (Fig. 1). The Mandi Granite had been subdivided into two types by Gupta (1974). The older one migmatitic (Gupta, 1970) and the younger one is an intrusive within the metasediments. Chatterjee (1976) classified the Mandi Granite into five types namely muscovite granite, biotite granite, tourmaline granite, aplitic granite and porphyritic granite. However, Purkayastha (1997) did not support this classification, as these varieties are not traceable in the field and hence, proposed a simple field classification with only two components i.e. porphyritic and non-porphyritic granites. The Mandi Granite is a mica granite with mylonitic foliation present everywhere. The age of this granite is dated to be 545 ± 12 Ma (Mehta, 1977).

The porphyritic and non-porphyritic granite are well exposed on most of the major road sections. Several apophysis of granites occur within the schists (Fig. 2). The attitude of the contact plane between granite and the schists is 60° due north. A pervasive foliation is present within both the metasediments and granite (Fig. 3) although degree of foliation varies from place to place. Purkayastha *et al.* (1999) show a photograph of an outcrop where the granite cuts across S_1 foliation of the metasediments (though erroneously inferred the Jugtoh metamorphics). This foliation is referred to as S_2 or major foliation. Mural joints are present within the granites (Fig. 4).

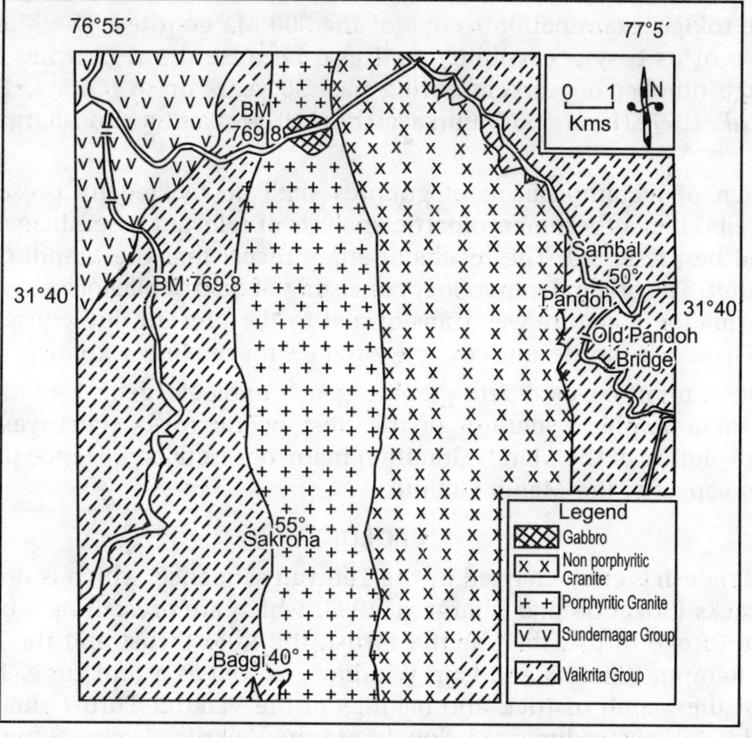

Fig. 1. Geological map of the Mandi area, Himachal Pradesh (After Chatterjee, 1976).

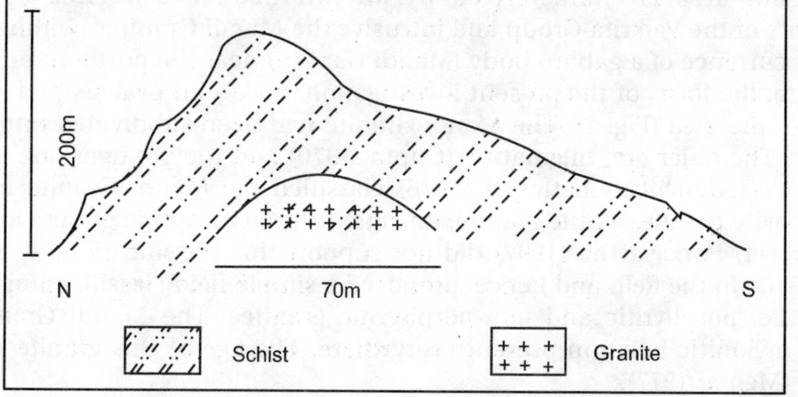

Fig. 2. Granite-schist contact. The granite forms a tounge within the schist. Location near Baggi bus stand.

Fig. 3. A photograph showing metapelites bed within the granite. Its trend is parallel to the foliation in the latter. Location 5 km east of Mandi on Mandi-Pandoh road section.

Fig. 4. Mural joints in the non-porphyritic granite. Location on the Baggi-Chail Chowk road section, north of the first kilometer stone-marker at Baggi.

METHODOLOGY

Routine petrography of the Mandi Granite has been done. For modal analyses the method given in Hutchinson (1974) was adopted. Oriented X-Z and Y-Z sections were prepared for this purpose along with routine sections. Both porphyritic granite and non-porphyritic granite were analysed. Out of a large number of thin sections from the Mandi Granite 18 were selected and examined in detail. Of these 4 sections were studied by electron probe micro analyzer for studying the chemical compositions of feldspar, muscovite and biotite. Mineral chemistry was carried out on JEOL JXA-8600M superprobe at Indian Institute of Technology, Roorkee. The operating voltage was 15 kV with 20 nA current. Beam diameter was kept at 2mm. The mineral standard used was SPI multi mineral Standard (Canada). In each grain, 3 to 4 close spaced points were analysed and averaged out to get a mean value. The deviation is noted to be less than 1%. Muscovite and K-feldspar were separated for X-ray diffraction. To separate muscovite, firstly the granite samples (1-2 Kg) were crushed to –80 to +120 mesh (ASTM) size. After crushing, the samples were washed and then magnetite was separated by a hand magnet. The remaining mass was subjected to gravity separation in a refined bromoform medium. Mica floats at the top and the mass that settled down in feldspar and quartz. This is repeated two to three times. Muscovite was separated from biotite using a isodynamic separator. For biotite, the current was kept at 0.5 amp, forward tilt is 25° and the sideward tilt is 15°. Muscovite separation requires 1 amp of current and side tilt was changed to 5°. X-ray diffraction was done on muscovite and feldspar using Philips X-ray diffractometer at Department of Geology, University of Delhi. Software programmes like Philips X-pert and CELREF were used for peak identification of X-ray diffraction pattern and mineral species studies respectively.

PETROGRAPHY

The granite contains quartz, K-feldspar, plagioclase as major minerals, muscovite, biotite as accessory minerals and zircon, tourmaline, apatite and garnet (in order of abundance) as minor minerals. Chlorite and sericite are present as secondary minerals. The rock is foliated and also shows mylonitic textures.

The Mandi Granite has been classified on the basis of modal analyses. The modal percentage data plotted on the IUGS igneous rock classification show that the compositions of the granite fall on quartz rich granitoids and syeno-granite area.

It is found in those samples, in which muscovite and biotite are in comparable amount, have preferential alignment of muscovite over biotite. In other words, the X-Z sections show higher biotite content than muscovite, while Y-Z sections reveal higher muscovite content.

Quartz occupies nearly half the volume of both porphyritic and non-porphyritic varieties. Feldspars account for nearly quarter to one third. Rest is mainly formed by micas. K-feldspar is dominant over plagioclase. Amongst micas, muscovite has a higher volume proportion than biotite.

Quartz is present as a matrix mineral or as inclusions and veins within feldspars. Fractures, granulation and undulose extinction are common and are very prominent where the rock is mylonitised. Fine grained, fractured and polygonised quartz grains are present in the porphyritic granites. The non-porphyritic granites have both fine as well as coarse grained quartz.

Feldspar is anhedral to subhedral in nature. It becomes fine grained where the granite is mylonitised otherwise, both medium and coarse grained varieties are present. The grains of K-feldspar observed as phenocrysts in handspecimen are actually porphyroclasts as fractures and quartz veins are commonly observed within feldspar. Inclusion of K-feldspar within plagioclase is common. Both carlsbad and cross hatched twinning are seen. K-feldspar and plagioclase commonly occur together and at places due to sectioning effect, the former appears to surround the latter giving an apparent poikilitic texture. The latter in one case has inclusions of subhedral muscovite. Inclusions of quartz are also present in some grains. Some grains exhibit perthitic texture. In the perthites the thickness of plagioclase lamellae vary. In some perthites, the plagioclase lamellae have fine albite twinning and in others, the plagioclase lamellae are very coarse. The host K-feldspar has cross-hatched twinning in some perthites. The K-feldspar grains are commonly sericitized in the core. In both the porphyritic and non-porphyritic types the K-feldspar is of microcline variety.

Plagioclase has K-feldspar and quartz inclusions in it. Various types of twins are present in it (carlsbad, pericline, cataclastic and polysynthetic). In some grains the twin lamellae are present from end to end. In others the twins are not complete. Antiperthitic texture is present. Sericitization at the cleavages with muscovite present in two cleavage directions are observed commonly.

Two feldspars in the Mandi Granite coexist which shows that the granite is a subsolvus granite (Morse, 1970).

Muscovite is subhedral, fine to medium grained. Muscovite in both porphyritic and non-porphyritic variety shows different types of habit: (1) Coarse grained, relict which is "fish" shaped (Fig. 5). (2) Fine grained and along with biotite at places define foliation (Fig. 6). (3) Found as inclusions within both types of feldspar (Fig. 7). (4) Fine grained grown at the grain boundary of feldspars (Fig. 8). Second type of occurrence is the most dominant type followed by the fourth type and then the third type. The first type of occurrence is not very common. The muscovite fish generally occurs in isolation along the S_2 plane; rarely muscovite fish is more than one, aligned along the prevailing schistosity plane. The optical properties of all these four varieties of muscovite are similar.

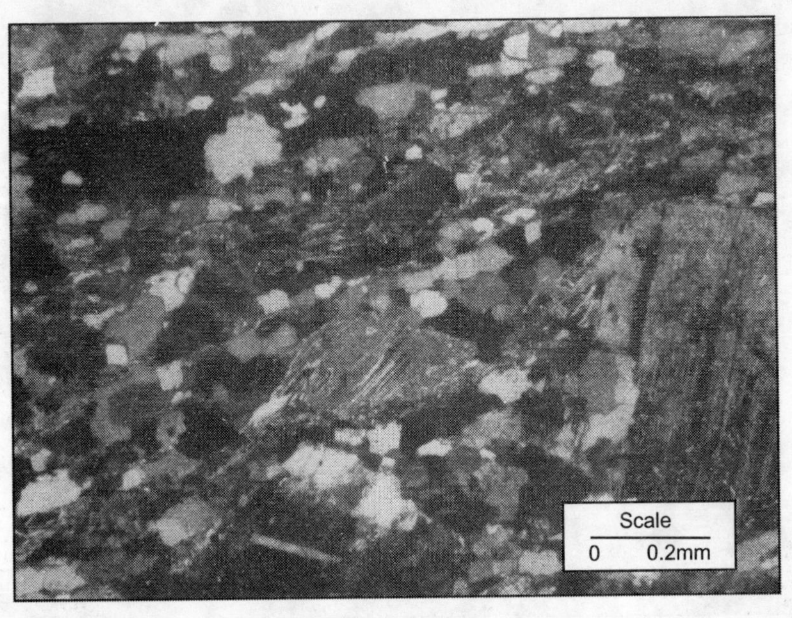

Fig. 5. Photomicrograph of muscovite fish in the Mandi Granite under X-nicols. Sample number MD 101/1.

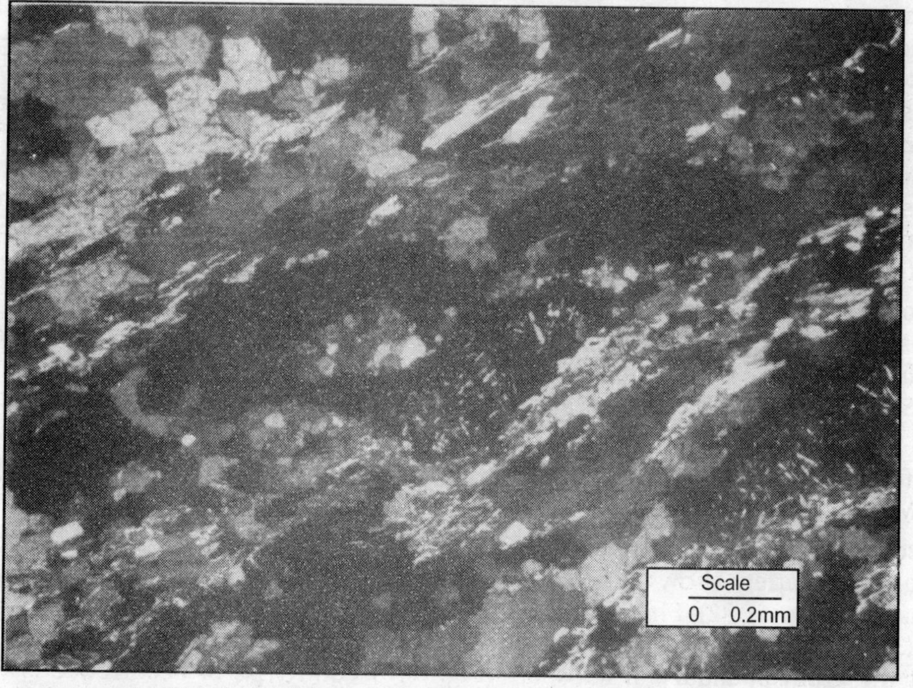

Fig. 6. Photomicrograph of biotite and muscovite defining a foliation plane parallel to S_2 under X-nicols. A large biotite fish parallel to the foliation and occurring at the eastern part of the photograph. Sample number MD 101/1.

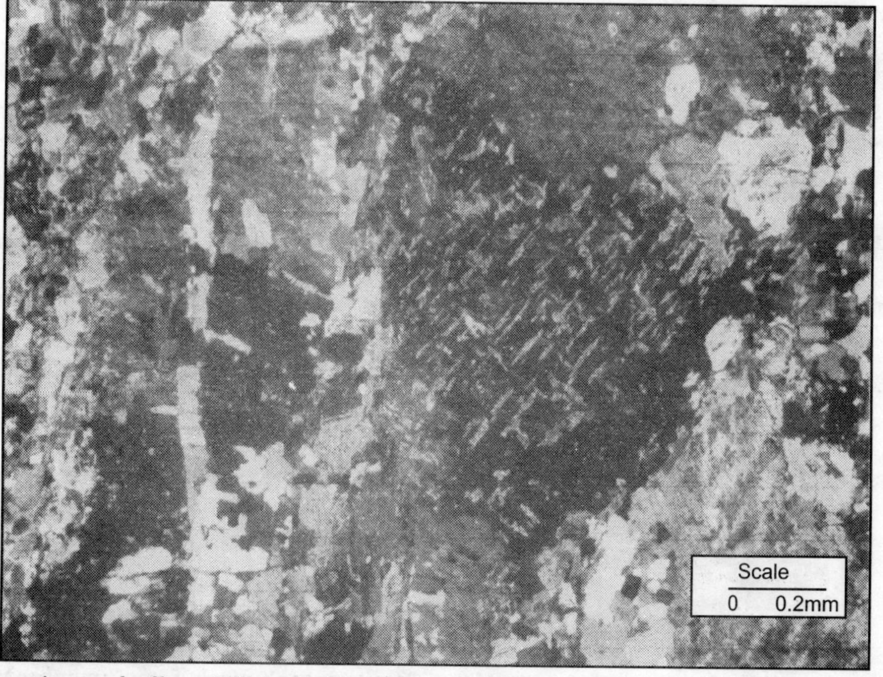

Fig.7. Photomicrograph of large K-feldspar with muscovite inclusion under X-nicols. Sample number MD 204/10.

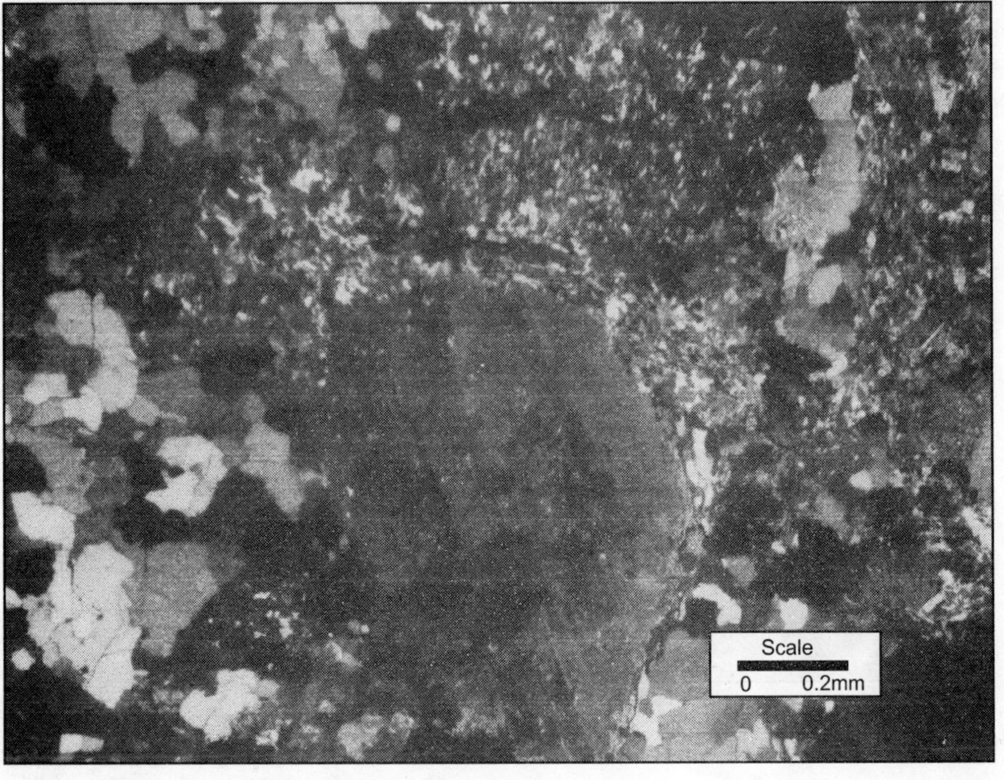

Fig. 8. Photomicrograph of small muscovite flakes occurring along the boundaries of large K-feldspar under X-nicols. Sample number MD 1/1.

Biotite is anhedral to subhedral and is fine to coarse grained. It is brown in color and is strongly pleochroic. Its pleochroic scheme is α = yellowish brown, β = brown and γ = dark brown. It is of two types. Some of the grains are coarse grained, "fish" shaped (Fig. 6), deformed and show undulose extinction. The fine grains are subhedral, intergrown with muscovite and define foliation. Along cleavage, the biotite is altered to chlorite.

The accessory minerals are zircon, tourmaline, garnet and chlorite. Zircon occurs only as inclusions within biotite. Garnet is euhedral in shape and occurs as individual grains (Fig. 9) with biotite, K-feldspar and quartz.

Porphyritic texture is present with profound inequality of grain size due to the presence of fine grained quartz, muscovite, biotite and medium to coarse grained feldspar. There are inclusions of quartz and plagioclase in K-feldspar and K-feldspar in plagioclase. Cooling textures are manifested in perthite and antiperthite. The rock has crude to strong foliation defined by annealed, polycrystalline and elongated quartz and aligned muscovite and biotite. This foliation is same as that observed at the scale of outcrop and hand specimen. In thin sections it is crenulated at places where foliation is well developed. Granulation of large grains is seen. Triple points are present at some places. The myrmekitic texture is present. The myrmekite occurs as colonies of lobes projecting into K-feldspar from the margins of the grains. The myrmekite is present in zones parallel to the foliation.

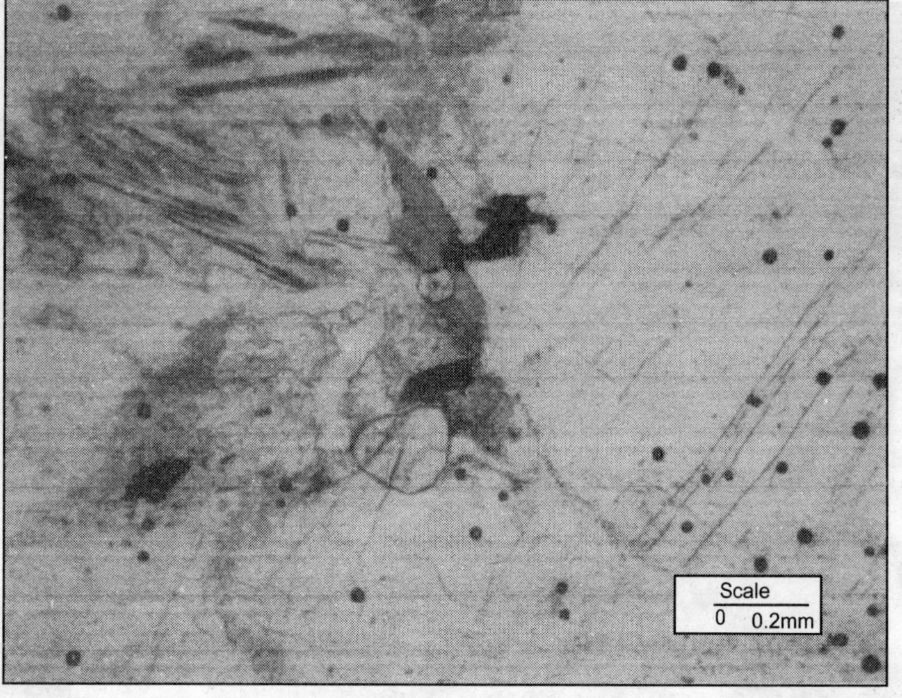

Fig. 9. Photomicrograph of Euhedral garnet occurring in reaction relation with biotite under plane polarized light. Sample number MD 101/53.

MINERAL CHEMISTRY

All the major minerals, *i.e.* K-feldspars, plagioclase, muscovite and biotite were analysed by electron microprobe analyzer. Analytical points were selected from the porphyroclasts, matrix and inclusions in porphyroclasts. The results are given in Tables 4 and 5.

Alkali Feldspar: In the Mandi granite, both independent and perthitic K-feldspar in both porphyritic and non-porphyritic varieties are Or rich ($Or_{0.94}$ to $Or_{0.96}$).

Plagioclase: In the Mandi granite, both independent grains and those present as perthite are Albite rich ($Ab_{0.69}$ to $Ab_{0.99}$).

X-ray diffraction of the feldspars from the Mandi Granite have been done on four porphyritic granite and four non-porphyritic granite (Table 1). The X-ray diffractograms of their powder were obtained. Figure 10 shows (060) versus ($\bar{2}04$) plot of feldspar of the Mandi Granite. It shows that K-feldspar is of low microcine variety. The cell dimensions and K-feldspar percentage are given in Table 2. The average K-feldspar percentage for porphyritic granite (89.79%) and non-porphyritic granite (90.71%) conform to the microprobe data.

Muscovite: The chemistry of the muscovite in the Mandi granite is given in Table 5. The muscovite falls on the muscovite phengite join with K as the predominant interlayer cation. In order to determine polytype of muscovite from the Mandi Granite, the authors selected samples from both porphyritic and nonporphyritic varieties. The hkl values of the present muscovite samples were compared with the muscovite polytypes of Baily (1988). It was found that out of all the hkl values which match with the samples of the Mandi Granite,

Table 1: *2 θ values and d spacing for (201), (002), (060), (204) of alkali feldspar from Mandi Granite*

Sample No.	2O (201)	d	2O (002)	d	2O (060)	d	2O (204)	d
PORPHYRITIC GRANITE								
MD 1/9	21.13	4.2005	27.48	3.2267	41.68	2.1651	50.7	1.799
MD 101/1	21	4.2267	27.4	3.2133	41.85	2.1567	50.75	1.7974
MD 101/52	21.024	4.2256	27.375	3.258	41.841	2.159	50.554	1.8055
MD 202/4	20.89	4.2487	27.5	3.2005	41.76	2.1611	50.65	1.8007
Average		4.225375		3.224625		2.160475		1.80065
NON-PORPHYRITIC GRANITE								
MD 2/6	20.982	4.233	27.442	3.2503	41.839	2.1591	50.56	1.8053
MD 3/9	20.95	4.2366	27.49	3.2005	41.8	2.1591	50.5	1.8057
PH 2/16	21.06	4.2133	27.6	3.2167	41	2.1493	50.7	1.799
PH 2/15	21.05	4.2167	27.8	3.2005	41.87	2.1057	50.5	1.8057
Average		4.2249		3.217		2.1433		1.803925

Table 2: *Cell dimensions and K-feldspar percentage in the alkali feldspar of the Mandi Granite*

Sample No.	a	b	c	Kf %
PORPHYRITIC GRANITE				
MD 1/3	8.543	13.001	7.2	83.8
MD 101/1	8.585	12.95	7.194	95.78
MD 101/52	8.59	12.966	7.233	n.d.
MD 202/4	8.648	12.997	12.977	n.d.
Average	8.5915	12.9785	8.651	89.79
NON-PORPHYRITIC GRANITE				
MD 2/6	8.59	12.966	7.233	n.d.
MD 3/9	8.621	12.965	7.225	n.d.
PH 2/16	8.574	12.906	7.2	90.25
PH 2/15	8.563	12.965	7.2255	91.17
Average	8.587	12.9505	7.220875	90.71

n.d. = not detected.

(11¯4), (114) and (115), are common to both 2 M_1 and 2 M_2 polytypes types. But (025) hkl value is specific for only 2 M_1 polytype. So, the muscovite of the Mandi Granite is of 2 M_1 polytype. This is confirmed by a software Philips programme "Xpert Philips Identifier".

The b_0 of muscovite from the Mandi Granite has also been calculated using chemical data (Guidotti *et al.*, 1989) as well as using software (CELREF). The b_0 of muscovite determined by the first method are based on substitution shown at the column heads of Table 4. The values from the first and third calibrations are almost similar to each other but different

Table 3: *Representative microprobe analysis of Feldspars in the Mandi Granite*

Sample No.	PH 2/16	PH 2/16	MD 101/1	MD 101/1	MD 2/6	MD 2/6
Mineral	K-feldspar	K-feldspar	K-feldspar	K-feldspar	K-feldspar	K-feldspar
points averaged	3	5	1	1	1	1
SiO_2	65.43 (0.17)	65.64 (0.41)	63.66	64.02	64.73	65
Al_2O_3	18.96 (0.19)	18.63 (0.28)	17.39	17.85	17.91	7.99
FeO						
MgO						
MnO						
CaO	0.01 (0.01)	0.02 (0.01)	0.04		0.01	
K_2O	15.09 (0.24)	15.44 (0.11)	17.25	17.99	15.3	15.4
Na_2O	0.63 (0.12)	0.51 (0.08)	0.41	0.4	0.57	0.72
TiO_2						
Total	100.12 (0.46)	99.24 (0.60)	98.76	100.26	98.63	99.1
No. of cations on the basis of 8 O						
Si	3 (0.01)	3 (0.01)	3	2.99	3.02	3.02
Al	1.02 (0.01)	1.02 (0.01)	0.97	0.98	0.99	0.99
Fe						
Mg						
Mn						
Ca						
K	0.88 (0.01)	0.91 (0.01)	1.04	1.07	0.91	0.91
Na	0.06 (0.01)	0.05 (0.01)	0.04	0.04	0.05	0.06
Ti						
Total	4.94 (0.01)	4.97 (0.01)	5.05	5.08	4.97	4.98

Sample No.	MD 105/65	MD 105/65	MD 101/1	MD 101/1	MD 2/6	PH 2/16
Mineral	Plag	Plag	Plag	Plag	Plag	Plag
points averaged	4	5	4	5	4	1
SiO_2	61.28 (0.18)	68.82 (0.03)	65.88	66.93	67.87	69.05 (0.34)
Al_2O_3	25.42 (0.32)	20.92 (0.17)	18.99	19.43	19.03	20.03 (0.28)
FeO						
MgO						
MnO						
CaO	6.07 (0.17)	0.93 (0.08)	0.94	0.73	0.06	0.21 (0.08)
K_2O	0.13 (0.02)	0.11 (0.02)	0.15	0.1	0.12	0.10 (0.01)
Na_2O	7.71 (0.17)	10.37 (0.21)	12.54	12.76	14.38	10.61 (0.21)
TiO_2						
Total	100.61 (0.61)	100.14 (0.79)	99.5	99.94	101.45	100.01 (0.69)
No. of cations on the basis of 8 O						
Si	2.70 (0.01)	2.96 (0.00)	2.95	2.95	2.96	3.00 (0.01)
Al	1.32 (0.01)	1.06 (0.00)	1	1.01	0.98	1.02 (0.01)
Fe						
Mg						
Mn						
Ca	0.29 (0.01)	0.04 (0.00)	0.05	0.03		0.01 (0.00)
K	0.01 (0.00)	0.01 (0.00)	0.01	0.01	0.01	0.01 (0.00)
Na	0.66 (0.01)	0.87 (0.02)	1.09	1.09	1.22	0.89 (0.02)
Ti		0.01				
Total	4.97 (0.01)	4.94 (0.01)	5.1	5.09	5.16	4.95 (0.01)

Note: Bracketed numbers are the standard deviations of the number of points analysed.

Table 4: *Representative microprobe analysis of mica in the Mandi Granite*

Sample No. Mineral	MD 105/65 Muscovite	MD 2/6 Muscovite	MD 101/1 Muscovite	MD 105/65 Biotite	MD 105/65 Biotite
points averaged	3	1	1	3	4
SiO_2	46.22 (0.24)	47.94	45.47	36.4 (0.18)	36.76 (0.30)
Al_2O_3	33.69 (0.27)	29.9	34.36	16.89 (0.06)	16.49 (0.15)
FeO	2.22 (0.06)	4.41	1.85	21.77 (0.39)	20.60 (0.35)
MgO	0.85 (0.06)	1.16	0.64	7.93 (0.46)	9.19 (0.04)
MnO	0.01 (0.02)	0.01	0.11	0.31 (0.01)	0.25 (0.05)
CaO	0.05 (0.06)	0.01		0.01 (0.02)	0.04 (0.03)
K_2O	11.02 (0.13)	0.99	10.24	9.79 (0.12)	9.48 (0.18)
Na_2O	0.26 (0.11)	0.06	0.28	0.18 (0.11)	0.08 (0.10)
TiO_2	0.37 (0.02)	0.42	0.34	1.60 (0.10)	1.64 (0.12)
Total	94.69 (0.36)	93.89	93.18	94.90 (0.51)	94.55 (0.46)
		No. of cations on the basis of 22 O			
Si	6.24 (0.03)	6.54	6.2	5.65 (0.00)	5.68 (0.03)
Al	5.54 (0.04)	4.81	5.5	3.09 (0.01)	3.00 (0.03)
Fe	0.85 (0.01)	0.5	0.21	2.82 (0.06)	2.66 (0.05)
Mg	0.17 (0.01)	0.24	0.13	1.83 (0.10)	2.12 (0.01)
Mn			0.01	0.04 (0.00)	0.03 (0.01)
Ca	0.01 (0.01)				0.01
K	1.90 (0.03)	1.74	1.78	1.94 (0.02)	1.87 (0.03)
Na	0.07 (0.03)	0.02	0.07	0.05 (0.03)	0.02 (0.03)
Ti	0.04 (0.00)	0.04	0.04	0.19 (0.01)	0.19 (0.01)
Total	14.03 (0.03)	13.89	13.94	15.62 (0.03)	15.57 (0.02)
X_K	0.96	0.99	0.96		
X_{Na}	0.04	0.01	0.04		
X_{Aliv}				2.35	2.32
X_{Alvi}				0.13	0.12
X_{Fe}				0.5	0.47
X_{Mg}				0.33	0.37
X_{Mn}				0.01	0.01
X_{Ti}				0.03	0.03

Note: Bracketed numbers are the standard deviations of the number of points analysed.

from the second one. The b_0 of muscovite determined by CELREF for porphyritic granite (MD 1/3) is 8.9 A^0 and that of non-porphyritic (MD 3/9) is 9.07 A^0.

Biotite: Biotite is an important mineral in the Mandi Granite. Its composition on the mica classification falls around phlogopite-annite join of the triangular plot. High Fe/Mg and low Ti are the characteristic of biotite in the Mandi Granite.

Table 5: *The b_0 of muscovite in the Mandi Granite using chemical data (Calibrations of Guidotti et al., 1989)*

Sample No.	Mineral formula	b=8.99+0.04 $(Mg+Fe_T)$	b=8.596+ 0.0666 Si	b=9.149 – 0.0258 $(Al^{iv} + Al^{vi})$
MD 2/6	(Na0.04,K1.86)(Mg0.24,Fe0.49,Al3.35)(Al1.46Ti0.05Si6.49)O20(OH)4	9.02	8.81	9.09
	(Na0.02,K1.74)(Mg0.24,Fe0.50,Al3.39)(Al1.42,Ti0.04,Si6.54)O20(OH)4	9.02	8.81	9.09
	(Na0.02,K1.79)(Mg0.23,Fe0.46,Al3.39)(Al1.45,Ti0.05Si6.50)O20(OH)4	9.02	8.81	9.09
	(Na0.04,K1.81)(Mg0.23,Fe0.51,Al3.39)(Al1.53,Ti0.03,Si6.44)O20(OH)4	9.02	8.81	9.09
	(Na0.05,K1.79)(Mg0.18,Fe0.51,Al3.39)(Al1.53,Ti0.03Si6.44)O20(OH)4	9.01	8.81	9.09
	(Na0.04,K1.83)(Mg0.23,Fe0.46,Al3.42)(Al1.52Ti0.04,Si6.44)O20(OH)4	9.02	8.81	9.09
	(Na0.04,K1.82)(Mg0.21,Fe0.39,Al3.49)(Al1.58,Si6.42)O20(OH)4	9.01	8.81	9.09
	(Na0.05,K1.86)(Mg0.24,Fe0.47,Al3.36)(Al1.46,Ti0.04,Si6.50)O20(OH)4	9.02	8.81	9.09
	(Na0.11,K1.72)(Mg0.1,Fe0.15,Al3.82)(Al1.77,Ti0.05,Si6.18)O20(OH)4	9	8.8	9.08
	(Na0.07,K1.78)(Mg0.22,Fe0.48,Al3.39)(Al1.47,Ti0.01Si6.52)O20(OH)4	9.02	8.81	9.09
MD 101/1	(Na0.08,K1.76)(Mg0.23,Fe0.54,Al3.31)(Al1.38Ti0.05,Si6.57)O20(OH)4	9.02	8.82	9.09
	(Na0.04,K1.78)(Mg0.21,Fe0.49,Al3.39)(Al1.55Ti0.04,Si6.41)O20(OH)4	9.02	8.81	9.09
	(Na0.04,K1.82)(Mg0.22,Fe0.53,Al3.38)(Al0.49,Ti0.06Si6.45)O20(OH)4	9.02	8.81	9.09
	(Na0.07,K1.78)(Mg0.13,Fe0.21,Al4.51)(Al0.99,Ti0.04,Si6.97)O20(OH)4	9	8.83	9.08
	(Na0.03,K1.86)(Mg0.31,Fe0.44,Al3.34)(Al1.46,Ti0.01,Si6.53)O20(OH)4	9.02	8.81	9.09
	(Na0.02,K1.79)(Mg0.30,Fe0.52,Al3.36)(Al1.49,Ti0.02,Si6.49)O20(OH)4	9.02	8.81	9.09
	(Na0.05,K1.69)(Mg0.12,Fe0.25,Al3.75)(Al1.8,Si6.2)O20(OH)4	9	8.8	9.08
	(Na0.02,K1.75)(Mg0.19,Fe0.42,Al3.47)(Al1.59,Ti0.01,Si6.4)O20(OH)4	9.01	8.81	9.08
	(Na0.03,K1.71)(Mg0.24,Fe0.54,Al3.37)(Al1.47,Ti0.03,Si6.50)O20(OH)4	9.02	8.81	9.09
	(Na0.03,K1.7)(Mg0.18,Fe0.35,Al3.6)(Al1.58,Ti0.18,Si6.24)O20(OH)4	9.01	8.8	9.08
	(Na0.03,K1.64)(Mg0.20,Fe0.49,Al3.48)(Al1.49,Ti0.05,Si6.46)O20(OH)4	9.01	8.81	9.08
	(Na0.08,K1.64)(Mg0.14,Fe0.25,Al3.76)(Al1.78,Ti0.02,Si6.2)O20(OH)4	9.01	8.81	9.08

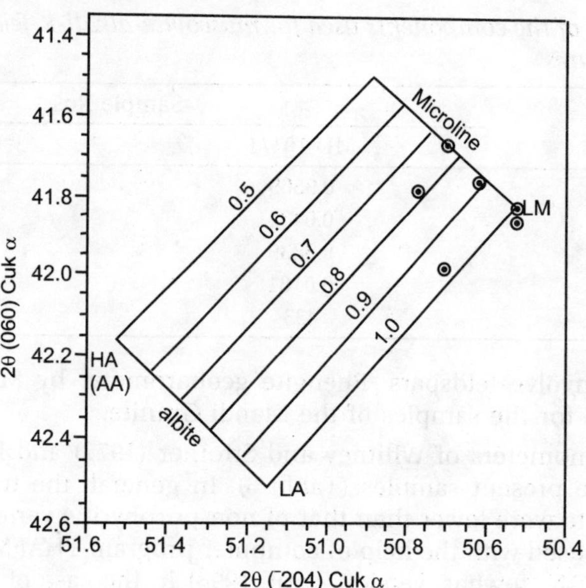

Fig. 10. (060) versus (¯204) plot of K-feldspar.

Geothermobarometry of the Mandi Granite

Well-calibrated thermobarometers can be applied to the Mandi Granite samples to decipher the condition of its formation. Two geothermometers, namely, two-feldspar-geothermometer and K-feldspar-muscovite geothermometer, have been used for this. Both

Table 6: *Two feldspar thermometry of the Mandi Granite*

Sample No.	Porphyritic/Non-porphyritic	Texture	S (1995); p = 5 Kb	Temperature (C) Calibration using H (1983); p = 5 Kb	W (1975); p = 5 Kb
MD 101/1	Porphyritic	Coexisting Plg & Kfeld	310	199.77	470.89
		Coexisting Plg & Kfeld	290	175	436.82
		Coexisting Plg & Kfeld	308	194.71	465.62
		Coexisting Plg & Kfeld	308	192.76	464.24
		Coexisting Plg & Kfeld	335	246.18	534.52
		Plg lamellae in K-feld	323	225.61	509.33
		Plg lamellae in K-feld	321	225.2	510.11
MD 2/6	Non-porphyritic	Coexisting Plg & Kfeld	307.01	201.96	476.71
		Coexisting Plg & Kfeld	323.1	223.98	507.08
		Coexisting Plg & Kfeld	323	216.88	497.45
		Coexisting Plg & Kfeld	350	256.67	551.99
		Plg lamellae in K-feld	335.71	238.17	524.83
		Plg lamellae in K-feld	351.09	263.26	561.85
		Plg lamellae in K-feld	342.57	248.42	540.077

Note: S-Stromer; H-Haselton *et al.*; W-Whitney and Stromer.

Table 7: *Mole fractions of the components used for muscovite-alkali-K-feldspar geothermometry and temperatures*

Mole fractions	Sample Nos.	
	MD 101/1	MD 2/6
X_{or}	0.9569	0.9198
X_{Ab}	0.0431	0.0802
X_{Mu}	0.9839	0.9909
X_{Pa}	0.0161	0.0091
TC	433	602

these thermometers involve feldspars. Phengite geobarometer by Massone and Schreyer (1987) have been used for the samples of the Mandi Granite.

Two feldspar thermometers of Whitney and Stromer (1977) and Haselton *et al.* (1983) have been used in the present samples (Table 6). In general, the temperatures obtained from porphyritic granite were lower than that of non-porphyritic variety. The calibration of Thompson (1976) was used with the help of computer program THAFMU.FOR (unpublished FORTAN program by T.N. Jowhar, version April, 1995) in the case of K-feldspar-muscovite geothermometer. The mole fractions of different components used are given in Table 7.

X-ray study shows that the alkali feldspar in the Mandi granite is of low structural state (as discussed in Mineral Chemistry section). As Whitney and Stormer (1977) have taken into account the low structural state of alkali feldspar, their calibration gives reasonable temperature for Mandi Granite. The second thermometer gives lower temperature for non-porphyritic granite and higher temperature for the porphyritic granite than two feldspar thermometry.

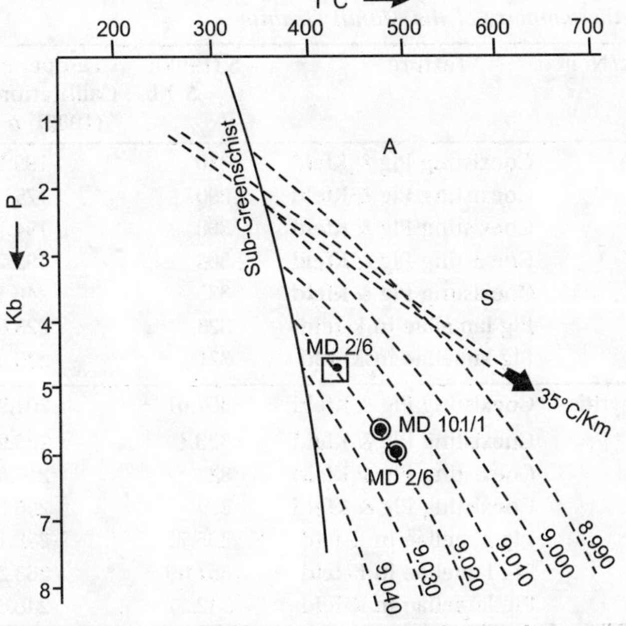

Fig. 11. The P estimate using b_0 from chemical data.

The b_0 values calculated using the method of Guidotti *et al.* (1989) (*see* Mineral Chemistry section) are plotted in Fig. 11. The base of this figure has been borrowed from Guidotti and Sassi (1986) which was prepared on the basis of muscovites from the rocks, whose pressures have been estimated by other means. This figure shows the effect of P and T on b_0. Conversely, if T and b_0 are known, P can be estimated. Here the b_0 values from the method following Guidotti *et al.* (1989) have been used. The geobarometer of Massone and Schreyer (1987) has also been applied to the present samples from the Mandi Granite. After consideration of all the aspects of the geothermobarometers finally the following steps are considered:

In the first step the temperature has been estimated from two feldspar thermometry of Whitney and Stormer (1977) for which assumed pressure was 5 Kb. On plotting these temperatures, the pressure obtained for the porphyritic granite is 6 Kb and that for non-porphyritic granite is 6.25 Kb. Again these pressures were in turn considered for the temperature calculation using Whitney and Stormer calibration. The temperature obtained was 571°C. When this temperature is used in Massone and Schreyer geobarometry, the pressure estimated is 6 Kb.

Metamorphism of the Mandi Granite

Systematic metamorphic studies in the Himalayas span the past fifty-year period (Ray, 1947; Le Fort, 1975; Verma, 1989, 2003; Searle and Treloar, 1993). Integrated studies involving geochronologic, microstructural and thermobarometric works (Chamberlain *et al.*, 1991; Vannay and Hodges 1996) are also available. Several authors (Jain and Manickavasagam, 1993; Hubbard, 1996) have stressed on the reverse metamorphism in the Himalayas while producing models on the geodynamic evolution of the region. A large number of studies have been concentrated on the Kashmir Sytaxis–Kohistan region as well as on the Central Crystallines of U.P. and Nepal (Hodges and Silverberg, 1988; Hodges *et al.*, 1988; Hubbard, 1996). Two distinct phases of Tertiary metamorphism have been postulated in these two regions. The western Syntaxis region shows the development of eclogite facies with more ubiquitous amphibolite assemblages. The Central Crystallines show higher grade than sillimanite grade. However, the present area Baggi-Mandi-Pandoh region does not belong to the Central Crystalline region (which could be its root) but is a nappe. Being southern most nappe, it possibly represents highest part of the root zone.

After compilation of database, called Himchron, on the radiometric ages from the Himalayas the first author observed two magmatic events in the Himalayas – one is Tertiary and the other around 500-640 Ma. The latter includes the Mandi Granite (545 Ma).

Textural and compositional criteria help to judge whether muscovite is igneous or metamorphic. Texturally muscovite in the Mandi Granite can be classified into four types and they are listed below according to the order of abundance:

(i) Fine grained and along with biotite defines a foliation (parallel to S_2)

(ii) Fine grained tracing the boundary of K-feldspar

(iii) Found as sericitic patch within K-feldspar

(iv) Muscovite 'fish'

The first kind of texture may have been formed due to metamorphism as the muscovite in this case is defining a foliation parallel to S_2 (formed at the peak of M_2). But the possibility of magmatic status of muscovite can also not be disregarded. It may so happen that the igneous muscovite got oriented parallel to S_2 due to the metamorphic episode. To support the fact that muscovite is formed later on due to metamorphism, compositional criteria can be brought in. Following the studies of (Spear, 1984) on Millevaches massif (France), it can be stated that late muscovite have Na/(Na+K) ratio less than 0.04. In the present case such ratio is varying from 0.011 upto 0.038. This can be also supported by the study in Appalachians (Spear, 1984); primary and secondary muscovites in granitoid rocks were plotted in ternary projection of Ti, Fe and Mg. Muscovites in the Mandi Granite also plot in the same position in the ternary diagram as the secondary muscovite of the Appalachians.

Fine grained muscovite occurring along the boundary of K-feldspar in the Mandi Granite suggest a reaction texture. Muscovite may have been produced by the alteration of K-feldspar. Such alteration can be due to hydrolytic H^+ metasomatism triggered by low T-fluids (Spear, 1984). In the present case, fluid influx from the neighbouring area altered the K-feldspar along its boundary. This fluid is produced due to progressive metamorphism in pelitic rocks hosting the granitic intrusion (Purkayastha *et al.*, 1999). This dehydration fluid invaded the granite through the fractures produced due to deformation in the area.

The weathering of feldspar produces sericitisation. Now whether the sericite within feldspar of the Mandi Granite is magmatic or metamorphic is hard to establish. But there is no doubt that this muscovite is secondary and is formed by the following hydrolysis reaction:

$$KAlSi_3O_8 + 2H^+ = KAlSi_3O_{10}(OH)_2 + 2K^+ + 6SiO_2$$

Muscovite 'fish' acquire its typical shape either due to shearing or pressure solution on pre-existing large muscovite flakes. This type of occurrence in the present rocks proves that some amount of muscovites in the granite were magmatic in nature (this type of texture is not very common in the Mandi Granite).

Thus the present authors tried to establish that muscovite in the Mandi Granite are generally secondary or metamorphic in nature. Only muscovite 'fish', which is very rare in this rock, points to the magmatic nature of the mineral. In this context it is necessary to point out that biotite 'fish' is also present and is more common than muscovite 'fish'. Hence magmatic biotite is abundant than muscovite of the same status. This leads the authors to think whether the original granite was two-feldspar biotite granite or not. For this the experimental results of Clarke (1992) and Thompson (1996) were studied. Crystallization of pure biotite granite starts at 765°C at 21 Km depth whereas the

Table 8: *The relation between different phases of deformation metamorphism and different generations of foliations*

Age		Deformation	Metamorphism	Foliation	Minerals
Tertiary		D_3	M_3	S_3	
					Gt3, Ky, Chl3
		D_2	M_2	S_2	Grt2
					Grt1, Bt
L. Cambrian	Granite intrusion				
Precambrian	Metasediments	D_1	M_1	S_1	Qz, Mus, Bt, Chl

crystallization of pure muscovite granite starts at 675°C, 17 Km depth. The geothermobarometric estimation for the Mandi Granite is 571°C and 5 Kb; the estimated temperature is much lower than the experimental values. Hence it can be said that the Mandi Granite has undergone some post-magmatic reactions.

The metapelites which host the granite show three sets of foliation: S_1, S_2 and S_3. The dominant equilibrium assemblage in metapelites is related to S_2. A foliation parallel to S_2 is related to M_2. Hence granite may have been affected by the same metamorphism. The relationship between tectonism and metamorphism in this area is given in Table 8. The peak of M_2 is related to garnet 3 of the metapelites Purkayastha (1997). The pressure and temperature are derived using garnet-muscovite-biotite-plagioclase geobarometer (Hodges and Crowley, 1985) and garnet-biotite thermometer (Ferry and Spear, 1978). The results estimated are 525°C and 5 Kb. The pressure estimated for the Mandi Granite is also 5 Kb. This raises a possibility that granite was affected by the same greenschist facies metamorphism.

The garnets in the Mandi Granite are euhedral and occur in a reaction relationship with biotite and quartz (Fig. 9). Such texture points to the fact that garnets are not igneous in origin but produced later on due to metamorphism. The reaction which can produce such a metamorphic garnet is as follows:

$$\text{Muscovite} + \text{Biotite} + \text{Quartz} \rightarrow \text{Garnet} + \text{K-feldspar} + H_2O$$

The above description establishes that the Mandi Granite has undergone a period of metamorphism synchronous with that of the country rocks. This metamorphism was of the Upper Green Schist facies. On the other hand, the post-tectonic granites of the Higher Himalayas are known to be the end result of the Himalayan metamorphism (Verma, 2003). The petrographic studies on the Tertiary granites indicate the occurrence of muscovite in them as a primary igneous phase along with quartz, feldspar, biotite, garnet etc. We may now extrapolate our results to the entire Himalayas, as the granites of each category are similar to each other. Therefore, we suggest that the granites of 500 Ma do not contain primary igneous muscovite as opposed to those of the Upper Tertiary age.

CONCLUSION

The Vaikrita Group of rocks was initially a pile of sediments, which got metamorphosed during the Precambrian. Then in the Cambrian period, there was melting at the bottom of the pile at 21 Km. There was ascent of the liquid upto a depth of about 15 Km where crystallization of biotite granite (Mandi Granite) took place around 545 Ma. A contact aureole is produced around the granitic body. Then there was a long period where no noticeable tectonism or metamorphism took place till the Tertiary period. During this time the whole area was affected by Himalayan metamorphism due to Himalayan orogeny. The progressive metamorphism in the metapelites produced dehydration fluid, which for obvious reason invaded the dry granitic mass through deformational fractures. In the granite, due to influx of the fluid hydrolytic metamorphism, K-feldspar was altered to sericite. Continuation of metamorphism changed the sericite into muscovite. The present mineralogy in granite is quartz + plagioclase + K-feldspar + biotite ± zircon + tourmaline ± apatite. Muscovite in the Mandi Granite is formed as a result of metamorphism and can be successfully used as a distinguishing factor for differentiating Tetiary and Paleozoic granite in the Himalayas.

ACKNOWLEDGMENT

The present work forms a part of a project awarded to the third author by the Department of Science and Technology, Government of India vide project number ESS/CA/A(-29/92). We are thankful to Prof. A.K. Jain, Prof. R.M. Manickavasagam and Dr. T.K. Ghosh of Indian Institute of Technology, Roorkee for their kind cooperation in making the probe facilities available. We are also thankful Dr. N.C. Pant of Electron Probe Microanalyser Laboratory of the Geological Survey of India, Faridabad for helping in microscopic photography.

REFERENCES

Bailey, S.W., 1988. X-ray diffraction indication of polytypes of mica, serpentine, and chlorite. Clay and Clay Minerals, **36** (3), 193-213.

Chamberlain, C.P., Zeitler, P.K. & Erickson, E., 1991. Constraints on the tectonic evolution of the northwestern Himalaya from geochronologic and petrologic studies of Babusar Pass, Pakistan. *Jour. Geol.*, **99**, 829-49.

Chatterjee, B.C., 1976. Petrology and structural geology of the Mandi Granite, Mandi district, Himachal Pradesh, Miscellaneous Publication, *Geological Survey of India*, **24**.

Clarke, D.B., 1992. Granitoid Rocks. Chapman and Hall, London, 283.

Ferry, J.M. & Spear, F.S., 1978. Experimental calibration of the partitioning of Fe and Mg between garnet and biotite. *Contrib. Min. Pet.*, **66**, 113-17.

Frank, W., Thoni, M. & Purtscheller, F., 1977. Geology and petrography of Kulu-S-Lahul area. *Colloques Internationauxn du C.N.R.S.*, **260**, 147-72.

Gansser, A., 1964. Geology of the Himalayas. Interscience Pub., John Wiley and Sons Ltd., 289.

Guidotti, C., 1984. Micas in Metamorphic rocks. *Rev. in Min.*, **13**, 357-67.

Guidotti, C.V. & Sassi, F.P., 1986. Classification and correlation of metamorphic facies series by means of muscovite b_0 data from low grade metapelites. *Neves. Jahr. Miner Abh.*, **153**(3), 363-80.

Guidotti, C.V., Sassi, F.P. & Blencoe, J.G., 1989. Compositional controls on the a and b cell dimensions of 2 M_1 muscovite. *Eur. Jour. Mineral.*, **1**, 71-84.

Gupta, L.N., 1974a. *Petrology of Mandi Granites and associated rocks*. Published by Centre of advanced studies in Geology, Punjab University, Chandigarh, **10**, 95-108.

Haselton, H.T., Horis, G.L., Hemingway, B.S. & Robie, R.A., 1983. Calorimetric investigation of the excess entropy of mixing in analbite-sanidine solid solutions: lack of evidence for Na, K short-range order implications for two feldspar thermometry. *Am. Min.*, **68**, 398-413.

Hodges, K.V. & Crowley, P.D., 1985. Error estimation and empirical geothermometry for pelitic systems. *Am. Min.*, **70**, 702-09.

Hodges, K.V., Hubbard, M.S. & Silverberg, D.S., 1988. Metamorphic constraints on the thermal evolution of the central Himalayan Orogen. In: *Tectonic evolution of the Himalayas and Tibet. Phil. Trans. Royal Soc. Lond.* **326**, 2577-80.

Hodges, K.V. & Silverberg, D.S., 1988. Thermal evolution of the Greater Himalayas, Garhwal, India. *Tectonics*, **7**, 583-600.

Hubbard, M.S., 1996. Ductile shear as a cause of inverted metamorphism: example from Nepal Himalaya. *Jour. Geol.*, **104**, 493-99.

Hutchinson, C.S., 1974. Laboratory handbook of petrographic techniques. John Wiley and Sons, New York, 577.

Jain, A.K. & Manickavasagam, R.M., 1993. Inverted metamorphism in the intracontinental ductile shear zone during Himalayan collision tectonics. *Geology*, **21**, 407-10.

Jain, A.K., Manickavasagam, R.M. & Singh, S., 1999. Collision tectonics in NW Himalaya: deformation, metamorphism, emplacement of leucogranite along Beas-Parbati valleys, Himachal Pradesh. *Gond. Res. Group Mem.*, **6**, 3-37.

Le Fort, P., 1975. Himalaya: the collide range present knowledge of the continental arc. *Amer. J. Sc.*, **275**, 7-44.

Massone, H.J. & Schreyer, W., 1987. Phengite geobarometry based on the limiting assemblage with K-feldspar, phlogopite and quartz. *Contrib. Min. Pet.*, **96**, 221-24.

Morse, S.A., 1970. Alkali feldspar with water at 5 Kb pressure. *Jour. Pet.*, **11**, 221-53.

Purkayastha, S., 1997. Constraining the metamorphic episodes in the Vaikrita polymetamorphites through textural and mineral chemical studies in Baggi-Mandi-Pandoh, H.P. Unpubl. *Ph.D. thesis, Univ. Delhi*, 188.

Purkayastha, S., Verma, P.K., Jain, A.K., Manickavasagam, R.M. & Ghosh, T.K., 1999. Growth and rotation of porphyroblasts and development of shear/extensional foliation in Baggi-Mandi-Pandoh region, Western Himalaya-Implications on the Himalayan tectonics. In A.K. Jain and R.M Manickavasagam (eds) *Geodynamics of the NW Himalaya. Gondwana Research Group Mem.*, **6**, 125-33.

Ray, S., 1947. Zonal metamorphism in the eastern Himalayas and aspects of local geology. *Quart. Jour. Geol. Soc. Ind.*, **14**, 117-40.

Searle, M.P. & Treloar, P.J., 1993. Introduction. Treloar, P.J. and Searle, M.P. (eds), Himalayan Tectonics, *Geol. Soc. Spl. Pub.*, **74**, 1-7.

Shams, F.A., 1983. Granites of Himalayas Karakoram and Hindukush. Burki Art Press, Lahore, 427.

Speer, J.A., Micas in igneous rocks. *Rev. in Min.*, **13**, 299-356.

Srikantia, S.V. & Bhargava, O.N., 1998. Geology of Spiti-Kinnaur Himachal Himalaya. Geological Society of India. *Geol. Surv. India, Mem.*, **124**, 210.

Thompson, A.B., 1976. Calculation of muscovite paragonite-alkali feldspar phase relation. *Contrib. Min. Pet.*, **44**, 173-94.

Thompson, A.B., 1996. Fertility of crustal rocks during anatexis. *Trans. Royal Soc. Edinb.*, **87**, 1-10.

Thoni, M., 1977. Geology, Structural Evolution and Metamorphic Zoning in the Kulu Valley (Himachal Himalayas, India) with Special Reference to the Reversed Metamorphism. *Mitt. Ges. Geol. Bergbaustud. Osterr.*, **24**, 125-87.

Vannay, J.C. & Hodges, K.V., 1996. Tectonometamorphic evolution of the Himalayan metamorphic core between the Annapurna and Dhaulgiri, *Central Nepal. J. Met. Geol.*, **14**, 635-56.

Whitney, J.A. & Stormer, J.C., 1977. Two-feldspar geothermometry, geobarometry in mesozonal granitic intrusions: three examples from the Piedmont of Georgia. *Contrib. Min. Pet.*, **63**, 51-64.

Verma, P.K., 1989. The Himalayan metamorphism. Evolution of metamorphic belts. Daly, J.S., Cliff, R.A., Yardley, B.W.D. (eds), *Geol. Soc. Pub.*, **43**, 377-83.

Verma, P.K., 2003. Presidential Address. Indian Science Congress Association, 90th Session.

11

MODELING OF P-T-t PATHS CONSTRAINED BY MINERAL CHEMISTRY AND MONAZITE DATING OF METAPELITES IN RELATIONSHIP TO MCT ACTIVITY IN SIKKIM, EASTERN HIMALAYAS

CHANDRA S. DUBEY*, E.J. CATLOS** AND B.K. SHARMA*

ABSTRACT

The metapelites of lesser Himalaya in Sikkim underwent two phases of prograde metamorphism, each culminating into a retrograde metamorphic episode. The first event was prograde regional metamorphism, M_1, during which S_1 and S_2 schistosity were developed. During this episode of metamorphism of chlorite, chloritoid, biotite, garnet, staurolite and kyanite were formed. Most of the prograde minerals (chlorite to staurolite) in this phase of metamorphism were formed by continuous reactions. Different geothermometers revealed temperatures upto 465 to 527°C in garnet zone and 555 to 576°C (Core parts) in staurolite zone. The maximum pressure in core of garnets in staurolite zone was inferred upto 6.6 Kbr.

The M_1 metamorphism was followed by a retrogressive phase M_{1a} due to which the prograde minerals from chlorite to kyanite showed alteration effects. The temperatures and pressure of this retrogressive event is inferred from outer core parts of garnets in staurolite zone from 492 to 522°C at 5.7 Kbr pressure.

M_2, metamorphism was again a prograde type during which garnet, staurolite, kyanite and sillimanite (with almandine) were formed. The temperature and pressure during M_2 metamorphic phase reached in the range of 630 to 695°C at 6.9-7.3 Kbr respectively in kyanite-sillimanite zone. The metamorphic episode M_2 again culminated into retrograde metamorphism M_{2a}, during which the rims of the multistage garnet were chloritised. The other evidences are replacement of k-rich feldspar by muscovite, reverse zoning of garnet at rim, pseudomorph of chlorite after idioblastic garnet, decrease in Zn content of staurolite.

In the Sikkim Himalaya, inverted metamorphism is the characteristics of rocks that underlie the MCT zone. Monazites dated in the thin section indicate that the Sikkim Himalayan MCT shear zone was active during 22-20 Ma and 14-10 Ma. The ages indicate inverted metamorphism generated via accretion of successive tectonic slivers of the Lesser Himalaya to the hanging wall. The paper outlines a model for the development of the Sikkim Himalayas based on P-T-t paths.

* Department of Geology, University of Delhi, Delhi-110007, India.

** School of Geology, Oklahoma State University, Stillwater, OK, USA.

KEY WORDS

Mineral Chemistry; Monazite dating, MCT and P-T-t paths.

INTRODUCTION

The Sikkim Himalaya is tectonically characterised by a domal structure on regional scale and the eastern half of domal structure in Eastern Himalaya (Sikkim) has been studied (Fig. 1A & 1B). Gansser (1964) opined that the gneisses were tectonically emplaced in the Dalings and were not related to the Darjeeling gneiss. Banerjee *et al.* (1980), based on landsat and Air photos, marked that the area witnessed differential uplift along a number of regional and local sub-vertical faults. Chloritic slates of the Dalings merged imperceptibly through biotite schists to garnet and staurolite schists of the Darjeeling group. Banerjee and Bhattacharya (1981) considered that there was an absence of prograde chloritoid and cordierite in the metamorphites of Sikkim and these minerals were locally formed as retrograde products of prograde assemblages.

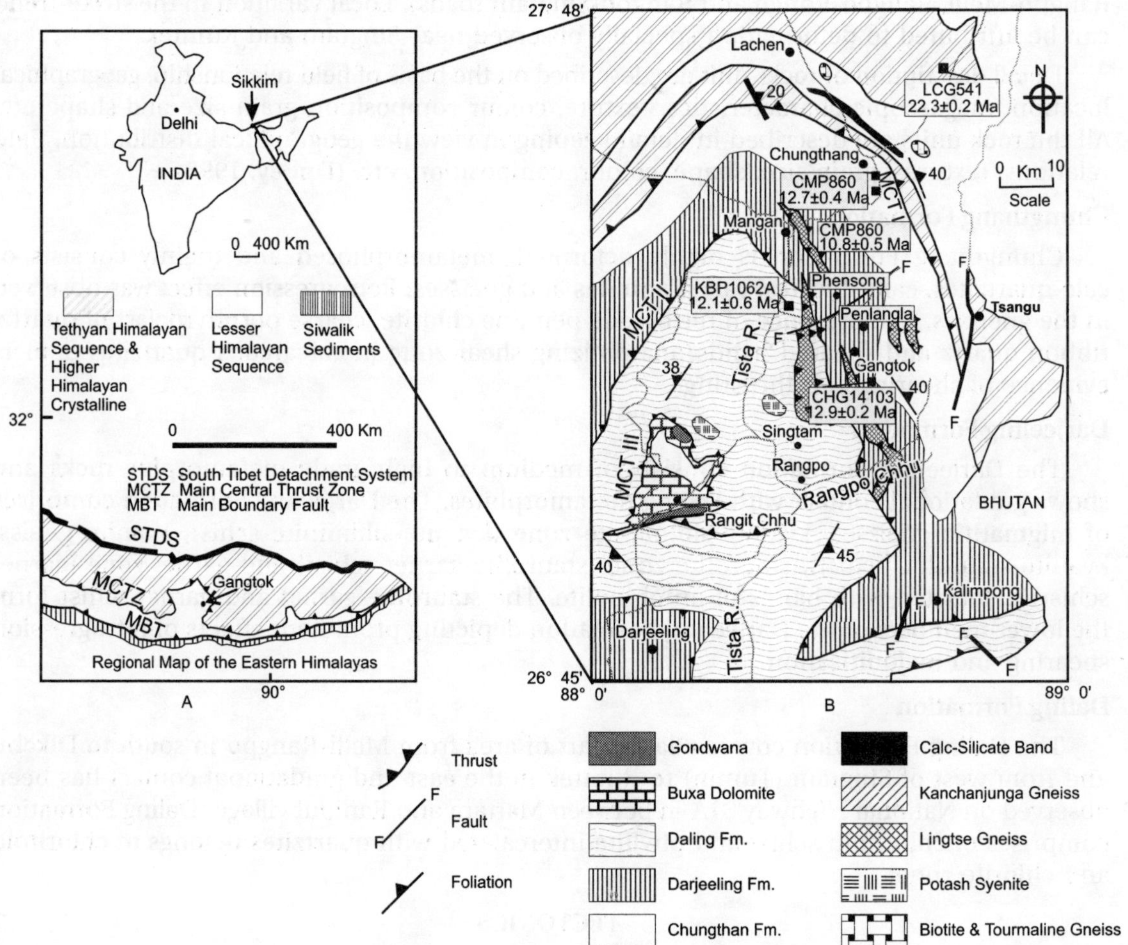

Fig. 1. Geological map of Sikkim (Eastern Himalaya).

Geological Setting

The Himalayan sequence of Sikkim consists of various lithostratigraphic units from low to high metamorphic grades. Igneous rocks occur as intrusives (potassic syenite, i.e. Sikkim Fm) in the Daling near Khamdong and Singtam. Stratigraphically these rocks have been grouped into Chungthang-Darjeeling-, Daling-, Lingtse- (Acharyya, 1989) and Sikkim igneous Formation (Dubey, 1993). The contact between Darjeeling and Daling near Deorali is a sheared one and the rocks are highly mylonitised along the contact especially near Deorali. The streaky biotite gneiss (Lingtse) occurs in the form of bands which are present within the low grade Daling and medium to high grade Darjeeling Formations. The Lingtse gneiss is characterized by sheared contacts with Darjeeling and Daling Formations. Chungthang and Darjeeling Formations are the part of Higher Himalayan Crystalline and Daling Formation is Lesser Himalayan low grade metamorphites (Fig. 1B).

The regional strikes of lithological units are NW-SE dipping 20-65° towards NE. However, in the southern part, gradually changes to NE-SW with 30-65° of dips in SE (e.g. along Rangpo-Melli, Rangpo-Rongli and Rangpo-Singtam roads). Local variation in the strike-trend can be attributed to deformation specially observed near Singtam and Rumtek.

Detail description of rocks unit are described on the basis of field relationship, geographical location, megascopic characteristics, texture, colour composition grain size and shape etc. All the rock units are described in detail keeping in view the geographical distribution, field relations, textures, grain size, shape, colour, composition, etc. (Dubey, 1993).

Chungthang Formation

Chungthang Formation is highly deformed, metamorphosed and mainly consists of calc-quartzites, calc-silicates, calc granulites and gneisses. Retrogression effect was observed in the gneisses. The presence of retrograde pennine chlorite, coarse porphyroclast of quartz, ribbon quartz and sheared bands/anatomizing shear zone in calcareous quartzites depicts evidence of shearing and thrusting.

Darjeeling Formation

The Darjeeling Formation consists of medium to high grade metamorphic rocks and shows gradational contact with Daling metamorphites. The Darjeeling Formation comprises of migmatitic gneisses, kyanite-sillimanite zone (kyanite-silliminite schist, kyanite gneiss, kyanite-staurolite schist), staurolite zone (staurolite-garnet-schist) and garnet zone (garnet schist) associated with bands of amphibolite. The staurolite-garnet and garnet schist form the lower members of the Darjeeling Formation depicting prominent effects of retrogression shearing and mylonitisation.

Daling Formation

The Daling Formation covers a large part of area from Melli-Rangpo in south to Dikchu and from west of Singtam (Tumin) to Rumtek in the east and gradational contact has been observed on National Highway 31A in between Martam and Ranipul village. Daling Formation comprises of chloritoid schist and phyllite intercalated with quartzites belongs to chloritoid and chlorite zones.

TECTONICS

The rock units exposed in Sikkim Himalaya have been multiply deformed and metamorphosed in due course of time. The three phase of deformation have been observed during the field investigations which are substantiated by the microstructural evidences. The Sikkim area is tectonically delineated by well defined thrusts i.e. the Main Central Thrust (shear zone) which brings the migmatites and gneisses over the medium to high grade schists and gneisses (Darjeeling Formation) and also depicts tectonic contacts with the Lingtse gneiss and low grade Daling phyllites. During the last stage of deformation (D_3), the rocks were affected by thrusting acquired domal structure.

Main Central Thrust (MCT-I) equivalent to the Vaikrita Thrust could be recognized in the northern part of the area where the Greater Himalayan Crystalline Complex is separated abruptly close to Chungthang by calc rich quartzites, calc-silicates and calc-granulites. Highly sheared and retrograded calcareous quartzites (Fig. 2A) calc-silicated and calc-granulites are observed 2 Km South of Chungthang (Fig. 1B). The shear fabrics L-S dominates, the calcareous quartzites and is clearly replaced by the flattening fabric of the Greater Himalayan Crystalline Complex in the north (Pecher, 1989). The position of MCT-II close to the MCT-I is marked in the south of Gangtok in the mylonitic Lingtse gneiss (Fig. 2B), retrograded staurolite and garnet bearing schists which overly the low grade Lesser Himalayan Formation at Martam. MCT-II of Sikkim Himalaya is recognized as Munsiari Thrust in the Garhwal Himalaya and could be related to MCT-I in Nepal (Arita, 1983). MCT-I and MCT-II is separated by the wide shear zone of strongly to weakly metamorphosed rocks of Darjeeling and Daling Formation along with thrust slice wedges of sheared Lingtse gneiss (Fig. 1B). MCT-III is recognized as Rangit Thrust by previous workers (Acharyya, 1989; Banerjee *et al.*, 1980; Sinha Roy, 1977). The dips of the foliation S_2 are recorded in one direction mainly i.e. 30-50° due NE, except and south of Rangpo where dips changes from 30-50° due SE. Tight isoclinal F_2 folds in Lingtse gneisses show intense flattening along the plane of S_2 foliation. The signatures of post-metamorphic thrusting are evidenced in the field by strong retrograde metamorphic effects (chloritization) after the F_3 folding movements (Dubey, 1993).

PETROGRAPHY

The relationship between S-surfaces, phases of folding and recrystallisation of mineral assemblages were established by several workers (Zwart, 1960, 1962; Vernon and Flood, 1979; Johnson, 1990; Bell and Johnson, 1992). On the basis of above works Table 1 correlates growth of various minerals in relation to deformation and metamorphism in metapelites on the basis of detailed petrographic studies in the area.

MINERAL CHEMISTRY

Analytical Techniques

Electron microprobe data of polished thin sections of key minerals in the samples were obtained by electron probe microanalyser (EPMA) and scanning electron microscope (SEM) at Roorkee University. For this purpose, standard operating conditions were 15-20 KV accelerating potential, 15 nA specimen current, 15-200 sec counting time, following the method of Bence and Albee (1968) and the correction factors of Albee and Ray (1970). Reproducibility of data is within 1 to 2% for major and minor elements.

Metapelites

Table 2 gives average representative chemical analysis of each of the minerals though a large number of point analyses have been obtained in different metamorphic grades.

Fig. 2. (A) Photomicrographs showing altered chlorite (peninnine) with obliquely elongated sheared quartz veins in calc-quartzites. (B) Photomicrograph showing sheared contact between Lingtse gneiss and garnet chlorite-schists. The feldspar porphoblast as well as the quartzose and micaceous layers show S – fabric bounded by strong C – plane.

Table 1: *Mineral growth in relation to deformation and metamorphism in Sikkim, Eastern Himalaya.*

	M_1 (Prograde) $S_1 - S_2$	M_{1a} (Retrograde) S_2	M_2 (Prograde) $S_2 - S_3$	M_{2a} (Retrograde) $S_3 - S_4$
Pelites				
Quartz	–	-	–	-
Plagioclase	–	Sericitization Zone	–	Sericitization and epidote formation
Chlorite	–	–	–	Penninine
Chloritoid	–	–	–	–
Biotite	–	Chlorite and Sigmoidal shapes	–	Chlorite
Muscovite	–	Sericite	–	Sericite
Garnet	– (Normal growth zoning)	Chloritization (inverse zoning)	Zoned garnet with inclusion free rims (normal growth zoning)	Chloritization (Penninine) and Chlorite pseudomorphs (reverse zoning)
Staurolite	– (Zn-rich)	Sericitization (decrease in Zn)	Zoned garnet inclusion free Staurolite (increase in Zn)	Sericitization (decrease in Zn)
Kyanite	–	Sericitization/ pyrophyllitization	–	Kinked and deformed cleavage and pyrophyllitization
Sillimanite	–	–	–	Sericitization
K-feldspar	–	Sericitization	Zoning	Muscovite Replacement
	Isoclinal Tight root folds	Sinistral shearing and less retrogression parallel or little oblique to S_2. Reclined and Isoclinal root-less coeval folds	Porphyroblast over growth and zoning open to tight 1B and 1C type folds	Retrogression phyllotization cross folding micropuckering
	F_1	F_2	F_3	F_{3a}
	D_1		D_2	

Table 2: *Selected representative chemical analysis of metamorphic rocks from Sikkim, Eastern Himalaya (Major trace elements in % and traces in ppm).*

Oxides	1	2	3	4	5	6	7	8	9	10	11	12	13
SiO_2	74.34	68.00	72.12	56.35	53.85	69.95	75.47	78.62	72.87	73.26	70.74	81.46	72.21
Al_2O_3	12.57	15.51	20.11	14.21	14.14	13.25	13.60	12.32	13.20	12.38	14.28	9.46	12.78
Fe_2O_3	4.27	12.89	0.05	10.94	17.07	4.80	3.42	2.70	4.20	5.74	5.21	3.62	7.46
CaO	0.14	0.00	4.57	0.05	0.07	0.17	0.19	0.52	1.02	2.00	1.96	0.00	0.06
MgO	1.05	0.00	0.82	9.07	5.40	0.63	1.46	0.75	0.78	0.50	0.89	1.20	1.36
MnO	1.07	0.20	0.08	0.14	0.25	0.09	0.02	0.06	0.06	0.10	0.08	0.07	0.09
K_2O	1.51	1.14	0.42	6.37	3.72	6.56	5.50	1.99	4.17	2.15	3.40	1.03	2.87
Na_2O	3.90	0.54	0.59	1.92	1.74	2.29	1.58	3.14	2.67	3.13	3.23	2.04	1.82
P_2O_5	0.11	0.57	0.91	0.03	0.36	0.09	0.09	0.68	0.66	0.22	0.61	0.76	0.60
TiO_2	0.53	0.58	0.29	0.86	0.70	0.67	0.98	0.25	0.38	0.52	0.48	0.36	0.74
Total	98.49	99.95	99.96	99.94	97.30	100.50	102.31	101.03	100.01	100.00	100.88	100.00	99.99
Fe^{+3}	3.56	11.74	3.73	7.95	17.15	5.64	3.46	2.40	3.85	4.74	4.69	3.12	5.54
Ni	29.62	68.93	15.15	68.93	82.81	31.88	46.89	8.30	19.22	14.91	11.12	14.55	17.77
Cu	15.68	45.86	9.70	45.86	137.50	33.53	30.16	00.00	2.11	33.30	1.78	13.53	25.02
Zn	45.66	13.62	5.35	138.62	138.45	55.21	44.74	27.23	46.64	55.91	60.02	8.37	29.35
Ga	17.48	17.48	17.01	17.48	19.44	22.44	20.48	15.94	16.21	16.67	16.68	16.62	17.91
Pb	21.46	2.75	28.04	2.75	26.60	41.17	36.71	28.02	23.44	27.31	24.85	32.95	25.48
Th	0.71	11.76	18.43	11.76	15.43	19.78	10.41	17.81	13.13	4.88	18.07	19.47	0.00
Rb	93.39	38.25	30.87	205.38	195.45	225.98	172.03	63.64	91.05	180.27	262.68	63.22	167.82
U	0.80	2.85	8.11	7.06	6.04	7.06	5.47	46.38	53.66	7.82	17.14	5.22	0.70
Sr	208.87	N.D	28.62	42.00	52.00	N.D.	35.61	57.38	60.33	87.36	87.03	12.78	21.81
Zr	236.5	356.97	205.48	15.25	15.72	68.28	34.81	81.52	82.14	83.01	50.58	193.80	270.19
Nb	8.92	13.29	1.60	18.92	N.D.	24.12	19.22	78.01	75.59	6.59	10.70	20.09	6.92
Y	20.40	55.07	4.46	N.D.	N.D.	N.D.	N.D.	65.00	68.00	16.61	20.98	29.75	28.93

1. Phyllite Schist; 2. Chloritoid Schist; 3. Garnet Schist; 4. Staurolite Garnet Schist; 5. Staurolite Garnet Schist; 6. Garnet Schist; 7. Kyanite Sillimanite Schist; 8. Migmatite; 9. Migmatite; 10. Feld. Gneiss; 11. Feld. Gneiss; 12. Lingtse Gneiss; Lingtse Gneiss.

Chlorite: The chlorite in low grade metamorphic rocks of Sikkim is ripidolite and is rich in iron and deficient in magnesium. It is enriched in Al_2O_3. The chlorite from garnet to staurolite zone varies from brunsvigite to pycnochlorite. The Fe/(Fe+Mg) ratio of chlorite in metapelites is usually in the range of 0.2-0.8 (Albee, 1962). The Fe/(Fe+Mg) ratio decreases from chloritoid to staurolite bearing rocks in which the FeO decreases but MgO increases with increasing metamorphic grades. The chloritoid bearing schist shows abundance of alumina and iron oxide as is observed in the chlorite of low grade rocks. The Al/(Al+Fe+Mg) versus Mg/(Mg+Fe) ratio (Laird, 1989) shows an inverse trend i.e. decrease of the ratio in increasing grade of metamorphism.

Chloritoid: Chloritoid is rich in aluminium and iron and shows normal chemical zoning. The abundance of (ferrous) iron, aluminium and low magnesium is a function of the bulk chemistry of chloritoid bearing schist. The recalculated analysis shows that chloritoid contains very little ferric iron. It shows negligible substitution of Al^{3+} by Fe^{3+} and Fe^{2+} by Mn. Na_2O and CaO also occur in negligible amounts.

Muscovite: The inferred percentage of muscovites (88-94% Mu) from Sikkim contains high K_2O (8.45-9.53), Al_2O_3 (33.33-34.5) while FeO (1.21-2.52), MgO (0.2-0.81) and Na_2O (0.97-1.75) are present in low percentage. These rocks show limited solid solution with paragonite (6-12% pa). The paragonite content increases in chloritoid to staurolite zones and decreases in kyanite-sillimanite zone. The highest sodium content is in staurolite zone. The increase in Ti content of muscovite is commonly inconsistent with the progressive increase of metamorphism upto staurolite grade (Guidotti, 1974). Al^{iv} content also show an increase from garnet to kyanite-sillimanite grade as observed by Das (1987) in Kishtwar area (Kashmir Himalayas).

Na substitutes for K upto 0.13-0.21 Na/(Na+K) in chloritoid and garnet zones. This content is upto 0.24 Na/(Na+K) in the staurolite zone. However the substitution decreases in (0.22 Na/Na+K) kyanite-sillimanite grade samples. The Na-content of muscovite is relatively higher in most of muscovites rich samples which may be due to muscovite co-existing with plagioclase (*cf.* Guidotti, 1969). The maximum amount of Si observed is 3.11/formula in garnet zone whereas it is about 3.09/formula in staurolite and kyanite-sillimanite zone. According to Labotka (1980) and Hyndman (1985), the phengitic component decreases and is substituted by Na with increasing metamorphic grade. However, in the staurolite zone significantly higher substitution of sodium content is noticed. According to Hyndman (1985) there is a lowering of Si, Fe^{2+} with increasing trend of metamorphic grade. In Sikkim area, Si, Ti, Mg, Na and EY in muscovite show an increase in slope from chloritoid to garnet grade zone while Al^{iv}, Al^{vi}, Fe, K and EX are characterised by a decreasing slope trend. This variation in staurolite and kyanite-sillimanite zone is different than the other rocks, except for a decrease of Al^{vi}, and an increase in Mg from chloritoid to kyanite-sillimanite zone. The decrease in Si content from garnet to staurolite zones is due to the effects of shearing and retrogression associated with staurolite zone (*cf.* Goff & Ballevre, 1990).

Biotite: The chemical variations in biotite during prograde metamorphism has been worked out by Dhana Raju *et al.* (1977), Ruiz *et al.* (1978), Labotka (1980), Schreurs (1985), Das (1987), Guidotti (1984) and Guidotti *et al.* (1988). The variations of Mg/Mg+Fe, Na/Na+K and Mn content largely depends on the various temperature controlled mineralogical reactions. Ti-content, EXII sites and EVI sites happen to be crystallochemically controlled responses to the variations caused by the metamorphic reactions (Guidotti *et al.*, 1988).

The chemical composition of biotite from Sikkim depicts Mg/Mg+Fe ratios as 0.48-0.49 in garnet zone; 0.49 to 0.53 in staurolite and 0.49-0.50 in kyanite-sillimanite zone respectively indicating the ratio being sensitive to an increasing trend of metamorphism from garnet to staurolite zone. It is consistent with the Mg/Mg+Fe ratio of chlorite for the above zones. The variation of Mg/Mg+Fe of biotite in these zones reflect that with the rise of temperature, the continuous reactions moved the assemblages in more Mg-rich positions field (Guidotti *et al.*, 1988). However, once the assemblages Al_2SiO_5 + staurolite + Biotite + garnet were formed, the continuous reactions made the phase to be more FeO rich in composition (Chinner, 1965). Similar feature is noticed in the Fe-rich biotite (lower Mg/Mg+Fe ratio) from kyanite-sillimanite zone of Sikkim.

In Sikkim area also the Na/Na+K ratio in biotite showed an increase from 0.03-0.04. to 0.082-0.091 in garnet and staurolite zone respectively. It decreased from 0.067 to 0.082 in kyanite-sillimanite zone. This relation was quite consistent in staurolite and kyanite-sillimanite bearing assemblages when the host rock was depleted in Na_2O and MgO (Dubey, 1993).

In Sikkim area an increase in Al^{iv}, Mg/Mg+Fe contents is observed with increasing grade of metamorphism whereas Al^{vi}, Ti, Si^{iv}, Na/Na+K and Mn showed a marked decrease in higher grade metamorphites (after Ruiz *et al.*, 1978). Guidotti (1984) is of the view that the increase in Ti content with metamorphic grade could be ascribed to biotite saturated with Ti and coexisting with intrinsic Ti phases such as ilmenite and rutile. Thus, Ti content is a function of prevailing intensive parameters. In Sikkim area, the biotite contains ilmenite and the Ti-content biotite (coexisting with garnet) of garnet and staurolite zone is concomitant with increasing metamorphic grade. However biotite of kyanite-sillimanite zone is not saturated with Ti. Dhana Raju *et al.* (1977) noted that the ratio MnO×100 TiO_2 was less than 5.0 (i.e., above 90%) in most of the biotites from high grade rocks (upper amphibolite and granulite facies), whereas it was more than 5.00 in low to medium grade rocks (epidote-amphibolite and greenschist facies). On the basis of above it can be inferred that the rocks of garnet, staurolite and kyanite-sillimanite zones show a transition from epidote-amphibolite to upper amphibolite facies on the basis of MnO versus TiO_2 ratio (*cf.* Dhana Raju *et al.*, 1977). Biotite show increase in Al^{iv} and decrease in Ti in medium to high grade metamorphic rocks (staurolite to kyanite-sillimanite). A marked decrease in MnO in biotite present in negligible amount in higher grades can be related with depletion of MnO in garnet of different metamorphic grades.

Plagioclase: The composition of plagioclase is in the range of An_{14-22}. The zoning is common and grain to grain compositional variation is very little and is marked by a decrease of co-existing anorthite content at the rims. The anorthite content of plagioclase increases from staurolite to kyanite-sillimanite zone. As noted by Labotka (1980); the Ca-rich garnet is more akin to co-exist with Ca-rich plagioclase and since these are the only Ca-phases present, the Ca-content is controlled to some extent by the occurrence of Ca in the total rock. In zoned garnet the Ca decreases from core to rim in all the metamorphic zones in Sikkim.

Garnet: Garnet from different metamorphic zones in Sikkim area exhibits a variable distribution of chemical elements *viz.* Fe, Mg, Ca and Mn and is characterized by chemical zoning. The zoned garnets are critical indicators of disequilibrium and are significant in application of increasingly sophisticated models of kinetics of metamorphic processes. The chemical evolution of garnet from Sikkim area is described below:

Garnets from the garnet zone is highly enriched in Mn-core showing a progressive decrease towards rim with a very slight inflection of relatively lower Mn near core. It also exhibits normal growth zoning i.e., Fe+Mg increasing and Ca+Mn decreasing from core to rim (Dempster, 1985). Garnet with normal growth zoning has a grain size varying from 1 to 3 mm. The smaller the size of garnets, the larger the volume of rock in which such garnets are formed (Hess, 1971; Dietvorst, 1982). Kinking in the core of such prograde garnet may be due to replacement of chlorite by biotite (Hollister, 1966; Thompson, 1976) overlapping T-X loops. The Mg/Mg+Fe(M/FM) ratio shows a slightly increasing trend varying near the rims (next to biotite) due to the effects of cation diffusion between garnet and biotite rather than any growth or resorption feature, which would mimic the porphyroblast shape (*cf.* Lasaga *et al.*, 1977). The deficiency of Mn at the rims of the garnet further supports this inference (*cf.* Dempster, 1985). The garnets of this zone contains inclusions of minute chlorite, muscovite and epidote along with rutile and $S_i(S_1)$ of quartz and ilmenite in core parts. The S_e in garnet is surrounded by S_2 of syntectonic origin at the rims which are affected by retrogression.

The garnet from staurolite zone shows complex textural and chemical patterns. The garnet is comparatively larger in size than that of the garnet zone (approx. 1-5 mm) and is dotted by clear inclusion-trails in its core which is smaller than inclusion free (ilmenite rich) idioblastic rim and is associated with biotite often showing retrogression.

The core of garnet consists rutile, ilmenite, muscovite and quartz as $S_i(S_1)$ followed by quartz, retrograded chlorite as $S_i(S_2)$ at outer part of core. It is followed by ilmenite and rarely retrogressed biotite near the idioblastic rim. Biotite coexisting with such garnet is mainly of prograde type and shows retrogression at the rims. It is observed that the inner inclusion trail is abruptly truncated in the outerpart of core and often the rim of the garnet is a pseudomorph after chlorite. A few grains show micropuckering or later foliation.

Such type of garnet characterized by an increase of Mn+Ca near the outerpart of core and with a decrease in Mg+Fe show an inverse zoning followed by a normal growth zoning from outer part of core to the inner part of rim. A reverse zoning near the outer part of the rim is also noticed. The inverse zoning i.e. enrichment of Mn content from inner core to the outer core is attributed to retrogressive effects (Grant and Weiblen, 1971; De Bethune *et al.*, 1975; Tracy *et al.*, 1976; Tracy, 1982; Dietvorst, 1982; Karabinos, 1984 and Dempster, 1985).

Lal *et al.* (1981) analysing the garnet from Takdah, Darjeeling area suggest that garnet from Zone C (Staurolite-biotite zone) is feebly zoned and is free from inclusions and lacks rotational synkinematic core. It was presumably formed during the post-kinematic phase, between the D_1 and D_2 deformation through a discontinuous reaction. Mohan *et al.* (1989) observe that Ca-increases from the core and then decreases towards the rim. The changes in garnet growth as suggested by them are in agreement with Crawford (1974) due to the garnet producing reactions. The garnet growth has also been attributed with contemporaneous increasing temperatures (Cygan and Lasaga, 1982). Continuous smooth zoning profiles and lack of textural zones in most of the grains of garnet was explained by them due to uninterrupted garnet growth in a single metamorphic event.

Some of the garnets from staurolite zone lacks distinct textural zones or inclusion free variety is post-kinematic and later possibly grew simultaneous to the overgrown inclusion-free rims. The reversal of chemical zoning i.e., an enrichment of Mn near the rim, has been discussed by several workers because of cation exchange between adjacent garnet and biotite (Hess, 1971; Tracy *et al.*, 1976; Robinson *et al.*, 1977; Lasaga *et al.*, 1977); resorption and reincorporation of Mn into the remaining garnet (De Bethune *et al.*, 1975; Kerr, 1981) and growth by continuous Fe-Mg-Mn exchange reactions between garnet and biotite (Tracy *et al.*, 1976; Thompson, 1976). The biotite inclusions near the rim of garnet show retrogression. Therefore, enrichment of Mn near the rim is due to another retrograde metamorphic event, which equilibrated rapidly with the co-existing minerals. The garnet from staurolite zone is marked by textural unconformities with varied chemical zoning due to multiple tectonometamorphic events and polymetamorphism (Thompson *et al.*, 1977).

The garnet in kyanite-sillimanite zone is coarser in size, varying from 2-7 mm texturally, some garnet contains inclusions while other are devoid of inclusions. The chemical analysis of these garnets supports the fact that the content of Mn increases with a decrease in Ca near the rim. Garnet shows an increasing trend of M/FM ratio from Darjeeling-Sikkim (Lal *et al.*, 1981; Mohan *et al.*, 1989). They interpreted the normal growth zoning with a reversed zoning near the rims of garnet due to resorption through a continuous prograde reaction. The zoning pattern of Ca in high grade metamorphic rocks, is complex and reflects the involvement of some calcic phases (*viz.* plagioclase in garnet producing reactions).

The garnet examined from kyanite-sillimanite zone shows no evidence of zoning in its interior parts as observed in garnet from low to medium grade metamorphics. The texture of most grains of garnet (high grade) reveals that it grew under peak metamorphic conditions. As a corollary of the preceding, complete recrystallisation of garnet of lower grade under high grade conditions possibly obliterated the evidences of low grade zoning.

All these types of garnet in the Sikkim depict an increasing M/FM ratio at rims sympathetic to other Fe-Mg phases and a rough trend of decreasing concentrations of Mn is exhibited in the garnet core with increasing metamorphic grades upto staurolite zone and occasionally, in kyanite-sillimanite zone.

Staurolite: The staurolite has been chemically analysed from staurolite and kyanite-sillimanite zones. The staurolite from these zones shows small content of Zn which increases progressively with metamorphic grade. Chemical zoning shows a slightly higher amount of X_{Mg} towards rim (Mohan *et al.*, 1989). Lal *et al.* (1981) described that X_{Mg} in staurolite increased from zone C (Staurolite-biotite) to D (Kyanite-biotite) and decreased from zone D to E (Sillimanite-biotite) and in respect of core and rim parts of staurolite there was a general decrease of X_{Mg}. The progressive increase of Zn-content was attributed to its resorption by the continuous (Fe-Mg-Zn) reaction ($T_{Zn} > T_{Fe} > T_{Mg}$). A progressive increase in Zn-content has also been noticed in staurolite with increasing metamorphic grade of Sikkim. The zoning is clearly related to retrogression of white mica shimmer aggregates (Zn-rich) formed by breakdown of staurolite which was incorporated in remaining staurolite and formed Zn-rich mid rim/outer core (similar to retrograde Mn of the garnet). The staurolite in this zone showed a decreasing trend of Ti and Zn at the rim. Al and Fe showed an increase at the rims (Al and Fe being maximum in mid rim/outer core). Mg decreased at the rim but was of increasing order at mid rim/outer core. The M/FM ratio decreased at the rim of staurolite. The appearance of such staurolites showing increase in Zn and low M/FM at the mid rim/outer core parts were possibly formed at the rims of first formed garnets with increasing grade of metamorphism. Such features were more commonly observed in the medium grade rocks where late deformations were accompanied by increasing hydrous retrogression (Dempster, 1985). The presence of staurolite with in staurolite in the medium grade rocks can be ascribed to its multistage growth.

The staurolite in kyanite-sillimanite zone is less abundant and shows a lower M/FM ratio as compared to the previously formed staurolite in staurolites zone. This staurolite shows a slightly high Zn-content from that of the previously formed staurolite. The increase in Zn content in higher grade metamorphic rocks can be explained due to decrease in modal staurolite characterized by staurolite consuming prograde reactions. This staurolite is richer in Fe than that from staurolite zone.

Metamorphism

The rocks of Sikkim area exhibit an inverted sequence of metamorphism. As a result, the grade of metamorphism increases from south to north specially northwards of the M.C.T.

The low grade Daling Formation is overlain by medium to high grade Darjeeling Formation. The prograde metamorphic rocks of the first event (M_1) consist low to high grade metamorphites from greenschists to upper amphibolite facies. The first appearance of chlorite porphyroblast is noticed in the chlorite schist belonging to Daling Formation (low grade). This pelitic rock shows a two phase assemblage of muscovite-chlorite. As the bulk rock composition (Dubey, 1993, *cf.* Table 2) shows very high alumina, the increase of modal chlorite (rich in Al_2O_3) is suggestive of the fact that chlorite was formed by the following reaction:

Pyrophyllite = Chlorite + Quartz ...(1)

or

Kaolinite = Chlorite + Quartz ...(2)

Guidotti (1984) observed that rotation of the lines on AFM implies the exchange reaction:

Caledonitic Muscovite + Al-rich Chlorite

= Less Caledonitic Muscovite + Al-poor Chlorite ...(3)

Miyashiro and Shido (1985) described the extent of Tschermark substitution (Mg, Fe) Si= Al_2 in chlorite, muscovite and biotite in low to medium, grade metapelites. They observed that distribution co-efficients of exchange reaction for Tschermak substitution between muscovite and chlorite varied greatly not only with temperature but also with the substitution of the two minerals because of their deviation from their ideal exchange equilibrium, not sensitive to pressure. The caledonite content of muscovite decreased with an increase in temperature.

The chloritoid bearing metamorphic assemblages show the presence of relatively caledonite deficient muscovite, aluminium rich Fe-chlorite i.e., Fe-ripidolite and Fe-chloritoid with magnetite. Hyndman (1985) observed that chloritoid formed with paragonite at the beginning or within biotite zone i.e.,

Paragonite + Chlorite + Quartz = Chloritoid + Albite ...(4)

Based on this reaction, low content of paragonite (9-15%) in muscovite and absence of albite commonly restrict the occurrence of chloritoid. Winkler (1979) opined that chloritoid persisted throughout the range of low grade metamorphism. The chlorite-chloritoid pair in this assemblage shows a consistent Fe-Mg partitioning. The relative thermometry suggests a temperature of 410 to 450°C. According to Frey (1974) the reaction responsible for first appearance of chloritoid is as following:

Pyrophyllite+Iron-rich Chlorite = Chloritoid + Quartz + H_2O ...(5)

Another reaction at slightly higher temperature described by Thompson and Norton (1968) is:

Haematite + Iron-rich Chlorite = Chloritoid + Magnetite + Quartz + H_2O ...(6)

They suggested it to be a dehydration reaction although on casual examination, it can be easily mistaken for deoxidation reaction. The presence of high alumina and iron oxide in bulk rock composition along with the abundance of magnetite with Fe-chlorite and chloritoid can be explained by reaction (6) which is mainly applicable for the formation of chloritoid in Sikkim. The low grade metamorphic chlorite schists show strong retrogressive effects with formation of penninite and sericite, during the last stage of second retrograde metamorphism. In the chloritoid schists sheaves of white mica with chloritoid and quartz are also present.

The formation of biotite in K-feldspar free metapelites can be explained by the following continuous Fe-Mg reactions:

Phengitic Muscovite + $Chlorite_1$ = less phengitic muscovite + biotite ± $chlorite_2$ ± Quartz ...(7)

3 Muscovite + 5 chlorite = Biotite + 4 Al - Chlorite ± 7 Quartz + H_2O ...(8)

The altered penninite due to shearing of biotite along the S_2 and alteration of biotite into chlorite later than S_2 can be ascribed to retrograde metamorphic effects.

The appearance of garnet in the rocks is first observed towards north of Ranipul. The garnet co-exists with biotite which shows retrogression to chlorite. The rims of the garnet often, partially or completely are surrounded by penninite. The garnet, very rarely, shows unaltered nature along with biotite and is associated with chlorite and muscovite in the metamorphic assemblages. Lal *et al.* (1981), Das (1987) and Saklani (1993) have reported this mineral assemblage in Lesser Himalayan region. According to Thompson and Norton (1968) garnet and biotite in aluminous rocks are formed due to:

Chlorite + Muscovite + Quartz = Almandine + Biotite + H_2O ...(9)

The other reactions responsible for the first appearance of garnet are:

Chlorite + $Biotite_1$ + Quartz = Almandine + $Biotite_2$ + H_2O ...(10)

Winkler (1979).

Chlorite + Muscovite + Epidote = Almandine rich Garnet + Biotite + H_2O ...(11)

Winkler (1979)

The rocks of garnet zone from Sikkim show Mn-rich and Ca-rich garnet-cores while their rims are rich in Fe and Mg. Such a Mn-rich almandine garnet core is formed by the following continuous reaction:

Muscovite + $Chlorite_1$ + Quartz = Almandine Garnet + Biotite + $Chlorite_2$ ...(12)

(Mn-richer) (Mg-richer)

The rocks of the garnet zone, north of Gangtok stadium, exposed close to staurolite zone, consists of garnet in which chlorite, muscovite and epidote occur as inclusions and also in separate phases in the matrix along with biotite. It is to be noted that reaction (11) also took place in calcium rich phases. According to Guidotti *et al.* (1988), the sodium enrichment of muscovite in low to medium grade rocks is due to the following continuous reaction:

Fe-Chlorite + Muscovite = Garnet + Mg-richer Biotite
+ Mg-richer chlorite + Na-richer Muscovite + H_2O ...(13)

This reaction explains the enrichment of sodium in muscovite and an increase in Mg-content of chlorite. Miyashiro and Shido (1985) inferred that these reactions with rising temperature caused a decrease in caledonite content in muscovite, chlorite and the phlogopite content in biotite in the low grade metamorphics. In Sikkim, the inferred reactions can be applied for the formation of phyllites, chlorite and chloritoid schists. The presence of a calcium phase in garnets during an early phase of evolution was completely used up to form either plagioclase or grossularite i.e.

Epidote + Quartz = Anorthite + Grossular + H_2O ...(14)

(Guidotti, 1974)

The distribution of CaO, dependent on temperature can be explained by the equation as described by Crawford (1974):

High Ca-Garnet + Na-plagioclase + Muscovite + Chlorite
= Low Ca-Garnet + Ca-plagioclase + Biotite ...(15)

Garnet from garnet zone depicts a small kink in the ternary plots. Hollister (1966) attributed this due to a break in garnet-chlorite tie line and a shift from garnet-biotite tie line to the pseudo-binary T-X loop. Other workers (De Bethune *et al.*, 1975; Dempster, 1985) suggested such a kink due to retrograde effects.

The staurolite zone is mainly exposed in the eastern proximity of the thrust wedge of Lingtse formation (Streaky biotite gneiss) near Gangtok. Staurolite is associated with biotite, chlorite, muscovite, garnet and oligoclase. The Mg-rich chlorite and Na-rich muscovite is present in the core of the garnet, as well as in matrix of the staurolite bearing schists.

The core parts of garnet in this zone consist of inclusions of chlorite, muscovite, rutile, ilmenite and quartz. The quartz inclusions (Si) of S_1 style (Stage 2-3 of Bell and Rubenbach 1983) are enveloped by S_2 inclusion trails of (syn. S_2 foliation) in small core followed by the retrograde chlorite pseudomorph after biotite and altered biotite in the outer parts of core from where the inclusion free overgrowth of garnet has taken place. The overgrown garnet rims are larger in size than the relict cores and occasionally, contain ilmenite inclusions near the outer rims. The biotite in the matrix, near the outer rim also shows retrogression to form chlorite. Often, the overgrown rims are prominently chloritised. A few garnet grains present in quartz and mica domains of staurolite garnet schist do not contain any inclusion and are later formed idioblasts. Small idioblasts of staurolite are enclosed as inclusions in coarse and fractured idioblasts of staurolite. The sheaves of coarse biotite with staurolite of later generation show alteration effects such as chloritization. Lal *et al.* (1981) and Mohan *et al.* (1989) described the following discontinuous reaction for the appearance of staurolite:

Chlorite + Muscovite = Staurolite + Garnet + Biotite + H_2O ...(16)

The garnets described above with a core consisting of inclusions of S_1 and S_2 were followed by the presence of retrograde chlorite in the outer core along with an inclusion free rim, depicting an inverse zoning with Mn-poor and Mg-rich core are followed by an increase of Mn and decrease of Mg-content in the outer part of the core where retrograde pseudomorphs of chlorite after biotite are present. Such cores are overgrown by normal growth zoned rims decreasing in Mn-content and increasing in Mg from the outer part of the core towards the inner part of rim. The inner part of outer rim again shows a reverse zoning followed by normal growth zoning. The low Mn-content and high Mg-content of inner core of garnet and the presence of inclusions of chlorite and muscovite suggests the first appearance of staurolite due to the following continuous reaction:

Chlorite + Muscovite = Staurolite + Garnet + Mg-rich Biotite
+ Mg-rich chlorite + Na-Muscovite + H_2O ...(17)
Guidotti (1974)

The appearance of altered biotite and chlorite in the outer parts of core; an increase of Mn and decrease of Mg; increase of Zn-content in the mid rim/outer core parts of staurolite, are due to retrograde metamorphic effects. This stage possibly enriched the fluids which might have played an important part in the formation of Mn-rich garnet by resorption or due to a continuous reverse reaction.

The presence of overgrown rim showing normal growth zoning over earlier formed core of garnet and inclusion of ilmenite in the outer rim part, presence of staurolite in staurolite; increase of modal biotite and its coarsening to sheaf like biotite flakes, indicate that the retrograde metamorphic stage was stabilized at a faster rate and another prograde metamorphic event formed garnets showing normal growth zoning. Later, coarser idioblasts of staurolite were formed. The staurolite of this generation also shows an increase in Zn content at its rims. The normal growth zoning of garnet rims were consistent with the predicted trend of garnet compositions involved in continuous Fe-Mg-Mn reaction with increasing temperature (*cf.* Thompson, 1976).

Staurolite + Biotite + Quartz = Muscovite + Garnet + H_2O ...(18)

Lal *et al.* (1981) and Mohan *et al.* (1989) described reaction (18) for the garnets of staurolite zone in Sikkim-Darjeeling area, and probably studied the later formed normal growth zoned garnets from the staurolite zone. The later formed staurolite and garnet without inclusions are also partially or prominently altered to sericite and chlorite. A few overgrown rims of garnet also show alteration product (penninite-chlorite) and the rock is strongly phyllonitized. The biotite shows alteration into chlorite. Such garnets with postectonically recrystallized partial pseudomorph of chlorite after garnet are enveloped by cross-micropuckers. This implies that the later prograde metamorphic event again culminated in a second retrograde metamorphic event. The garnet from staurolite zone show a reversal of zoning near its outer margin. The increase of Mn and decrease of Mg-content in this zone, suggests resorption or effects of later retrograde metamorphic episode.

Kyanite is first observed associated with biotite and quartz along with staurolite, chlorite and muscovite. Lal *et al.* (1981) concluded that staurolite-chlorite join of staurolite zone is replaced by kyanite-biotite join. Mohan *et al.* (1989) did not observe a direct textural evidence that kyanite grew at the expense of staurolite and muscovite. However, an assemblage consisting of abundant staurolite and muscovite with lesser amount of chlorite and kyanite and plagioclase was followed by abundant kyanite, biotite and quartz in rocks exposed at Dikchu road. Lal *et al.* (1981) suggested a discontinuous (Fe-Mg) reaction mentioned below:

Staurolite + Chlorite + Muscovite = Kyanite + Biotite + Quartz ...(19)

The kyanite in the kyanite-staurolite bearing rock shows strong retrogression in which kyanite grains (parallel to S_2) are bent, broken and altered to pyrophyllite. Staurolite alters into sericite and biotite into chlorite. The presence of plagioclase along with kyanite in these schists, in significant modal amount, is attributed to a continuous reaction A-Na-K system as described by Lal *et al.* (1981):

Na-rich Muscovite + Quartz = Kyanite + Albite + K-rich Muscovite + H_2O ...(20)

However, a better discontinuous reaction proposed by Guidotti (1984) can be applied for the mentioned mineral assemblage, on the basis of petrographic observations:

Staurolite + Mg-chlorite + Na-Muscovite =
Al_2SiO_5 + Biotite+K-rich Muscovite + Albite + quartz + H_2O (Kyanite) ...(21)

In most of the metamorphic assemblages the prograde chlorite becomes negligible and so it can be inferred that the reaction (20) possibly continued to increase the modal biotite and kyanite in the late stages due to the following reaction (Mohan *et al.*, 1989):

Staurolite + Na-Muscovite + Quartz = Kyanite + Biotite + H_2O ...(22)

A few biotite laths in these gneissic rocks show alteration into chlorite of later metamorphic phase. Plagioclase in these gneissic rocks is more Ca-rich and K-feldspar shows perthitic intergrowth. So, the continuous reaction (21) consisting of Ca-bearing phases continued in the higher grade metamorphic rocks (Guidotti, 1984):

Muscovite + Plagioclase + Quartz = Al_2SiO_5 + Na bearing K-Feldspar
+ Ca-rich Olgioclase + H_2O ...(23)

In the kyanite-sillimanite zone, the assemblages consisting of garnet, along with kyanite, biotite with a decrease of modal muscovite and staurolite, were aided by the kyanite and/or sillimanite; decrease in M/FM ratio of biotite and staurolite; increase in Zn-content of Fe-rich staurolite; K-richer muscovite, Ca-rich plagioclase and zoned K-feldspar. The reaction attributing to the increase in modal amount of kyanite and biotite along with garnet as an extra phase has been discussed by Dempster (1985). He observed that an increase in content of Zn in staurolite possibly acted as an internal buffer to the kyanite forming reaction (22). Hence, the staurolite breakdown reaction:

Staurolite + Na-Muscovite + Quartz = Kyanite + Biotite + Garnet + H_2O ...(24)
(Dempster, 1985)

(24) was important i.e., for the production of kyanite and garnet in the retrograde zones. The alteration of staurolite during the reaction (21) or (22) might have produced garnet in the prograde kyanite-biotite assemblage due to the first retrograde metamorphic stage. It is observed that garnet present in kyanite-sillimanite bearing schists often occurs in broken pieces recrystallised in granoblastic polygonal to platy quartz in the prograde assemblages. The co-existence of kyanite and sillimanite in the higher grade metamorphic rocks can be attributed to the polyphase transformation as exemplified by Carmichael (1969):

Kyanite = Sillimanite ...(25)

Banerjee and Bhattacharya (1981) inferred this reaction as a result of breakdown of garnet and staurolite bearing assemblages, instead of the breakdown of muscovite in the presence of quartz. The occurrence of sillimanite along with kyanite and increase in modal amount of Fe-rich biotite and increase in Mn, Fe of garnet, K-rich muscovite can also be interpreted by the following continuous reaction (Guidotti, 1974):

Mg-Biotite + Muscovite + Garnet + Mg-Staurolite + Quartz = Sillimanite
+ Fe-rich biotite + Fe-rich staurolite + Mn-Fe Garnet + Albite + H_2O ...(26)

The presence of Ca-rich plagioclase and zoned K-feldspar point towards the continuous reaction (23) which can be applicable to this zone.

The occurrence of chlorite formed due to alteration of biotite and epidote suggests for the extent of another retrograde metamorphism in the area. Biotite readily alters to chlorite at low to moderate temperatures. Eggleton and Banfield (1985) have mentioned the following coupled reaction for chloritization at 340°C:

1 Biotite + 0.21 Anorthite + 1.64 H_2O + 1.46 H + 0.08 O_2 = 0.46 Chlorite + 0.11 titanite
+ 0.10 Epidote + 0.56 Muscovite + 0.63 Quartz + 0.02 Magnetite + 1.46K^+ ...(27)

The K^+ released from biotite, sericitized plagioclase and the calcium released from the plagioclase was used to produce epidote and titanite combined with sheet silicates. The presence of sericite requires the addition of water and K^+ and it possibly existed during the chloritization of biotite. This K^+ also reacted with anorthite component of plagioclase to free Ca^{2+}. It is observed that in vigorous hydrothermal systems, there can be a general flux of ions so that intense sericitization is commonly accompanied by K-metasomatism.

Epidote is also formed as an alteration product of Ca-rich plagioclase due to addition of water i.e.,

Oligoclase + tschermakite = Epidote + Chlorite + Albite ...(28)
(Liou, 1973)

In the high grade migmatitic zone, the partial melting might have proceeded in a progressive order i.e.,

Higher grade regional metamorphic rocks → Migmatic (Tronjhemitoids) gneisses → Granitic gneisses → foliated granites.

The absence of staurolite from the migmatic tronjhemitoids suggest an increase of temperature to about 700°C as inferred by Winkler (1979):

Staurolite + K-feldspar component in melt = Al_2SiO_5 + Garnet + Biotite ...(29)

The anatexis of a muscovite-plagioclase-quartz gneiss was examined by Winkler (1979). At P_{H_2O} = 5 kb and 680°C an anatectic melt begins to form. At this stage, it is observed that muscovite disappears and the melt is enriched by K-feldspar component. This component, together with albite, quartz and a small amount of anorthite forms a granitic melt. The reaction (23) proposed after Guidotti (1984) in its modified form representing this assemblage is as follows:

Muscovite + Quartz + Plagioclase = K-feldspar + Plagioclase + Quartz component in anatectic melt + An-rich Plagioclase + Sillimanite + H_2O ...(30)

At higher pressures Fe-biotite in the high grade metamorphic rocks of Sikkim without cordierite, with a slight increase in temperatures, reacted with Al_2SiO_5 (Winkler, 1979):

Biotite + Al_2SiO_5 + 2 Quartz = K-feldspar component + Almandine garnet + H_2O ...(31)
Sillimanite
Kyanite

and

Biotite + Al_2SiO_5 = Quartz + Albite + Garnet + Melt ...(32)

Ghose (1979) working on the melting relationship between schists and gneisses of Darjeeling area experimentally noticed the disappearance of muscovite at 715°C-725°C at 7 kb with an approximate 30 per cent of melt and the appearance of the new phases of cordierite, ortho-amphibole, ortho-pyroxene and sillimanite at the expense of quartz and biotite in sericite-chlorite schist. The melting of sillimanite gneiss began at 677°C at 5 kb with 5 per cent melt and muscovite disappeared between 680°C-700°C. The crystalline phases—biotite, quartz and sillimanite remained stable with 25 per cent rock-melt upto 730°C. Commonly the plagioclase (sodic) gets melted at a few degrees centigrade higher than muscovite in the gneisses.

P-T conditions of metamorphism

Geothermometry: The temperatures have been estimated on the basis of Fe-Mg partitioning in garnet-biotite pairs for medium to high grade assemblages and garnet-phengite for low grade assemblages.

Table 3: *Representative microprobe analyses of minerals from Sikkim metapelites, Eastern Himalaya.*

Mineral	Chlorite			Chloritoid								Muscovite				
Zones	Chloritoid	Garnet	Staurolite	1		2		3		4		Chloritoid		Garnet	Stauro-lite	Kyanite-Sillimanite
				Rim	Core	Rim	Core	Rim	Core	Rim	Core	a	b			
SiO_2	22.62	23.50	25.82	23.53	23.56	24.08	24.19	23.61	23.68	23.47	23.84	44.02	44.81	44.54	45.34	45.18
TiO_2	0.05	0.06	0.14	-	-	-	-	-	-	-	-	0.14	0.17	0.27	0.46	0.38
Al_2O_3	22.10	23.60	22.12	38.90	38.39	39.64	38.34	39.10	38.83	38.43	39.60	33.80	34.52	33.26	34.16	34.11
FeO	35.09	29.40	24.88	25.61	24.06	24.84	24.40	25.22	23.83	24.24	24.70	2.52	1.74	1.11	1.21	1.32
MnO	0.04	0.12	0.06	0.09	0.08	0.16	0.10	0.04	-	0.03	0.02	0.01	0.03	0.07	-	0.02
MgO	5.58	11.60	16.72	1.10	0.85	1.44	1.16	1.83	1.61	1.76	1.76	0.25	0.19	0.72	0.80	0.81
CaO	0.02	-	-	-	-	0.04	0.02	-	-			0.03	-	0.04	0.05	0.02
Na_2O	-	-	-	0.04	0.02	-	0.02'	-	-	0.03	-	0.97	1.56	1.51	1.75	1.62
K_2O	0.01	-	-	0.01	0.001	0.006	0.02	-	-	-	-	9.53	8.85	9.14	8.45	8.85
Cr_2O_3	-	-	-	-		0.008	0.039	-	0.03			-	-	-	-	-
NiO	-	-	-	-		0.12				0.018				-	-	-
ZnO	-	-	-	-	-	-	-	0.08	-	-		0.140	0.01	-	-	0.08
Total	85.49	88.28	89.74	89.37	86.97	90.51	88.35	89.34	88.22	88.05	90.54	94.58	92.24	90.79	92.24	93.05
	Cations based on 28 oxygens			Cations based on 12 oxygens								Cations based on 28 oxygens				
Si	5.18	5.99	6.15	2.02	2.07	2.03	2.10	2.02	2.07	2.05	2.04	6.19	6.16	6.22	6.19	6.18
Al IV	2.82	2.01	1.85		-							1.81	1.84	1.78	1.81	1.82
Al				3.94	3.97	3.94	3.92	3.98	3.94	3.95	3.95					
Al VI	3.14	3.29	2.79	-	-	-	-	-	-	-	-	3.69	3.76	3.69	3.68	3.68
Ti	0.01	-	-	-	-	-	-	-	-	-	-	0.01	0.01	0.03	0.05	0.04
Fe	6.72	6.27	4.96	1.84	1.76	1.75	1.76	1.75	1.74	1.76	1.76	0.29	0.20	0.13	0.14	0.15
Mn	0.01		0.01	0.006	0.005	0.01	0.007	-	-	-	-	-	-	0.01	-	-
Mg	1.90	4.40	5.94	0.14	0.11	0.18	0.19	0.20	0.18	0.20	0.20	0.05	0.03	0.15	0.16	0.16
Ca	-	-	-	-	-	-	-	-	-	-	-	-	-	-	-	-
Na	-	-	-	0.006	0.003	-	-	-	-	-	-	0.26	0.41	0.41	0.46	0.43
K	-	-	-	-	-	-	-	-	-	-	-	1.71	1.55	1.63	1.47	1.54
Zn	-	-	-	-	-	-	-	-	-	-	-	0.01	-	-	-	0.01
Total	19.78	21.96	21.40	7.95	7.92	7.91	8.00	7.95	7.93	7.96	7.98	14.02	13.96	14.05	13.96	14.11

(*Contd...*)

Table 3: *Contd...*

Mineral	Garnet								Staurolite								Kyanite-Sillimanite	
Zone	Garnet		Staurolite						Kyanite-Sillimanite					Staurolite			Kyanite-Sillimanite	Kyanite-Sillimanite
	a	b	a	b	c	d	e	f	a	b	c	d	e	Rim	Core	Rim	Core	
SiO_2	35.68	35.20	35.65	36.71	35.20	35.07	34.22	34.91	34.54	34.82	34.45	35.01	34.47	63.61	63.68	62.50	62.36	62.04
TiO_2	1.45	1.38	1.28	1.22	1.31	1.32	1.31	1.30	1.30	1.45	1.68	1.49	1.37	0.04	0.04	0.03	0.04	0.02
Al_2O_3	18.88	18.87	19.08	19.53	18.96	19.30	18.57	18.88	19.33	19.40	19.17	19.78	19.71	21.39	21.96	22.96	22.86	22.80
FeO	18.97	18.98	16.69	18.69	15.85	18.01	16.87	17.87	18.55	17.97	17.43	17.75	17.91	0.07	0.03	0.07	-	0.06
MnO	0.10	0.04	0.07	-	0.04	-	0.05	-	-	0.01	0.01	0.04	0.04	-	-	-	-	0.01
MgO	9.86	10.41	7.79	10.44	10.42	10.32	9.92	9.93	10.06	10.09	9.71	10.31	10.53	0.00	0.02	0.02	0.01	
CaO	0.03	0.10	0.03	0.05	-	-	0.03	0.05	0.02	0.06	0.05	0.03	0.06	3.08	3.21	4.45	4.51	4.64
Na_2O	0.25	0.21	0.57	0.53	0.53	0.50	0.61	0.54	0.39	0.48	0.49	0.44	0.50	10.63	10.60	9.54	9.14	9.40
K_2O	8.98	8.32	8.35	8.77	8.77	8.61	8.36	8.48	8.31	8.51	8.59	8.59	8.67	0.21	0.14	0.09	0.12	0.08
ZnO	0.06	0.07	0.09	-	-	0.11	0.05	0.05	-	-	0.01	0.12	-	-	-	-		-
Total	94.48	93.69	90.85	95.97	91.87	93.39	90.35	92.06	93.53	93.67	91.60	93.57	94.66	99.04	99.68	99.74	99.04	99.06
	Cations based on 22 Oxygens													Cations based on 8 Oxygens				
Si	5.46	5.42	5.46	5.49	5.50	5.40	5.45	5.45	5.37	5.39	5.40	5.37	5.39	2.84	2.83	2.78	2.79	2.78
Al IV	2.54	2.58	2.54	2.41	2.50	2.60	2.55	2.55	2.63	2.62	2.60	2.64	2.61	-	-	-		
Al														1.13	1.15	1.20	1.20	1.20
Al VI	0.87	0.84	1.00	1.03	0.99	0.91	0.94	0.92	0.91	0.91	0.86	0.93	0.92					
Ti	0.17	0.16	0.15	0.14	0.15	0.15	0.16	0.15	0.15	0.17	0.20	0.17	0.16	0.00	0.00	0.00	0.00	0.00
Fe	2.43	2.44	2.20	2.34	2.07	2.32	2.25	2.41	2.41	2.32	2.28	2.28	2.28	0.00	0.00	0.00	-	0.00
Mn	0.01	0.00	0.01	0.00	0.00	0.00	0.01	-	-	0.00	0.00	0.00	0.00	-	-	-	-	-
Mg	2.25	2.39	2.30	2.32	2.42	2.37	2.33	2.34	2.34	2.33	2.27	2.35	2.39	0.00	0.00	0.00	0.00	-
Ca	-	0.01	-	0.01	-	-	-	-	-	0.01	0.01	-	0.01	0.15	0.15	0.21	0.22	0.22
Na	0.07	0.06	0.17	0.15	0.16	0.15	0.19	0.12	0.12	0.15	0.15	0.13	0.15	0.92	0.91	0.82	0.79	0.82
K	1.75	1.63	1.68	1.67	1.75	1.69	1.70	1.65	1.65	1.68	1.69	1.68	1.68	0.01	0.01	0.00	0.01	0.00
F	0.07	-	0.22	-	-	0.10	0.36	-	-	0.71	0.07	-	0.30	-	-	-	-	-
Zn	0.01	0.01	0.01	-	-	0.01	0.00	0.00	-	-	0.00	0.01	-	-	-	-		-
Total	15.69	15.60	15.76	15.57	15.56	15.72	15.96	15.60	15.59	16.31	15.63	15.58	15.91	5.06	5.05	5.03	5.01	5.03

(Contd...)

Table 3: *Contd...*

Mineral	Garnet														Staurolite							Kynanite-Sillaminite	
Zone	Garnet							Staurolite					Kyanite-Sillimanite		Staurolite		MR/ Outer		Kyanite-Sillamanite			Kyanite-Sillimanite	Kyanite-Sillimanite
	Rim						Core	Rim			Core	Rim											
	a	b	c	d	e	f	g	a	b	c	d	e	Rim	Core	Rim	Core	Rim	Core	Core	Rim	Core		
SiO_2	36.16	36.04	36.17	36.27	36.46	36.12	36.87	36.47	35.86	36.00	37.85	36.17	36.18	37.03	27.10	27.30	26.79	27.87	26.96	26.98	27.55	37.38	37.67
TiO_2	-	-	-	-	-	-	-	-	-	-	-	-	-	-	0.46	0.49	0.38	0.52	0.64	0.42	0.50	0.00	-
Al_2O_3	20.40	20.55	20.59	20.32	20.67	20.54	20.96	21.01	20.17	2.55	21.32	20.28	20.35	20.42	51.78	50.90	51.46	52.83	50.45	50.77	51.00	16.25	58.80
FeO_2	36.63	36.64	35.69	33.42	34.10	33.36	31.95	35.46	34.69	35.10	32.66	33.48	35.19	34.20	13.19	12.99	11.63	13.10	10.92	13.25	13.37	0.16	0.21
MnO	0.38	1.00	2.30	2.71	2.33	3.69	5.06	0.80	0.57	1.39	2.60	1.51	4.12	3.70	0.16	0.14	0.09	0.10	0.09	0.02	0.08	0.04	-
MgO	2.61	2.33	2.27	2.27	2.55	2.58	1.97	3.43	3.42	2.70	2.47	3.08	0.45	0.29	1.64	1.71	1.56	1.89	1.85	1.76	1.81	0.01	0.04
CaO	2.29	2.30	2.17	2.25	2.38	2.48	2.94	1.54	2.26	2.73	3.08	2.23	2.82	3.61	-	-	-	-	0.01	0.03	0.00	-	0.07
ZnO	-	-	-	-	-	-	-	-	-	-	-	-	-	-	0.16	0.17	0.13	0.16	0.11	0.20	0.25	-	-
Total	98.48	98.87	99.20	97.24	98.50	98.77	99.75	98.68	96.97	98.46	99.99	96.75	99.11	99.25	94.51	93.71	92.05	96.48	91.03	93.43	94.56	98.84	96.79
Cations based on																							
12 Oxygens															48 Oxygens							20 Oxygens	
Si	2.98	2.96	2.97	3.01	2.99	2.97	2.99	2.97	2.98	2.96	3.03	3.00	2.99	3.03	8.12	8.24	8.17	8.16	8.29	8.18	8.26	4.21	4.19
Al	1.98	1.99	1.99	1.99	1.20	1.99	2.00	2.02	1.97	1.99	2.01	1.98	1.97	1.97	18.29	18.11	18.50	18.24	18.28	18.15	18.01	7.62	7.72
Ti	-	-	-	-	-	-	-	-	-	-	-	-	-	-	0.10	0.11	0.08	0.11	0.15	0.09	0.11	0.00	0.02
Fe	2.52	2.52	2.45	2.32	2.34	2.29	2.17	2.42	2.41	2.42	2.19	2.32	2.43	2.34	3.31	3.28	2.97	3.21	2.81	3.36	3.35	0.01	-
Mn	0.03	0.07	0.16	0.19	0.16	0.26	0.35	0.05	0.04	0.09	0.17	0.10	0.03	0.02	0.04	0.04	0.02	0.03	0.02	0.01	0.02	-	-
Mg	0.32	0.29	0.28	0.28	0.31	0.32	0.24	0.42	0.42	0.33	0.29	0.38	0.35	0.29	0.73	0.77	0.71	0.82	0.65	0.79	0.81	-	-
Ca	0.20	0.20	0.19	0.20	0.21	0.22	0.25	0.13	0.20	0.24	0.26	0.20	0.25	0.32	-	-	-	-	0.00	0.01	0.00	-	-
Zn	-	-	-	-	-	-	-	-	-	-	-	-	-	-	0.03	0.03	0.03	0.03	0.02	0.04	0.05	-	-
Total	8.03	8.04	8.04	8.00	8.01	8.04	8.01	8.01	8.03	8.03	7.96	8.00	8.02	7.98	30.63	30.59	30.49	30.61	30.42	30.65	30.62	11.84	11.93

Garnet-biotite pairs: As already mentioned, the garnet from garnet and staurolite zone is characterized by normal growth, inverse and reverse zoning. Garnet in the kyanite-sillimanite zone was more homogenised. For the estimation of temperature of garnet-biotite pairs, the models of Ferry and Spear (1978), Hodges and Spear (1982) and Ferry and Spear versus Berman (1990) were applied (Table 3). The temperatures and pressures have been calculated by using a composite computer programme 'Thermobarometry' (Spear and Peacock, 1990) inclusive of a new model Ferry and Spear versus Berman (1990). The author aimed the results by examining the microprobe data of garnet-biotite pairs from different metamorphic zones.

The results obtained by the three models for garnet-biotite pairs from different metamorphic zones are quite consistent. The temperatures estimated lie in range of 490 to 527°C for rim, 465 to 490°C/4-5 kb for core of garnet zone; 546 to 575°C for rim, 492 to 522°C for outer core and 555-576°C/5-6 kb for core parts of staurolite zone; 658 to 695°C for rim and 630 to 675°C/7-7.5 kb for core of kyanite-sillimanite zone, respectively. Neogi *et al.*, 1998; Catlos *et al.*, 2000; Ganguly *et al.*, 2000 have inferred the pressure and temperature for sillimanite-k-feldspar zone (K-feldspar in) in the HHC, Eastern Himalaya ranging from 820°-850°C and pressure 8.8 - 11.8 kbs. This higher variation in pressure and temperature in comparison to 658°-770°C temperature and 7-7.5 kbs pressure given by other workers (Mohan *et al.*, 1989 and Dubey, 1993) which is different as the later P-T are in sillimanite-muscovite zone (muscovite in).

Garnet-muscovite pairs: The Fe-Mg ratio between co-existing garnets and phengite were studied by Krogh and Raheim (1978). The thermometer was applied to garnet-phengite pairs from garnet zone of the area and temperatures inferred are 485°C for the core and 510°C for rim-parts of garnet under 4-5 kb. These values are also consistent with the temperatures estimated by Hodges and Spear (1982).

Geobarometry: Different mineral equilibria involving high volume change (V) of the reactions are commonly applied for the estimation of pressure conditions in relation to temperatures inferred by various models of geothermometers. The model given by Newton and Haselton (1981), which is based on reaction (garnet-plagioclase-sillimanite/kyanite-quartz) has been applied to infer the pressure experienced by the kyanite-sillimanite zone in Sikkim.

The most significant limitation of this geobarometer is that the end member equilibrium is known within ± 2 kbrs. The uncertainties are generated by the long extrapolation of pressure and evaluation of grossular activity in garnet solutions at 95% dilution. Thus, a comparative geobarometry based on Ghent and Stout (1981) model was also applied to the kyanite-sillimanite assemblage.

Following the method by Newton and Haselton (1981), the pressure for kyanite-sillimanite zone is estimated as 7-7.2 kb but according to the model of Ghent and Stout (1981) it ranges from 6.9-7.3 kb. These estimates are compatible with one another and are much closer to the estimation of Lal *et al.* (1981). Pressures were also calculated for Staurolite zone by applying the model of Hoisch (1990). It is based on empirical calibration of six different geobarometers for the mineral assemblage:

Quartz + muscovite + biotite + plagioclase + garnet

Based on this geobarometer a pressure of 6.6 to 6.7 kb is inferred for rims and 5.7 kb for the core parts of garnet in staurolite zone. The temperatures and pressures estimates of

Mohan *et al.* (1989) are too high for garnet zone and these in fact correspond to the pressure estimated for staurolite zone of the area. The pressure estimated for this zone is low as revealed by the retrogressed outer core parts of garnet. Fig. 3 shows P-T-t paths (arrow), AFM topology and inferred important mineral reactions and assemblages in Sikkim.

GEOCHRONOLOGY

Analytical Procedure

Ion Microprobe Th-Pb dating of monazite has been carried out at UCLA is briefy summarized. 15-25 μm diameter oxygen (O^-) primary beam is used to sputter positive ions from the surface of the sample. The intensity of the primary beam ranges from 2 to 13 nA during different analysis sessions. A mass resolving power of ~5000 is sufficient to separate molecular interferences and distinguish between Pb isotopes in the 204-208 mass range (Harrison *et al.*, 1999). Specially, a 50eV energy window with a 10 to 15eV offset for $^{232}Th^+$ is used. The magnet is steeped through six different mass species: $^{204}Pb^+$, $^{207}Pb^+$, $^{208}Pb^+$, $^{232}Th^+$, ThO_2^+ and U^+. This cycle was repeated 15 times to correct for variations in beam ability and average values were used to calculate ages. Determination of the $^{208}Pb^+/^{232}Th^+$ sensitivity factor is accomplished by dividing the measured magnet $^{208}Pb^+/^{232}Th^+$ value of a standard monazite at a reference $^{208}Pb^+/^{232}Th^+$ value by standard's known daughter to parent ratio. The age of an unknown, measured under identical operating conditions, is determined by applying this sensitivity factor to the measured $^{208}Pb^+/^{232}Th^+$ value of the unknown. The reproducibility of the calibration line is the limiting factor on the precision of ion microprobe age determinations.

All *in situ* Th-Pb age analyses utilized monazite standard 554, a peraluminous granodiorite from Santa Catalina Mountains in Arizona (Force, 1997). Monazite grains dated *in situ* are typically analyzed with one spot, limited by the size and irregular shape of the grain (typically <60 μm diameter) and primary beam (typically ~25-30 μm diameter), as well as the optical capabilities of the reflected light microscope attached to the ion microprobe. Two analyses are performed on each spot to identify any variation in age as the ion beam drilled deeper into the grain. Reported ages are corrected from common ^{204}Pb using and assumed $^{208}Pb/^{204}Pb = 38.6$ (Stacy and Kramers, 1975). Himalayan monazite grains typically yield >95% radiogenic ^{208}Pb (Catlos, 2000; Gilley, 2001).

Monazite Dating

A detailed geochronological studies and their metamorphic relationship has been described by Catlos *et al.*, 2003. Lesser Himalaya samples have garnet + biotite + staurolite + muscovite + plagioclase + chlorite + ilmenite + apatite + monazite + zircon + xenotime + quartz. MCT shear zone rock (CMP860) has garnet + biotite + sillimanite + muscovite + plagioclase + chlorite + ilmenite + monazite + tourmaline + zircon + apatite + allanite + quartz, whereas samples CH14103 have garnet + biotite + staurolite + muscovite + plagioclase + chlorite + ilmenite + zircon + monazite ± xenotime ± apatite ± allanite + quartz. Greater Himalayan Crystallines (Catlos, 2002c). All mineral grains were dated using an ion microprobe and the reader is referred to Catlos *et al.* (2002b) for technical details of *in situ* Th-Pb ion microprobe monazite analysis, and Schneider *et al.* (1999) for U-Pb ion microprobe dating of zircon. The errors for all ages reported in this paper are 1σ. MCT shear zone rock CMP860 (650±10°C, 6±1 kbar; Fig. 5b, c, d) has monazite inclusions in garnets that range from 20.5±0.6 Ma to 10.8±0.3 Ma (Fig. 5b, c, d). The garnet also contains inclusions of quartz, apatite, allanite,

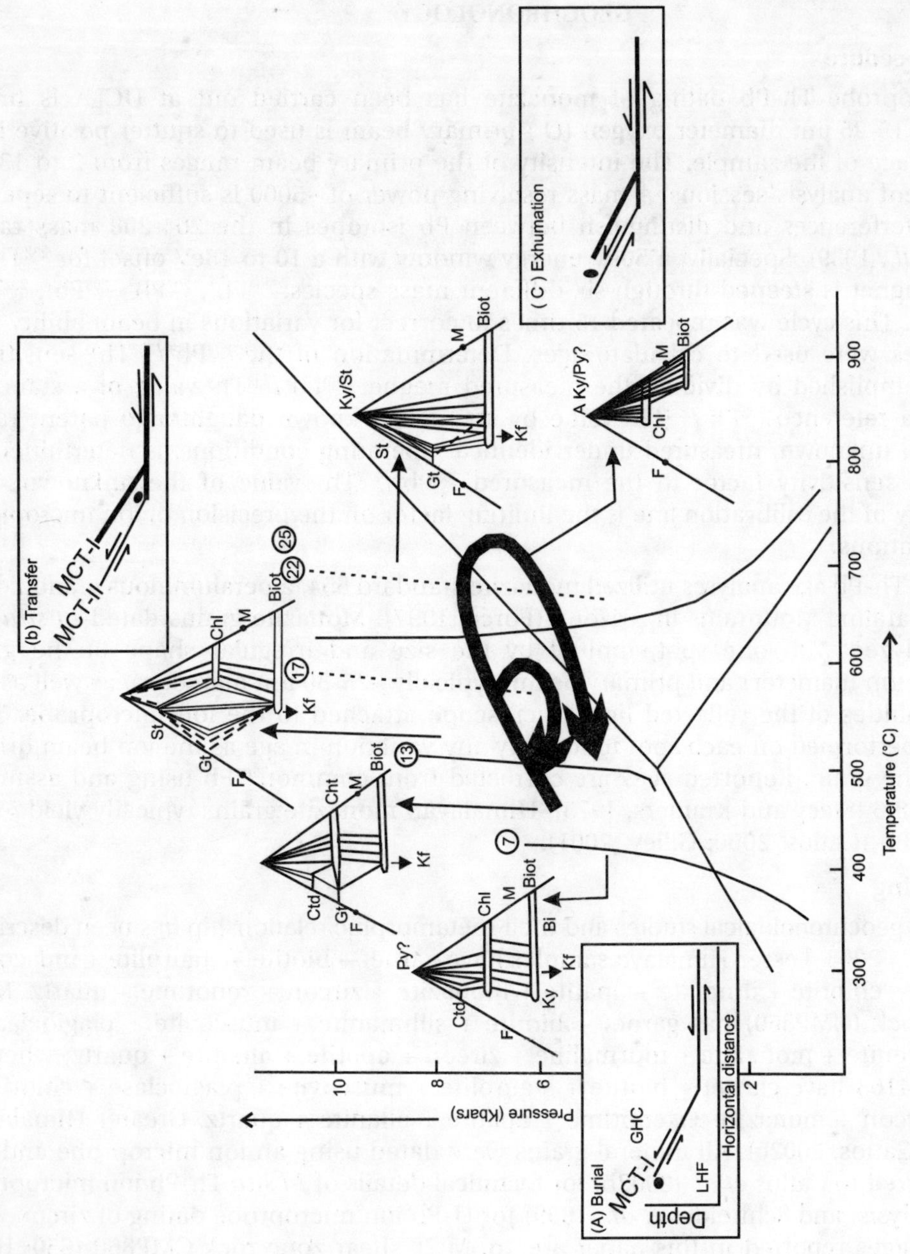

Fig. 3. P-T-t paths (arrow), AFM topology and inferred important mineral reactions and assemblages in Sikkim, Eastern Himalaya.

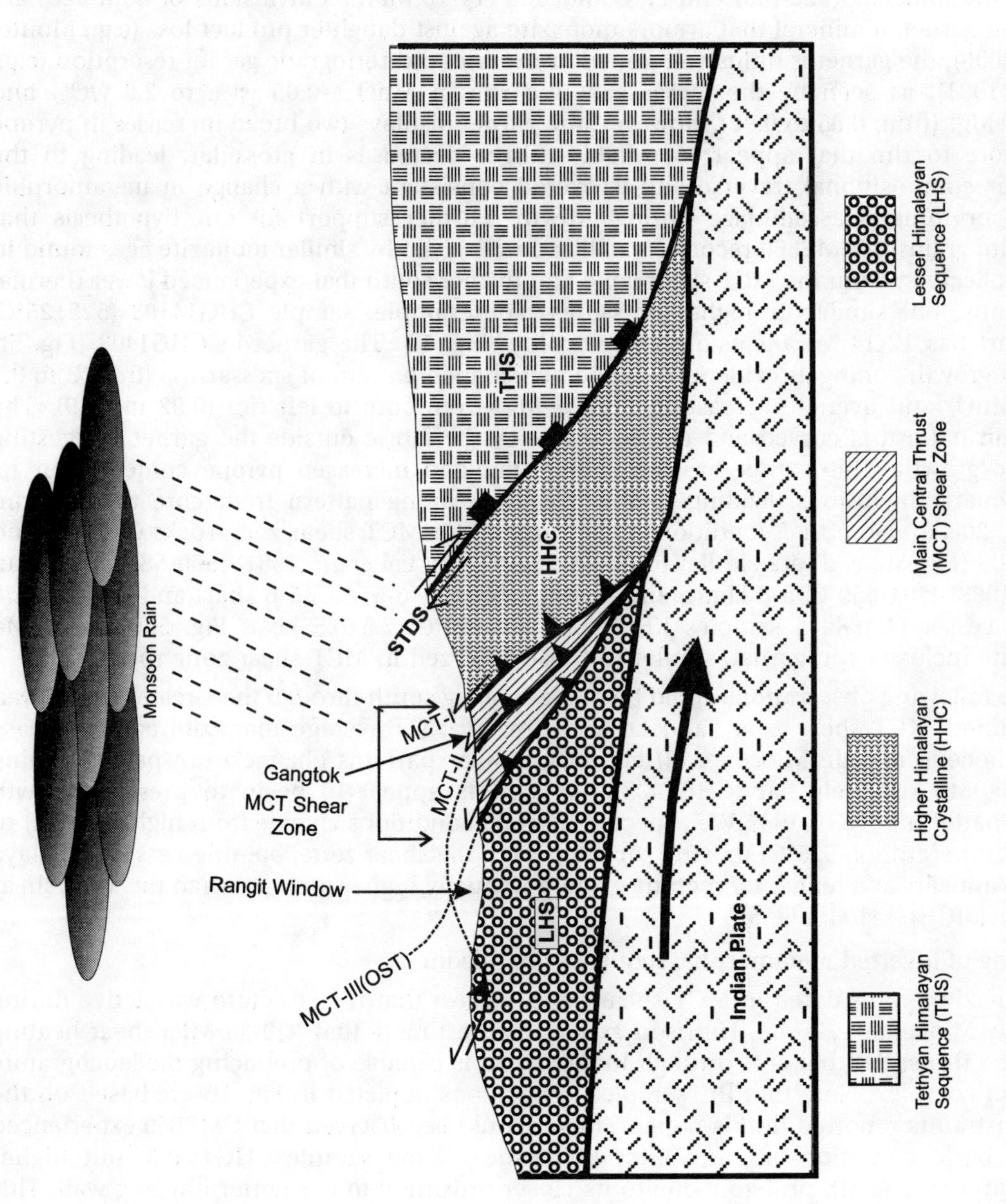

Fig. 4. Schematic cross section showing the positions of MCT-I, MCT-II and OST (MCT-III) in Sikkim, Eastern Himalaya.

monazite, biotite, quartz, and plagioclase and records peak conditions below the closure temperature of Pb in monazite (e.g., Smith & Giletti, 1997). Thus the age distribution in this rock suggests the sample went through a minimum of two monazite forming reactions during the Miocene (~20 Ma) and Late Miocene (14-10 Ma). As inclusions of both ages are found in garnet, a mineral that armors monazite against daughter product loss (e.g., Montel *et al.*, 2000), the garnet is diffusively zoned or experienced retrograde garnet resorption (e.g., Spear, 1991), as seen by the increase at the rim in MnO (~0.05 wt% to 2.8 wt%) and Fe/(Fe+Mg) (from 0.85 to 0.90). However, the garnet displays two broad increases in pyrope from core to rim that appear to roughly mirror decreases in grossular, leading to the irregular compositional traverse and is overall consistent with a change in metamorphic history or mineral assemblage during growth. Further support for our hypothesis that monazite grains in CMP860 record crystallization is given by similar monazite ages found in rocks collected within the MCT shear zone and further south that experienced lower thermal conditions, but similar or higher pressures. For example, sample CHG14103 (525±25°C; 6±1 kbar) has 12-14 Ma inclusions in staurolite (Fig. 5f). The garnet in CHG1403 (Fig. 5f) records growth zoning as evidenced by the bell-shaped pattern of spessartine (from 3 to 0.1 wt % MnO) and overall decrease in Fe/(Fe+Mg) from core to left rim (0.92 to 0.89). The inclusion pattern is curved and continuous with the fabric outside the garnet, suggesting tectonic growth. This garnet has lower grossular and increased pyrope content near its larger quartz inclusions, leading to the irregular zoning pattern from core to right rim (Catlos, 2002a). Note that conditions for estimated for MCT shear zone rocks CMP860 and CHG1403 (Fig. 5b, c, d) are similar to those reported by Lal *et al.* (1981) (560-583°C), Mohan *et al.* (1989) (594-659°C, 5-8 kbar), and Dubey (1993) (546-575°C, 5-6 kbar) and Catlos *et al.* (2003). Lesser Himalaya samples KBP1062A (610±10°C; 7.5±0.5 kbar, Fig. 5a) ~12-13 Ma monazite inclusions in garnet, similar to those analyzed in MCT shear zone rocks.

The following observations could be made moving south through the Greater Himalayan Crystallines, MCT shear zone, and Lesser Himalaya: (1) average monazite ages decrease from Miocene to Late Miocene, (2) garnet grossular patterns change from patchy zoning (CMP860) to relatively flat (CHG1403), (3) garnets appear to begin to preserve growth zoning patterns (CHG14103). We expected thermal conditions change from high (~700°C) to lower temperature (~525°C) moving south through the shear zone, but one Lesser Himalaya sample appears at a higher temperature and potentially higher pressure than the MCT shear zone rock (CHG14103; Fig. 5f).

Modeling of inverted metamorphic sequences in Sikkim

Monazite grains dated in MCT shear zone suggest that the structure was active during the early Miocene (~22 Ma). Harrison *et al.* (1998) estimate that ~20-30 MPa shear heating along the thrust flat that cuts through Indian rocks is capable of producing the leucogranite injection complex. Note that the sample's 4 positions depicted in Fig. 1b are based on the P-T constraints reported in this paper. Thus it could be observed that CMP860 experienced similar baric conditions as the other MCT shear zone sample CHG14103, but higher temperatures (Fig. 5f), probably due to its closer proximity to the hotter hanging wall. This age may be the result of: (1) Pb loss from ~22 Ma grain, (2) a crystallization event related to MCT movement at this time, or (3) a crystallization event not related to MCT movement but facilitated by Miocene emplacement of the Greater Himalayan Crystallines. Thus the rocks experienced conditions not only conducive to monazite growth during Miocene MCT movement,

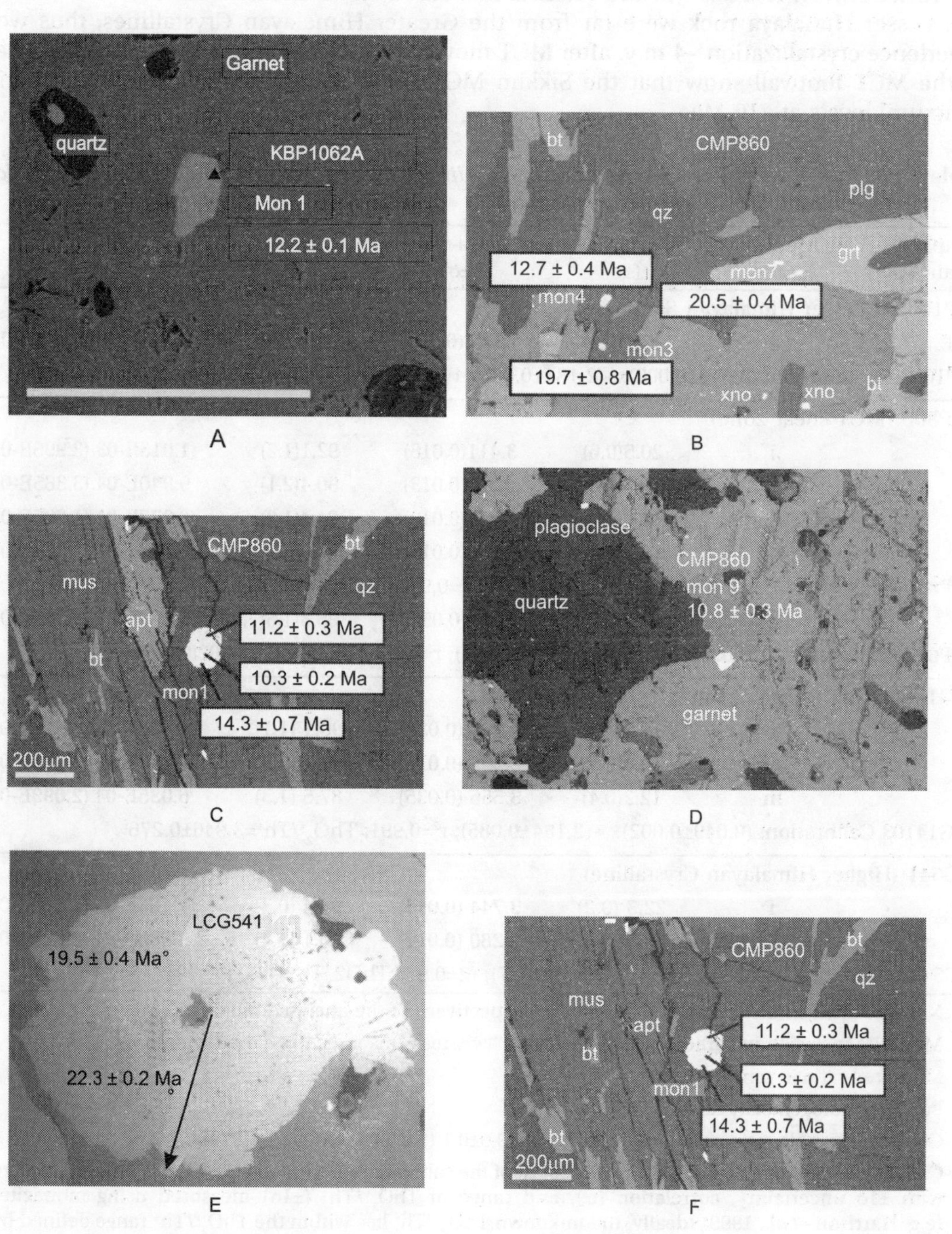

Fig. 5. Ages of Sikkim Himalaya monazite inclusions in Garnet and Staurolite (modified after Catlos *et al.*, 2003).

but also hot enough to experience huge amount of Pb loss. However, garnet crystallized in this rock between 12.2±0.1 Ma and 11.5±0.2 Ma, based on its monazite inclusions of (Table 4), The Lesser Himalaya rock were far from the Greater Himalayan Crystallines, thus would experience crystallization ~4 m.y. after MCT movement. The youngest monazite grains dated in the MCT footwall show that the Sikkim MCT shear zone continued activity at various structural levels at ~10 Ma.

Table 4: *In situ monazite age results from Lesser Himalaya, MCT shear zone and Higher Himalayan Crystalline, Sikkim, Eastern Himalaya.*

Sample[a] (Grain spot)	Monazite location[b]	Age (Ma) (±σ)	ThO_2^+/Th^+[e] (±σ)	^{208}Pb(%)[d] (±σ)	$^{208}Pb^*/Th^+$[e] (±σ)
KBP1062A (Lesser Himalayan Sample)					
10-1[f]	i	12.1(0.6)	3.343 (0.034)	64.9 (2.8)	5.973E-04 (3.084E-05)
KBP1062A Calibration: (0.049±0.002)x+(2.184±0.085); r^2=0.991; ThO_2^+/Th^+=3.946±0.276[f]					
CMP860 (MCT shear zone)					
7-1	i	20.5(0.6)	3.411(0.018)	92.1(1.2)	1.015E-03 (2.996E-05)
8-1	m	19.7(0.8)	3.299(0.013)	80.4(2.1)	9.940E-04 (3.385E-05)
4-1	m	12.7(0.4)	3.557(0.019)	84.4(1.8)	6.277E-04 (1.885E-05)
1-2	m	10.3(0.2)	3.800(0.018)	87.5(1.7)	5.117E-04 (1.224E-05)
CMP860 Calibration 1: (0.049±0.001x +(2.184±0.085;r^2=0.991; $ThO_2^+Th^+$=3.946±0.276f[f]					
9-1	i	10.8(0.3)	3.558(0.054)	85.3(1.6)	5.362E-04 (1.359E-05)
CMP860 Calibration 2: (0.099±0.001)x +(0.0881±0.037); r^2=0.948; ThO_2^+/Th^+=4.028±0.668[f]					
CHG14103 (MCT shear zone)					
2-1	m	13.8(0.5)	3.328 (0.020)	90.1 (1.1)	6.826E-04 (2.310E-05)
1-1	m	12.9(0.5)	3.380 (0.015)	92.1 (0.9)	6.380E-04 (1.866E-05)
3-1	m	12.2(0.4)	3.556 (0.035)	87.5 (1.3)	6.035E-04 (2.099E-05)
CHG14103 Calibration: (0.049±0.002)x +(2.184±0.085); r^2=0.991; ThO_2^+/Th^+=3.946±0.276[f]					
LCG541 (Higher Himalayan Crystalline)					
3-1	i	22.3 (0.2)	3.744 (0.015)	94.4 (0.3)	1.105E-03 (1.108E-05)
4-1	i	19.5 (0.4)	2.280 (0.012)	90.1 (1.4)	9.668E-04 (2.210E-05)
LCG541 Calibration: (0.085±0.002)x +(1.803±0.047); r2=0.976;ThO2/Th^+=4.000±0.491[f]					

(a) Nomenclature indicates the grain and spot, respectively, of the analysed monazite.

(b) Monazite inclusion in garnet is designated as "i" whereas "m" indicates a matrix grain.

(c) Measured ratio in sample.

(d) Per cent radiogenically derived ^{208}Pb.

(e) Corrected sample ratio assuming $^{208}Pb/^{204}Pb$=39.5±0.1 (Stacey & Kramers, 1975).

(f) Calibration information: Sample name, best fit of the calibration to the equation of a line (slope*x+intercept) with ±1σ uncertainty, correlation (r^2), and range of ThO_2^+/Th^+ (±1σ) measured using monazite 554 (e.g. Harrison *et al.*, 1999) Ideally, the unknown ThO_2^+/Th^+ lies within the ThO_2^+/Th^+ range defined by the standard.

CONCLUSION

It can be summarised that the Daling and Darjeeling Formations of Sikkim area underwent two phases of prograde metamorphism, each culminating into a retrograde metamorphic episode. The first event was prograde regional metamorphism (M_1) during which S_1 and S_2 schistosities were formed. During this episode of metamorphism chlorite, chloritoid, biotite, garnet, staurolite and kyanite were formed. Distinct evidences have not been observed to prove whether the inverted metamorphic sequence coincides with an inversion of stratigraphic sequence. Based on metamorphic reactions, it can be stated that M/FM ratio increased during the prograde metamorphism. Most of the prograde minerals (chlorite to staurolite) in this phase of metamorphism were formed due to continuous reactions. The chemical zoning of garnets in garnet zone is normal growth type. The chemical zoning of garnets in staurolite zone show a normal growth zoning followed by an inverse chemical zoning in the outer part of core.

The results obtained by applying different geothermometers revealed temperatures upto 465 to 527°C in garnet zone and 555 to 576°C (Core parts) in staurolite zone. The maximum pressure in core parts of garnets in staurolite zone was inferred upto 6.6 kb.

The M_1 metamorphism was followed by a retrogressive phase M_{1a} due to which the prograde minerals from chlorite to kyanite were marked by alteration effects. The retrogressive effects were more dynamic (brittle fracturing of garnet and kyanite) while at some places mylonitisation and phyllonitisation was common. The other evidences of retrogressive phase is supported by inverse chemical zoning in the outerparts of core alongwith presence of retrograded chlorite in garnet from staurolite zone; decrease of Zn in outer core parts of staurolite; penninization of chlorite; chloritisation of biotite, sericitization of feldspar and pyrophyllitisation of kyanite. The temperature and pressure of this retrogressive event are inferred from outer core parts of garnets in staurolite zone ranging from 492 to 522°C and 5.7 kb.

The prograde M_2 phase of metamorphism was primarily of burial type which later acquired a regional nature with formation of garnet, staurolite, kyanite and sillimanite (with almandine). The garnets in staurolite zone also show another prograde normal zoning varying from outer part of cores to inner parts of rim. Complete recrystallisation of garnet of low grade under high grade conditions possibly obliterated the evidences of low grade zoning in the kyanite-sillimanite zone.

Some other evidences supporting prograde metamorphism (M_2) are: (i) weak penetrative deformation and inclusion free rims of earlier multistage garnets, (ii) inclusion free and later formed idioblastic garnets and staurolite, (iii) presence of staurolite inclusion in staurolite, (iv) increase of Zn content in Fe-rich staurolite and (v) presence of K-richer muscovite, Ca-rich plagioclase and zoned K-feldspar in high grade schists and gneisses. The presence of garnet in kyanite-sillimanite zone formed during this metamorphic phase is attributed to an increase in content of Zn in staurolite which possibly acted as an internal buffer.

The temperature and pressure during M_2 phase varied from 630 to 695°C at 6.9-7.3 kb as inferred by geothermobarometric studies of kyanite-sillimanite zone. This metamorphic episode M_2 again culminated into retrograde metamorphism M_{2a}, which largely remained confined to alteration products. During this retrogressive phase the rims of the multistage garnet were chloritised. The biotite as inclusion in the rim part of garnet showed alteration to chlorite. The other evidences are replacement of K-rich feldspar by muscovite; reverse zoning of garnet at rim, pseudomorph of chlorite after idioblastic garnet; decrease in the Zn content of staurolite and kinked and deformed cleavages in kyanite.

The studies in the Sikkim area, therefore, reveals polymetamorphic conditions of rocks. According to Staubli (1989) the Himalayan metamorphism was of Barrovian type and was of three phases. The earlier stages of metamorphism were hampered due to retrogression which affected the then formed equilibrium conditions. It records only a late retrogressive phase of polymetamorphic evolution. The polymetamorphic events in relation to deformative phases of the Daling-Darjeeling Formations of Sikkim area are suggestive of two major periods of movement of thrusting episode. The M_1 metamorphism (Barrovian) was associated with the earlier thrusting episode while M_2 phase (burial to regional) was associated with the vertical uplift followed by thrusting in later stages.

According to Sorkhabi and Stump (1993) the two phases of prograde metamorphism were Early Eocene and Early Miocene in age. It may be inferred that these two prograde metamorphic episodes in Sikkim must be related to an early Eocene and early Miocene period respectively (*cf.* Sorkhabi & Stump, 1993) whereas the retrograde metamorphic events could have occurred during early middle Miocene and late Miocene period respectively (*cf.* Macfarlane *et al.*, 1992).

The Late Miocene ages of Sikkim monazite grains are wholly consistent with a MCT shear zone that is active at the same time as that documented by the MBT (e.g., Meigs *et al.*, 1995; Upreti, 1999). In this way, the Himalayan range does not solely follow a sequence of deformation in which convergence shifts towards the foreland as mountain building progressed, but instead records out-of-sequence thrusting. Pressure conditions recorded by the garnet rim in sample CMP860, the precise age of the monazite inclusion, and a lithostatic gradient suggest the exhumation rate of the MCT shear zone in the Sikkim region is ~2 mm yr-1 [= 6 kbar / (0.27 kbar km-1 * 10.8 Ma)], similar to that calculated for eastern Nepal (Catlos *et al.*, 2002a). The evidence suggests synchronous thrusting and out-of-sequence imbrication are mechanisms in which the Himalayan wedge maintains a critical taper (e.g., Davis *et al.*, 1983, Seeber & Gornitz, 1983, Boyer, 1992). Documentation of large-scale out-of-sequence events in other contractional terrains lends further support for the feasibility of the mechanism for slip accommodation in convergent plate-tectonic settings (see Gessener *et al.*, 2001). In the Sikkim, the MCT was active at 22-20 Ma and 14-10 Ma. The *in situ* monazite ages from the Sikkim region are similar to ~14 Ma monazite ages found in several rocks collected ~150 km west along the Dudh Kosi-Everest transect in eastern Nepal (Catlos *et al.*, 2002a). For example, the CMP860 inclusion age of 10.8±0.3 Ma is within error of a 10.3±0.8 Ma monazite inclusion in garnet found at similar structural levels in eastern Nepal. These geochronologic results indicate the lateral continuity of orogenic events in eastern Nepal and NE India.

The MCT footwall increases in metamorphic intensity towards higher structural levels (e.g., Ray, 1947), and the presence of 14-10 Ma monazite inclusions in garnet from rocks collected from the inverted sequence suggest the thrusts within the footwall shear zone juxtaposed the metamorphic sequences into their inverted position (e.g., Harrison *et al.*, 1997a; 1998). Although rocks collected from the Sikkim region did not record Pliocene monazite ages as seen in central Nepal (Catlos *et al.*, 2001) and Garhwal and H.P. (Catlos, 2003), the nappe structures in this area may obscure the reactivated ramp observed by the geometrical and structural relationship to MCT-II and MCT-III (Fig. 4), Eastern Himalaya, Sikkim. Dubey (2003) indicates that the region between the MCT-II and MCT-III is the locus of 3.5-6.8 (Mb) magnitude earthquakes since 1973-2003, leaving open the possibility that the Sikkim MCT shear zone is presently active.

ACKNOWLEDGMENTS

Authors are grateful to Dr. R.M. Manikavasagam, University of Roorkee for providing EPMA laboratory facilities and thankful to the Department of Earth and Space Sciences, University of California, Los Angeles for providing Ion Microprobe facilities for Monazite dating. Authors are also grateful to Prof. S.K. Tandon and Prof. P.S. Saklani, Department of Geology, University of Delhi for valuable guidance while writing the manuscript.

REFERENCES

Acharya, S.K., 1989. The Daling group, its nomenclature, tectono-stratigraphy and structural grains: with notes on their possible equivalents. In Daling group and related rocks. *Geol. Surv. India, Spl. Publ.*, **No. 22**, 5-14.

Albee, A.L., 1962. Relationships between the mineral association, chemical composition and physical properties of chlorite series. *Am. Miner.*, **47**, 851-70.

Albee, A.L. & Ray, L., 1970. Correction factors for the electron probe microanalysis of silicates, oxides, carbonates, phosphates and sulfates. *Anal. Chem.*, **42**, 1408-14.

Banerjee, H. & Bhattacharya, P.K., 1981. Concurrent granitisation under different metamorphic facies conditions in the lesser Himalayas of the Sikkim-Darjeeling region. *N. Jb. Miner. Abh.*, **142**, 199-222.

Banerjee, P.K., Guha, P.K. & Dhiman, L.C., 1980. Inverted metamorphism in the Sikkim-Darjeeling, Himalaya. *Jour. Geol. Soc. India*, **21**, 330-42.

Bell, T.H. & Johnson, S.E., 1992. Shear sense: a new approach that resolves conflicts between criteria in metamorphic rocks. *Jour. Met. Geol.*, **10**, 99-124.

Bell, T.H. & Rubenbach, M.J., 1983. Sequential porphyroblast growth and crenulation cleavage development during progressive deformation. Tectonophysics, **92**, 171-94.

Bence, A.E. & Albee, A.L., 1968. Empirical correction factors of electron microanalysis of silicates and oxides. *Jour. Geol.*, **6**, 382-403.

Berman, R.G., 1990. Mixing properties of Ca-Mg-Fe-Mn garnets. *Amer. Miner.*, **75**, 328-44.

Carmichael, D.M., 1969. On the mechanism of prograde metamorphic reactions in quartz-bearing pelitic rocks. *Contr. Miner. Petrol.*, **20**, 244-67.

Catlos, E.J., Dubey$_2$, C.S., Harrison$_3$, T.M. & Edwards, M.A., 2003. Late Miocene Movement within the Himalayan Main Central Thrust Shear Zone, Sikkim, NE India. *Journal of Metamorphic Petrology (Communicated)*.

Catlos, E.J., Gilley, L.D. & Harrison, T.M, 2002b. Interpretation of monazite ages obtained via *in situ* analysis. *Chemical Geology*, **188**, 193-215.

Catlos, E.J., Harrison, T.M, Kohn, M.J., Grove, M., Ryerson, F.J., Manning, C.E. & Upreti, B.N., 2001. Geochronologic and thermobarometric constraints on the evolution of the Main Central Thrust, central Nepal Himalaya. *Journal of Geophysical Research*, **106**, 16177-204.

Catlos, E.J., Harrison, T.M., Manning, C.E., Grove, M., Rai, S.M., Hubbard, M.S., & Upreti, B.N., 2002a. Records of the evolution of the Himalayan orogen from *in situ* Th-Pb ion microprobe dating of monazite: Eastern Nepal and Garhwal. *Journal of Asian Earth Sciences*, **20**, 459-79.

Catlos, E.J., Harrison, T.M., Dubey, C.S. & Edwards, M.A., 2002c. P-T-t constrains on the evolution of the Sikkim Himalaya. *Jour. Asian Earth Sc.* (Special Abstract Issue, 17th HKT Workshop, Gangtok, Sikkim, India) **20**, No. 4A, 6-7.

Chinner, G.A., 1965. The kyanite isograd in Glen Clova, Angus, Scotland. *Miner. Mag.*, **34**, 132-43.

Crawford, M.L., 1974. Calcium zoning in almandine: a model based on plagioclase equilibria. In: The Feldspars (Ed. Mackenzie, W.S. & Zussman, J.), Manchester University Press, Manchester, 629-44.

——. 1977. Calcium zoning in almandine garnet, Wissahickon Formation, Philadelphia, Pennsylvania. *Canadian Miner.*, **15**, 243-49.

Cygan, R. & Lasaga, A.C., 1982. Crystal growth and formation of chemical zoning in garnets. *Contr. Miner. Petrol.*, **79**, 187-200.

Das, B.K., 1987. Petrology of Barrovian type regional metamorphism in pelitic schists of Kishtwar, Kashmir Himalayas, India. *Proc. Nat. Sem. Tert.*, Orogeny, 153-87.

De Bethune, P., Laduron, D. & Bocquet, J., 1975. Diffusion processes in resorbed garnets. *Contr. Miner. Petrol.*, **50**, 197-204.

Dempster, J., 1985. Garnet zoning and metamorphism of Barrovian type area, Scotland. *Contr. Miner. Petrol.*, **89**, 30-38.

Dhana Raju, R., Satyanarayana, B. & Krishna Rao, J.S.R., 1977. Variation of MnO and TiO_2 contents of biotite in relation to metamorphic grade. *Ind. Jour. Earth Sci.*, **4**, 13-19.

Dietvorst, E.J.L., 1982. Retrograde garnet zoning at low water pressure in metapelitic rocks from Kemio, SW Finland. *Contr. Miner. Petrol.*, **79**, 37-45.

Dubey, C.S., 1993. Study of the lesser Himalayan metamorphics, Eastern Sikkim, India. Ph.D. Thesis (unpublished), 151.

Eggleton, R.A. & Banfield, J.F., 1985. The alteration of granitic biotite to chlorite. *Am. Miner.*, **70**, 902-10.

Ferry, J.M. & Spear, F.S., 1978. Experimental calibration of the partitioning of Fe and Mg between biotite and garnet. *Contr. Miner. Petrol.*, **66**, 113-17.

Frey, M., 1974. Alpine metamorphism of pelitic and marly rocks of the Central Alps. *Schweiz. Mineral. Petrogr. Mitt.*, **54**, 489-506.

Ganguly, J., Dasgupta, S., Cheng, W. & Neogi, S., 2000. Exhumation history of a section of the Sikkim Himalayas India: records in the metamorphic mineral equilibria and compositional zoning of garnet. *Earth and Planetary Science Letters*, **183**, 471-86.

Ghent, E.D. & Stout, M.Z., 1981. Geobarometry and geothermometry of plagioclase-biotite-garnet-muscovite assemblages. *Contr. Miner. Petrol.*, **76**, 92-97.

Ghose, N.C., 1979. Hydrothermal melting relationship of some high grade metamorphic rocks of Darjeeling. In: *Metamorphic rock sequences of the Eastern Himalaya*, (Ed. P.K. Verma) K.P. Bagchi and Co., Calcutta, 139-49.

Goff, E.L. & Ballevre, M., 1990. Geothermobarometry in Albite-garnet orthogneisses: a case study from Gran Paradiso nappe (Western Alps). *Lithos*, **25**, 261-80.

Grant, J.A. & Weiblen, P.W., 1971. Retrograde zoning in garnet near the 2nd sillimanite isograd. *Amer. Jour. Sci.*, **270**, 281-96.

Guidotti, C.V., 1969. A comment on 'chemical study of minerals from the Moine schists of the Ardnamurchan area, Argyllshire, Scotland and its implications for the phengite problem. *Jour. Petrol.*, **10**, 164-70.

——, 1974. Transition from staurolite to sillimanite zone, Rangeley Quadrangle, Maine. Bull. *Geol. Soc. Amer.*, **85**, 475-90.

——, 1984. Micas in Metamorphic rocks. In: Micas: Mineralogical Society of America, (Ed. S.W. Bailey), *Rev. in Miner.*, **13**, 435-67.

Guidotti, C.V., Cheney, J.T., & Henry, D.J., 1988. Compositional variation of biotite as a function of metamorphic reactions and mineral assemblage in the pelitic schists of Western Maine. *Amer. Jour. Sci.*, **288-A**, 270-92.

Harrison, T.M., Grove, M., Lovera, O.M. & Catlos, E.J., 1998. A model for the origin on Himalayan anatexis and inverted metamorphism. *Journal of Geophysical Research*, **103**, 27017-32.

Harrison, T.M., Grove, M., McKeegan, K.D., Coath, C.D., Lovera, O.M. & Le Fort, P., 1999. Origin and emplacement of the Manaslu intrusive complex, Central Himalaya. *Journal of Petrology*, **40**, 3-19.

Harrison, T.M., Ryerson, F.J., Le Fort, P., Yin, A., Lovera, O.M. & Catlos, E.J., 1997a. A Late Miocene-Pliocene origin for central Himalayan inverted metamorphism. *Earth and Planetary Science Letters*, **146**, E1-E8.

Hess, P.C., 1971. Prograde and retrograde equilibrium in Garnet-cordierite gneisses in the south-central Massachussets. *Contr. Miner. Petrol.*, **30**, 177-95.

Hodges, K.V. & Spear, F.S., 1982. Geothermometry, geobarometry and the Al_2SiO_5 triple point at Mt. Moosilauke, New Hampshire. *Amer. Miner.*, **67**, 1118-34.

Hoisch, T.D., 1990. Empirical calibration of six geobarometers for the mineral assemblage quartz + muscovite + biotite + plagioclase + garnet. *Contr. Miner. Petrol.*, **104**, 225-34.

Hollister, L.S., 1966. Garnet zoning and interpretation based on the Rayleigh fractionation model. *Science*, **154**, 647-51.

Hyndman, D.W., 1985. Petrology of igneous and metamorphic rocks. McGraw Hill, New York.

Johnson, S.E., 1990. Deformational history of the Otago schists, New Zealand, from progressive developed phorphyroblast/matrix microstructures: cyclic uplift-collapse orogenesis and its implications. *Jour. Struct. Geol.*, **12**, 727-46.

Karabinos, P., 1984. Polymetamorphic garnet zoning from Southeastern Vermount. *Amer. Jour. Sci.*, **284**, 1008-25.

Kerr, A., 1981. Zoning in Garnet from the mainland Lewisian. *Miner. Mag.*, **44**, 919-94.

Krogh, E.J. & Raheim, A., 1978. Temperature and pressure dependence of Fe-Mg partitioning between garnet and phengite, with particular reference to eclogites. *Contr. Miner. Petrol.*, **66**, 75-80.

Labotka, T.C., 1980. Petrology of a medium pressure regional metamorphic terrain, Funeral Mountains, California. *Amer. Miner.*, **65**, 670-89.

Laird, J., 1989. Chlorites: Metamorphic petrology. In: Hydrous phyllosilicates, Mineralogical Society of America. *Rev. in miner.*, **19**, 405-53.

Lal, R.K., Mukherjee, S. & Ackermand, D., 1981. Deformation and Barrovian metamorphism at Takdah, Darjeeling (Eastern Himalaya). In: *Metamorphic tectonites of the Himalaya* (Ed. P.S. Saklani), 231-78.

Lasaga, A.C., Richardson, S.M. & Holland, H.D., 1977. The mathematics of cation diffusion and exchange between silicate minerals during retrograde metamorphism. In: *Energetics of geological processes.* (Ed. S.K. Saxena & S. Bhattacharji), Springer Verlag, New York, 353-88.

Liou, J.G., 1973. Synthesis and stability relations of epidote, $Ca_2Al_2FeSi_3O_{12}(OH)$, *Jour. Petrol.*, **14**, 381-413.

Macfarlane, A.M., Hodges, K.V. & Lux, D., 1992. A structural analysis of the Main Central Thrust zone, Langtang National Park, central Nepal Himalaya. Bull. *Geol. Soc. Amer.*, **104**, 1389-402.

Miyashiro, A. & Shido, F., 1985. Tschermak substitution in low- and middle-grade pelitic schists. *Jour. Petrol.*, **26**, 449-87.

Mohan, A., Windley, B.F. & Searle, M.P., 1989. Geothermobarometry and development of inverted metamorphism in Darjeeling-Sikkim region of the Eastern Himalaya. *Jour. Met. Geol.*, **7**, 95-110.

Neogi, S., Dasgupta, S. & Fukuoka, M., 1998. High P-T polymetamorphism, dehydration melting, and generation of migmatites and granites in the Higher Himalayan crystalline complex, Sikkim, India. *Journal of Petrology*, **39**, 61-99.

Newton, R.C. & Haselton, H.T., 1981. Thermodynamics of the garnet - plagioclase - Al_2SiO_5 quartz geobarometer. In: *Thermodynamics of mineral and melts* (Ed. R.C. Newton *et al.*), Springer-Verlag, New York, 131-47.

Robinson, P., Tracy, R.J. & Pomeroy, J.S., 1977. High pressure sillimanite garnet biotite assemblage formed by recrystallisation of mylonite in sillimanite-cordierite-orthoclase schist, Central Massachusetts. *Geol. Soc. Amer. (Abst. with progr.)*, **9**, 1144-45.

Ruiz, J.L., Apparicio, A. & Cacho, L.G., 1978. Chemical variation in biotites during prograde metamorphism, Sierra De Guadarrama, Sistema Central, Spain. *Chem. Geol.*, **21**, 113-29.

Saklani, P.S., 1993. Geology of the Lower Himalaya (Garhwal). *International Book and Periodicals*, Delhi, 241.

Schreurs, J., 1985. Prograde metamorphism of metapelites garnet-biotite thermometry and prograde changes of biotite chemistry in high grade rocks of West Uusimaa, Southwest Finland. *Lithos*, **18**, 69-80.

Sinha Roy, S., 1977. Metamorphism and tectonics of the Himalayas, as illustrated in the Eastern Himalayas. *Geotectonics*, **11**, No. 2, 120-24.

Sorkhabi, R.B. & Stump, E., 1993. Rise of the Himalaya: A geochronologic approach. *GSA Today*, **3**, 85-92.

Spear, F.S., 1981. Amphibole-plagioclase equilibria: An empirical model for the relation Albite + Tremolite = Edenite + Quartz. *Contr. Miner. Petrol.*, **77**, 355-64.

Spear, F.S. & Peacock, S.M., 1990. Metamorphic P-T-t paths. Program manual and computer exercises for the calculation of metamorphic phase equilibria, pressure-temperature-time paths and thermal evolution of orogenic belts. Deptt. of Geology, Arizona State University, Tempe, U.S.A.

Staubli, A., 1989. Polyphase metamorphism and the development of the Main Central Thrust. *Jour. Met. Geol.*, **7**, 73-94.

Thompson, A.B., 1976. Mineral reactions in pelitic rocks: 1. Prediction of P-T-X (Fe-Mg) phase relations. *Amer. Jour. Sci.*, **276**, 401-24.

Thompson, A.B., Tracy, R.J., Lyttle, P. & Thompson, J.B. Jr., 1977. Prograde reaction histories deduced from compositional zonation and mineral inclusions in garnets from Gassette schist. Vermont. *Amer. Jour. Sci.*, **277**, 1152-67.

Thompson, J.B. Jr. & Norton, S.A., 1968. Paleozoic regional metamorphism in New-England and adjacent areas. In: *Studies of Appalachian Geology* (Ed. Zen, E-an *et al.*). Northern and Maritime New York. Interscience, 319-29.

Tracy, R.J., 1982. Compositional zoning and inclusions in metamorphic minerals. In: Characterisation of metamorphism through mineral equilibria (Ed. J.M. Ferry), Mineralogical Society America, *Rev. in Miner.*, **10**, 355-97.

Tracy, R.J., Robinson, P. & Thompson, A.B., 1976. Garnet composition and zoning in determination of temperature and pressure of metamorphism, Central Massachusetts. *Amer. Miner.*, **61**, 762-75.

Vernon, R.H. & Flood, R.H., 1979. Microstructural evidence of time relationships between metamorphism and deformation in the metasedimentary sequence of the Northern Hill End Trough, New South Wales, Australia. *Tectonophysics*, **50**, 127-37.

Winkler, H.G.F., 1979. Petrogenesis of metamorphic rocks. Springer-Verlag, New York.

Zwart, H.J., 1960. The chronological succession of folding and metamorphism in the Central Pyrenees. *Geol. Rundsch.*, **50**, 203-18.

Zwart, H.J., 1962. On the determination of polymetamorphic mineral associations and its application to the Bosost area (Central Pyrenees). *Geol. Rundsch.*, **52**, 38-65.

SECTION–IV

UHP/UHT METAMORPHISM

12

UHP METAMORPHISM AND CONTINENTAL SUBDUCTION/COLLISION

J.G. LIOU*, T. TSUJIMORI*, I. KATAYAMA** AND S. MARUYAMA**

ABSTRACT

Continental crust (density ~2.8 g.cm^{-3}) resists subduction into the earth's mantle (~3.3 g.cm^{-3}) because of buoyancy forces. However, more than 20 recognized ultrahigh-pressure (UHP) terranes have been documented; their occurrences demonstrate that not only is continental crust subducted to depths as great as 150 km, but also that some supracrustal rocks were subsequently exhumed to the earth's surface. UHP terranes are composed of mainly supracrustal rocks that contain minerals such as coesite or diamond, indicative of $P > 2.5$ GPa. In general, quartzofeldspathic units are thoroughly back reacted, and only mafic eclogite lenses and boudins retain scatter UHP phases. These index minerals are restricted to micron-scale inclusions in chemically and mechanically resistant zircon, garnet and a few other strong container minerals, and are difficult to identify by conventional petrologic studies. The continental rocks were subjected to UHP metamorphism at T ranging from ~700 to 950°C and $P > 2.8$ to 5.0 GPa, corresponding to depths of ~100 to 150 km. These UHP units were subsequently exhumed to crustal depths and subjected to intense hydration and amphibolite-facies overprint. Widespread Barrovian-type metamorphism in many collisional orogens may mask an earlier, higher pressure metamorphic history. We suspect that coesite-bearing UHP rocks were once generated in the majority of exhumed collisional orogens.

The recent finding of coesite inclusions in rare Himalayan eclogites and country rock gneisses is a typical example. We use the Himalayan model to illustrate UHP metamorphism and subduction of continental crustal rocks to mantle depths and the subsequent Barrovian-type overprint during exhumation. Himalayan UHP eclogites and adjacent gneisses were formed at mantle depths > 100 km at 46 to 52 Ma. These rocks were extruded to crustal depths and subjected to Barrovian amphibolite- to granulite-facies metamorphism; associated magmatism occurred at 30 to 15 Ma. The Himalayan metamorphic belt was domally uplifted and the mountain-building process initiated since 11 Ma, when underthrusting of the Indian tectosphere beneath the Lesser Himalayas occurred.

* Department of Geological and Environmental Sciences, Stanford University, Stanford, CA 94305-2115.

** Department of Earth and Planetary Sciences, Tokyo Institute of Technology, Meguro-ku, Tokyo 152-8551, Japan.

KEY WORDS

UHP Metamorphism; Continental Subduction/Collision; Supracrustal; exhumed; Coesite or Diamond and eclogites.

INTRODUCTION

Ultrahigh-pressure (UHP) metamorphism refers to the metamorphism of crustal rocks (both continental and oceanic) brought to pressures high enough to crystallize the index minerals coesite at a minimum $P > 2.7$ GPa at $T > 600°C$ and/or diamond. Prior to the initial discoveries of coesite in supracrustal rocks in 1984 (Chopin, 1984; Smith, 1984), coesite and diamond were thought to occur only in meteorite impact craters and mantle xenoliths. Fig. 1 shows the relevant *P-T* conditions defining UHP, high-pressure (HP), and low-pressure (LP) metamorphism; in addition, geotherms of about $5°C.km^{-1}$ (extreme high *P*/*T*) and $20°C.km^{-1}$ (old plates) are illustrated.

The objective of the present article is to present a summary to the literature of continental crust subduction and UHP metamorphism and illustrate recent petrochemical research of various UHP rocks. To honor Professor R.S. Sharma, who has made many contributions in the study of granulite-facies metamorphism and growth of continent, it is appropriate to emphasize the interactions between metamorphism and tectonics in continental collision belts using the Himalayan Orogen as an example. Distribution and mineralogical and petrological characteristics of other UHP terranes in the world are also described.

The stabilities of coesite and other UHP minerals in metamorphic regime require abnormally low geothermal gradients less than approximately $7°C.km^{-1}$. Such environments can be attained only by the subduction of old, cold, oceanic crust-capped lithosphere ± pelagic sediments or an ancient continental crust. Eclogites with compositions of the mid-oceanic ridge basalt (MORB) + H_2O system have been experimentally studied (e.g. Schmidt & Poli, 1998; Okamoto & Maruyama, 1999); the results shown in Fig. 1 subdivide *P-T* regimes for the eclogite-facies into those for amphibole eclogite, epidote eclogite, lawsonite eclogite, and dry eclogite. UHP and HP metamorphism can be separated conveniently by the *P-T* boundary for the quartz-coesite equilibrium. Eclogitic rocks formed at pressures greater or less than the transition boundary have been called 'hot' or 'cold' eclogite respectively. The equilibrium boundary for the graphite-diamond transition further subdivides the UHP regimes into diamond-grade and coesite-grade; with appropriate composition and fO_2 and X_{CO_2} condition, microdiamond (+ coesite) occurs in diamond-grade UHP rocks.

We have classified the global HP and UHP belts into two types according to the protoliths (Maruyama *et al.*, 1996). In active margins of the Pacific type, protoliths for HP metamorphism consist of rock assemblages at an accretionary complex including bedded cherts, MORB-origin greenstones, seamount fragments, and enclosing trench turbidites. On the other hand, protoliths of the collision-type HP-UHP rocks include continental basement complexes and their overlying sediment + volcanoclastic rocks of a variety of tectonic settings. Fig. 2 is a schematic diagram showing the contrasting tectonic settings for accumulation of A-type protoliths in a rifted continental margin, and B-type protoliths in an active continental margin.

Most UHP rocks have A-type protoliths and are characterized by occurrences of continental shelf carbonates, bimodal volcanics, peraluminous sediments, and granite-gneiss basement rocks. Eclogites and garnet peridotites, although volumetrically small, are widespread in all

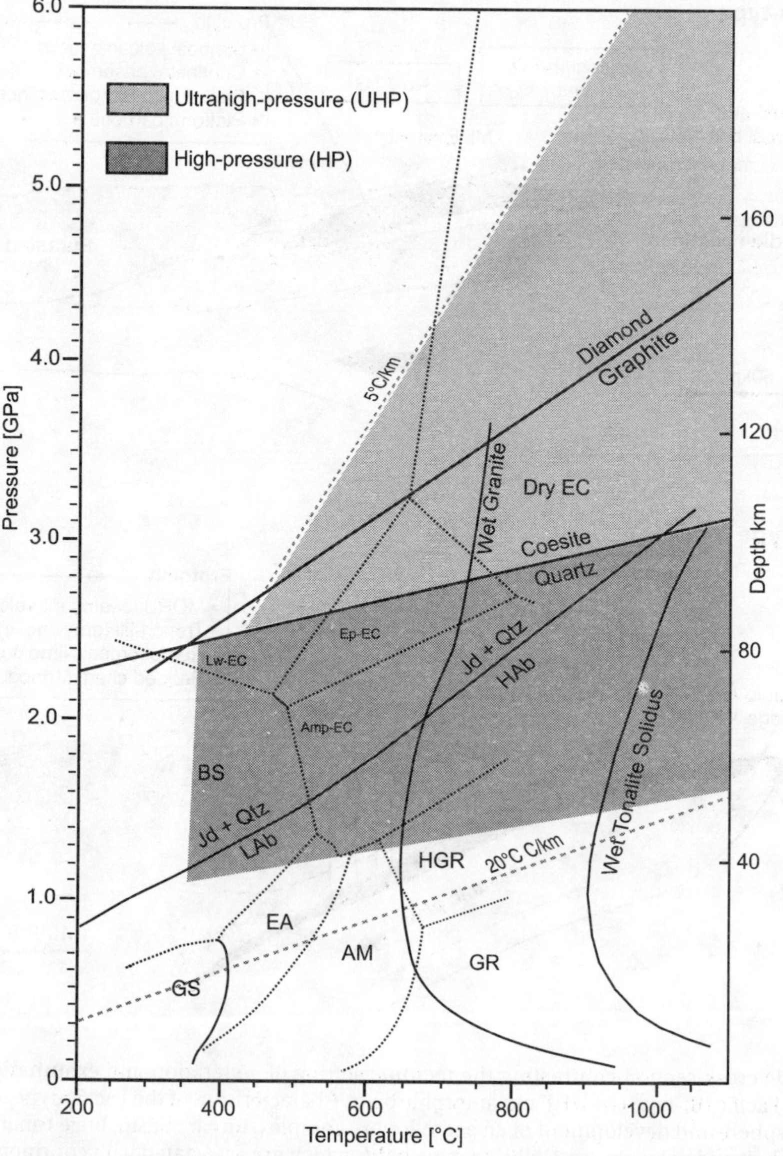

Fig. 1. P-T regimes assigned to various metamorphic types: (1) Ultrahigh-P, (2) High-P, and (3) Low-P. Geotherms of 5°C/km and 20°C/km are indicated. Stabilities of diamond, coesite, glaucophane, jadeite + quartz, aragonite, kyanite, sillimanite, andalusite, paragonite and the minimum meltings of granite and tonalite are shown. P-T boundaries of various metamorphic facies [granulite, amphibolite, epidote amphibolite, greenschist, and subgreenschist facies] and subdivision of the eclogite field into amphibole (Amp) eclogite, epidote (Ep) eclogite, lawsonite (Lw) eclogite and dry eclogite are indicated (after Liou *et al.*, 2000).

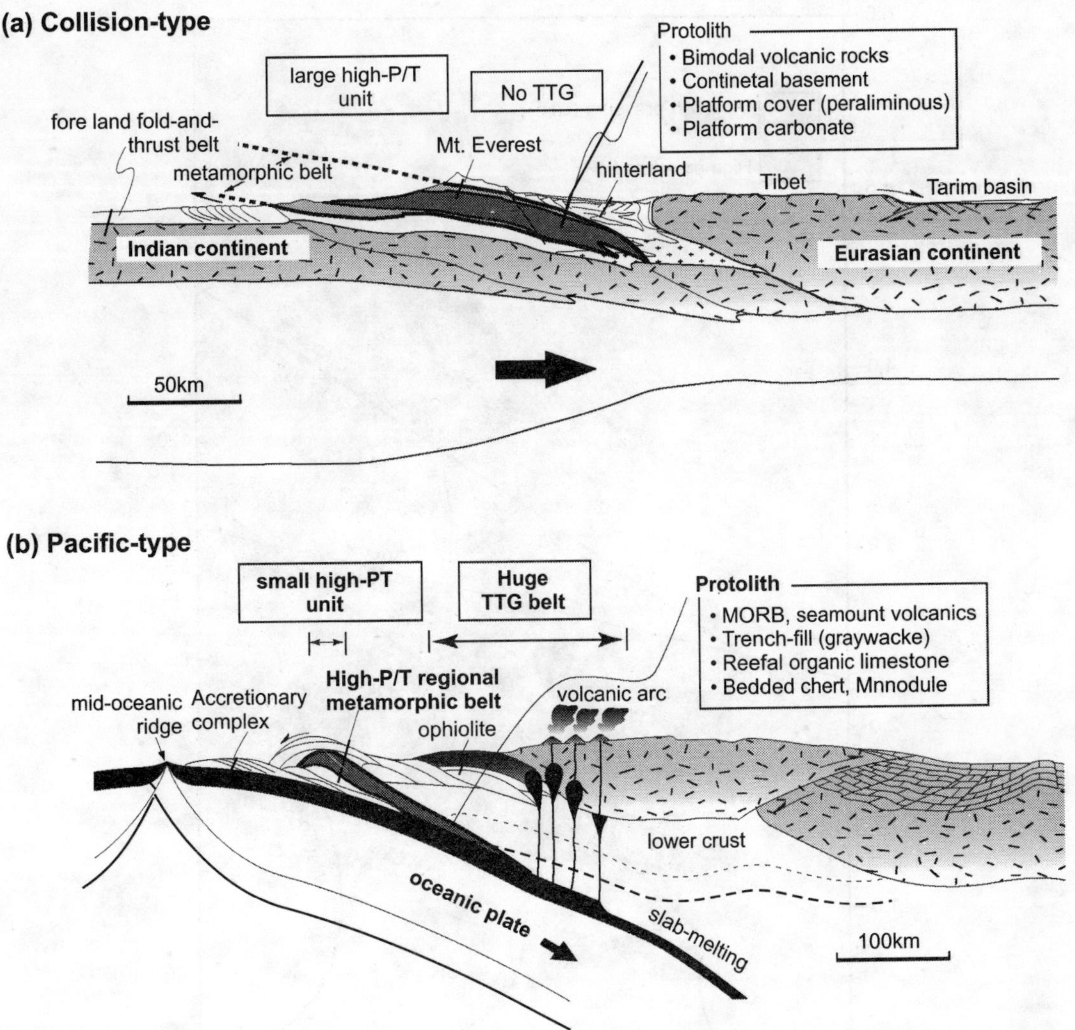

Fig. 2. Schematic cross-section contrasting the tectonic setting of generation and exhumation of (a) Collision (A)-type and (b) Pacific (B)-type HP-UHP metamorphic belts. Characteristics of the Pacific-type include subduction of oceanic lithosphere and development of an accretionary complex, forearc basin, huge tonalite-trondhjeimite-granitoid (TTG) belt and volcanic arc. Collision-type belts, which are associated with continental collision, lacks those characteristics, and the high-P belt is generally of a much large extent (after Maruyama *et al.*, 1996).

recognized UHP terranes of the world. They are the most significant components inasmuch as they preserve most completely UHP index minerals such as coesite and microdiamond. These mafic-ultramafic rocks may have originated from different tectonic settings and were subjected to similar UHP metamorphism, deformation, and retrogression with their enclosing granitic gneisses and supracrustal rocks. Because of better preservation of the effect of UHP metamorphism compared with adjacent gneissic rocks, eclogites and garnet peridotites have received the most intensive study; accordingly, they provide important petrochemical and isotopic constraints on tectonic models of continental subduction, collision and subsequent exhumation for an UHP terrane.

GLOBAL DISTRIBUTION OF UHP METAMORPHIC TERRANES

Occurrences of UHP rocks have been increasingly recognized and extensively reviewed (e.g. Liou *et al.*, 1999, 2002). Thus far, more than 20 UHP terranes shown in Fig. 3 have been documented. These UHP terranes lie within major continental plate collision belts and extends several hundred km or more; many are in Eurasia and few in Africa, South America and Antarctica. They share common structural and lithological characteristics. (1) UHP records are preserved mainly in eclogites and garnet peridotites, enclosed as pods and slabs within gneissic units. A few of those rocks contain minute inclusions of coesite in zircon, garnet and omphacite, and microdiamonds in garnet and zircon. (2) Most lithologies are continental in chemical compositions. (3) Exhumed UHP units are now present in the upper continental crust as thin sub-horizontal slabs, bounded by normal faults on top, and reverse faults on bottom, and sandwiched in amongst HP or lower grade metamorphic units. (4) Coeval island-arc volcanic and plutonic rocks do not occur, whereas post-collisional or late-stage A-type granitic plutons are common in some occurrences.

In this section, classical (Ernst & Liou, 2000), less intensively studied (Liou, 1999) and recently recognized UHP terranes are described below. Lithological and tectonic characteristics including index minerals, *P-T* conditions, size and peak metamorphic ages of these UHP terranes are summarized in Table 1.

Classical UHP Terranes

Several classical UHP terranes including (1) the Dora Maira massif of the Western Alps (e.g. Chopin, 1984), (2) the Western Gneiss Region of Norway (e.g. Smith, 1984), (3) the Dabie-Sulu terrane of east central China (Wang *et al.*, 1989), (4) the Kokchetav massif of northern Kazakhstan (e.g. Sobolev & Shatsky, 1990), and (5) the Bohemian massif (Bakun-Czubarow, 1991) have been well studied and extensively reviewed (e.g. Ernst & Liou, 2000). Except for the Dora Maira massif, occurrences of microdiamond in these UHP terranes have been reported; petrochemical characteristics of these terranes are summarized below.

The Dora Maira massif, Western Alps

The Dora Maira massif consists of Late Paleozoic and older continental basement rocks, metamorphosed under UHP and HP conditions as a result of Mesozoic-Cenozoic convergence of the European and African plates. The HP + UHP belt is a stack of four major units separated by low-angle faults. These units have been affected by Cenozoic HP metamorphism and later re-equilibration at lower pressures. The UHP unit consists of dark- and light-colored eclogites, pyrope quartzite with phengite schist and jadeite-rich rock inclusions, as well as orthogneiss country rocks and undeformed metagranites. Boudins of pyrope quartzite contain UHP relics such as coesite and other UHP minerals. This entity was subdivided into two metamorphic subunits. (a) A coesite-bearing, UHP subunit contains kyanite-eclogites with relict coesite and coesite pseudomorphs, and has an estimated peak *P-T* conditions of 3.7 GPa and 790°C, within the diamond stability field. (b) A lower-P eclogite subunit was metamorphosed at 1.5 GPa and 500°C. The time of UHP/HP metamorphism has been considered to have culminated between 90 to 125 Ma, and retrograded between 35 to 41 Ma. However, recent SHRIMP dating of zircons extracted from gneiss, schist, pyrope inclusion, and pyrope quartzite indicates the occurrence of 240-275 Ma-old zircon cores and newly

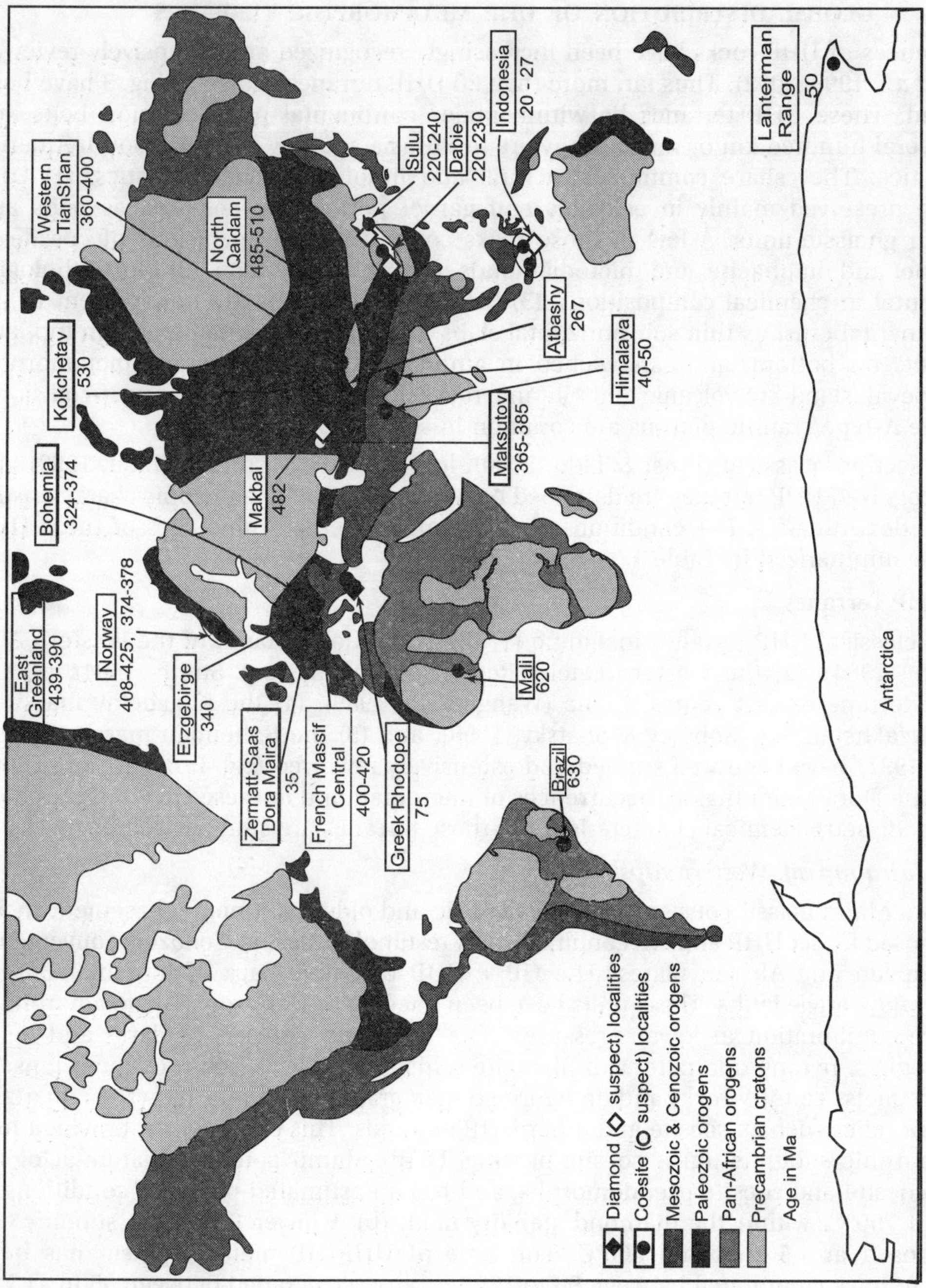

Fig. 3. Distribution and peak metamorphic age of recognized UHP terranes in the world (modified after Liou *et al.*, 2000).

formed 35 Ma rims for some zircons. Thus, UHP metamorphism must have taken place during Oligocene time, with an estimated exhumation rate of 2 to 2.4 $cm.yr^{-1}$.

The Western Gneiss Region, Norway

The Western Gneiss Region lies within the Early Paleozoic collision zone; the gneissic unit, about 300 km long and 150 km wide, consists of interlayered pelite and migmatite, marble, quartzite, and amphibolite, with tectonic inclusions of gabbro, and peridotite. The gneissic unit exhibits mainly amphibolite-facies assemblages, but relics of HP assemblages also occur. Eclogite boudins are widespread. Coesite was first reported from Grytting (Smith, 1984); several new localities of coesite and coesite pseudomorphs have been recognized recently in both eclogites and the adjacent gneissic rocks (Wain, 1997; Cuthbert *et al.*, 2000; Wain *et al.*, 2000). Microdiamond grains 20-50 microns across have been described from residues separated from two gneisses (Dobrzhinetskaya *et al.*, 1995); the associated kyanite-eclogites contain inclusions of coesite pseudomorphs in garnet. Relict majoritic garnets showing exsolution of pyroxene lamellae in peridotite bodies have been recently discovered on the islands of Otrogy and Flemsoy (van Roermund, 1998, 2000). These peridotite bodies originally have an even higher-pressure, deeper mantle origin, probably within a rising mantle diapir at ~1700 Ma. In these garnet peridotites, diamond inclusions in sympletetic spinel after garnet have been recently discovered (van Roermund *et al.*, 2002). Radiometric ages for eclogites of the West Gneiss region are 408-425 Ma for the peak metamorphism, and 374-378 Ma for the retrogressive mid-crustal metamorphic overprinting (for review, see Carswell *et al.*, 1999). Carswell *et al.* (2003) provide new petrographic evidence and a review of the latest radiometric age data, and conclude that the UHP metamorphism of eclogites occurred at 400-410 Ma, significantly younger than the previous, widely accepted age of 425 Ma. A two-stage exhumation process is suggested: an initial exhumation to about 35 km depth by about 395 Ma at a mean rate of about 10 $m.Ma^{-1}$, and subsequent exhumation to 8-10 km by about 375 Ma at a much slower rate of about 1.3 $m.Ma^{-1}$.

The suture zone occupied by UHP/HP rocks of the Western Gneiss Region reflects collision of the eclogite-bearing Greenland sialic crust-capped lithospheric plate with Fennoscandia during the Caledonian orogeny (Gilotti, 1993; Krogh & Carswell, 1995; Brueckner *et al.*, 1998). Inclusions of coesite pseudomorph in garnet and omphacite from kyanite eclogites have been recently documented in the East Greenland eclogite province (Gilotti & Krogh Ravna, 2002); this new UHP/HP terrane is remote, but has excellent exposures and requires close examination.

The Dabie-Sulu terrane of east-central China

The Dabie-Sulu terrane occupies a Triassic collision zone between the Sino-Korean and Yangtze plates. The central UHP coesite-bearing eclogite belt (P = 2.7-5 GPa) is flanked to the north by a migmatite zone (P < 2 GPa), and to the south by a belt of blueschist, epidote amphibolite and eclogite-facies rocks (P = 0.5-1.2 GPa). Prevalent rock types include felsic gneiss, orthogneiss, marble, and quartzite. Protolith ages range from Proterozoic to Ordovician but the majority of reported ages are Neoproterozoic (Li *et al.*, 1993; Hacker *et al.*, 2000). Blocks, boudins and layers of eclogites and garnet peridotites occur as enclaves in gneisses in the UHP unit. The Dabie-Sulu UHP rocks show several characteristics. (1) Widespread occurrences of coesite with hydrous minerals such as talc, zoisite/epidote, and phengite are found in eclogites. (2) Several garnet peridotites of mantle origin occur. (3) Abundant exsolution textures were identified in UHP minerals from garnet peridotite and eclogite.

Diamond separates from both eclogite and garnet peridotite have been reported, although these findings are not confirmed by thin section observation and extensive studies of mineral inclusions in zircon (for review, see Liou *et al.*, 2002). Some garnet peridotites record much higher pressure than associated coesite-bearing eclogites, reaching 5-6 GPa (Yang *et al.*, 1993; Zhang *et al.*, 2000, 2003). Various isotopic age dating methods for UHP minerals and rocks from Dabie-Sulu terrane gives 220-240 Ma (Li *et al.*, 1993; Hacker *et al.*, 2000).

Eclogites in the Huwan shear zone from the north-western Dabie Mountains have SHRIMP U/Pb dates of 309 ± 3 Ma (Sun *et al.*, 2002) suggesting a discrete Carboniferous subduction/collision between the North and South China. Further to the west in North Qinling, inclusions of microdiamond were discovered in zircons from both eclogite and its gneissic country rocks (Yang *et al.*, 2002); SHRIMP U/Pb dating of zircons from granitic gneiss yielded 507 ± 38 Ma for metamorphic rims and 1200-1800 Ma for relic magmatic or old cores suggesting an Early Paleozoic UHP metamorphic event. These new data suggest episodic subduction and collision of continents occurred during the closures of the paleo Tethys prior to the well-documented major Triassic continent-continent collision for the Dabie-Sulu terrane.

The Kokchetav Massif

The Kokchetav Massif, northern Kazakhstan, is the type discovery site for the UHP diamond-eclogite regional metamorphic facies (Sobolev & Shatsky, 1990; Parkinson *et al.*, 2002). Diamonds occur as minute inclusions in garnet, diopside, and zircon porphyroblasts from marble, pyroxene-carbonate-garnet rock, and garnet-biotite gneiss and schist. Inclusions of coesite and coesite pseudomorphs are found in garnet and zircon from eclogite and diamond-bearing gneiss. The UHP/HP unit extends NW-SE for at least 80 km and is about 17 km wide. The UHP slab is structurally overlain by a weak- to low-grade metamorphic unit and is underlain by a low-*P* metamorphic andalusite-staurolite facies unit (Kaneko *et al.*, 2000). The UHP slab is composed of felsic gneiss with locally abundant eclogite lenses and minor orthogneiss, metacarbonate, and rare garnet peridotite with a range of Proterozoic protolith ages. A well-constrained P-T-time path deduced from mineral inclusions and SHRIMP geochronology of zircons is illustrated in Fig. 4. Zoned zircons from diamond-bearing gneiss yield the following ages: the quartz-bearing inherited core with 1100-1400 Ma as protolith age, coesite-bearing mantle with 537 ± 9 Ma for UHP metamorphism and the plagioclase-bearing rim with 507 ± 8 Ma for amphibolite-facies overprint (Katayama *et al.*, 2001). Many diamond-grade marbles contain clinopyroxene with exsolution lamellae of quartz, K-feldspar, phengite and phlogopite (Katayama *et al.*, 2001; Zhu & Ogasawara, 2002), and titanite with coesite lamellae (Ogasawara *et al.*, 2000); their occurrence suggest the UHP metamorphism occurred at $P > 6$ GPa and 1000°C.

C^{-13} isotopic data of diamond suggest biogenic sources; inclusions of water, carbonate and nanometric oxide in diamonds indicate that metamorphic microdiamonds in this and other UHP terranes were precipitated from C-O-H supercritical fluids (e.g., Stockhert *et al.*, 2001). Ogasawara *et al.* (2002) further suggests a two-stage growth mechanism for microdiamonds. Masago *et al.* (2002) report negative ^{18}O values (-3.9 per mil) for minerals in eclogite and whiteschist of the Kokchetav Massif, similar to the Chinese UHP rocks first described by Yui *et al.* (1995). The Kokchetav Massif is the second recognized UHP region that preserves a significant effect of the interaction of cold meteoric water with the protolith prior to subduction.

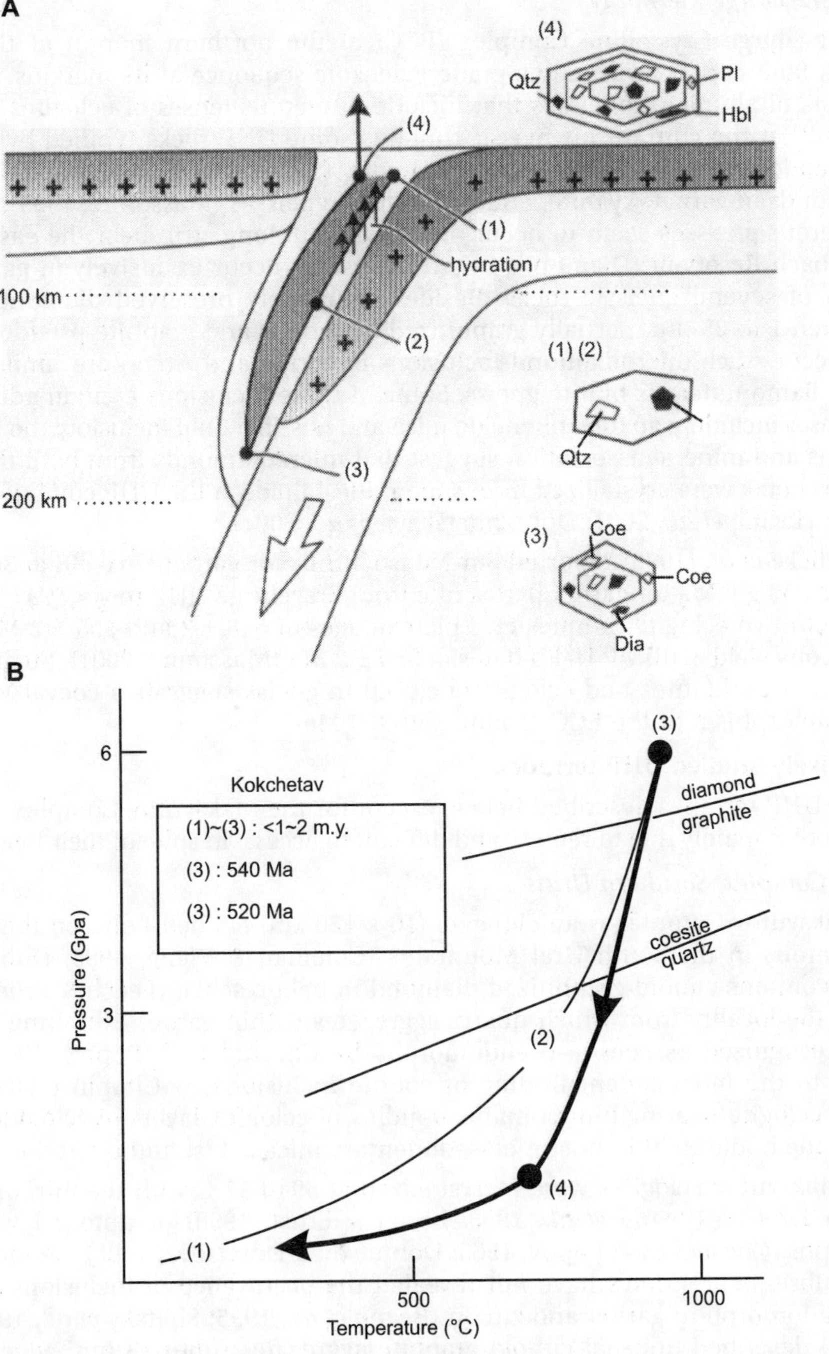

Fig. 4. A. Schematic diagram showing the growth of detrital zircon (stage 1) through subduction (stage 2 and peak stage 3) and exhumation (stage 4) of supracrustal rocks. Mineral inclusions in various stages of zircon growth are shown. B. SHRIMP U-Pb dating of zoned zircons from a Kokchetav diamond-bearing gneiss yields a well constrained P-T-time path (For details, see Katayama *et al.*, 2002).

Saxonian Erzgebirge, Germany

The Erzgebirge Crystalline Complex (ECC) at the northern margin of the Bohemian massif is in fault contact with a low-grade Paleozoic sequence at its margins. The Variscan ECC consists of abundant gneisses that include numerous lenses of eclogites. The 'Gneiss-Eclogite Unit' at the core of this massif contains some UHP rocks typified by inclusions of coesite pseudomorph in garnets from eclogite (Schmadicke, 1991) and inclusions of microdiamonds in garnet, kyanite, and zircon from gneisses (Massonne, 1999). Thus far, the diamondiferous gneisses seem to occur only in a 1 km long strip near the eastern shore of the Saidenbach Reservoir. Diamonds micro-inclusions occur exclusively in garnet, kyanite, and zircon of several gneissic rocks. Besides completely preserved diamonds with grain sizes between 1 to 25 μm, partially graphitized diamonds and graphite pseudomorphs after diamond occur. Such microdiamond inclusions in garnet and zircon are similar to those in Kokchetav diamondiferous biotite gneiss. Some of these inclusions contain additional fluid-bearing phases including apatite, phengitic mica and possible fluid inclusion; the morphologies of inclusions and mineral association suggest that microdiamonds from both the Erzgebirge and the Kokchetav were crystallized from supercritical fluids at the UHP conditions (Stockhert *et al.*, 2001; Hwang *et al.*, 2001; Dobrzhnetskaya *et al.*, 2001).

Schmadicke *et al.* (1995) reported Sm-Nd isochrons for garnet-Cpx-WR at 360 ± 7 Ma for eclogite and 353 ± 6 Ma for garnet pyroxenite from Erzgebirge UHP rocks. $^{40}Ar/^{39}Ar$ spectra of phengite from two eclogite samples give plateau ages of 348 ± 2 and 355 ± 2 Ma. Diamond-bearing zircons yield SHRIMP U/Pb dates at 336 ± 2 Ma (Massonne, 2001). Such similarity in ages for garnet peridotites and eclogite enclosed in gneiss suggests a coeval Variscan UHP *in situ* metamorphism of the ECC around 340-360 Ma.

Less Intensively Studied UHP terranes

Several UHP terranes described below, except for the Maksyutov Complex, received less intensive studies mainly due to remote and difficult to access, in spite of their long recognition.

Maksyutov Complex, Southern Urals

The Maksyutov Complex is an elongate (10 × 120 km) NS belt between the Russian and Siberian cratons in the south Ural Mountains (Coleman & Wang, 1995; Dobretsov *et al.*, 1996), and contains cuboid graphitized diamond in pelitic schist (Leech & Ernst, 1998). The complex is the locality from which quartz aggregates within garnet exhibiting radial cracks were first recognized as coesite-pseudomorphs by Chesnokov & Popov (1965), nearly 20 years prior to the independent finding of coesite inclusions by Chopin (1984) and Smith (1984). The eclogite-bearing unit contains boudins of eclogite, layers of eclogitic gneiss, and rare ultramafic bodies within host metasedimentary mica schist and quartzite.

Some Maksyutov eclogites were recrystallized at 594-637°C with the minimum pressure from 1.5 to 1.7 GPa (Beane *et al.*, 1995; Leech & Ernst, 1998) or upto 2.7 GPa if coesite pseudomorphs (Chesnokov & Popov, 1965; Dobretsov & Doretsova, 1988) are indeed present. However, subsequent studies have not revealed the occurrence of inclusions of coesite or coesite pseudomorph in garnet and zircon (Beane *et al.*, 1995; Shatsky *et al.*, 1995). Leech & Ernst (1998) described unusual cuboid graphite aggregates (upto 13 mm edge length) that may be diamond pseudomorphs. This suggestion was based on the blocky morphology of the cuboid graphite aggregates cross-cutting rock foliation, pressure shadows, biogenic isotopic signatures and other spectroscopic characters similar to those graphitized diamonds from the Beni Bousera massif. This study has yielded the suggestive, though unproven, possibility that the eclogitic unit of the Maksyutov Complex experienced diamond-grade UHP metamorphism. The UHP/HP metamorphism with peak metamorphic ages of 370-374 Ma (Shasky *et al.*, 1995; Beane *et al.*, 1998) may be related to the collision of the Russian platform with a fragment of the Siberian craton during Devonian time, but was affected later by the Triassic collision of Kazakhstan block with the Russian platform (e.g. Matte, 1998).

Zermatt-Saas area, Western Alps

In contrast to most other UHP terranes where the protolith is continental crust, the eclogites and metasediments at Lago di Cignana of the Western Alps have been derived from oceanic crust and constitute an ophiolite sequence of the Zermatt-Saas zone. Inclusions of coesite and coesite pseudomorphs occur in tourmaline and garnet of manganiferous quartzite and in omphacite and garnet of the underlying eclogite (Reinecke, 1991). The metasediments are mainly garnet-phengite-quartz schist, with variable amounts of garnet-clinopyroxene quartzite, and piedmontite-phengite-quartz schist. Eclogites and retrograded eclogites with preserved UHP mineral parageneses consist of zoned garnet, omphacite, glaucophane, zoisite, phengite, and dolomite. Zoned garnets preserve prograde quartz crystals in their cores and coesite inclusions in mantles and rims. Coesite inclusions are much more common in omphacite than in garnet. UHP conditions were estimated at 2.6-3.0 GPa and 590-630°C. Sm-Nd isotopic analyses of eclogites yield UHP metamorphism at 52 ± 18 Ma (Bowtell *et al.*, 1994), consistent with a Tertiary age of eclogitization recently established for Alpine metamorphism in the Western Alps (Gebauer *et al.*, 1997).

Thus far, the recognized areal extent of the coesite-bearing rocks is ca. 2 km^2. The apparent thickness of the UHP slice does not exceed a few hundred meters. This area is the best example for subduction of oceanic lithosphere to mantle depth, and later exhumation as a fragmented tectonic slab. Similarly, the Allalin metagabbro, part of the oceanic Piemonte zone, contains magnesite + talc + kyanite + garnet recrystallized at about 600°C and 3.5 GPa, with an estimated geothermal gradient of 5-6°C.km^{-1}; however, inclusions of coesite or coesite pseudomorphs have not been reported for the Allalin metagabbro.

Mali, Africa

Eclogitic rocks in the Pan-African collision zone cropping out in northern Mali contain inclusions of coesite in omphacite from mafic nodules within a calc-silicate layer enclosed in impure marble, and coesite pseudomorphs in garnets from the surrounding eclogitic metasediments (Caby, 1994). Theses UHP eclogitic mica schists have P-T estimates of 700-750°C and > 2.7 GPa; phengite yields a well-defined plateau $^{39}Ar/^{40}Ar$ age of 1045 ± 9 Ma. Recent Sm/Nd ages of 620 Ma were obtained by Jahn *et al.* (2001) for peak UHP metamorphism. This UHP terrane may be the oldest one in the world.

This coesite-bearing UHP eclogitic unit about 3 km thick lies within an internal nappe as a flat, thin slice bounded on the top by a thick passive-margin shaly formation, and on the bottom by low-grade greenschist-facies phyllites. Caby (1994) suggested that a large portion of the terrigenous metasediments of passive-continental-margin affinity of the West African plate was subducted eastward to mantle depths (> 90 km) beneath an oceanic domain. Collision of the West African plate with an island arc resulted in a low-angle, westward sub-horizontal extrusion of this subducted slice of sialic materials to their present-day position. The coesite-bearing eclogite unit may also exist 1500 km to the south in Togo, where kyanite-bearing eclogites occur as part of the passive-margin assemblage.

Makbal (480 Ma) and Atbashy (270 Ma), Kazakhstan

The Tien-Shan is divided into three mountain ranges — the northern, the middle and the southern Tien-Shan. Within the Kyrghyzstan Tien-Shan, UHP eclogite localities with coesite pseudomorph inclusions in garnet and omphacite from northern and southern Tien-Shan represent Caledonian and Hercynaian orogenic belts respectively (Tagiri & Bakirov, 1992; Tagiri *et al.*, 1995). The Makbal Formation in the western Kyrghyz Ridge of northern Tien-Shan consists of quartzose schist alternating with pelitic schists intercalated with thin marbles and layers or lenses of eclogites and amphibolitized eclogite. Tagiri & Bakirov (1992) found inclusions of coesite pseudomorphs in garnet from a garnet-chloritoid-talc schist with peak metamorphic assemblage of coesite + almandine + chloritoid + talc + phengite (Si = 3.42 p.f.u) + phlogopite + rutile. Makbal coesite-grade eclogite has a paragonite K/Ar age of 480 Ma.

The eclogite-bearing complex of Atbashy Ridge from the southern Tien-Shan is composed of pelitic and siliceous schists alternating with thin UHP eclogitic layers, and is unconformably overlain by Upper Paleozoic molasse and silicic volcanics. Inclusions of coesite pseudomorphs occur in omphacite cores and in garnet. Eclogites from the Atbashy area preserve prograde and retrograde paths with a peak metamorphic condition at 660°C, 2.5 GPa. The Atbashy coesite-eclogites yielded a Rb/Sr mineral isochron age of 270 Ma (Tagiri *et al.*, 1995).

Central Indonesia UHP terrane

The pre-Tertiary basement of the Indonesian region comprises a variety of imbricate terranes, melange, ophiolite, and variably metamorphosed accretionary complexes; some contain HP to UHP metamorphic rocks resulting from the collision of an Australia-derived continent with Eurasia (Parkinson, in press). HP rocks including eclogites and garnet peridotites are widely distributed in Cretaceous accretionary complexes. Many of these rocks occur as imbricate slices of carbonate, quartzite, and pelitic schist of shallow-marine or continental margin parentage, interthrust with subordinate mafic schist and serpentinite; some yield mica K/Ar ages of 110-120 Ma (Parkinson, in press).

HP and UHP rocks are sporadically exposed as tectonic blocks throughout the Cretaceous accretionary complexes. They include eclogite, garnet-glaucophane rock (P = 1.8-2.4 GPa, T = 580-620°C), and jadeite-garnet quartzite ($P > 2.7$ GPa, T = 720-760°C) in Bantimala, southwest Sulawesi, eclogite and garnet granulite in west-central Sulawesi, eclogite and jadeite-glaucophane-quartz rock ($P \sim 2.2$ GPa, $T \sim 530$°C) in central Java, Mg-chloritoid-bearing whiteschists ($P \sim 1.8$ GPa) in southeast Kalimantan, garnet lherzolites in east-central Sulawesi (P = 2.2-2.8 GPa, T = 1000-1100°C), west-central Sulawesi (P = 1.6-2.0 GPa, T = 1050-1100°C), and garnet pyroxenite ($P \sim 2$ GPa, $T \sim 850$ °C) in Sabah, northeast Borneo. Evidence for UHP rocks include: (1) inclusions of coesite in zircon and coesite pseudomorph in jadeite from jadeite quartzite and eclogite of the Bantimala Complex of south Sulawesi (Parkinson, in press); and (2), P-T estimates of peak-stage recrystallization at 2.7-3.5 GPa and 1000-1100°C for most garnet peridotites (Kardarusman & Parkinson, 2000). Many of these rocks were probably recrystallized in a north-dipping subduction zone at the margin of the Sundaland craton in the Early Cretaceous. Exhumation may have been facilitated by the collision of a Gondwana continental fragment with the Sundaland margin at ~120-115 Ma.

Recently recognized UHP terranes

Several new UHP terranes were recently identified as they have partially preserved trace index minerals in strong containers such as zircon or garnet. In fact, zircon has been considered to be the best container and many new terranes were discovered through positive identification of inclusions of coesite or coesite pseudomorphs in zircons. Detailed examination of mineral inclusions in core, mantle and rims of zircon separates from eclogites and their enclosing country rock gneisses have yielded both prograde and retrograde P-T-time paths for various UHP terranes mentioned above (see Katayama *et al.*, 2001 for details; also see the Himalayan eclogite for example).

The latest recognized UHP terranes include the followings: (1) coesite in the French Massif Central (Lardeaux *et al.*, 2001), (2) coesite and diamond from the Greek Rhodope metamorphic province (Mposkos & Kostopoulos, 2001), (3) coesite pseudomorph in the Northeast Greenland Eclogite Province (Gilotti & Ravna, 2002), (4) coesite in granulite-facies overprinting eclogites of SE Brazil (Parkinson, in review), (5) coesite in Himalayan eclogite from the Upper Kaghan Valley, Pakistan (O'Brien *et al.*, 2001; Kaneko *et al.*, 2003) and coesite from the Tso Morari crystalline complex of India (Mukherjee *et al.*, 2002), (6) coesite inclusions in gneissic rocks from the North Qaidam belt, western China (Yang *et al.*, 2001; Song *et al.*, 2003), (7) diamond inclusions in garnets of eclogite and gneiss from North Qinling (Yang *et al.*, 2002), (8) inclusions of quartz pseudomorphs and minor coesite in garnets of mafic eclogites from the Lanterman Range in Antarctica (Ghiribelli *et al.*, 2002), and (9) inclusions of calcite pseudomorphs after aragonite + magnesite in dolomite in metapelites from the western Tianshan inferred its metamorphism in a UHP region of the dolomite decomposition (Zhang L. *et al.*, 2002). The results of petrochemical and geochronological data for these new terranes are summarized below. The Himalayan UHP terrane in the Upper Kaghan Valley and the Tso-Morari complex is used to illustrate the processes of continental subduction and collision in the later session.

The French Massif Central

The French Massif Central in the western part of the Variscan Belt has experienced Late Silurian-Early Devonian up to Late Carboniferous orogenic events (e.g., Matte, 1986). High-pressure metamorphic rocks occur mainly in the upper gneissic unit within the leptyno-amphibolite group. The latter is an association of metagreywacke, mica schist, metabasalt, leptynite, metagranite and peridotite. Eclogites are associated with silicic and mafic HP granulites (Lardeaux *et al.*, 2001) and also with spinel and/or garnet lherzolites (Gardien *et al.*, 1990). Inclusions of coesite and coesite pseudomorphs were discovered in garnets of kyanite-bearing eclogite (Lardeaux *et al.*, 2001). Only two coesite grains are preserved as relicts; most coesite grains are completely transformed into polycrystalline radial quartz (palisade texture) or into polygonal quartz surrounded by radial cracks. Metamorphic temperatures were estimated to be 740-780°C, and subsequent amphibolite facies overprint at 750°C at 1.5-1.7 GPa during decompression. UHP metamorphism occurred between 420-400 Ma (Paquette *et al.*, 1995); $^{40}Ar/^{39}Ar$ data from amphibole separates from retrogressed eclogites yielded 339 ± 4 Ma (Costa *et al.*, 1993).

The North-East Greenland Caledonides

The first evidence for UHP metamorphism in the Greenland Caledonides was reported from kyanite eclogites and associated host gneisses on an island in Jokelbugt (Gilotti & Krogh Ravna, 2002). Inclusions of quartz pseudomorph exhibiting palisade structure and radiating fractures occur in garnet and omphacite of eclogites and in garnet of the host gneisses. *P-T* estimates of peak stage metamorphism of eclogite at ~972°C and 3.6 GPa are well within the coesite stability field. SHRIMP U/Pb dates of 403 ± 5 Ma for zircon rims from HP metapsammite and 404 ± 4 Ma for zircons from anatectic melt derived from metapelite coeval with the estimated ages of exhumation of UHP terranes in the Scandinavian Caledonides described in the previous sections (McClelland & Gilotti, 2003).

Neoproterozoic nappes in SE Brazil

Neoproterozoic nappes of SE Brazil consist mainly of coarse-grained kyanite-garnet granulite with intercalations of impure quartzite, cal-silicate rocks and minor lenses and sills of mafic-ultramafic rocks, including eclogite. Early eclogitic phases are strongly obliterated during granulite-facies overprint during a major Pan-African continent collision at 640-630 Ma (Campos Neto & Caby, 1999). Rare ultramafic lenses consist of phlogopite-orthopyroxene rocks and garnet clinopyroxenite. Cores of granulite boudins retain eclogitic assemblages; some zoned garnet porphyroblasts occur quartz inclusion within the mantle regions of the garnets are surrounded by intense radial fractures, suggestive of the former presence of coesite. In fact, coarse-grained zircons in some granulites contains numerous micro-inclusions of K-feldspar + kyanite + coesite confirmed by Raman (Parkinson *et al.*, 2001). Many exsolution microstructures observed in other UHP terranes also occur. This area occurs the first coesite-grade UHP rocks in the Americas and it can be correlated with the Mali UHP terrane of South Africa (Caby, 1994). It was suggested that eclogitic assemblages and probable UHPM components may be disposed in several other correlative nappes systems at the margins of San Francisco craton in Brazil.

The Greek Rhodope

The Greek Metamorphic Province (RMP) at the border between Greece and Bulgaria represents a synmetamorphic nappe-system of Alpine age that was formed during the Cretaceous to Mid-Tertiary collision of Apulia and paleo-Europe. The Kimi complex of the intermediate nappe comprises crustal eclogites, leucocratic orthogneisses, pelitic gneisses and minor mantle-derived ultramafics; these rocks occur many mineralogical indicators for UHP metamorphism (Mposkos *et al.*, 2001). The ultramafics include garnet-spinel-bearing lherzolite and layers of spinel-garnet clinopyroxenite, garnet pyroxenite and clinopyroxene garnetite. These rocks were subjected to UHP recrystallization at P-T conditions of ~1200°C and > 3.0 GPa. The pelitic gneiss occurs microdiamond inclusions in garnet porphyroblasts as well as rods or needles of silica, rutile, biotite and apatite in sodic garnet suggesting the occurrence of majoritic garnet. The growth of such supersilisic garnet was suggested to be at P of ~7 GPa. Inclusions of polycrystalline quartz aggregates surrounded by radial cracks occur in eclogitic garnet. Exsolution lamellae of quartz in eclogitic clinopyroxene also indicate an UHP precursor of supersilicic clinopyroxene.

The Lanterman Range, Antarctica

The Lanterman Range in northern Victoria Land of Antarctica occurs several metamorphic complexes. The Gateway Hills Metamorphic Complex is a thin discontinuous belt of more than 50 km long consisting of mafic and ultramafic rocks including lenses and pods of eclogites within felsic gneisses. Some less foliated eclogites have inclusions of coesite and coesite pseudomorph in garnet porphyroblasts confirmed by *in situ* Raman spectroscopy and the occurrence of radial fractures around the inclusions (Ghiribelli *et al.*, 2002). The eclogites were subjected to UHP metamorphism at $T > 850$°C and $P > 2.9$ GPa and amphibolite-facies overprint during isothermal decompression. The UHP event was dated by a Sm/Nd and $^{238}U/^{206}Pb$ age of 500 Ma and the $^{40}Ar/^{39}Ar$ ages of Ca-amphibole of 490-486 Ma for amphibolite overprint suggesting a fast cooling with an average exhumation of 3-4 km/Ma for the UHP terrane (Di Vincenzo & Palmeri, 2001).

The Western Tianshan, China

Caledonian eclogites of the western Tianshan, China have been classified into 3 types (Zhang *et al.*, 2002). Type I eclogite pods are layered with mafic blueschists and type II eclogites preserved pillow structures, whereas Type III eclogites are banded with marbles. This eclogite-bearing belt of 200 km long in China Tianshan extends westward to connect with the Atbashy UHP belt in Kazakhstan (Tagiri *et al.*, 1995). Reported UHP evidences include inclusions of coesite pseudomorphs in garnets from Type I and III eclogites, quartz exsolution lamellae in omphacites from Type II eclogites and relict magnesite within matrix dolomite in Type III eclogites (Zhang *et al.*, 2002). *P-T* estimates of the peak UHP stage metamorphism are T = 500-600°C and P = 2.6 to 2.7 GPa (Wei *et al.*, 2003).

The Northern Qaidam Mountains, western China

Many lenses and layers of garnet peridotite and eclogite occur in an amphibolite-facies terrane in the North Qaidam Mountains along the NE rim of the Qaidam. This belt occurs mainly felsic gneiss, quartz schist, garnet-bearing gneiss, amphibolite, and minor eclogite and garnet peridotite and extends westward for more than 1000 km to the Altun Mountains and was displaced by a left-lateral Altyn-Tagh fault. At the eastern end of this UHP belt at Dulan, occurrences of coesite inclusions in zircon separates from paragneiss, inclusions of coesite pseudomorph in garnet and omphacite of eclogites, quartz rods in omphacite of eclogite, and P-T estimates of eclogites (P = 2.87-3.17 GPa, and T = 631-687°C) demonstrate that this is a typical UHP metamorphic terrane (Song *et al.*, 2003). The Dulan eclogites have Sm-Nd mineral isochron age of 497 ± 87 Ma and SHRIMP U/Pb zircon age of 495 Ma (Yang *et al.*, 2002).

The Altun Mountains, western China

This UHP terrane, about 200 km long in the southern margin of the Altun Mountains, is a western extension of the North Qaidam belt. Eclogitic lenses and blocks of various sizes together with garnet-lherzolite and clinopyroxenite are enclosed within Early Proterozoic amphibolite, amphibole schist, garnet-bearing granitic gneiss, and pelitic schist sequence. Garnet lherzolite contains magnesite that reacted to form dolomite and *P-T* estimate of the peak magnesite-bearing assemblage yield P > 3.6 GPa and 850°C (Liu *et al.*, 2001). Garnet porphyroblasts include exsolution lamellae of clinopyroxene and rutile similar to that of eclogite reported by Ye (2000) as majoritic garnet. The associated eclogites contain inclusions of coesite pseudomorphs in garnet with characteristic radial fractures; this together with the exsolved quartz rods in omphacite and P-T estimates of 820-850°C and 2.8 to 3.2 GPa provide additional evidence for UHP (Zhang *et al.*, 2001). A Sm/Nd mineral isochron for eclogitic garnet, omphacite, and whole rock yields 500 ± 10 Ma, whereas an U/Pb isochron for zircon separates from eclogite gives an age of 504 ± 5 Ma (Zhang *et al.*, 2000).

The North Qinling Mountains, central China

Lenses and blocks of eclogitic rocks in the northern Qinling are enclosed in amphibolite-facies gneiss and garnet-bearing quartz and phengitic mica schist. Inclusions of coesite in eclogitic garnet (Hu *et al.*, 1995) and microdiamonds in zircon separates from both eclogite and gneiss (Yang *et al.*, 2002) have been recently reported. While the individual outcrops are no more than a few meters wide, the belt of eclogites extends more than 10 km. P-T estimates of amphibolitized coesite-bearing eclogite yield T = 590-758°C and P > 2.6 GPa. A Sm-Nd isochron for garnet, omphacite, rutile, amphibole and whole rock yields 400 ± 16 Ma. Microdiamonds were recently discovered as inclusions in zircon from both eclogite and its gneissic country rocks; SHRIMP U/Pb dating of zircons from granitic gneiss yields 507 ± 38 Ma for peak UHP metamorphism (Yang *et al.*, 2002). Such discovery indicates that two discrete collision events occurred between the North China and Yangtze cratons: an Early Paleozoic North Qinling UHP belt and to the Triassic collision suture along the Dabie-Sulu UHP-HP belt. This Early Paleozoic belt extends westward to the North Qaidam-Altun Mountains for more than 4000 km (Yang *et al.*, 2002).

HIMALAYAN UHP ECLOGITES

The Himalayan orogen has long been considered as a classic locality of continental collision between the Indian and Eurasian continents (Fig. 5). Several eclogite-facies HP rocks in the Himalayans have been reported since late 80s (e.g. Chaudary & Ghazanfar, 1987; Pognante & Spencer, 1991; de Sigoyer *et al.*, 1997; Lombardo & Rolfo, 2000). Inclusions of coesite were recently discovered in omphacite and garnet for a few Himalayan eclogites from the Upper Kaghan Valley, Pakistan (O'Brien *et al.*, 2001; Kaneko *et al.*, 2003). Subsequent findings of additional coesite localities in the Tso Morari area have established the presence of an UHP belt in the Himalayas, an active ongoing collision since the Eocene. However, UHP relicts are volumetrically very minor constituents of the metamorphic belts dominated mostly by Barrovian assemblages. In these rocks, hydration related to Barrovian-zone metamorphism has resulted in remarkably pervasive recrystallization during the later exhumation; this event has almost entirely masked the preceding UHPM record.

The finding of HP-UHP evidence has led to new insights into the Himalayan tectonic model. England & Houseman (1986) showed that, in a model of homogeneous thickening of the Himalayan-Tibet boundary, the maximum crustal thickness should be less than 70 km. However, the occurrences of coesite-bearing eclogites and gneisses in the Himalaya cannot be explained by a homogeneous thickening of the crust, but rather by continental subduction as described below. According to geochronological data, the exhumation rate of the Himalayan UHP rocks has been estimated to be ~1 $cm \times yr^{-1}$ (O'Brien *et al.*, 2001), which lies in the same range with those calculated from other UHP-HP terranes. This requires tectonic emplacement of the UHP-HP Himalayan slab into shallow crust, following isostatic rebound due to slab decoupling between the Indian continent and the oceanic crust initiated during the Middle Eocene. UHP rocks from both the Kaghan Valley and Tso-Morari are described below.

Kaghan Valley, Pakistan

The Kaghan nappe in the Northwestern Himalayan syntaxis, lies between the Indus suture (also called main mantle thrust: MMT) to the north and the main central thrust (MCT) to the southwest (Pognante & Spencer, 1991; Spencer, 1993) (Fig. 5). The Cretaceous Kokistan island arc sequence lies north of the Indus suture. It developed as a result of intra-oceanic subduction. The main central thrust separates an underlying Salkhala unit from an overlying higher Himalayan crystalline (HHC) unit. The Kaghan nappe consists mainly of granitic gneiss and paragneiss, with minor intercalated layers of amphibolite, marble and quartzite. Eclogites are preserved in the cores of mafic boudins (less than 1 m thick) in gneisses, and are significantly overprinted by amphibolite assemblages. They contain

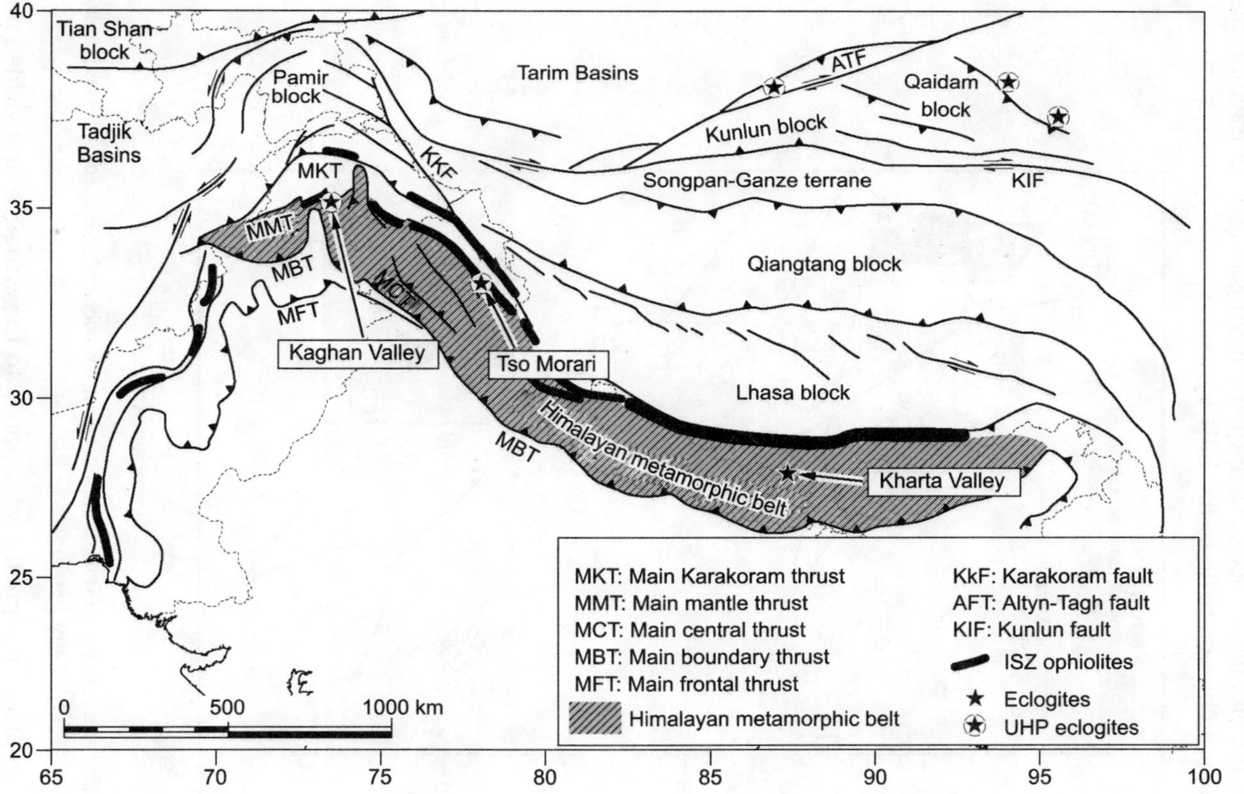

Fig. 5. A simplified tectonic map of the western Himalaya, Hindu Kush, Pamir, Karakoram, and western Tibet modified after Fig. 3 of Searle *et al.* (2001). This map shows the locations of eclogite and UHP eclogites from both Himalayan and North Qaidam, Indus Suture Zone (ISZ) ophiolites and major structures.

fine-grained (< 1 mm) garnet and omphacite, together with aggregates of quartz, chains of rutile rimmed by titanite, and mm-sized phengitic micas overgrown by randomly oriented amphiboles upto 1 cm in length. Coesite and coesite pseudomorphs occur as inclusions in omphacite, and show the characteristic palisade-quartz aggregate texture; fractures radiating from these inclusions are characteristic in the host pyroxene.

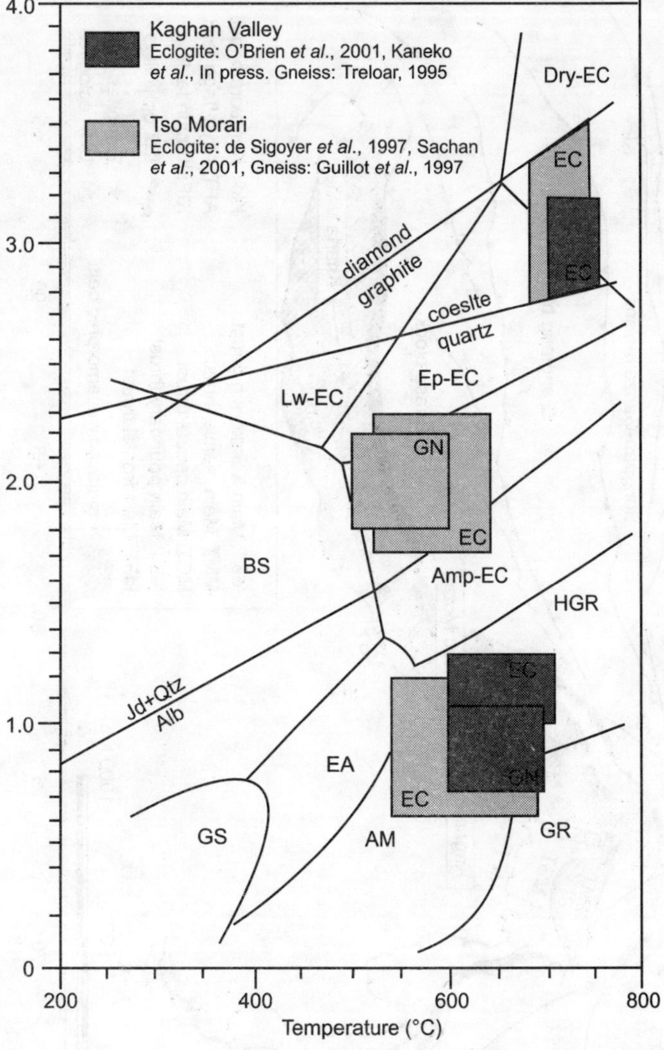

Fig. 6. P-T estimates of peak- and retrograde stage metamorphism for coesite-bearing eclogites and their country rock gneisses from the Kaghan Valley and Tso Morari.

Garnet-pyroxene-phengite barometry was applied to the eclogites, and yielded peak *P-T* conditions of 2.7-2.9 GPa and 690-750°C (O'Brien *et al.*, 2001) (Fig. 6). Retrograde textures are common in the eclogites, including symplectic intergrowths of augite, amphibole and plagioclase after omphacite; thermobarometric analyses of these assemblages yielded 1.0-1.3 GPa and 600-710°C (Kaneko *et al.*, 2003). On the other hand, the dominant granitic gneiss and metapelites of the higher Himalayan crystalline unit contain typical Barrovian-zone amphibolite-facies assemblages, represented by kyanite- and staurolite-bearing garnetiferous metapelites (Treloar, 1995). Extensive overprinting has mostly obliterated the peak assemblages; matrix minerals of the gneisses yielded *P-T* conditions of 0.7-1.1 GPa and 600-700°C (Treloar, 1995), consistent with the retrograde conditions of the eclogites. However, zircon preserved relict coesite as inclusions, which are ovoid and approximately 10 μm in diameter (Fig. 7A). This discovery of coesite-bearing zircons from the Himalayan quartzofeldspathic gneisses demonstrates the subduction of Indian continental crust to the depths of UHP conditions (Kaneko *et al.*, 2003).

Geochronological studies using Sm/Nd (garnet-omphacite), Rb/Sr (phengite), and U/Pb (rutile, zircon) methods suggest that the eclogite facies event took place at 40-50 Ma (Tonarini *et al.*, 1993; Spencer & Gebauer, 1996). $^{40}Ar/^{39}Ar$ cooling ages for hornblende and mica from gneisses were reported to be ca. 43 Ma and ca, 25 Ma, respectively (Chamberlain *et al.*, 1991). Recent SHRIMP U/Pb dates of zoned zircons with mineral inclusions formed at different stages indicate the ages of non-UHP mineral-bearing mantle domains of zircon and UHP mineral-bearing rims, are at about 50 and 46 Ma, respectively (Kaneko *et al.*, 2003).

Tso Morari

The Tso Morari crystalline dome is located between the Indus suture zone on the north bordered by the Zildat detachment fault, and the Zanskar sedimentary unit on the south (de Sigoyer *et al.*, 1997). The dome represents an internal massif of the northern part of the higher Himalaya. The Tso Morari area has been considered as a typical Indian Tethyan margin because of a stratigraphic succession similar to the Nimaling antiform in the core of the dome, a metamorphic basement consists of Cambro-Ordovician augen gneisses (Stutz, 1988). This complex is covered by a metasedimentary series of Cambrian to Devonian quartzite, schist and conglomerate, overlain by Lower Carboniferous to Triassic marble and metapelite. Eclogites occur as lenses or irregular bodies within the basement and cover gneisses, and are strongly overprinted by amphibolite-facies assemblages. de Sigoyer *et al.* (1997) reported peak *P-T* conditions of the eclogites at 1.7-2.3 GPa and 520-640°C (Fig. 7). However, relics of coesite 30-80 μm in size were recently found as inclusions in garnet (Sachan *et al.*, 2001) typified by radial fractures (Fig. 7B,C); for the SiO_2 inclusions, a characteristic Raman spectrum is centred at 523 cm^{-1} (Fig. 7D). The country rock metasediments with jadeite + chloritoid + paragonite + garnet assemblages have peak *P-T* estimates of 1.8-2.2 GPa and 500-600°C, and record retrogression at the eclogite-blueschist facies transition of 1.3-1.8 GPa and 490-590°C (Guillot *et al.*, 1997).

The geochronology of the Tso Morari eclogites has been extensively studied and subdivided into three different timings (de Signoyer *et al.*, 2000). Eclogitization at ca. 55 Ma was obtained by Lu-Hf and Sm-Nd methods on garnet, omphacite, glaucophane and the whole rock. An amphibolite-facies overprint occurred at 47 Ma, judging by Rb/Sr dating on phengite, apatite and whole rock from metapelites. $^{40}Ar/^{39}Ar$ ages of biotite and muscovite at ca. 30 Ma suggest that the Tso Morari unit rose to upper crustal levels and recrystallized at the end of the exhumation (de Signoyer *et al.*, 2000). SHRIMP U/Pb dating of zircons from the Tso Morari has not been completed; available data indicate that the peak metamorphic age from Tso Morari eclogites is almost 10 Ma older than the 46 Ma U/Pb age for Kaghan eclogites (Kaneko *et al.*, 2003). Nevertheless, similarity in the tectonic setting and lithologies suggest that they probably belong to a single UHP belt.

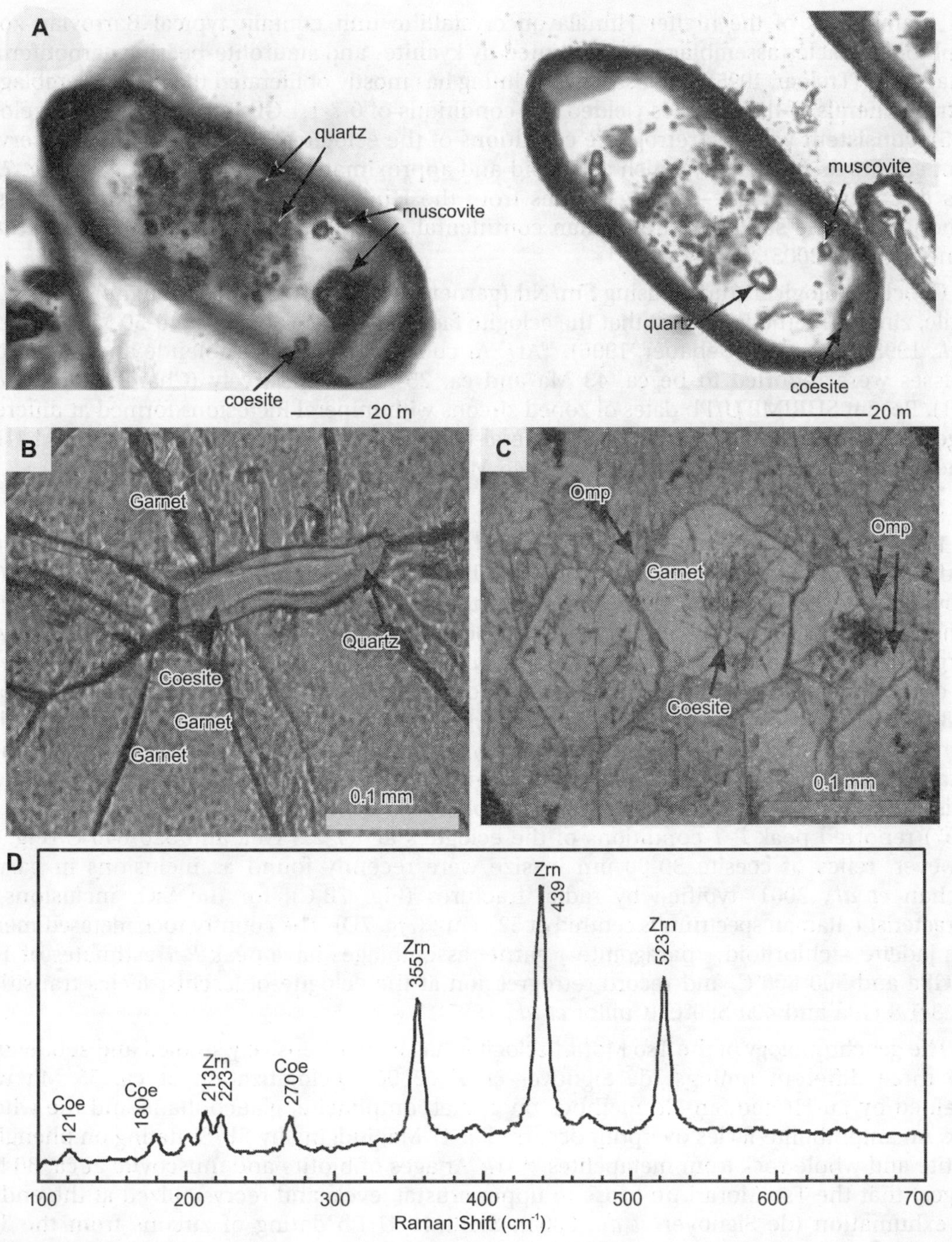

Fig. 7. (A) Inclusions of low-P phases (quartz & muscovite) at core and coesite at rim from zircon separates of gneissic rocks from Kaghan Valley (Kaneko *et al.*, 2003); (B) Photomicrograph of coesite inclusion in garnet from Tso Morari eclogite showing thin palisade quartz around coesite and radial fractures in host garnet (from Fig. 2 of Mukherjee *et al.*, 2002); (C) Inclusion of coesite in eclogitic garnet from Tso Morari (from Fig. 2 of Mukherjee *et al.*, 2002); (D) Laser Raman spectra of coesite inclusion in zircon separates of Fig. H-3A.

CONTINENT SUBDUCTION AND COLLISION (HIMALAYAN MODEL)

Continental subduction and UHP metamorphism at 45 to 52 Ma

The Himalayan orogeny was controlled principally by subduction of the Indian continental crust beneath Asian continental lithosphere; this may provide a present-day analog for UHP metamorphism. So far, numerous tectonic models for the Himalayan orogeny have been proposed (for review see Hodges, 2000); most contributions have focused on the dynamics of unusually crustal thickening (upto 75 km) that resulted in a regional Barrovian-zone sequence. In general, the closure of Neo-Tethys to form the Indus suture was followed by India-Asia continental collision at 52-50 Ma (Fig. 8). Both Asian and Indian margins were thickened and metamorphosed, and the intra-oceanic convergence generated a continental suture (e.g. Rowley, 1996). During the infant stage of the Himalayan orogeny (~80-50 Ma), a Pacific-type arc-trench system occurred, and huge, multiple volcanic/plutonic calc-alkaline arc formed the Kohistan-Ladakh and Trans-Himalayan batholiths. The Pacific-type subduction formed a Cretaceous accretionary complex at the southern margin of the Kohistan-Ladakh arc; the accretionary complex contains 80 Ma blueschists (e.g. Anczkiewicz *et al.*, 2000). Granitic arc-magmatism continued until 50 Ma at Ladakh (Weinberg & Dunlap, 2000), when collision was initiated at the western syntax and propagated to the SE. Since then, subduction of the Indian continent occurred, and the India-Asia convergence rate drastically decreased from 19 to 5 $cm.yr^{-1}$ at about middle Eocene time (e.g. Klootwijk *et al.*, 1992). The continental collision closed the intervening ocean and terminated the arc magmatism (Fig. 8); hence 45 Ma lacustrine deposits of the Indian continent contain detritus derived from the Asian continent.

Recent seismologic studies of the ongoing Himalayan orogeny have provided important constraints on its tectonic evolution. In the western Himalaya, the Hindu Kush region is seismically most active, with abundant intermediate-depth seismicity (Regler, 1995; Pegler & Das, 1998). The seismic zone extends to a depth of 100-250 km, and steepens with increasing depth. A seismic reflection profile through the region shows the presence of subducted continental crust and a trapped oceanic basin at that depth. On the basis of the available seismic data, Searle *et al.* (2001) proposed a continental subduction model for the formation of UHP metamorphic rocks beneath the Hindu Kush region. Their model suggests that a cold lower crust of the Indian block comprising Precambrian granulite-facies basement and a cover sequence of Paleozoic-lower Mesozoic supracontinental sediments was subducting beneath the Hindu Kush; subsequently the crustal rocks underwent UHP metamorphism at about 150 km depth and ca. 800-900°C at 4-6 GPa. On the other hand, Kaneko *et al.* (2003) obtained a sinking rate of 1.2-1.6 $cm.yr^{-1}$ for the downgoing Indian crust to ca. 110 km depth, based on SHRIMP U/Pb ages of zoned zircon in UHP rocks from the Kaghan Valley. These data also suggest a time-integrated subduction angle of 18-22°, similar to the estimated angle of the present Moho between the Himalaya foreland and southern edge of the Tethyan Himalaya (Zhao & Nelson, 1993). The timing of UHP metamorphism at 46 Ma (Kaneko *et al.*, 2003) is synchronous with that of the rapid cooling (Weinberg & Dunlap, 2000) of the youngest plutons of the Ladakh batholith.

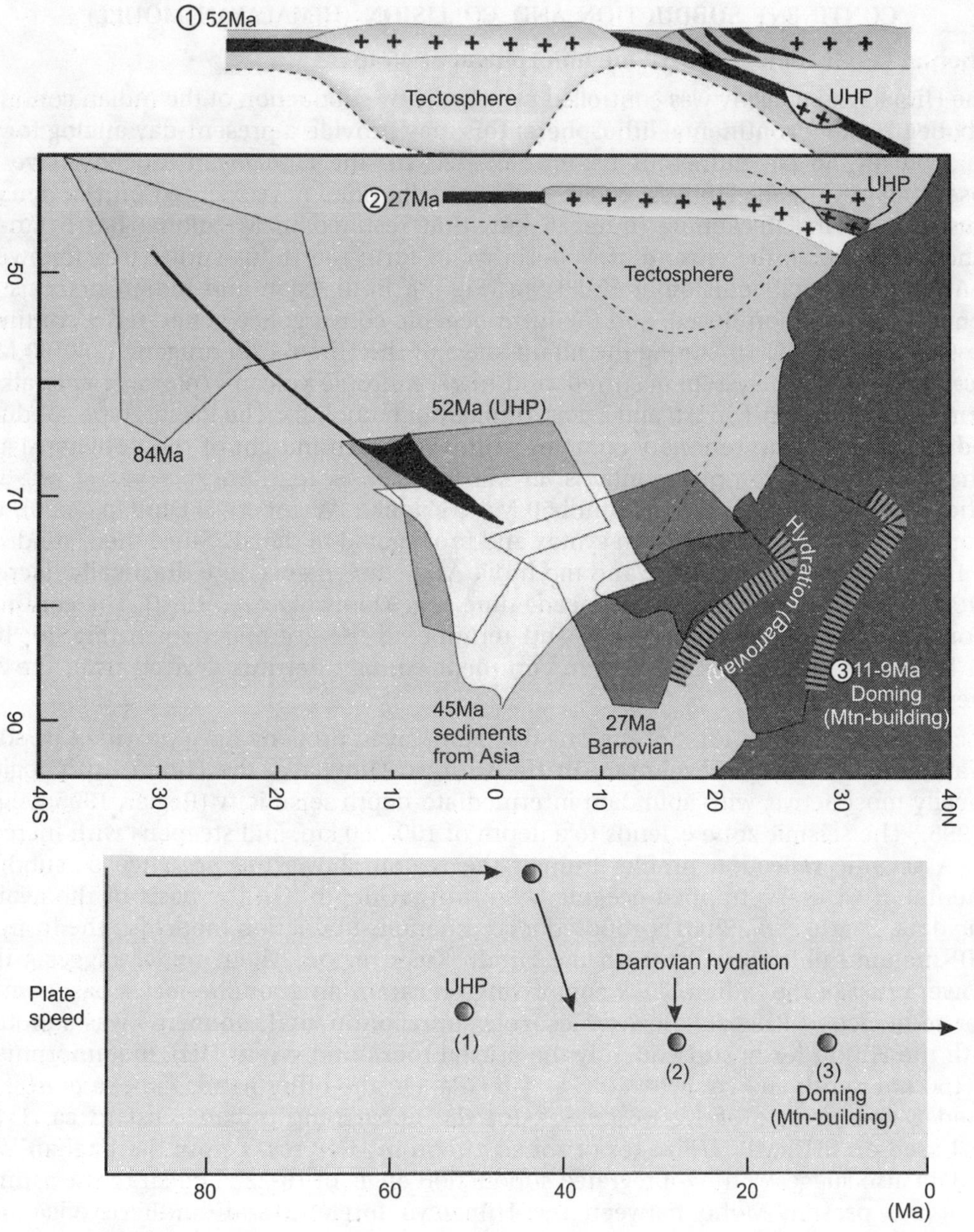

Fig. 8. Subduction and collision of the Indian and Eurasian continents showing the subduction of microcontinent for formation of UHP rocks at 52 Ma (1) and exhumation of UHP rocks from the Himalayan metamorphic belt. Exhumation of the UHP slab includes (2) wedge extrusion from mantle depth to crustal level at 27 Ma and (3) doming event at 11-9 Ma. Paleogeographic positions of the Indian continent at 84, 52, 45, 27 and 11-9 Ma are shown together with the rate estimates of northeastward motion of the Indian plate. The doming event was responsible for the mountain-building processes for high Himalayan mountain chain.

Barrovian Metamorphic Overprint at 30-15 Ma

Himalayan orogeny involves crustal thickening that resulted in a regional Barrovian-type kyanite-sillimanite series metamorphism. This metamorphism is well-documented in metapelites of the greater Himalayan series (including the Zanskar Himalaya series) and the north Himalayan gneiss domes. Most Oligocene kyanite-grade rocks were metamorphosed at P = 0.6-1.0 GPa) and T = 450-600°C, and later Early Miocene sillimanite-grade rocks were formed at higher T (650-770°C) but lower P (0.4-0.7 GPa). The U/Pb and Sm/Nd geochronology in the Zansker and Lahoul region yields ca. 37-29 Ma for regional Barrovian-type kyanite-grade metamorphism (e.g. Vance & Harris, 1999; Walker *et al.*, 1999). The $^{40}Ar/^{39}Ar$ cooling ages of muscovite in the metapelites yield 22-21 Ma (Walker *et al.*, 1999) and overlap those in the Himalayan leucogranite (24-17 Ma: Searle *et al.*, 1988). Monazite from sillimanite-bearing metapelites of the greater Himalayan series in the Everest region has U/Pb ages of 32 Ma (Simpson *et al.*, 2000); the monazite U/Pb ages of 23 Ma from the sillimanite-cordierite-bearing granitic gneiss represent timing of cordierite-grade low-P metamorphism related to crustal melting.

The kyanite- and staurolite-bearing mineral assemblage occurs as retrograde phase of Tso Morari eclogitic metapelites (Guillot, 1997) in the largest gneiss dome of the Himalayan chain. $^{40}Ar/^{39}Ar$ plateau ages of 31-29 Ma from low-Si muscovite and biotite of retrograded metapelites represent thermal relaxation at upper crustal levels (de Sigyor *et al.*, 2000). A series of gneiss dome (north Himalayan gneiss domes) that are mantled by Barrovian-type kyanite-grade metamorphic rocks can be traced for at least 2000 km; their cooling age is similar to the Barrovian overprint of the Tso Morari UHP eclogites.

Domal Uplift and Mountain-Building since 11 Ma

The uplift of the higher Himalayas apparently occurred in Late Miocene, as documented by ODP Leg 116 and on-land geology. Drilling of the distal Bengal fan in the central Indian Ocean revealed that the sedimentation rate drastically increased at 10.9 Ma (Amano & Taira, 1992). Paleontological evidence suggests that a glacial paleoclimate began in the Higher Himalayas at about 7 Ma (Lakhanpal *et al.*, 1983). The doming event at 15-11 Ma is also suggested by mica $^{40}Ar/^{39}Ar$ thermochronology of the Kangmar Dome (Maluski *et al.*, 1988; Lee *et al.*, 2001). This doming event raised the Himalayan Mountain chain more than 5 km in elevation; active erosion has continued and thick flesh sequences have accumulated in the Bengal fan. The major driving force for the domal uplift is not well understood. The Miocene underthrusting of the Indian tectosphere beneath the Lesser Himalayas may have elevated the Higher Himalaya.

These data clearly show a two-step exhumation process for the Himalayan UHP and high-grade metamorphic terrane. (1) Emplacement of the Himalayan high-grade metamorphic rocks at shallow crustal levels occurred at > 20 to 30 Ma, when no mountain building was apparent, and (2) Himalayan mountain building initiated at 10.9 Ma, when extensive surface erosion commenced. Fig. 8 schematically shows such events together with the paleogeography of the Indian and Eurasian continents since 80 Ma.

ACKNOWLEDGMENTS

This manuscript was initiated in 2000-01 while Maruyama held a sabbatical position as a Cox Professor at Stanford; both T. Tsujimori and I. Katayama were supported by JSPS Research Fellowship for Young Scientists, and Grant-in-Aid for JSPS Fellows. It represents a product of the U.S.-Japan project supported by the NSF Global Partnership Fellowship INT-9820171, and the NSF US-China project EAR 0003355. We thank Y. Kaneko and Himanshu Sachan for preprints and photomicrograph of coesite inclusions in UHP rocks respectively from the Kaghan Valley of Pakistan and Tso Morari. We appreciate a constructive review of the manuscript by Gary Ernst.

REFERENCES

Amano, K. & Taira, A., 1992. Two-phase uplift of Higher Himalayas since 17 Ma. *Geology*, **20**, 391-94.

Bakun-Czubarow, N., 1991. On the possibility of occurrence of quartz pseudomorphs after coesite in the eclogite-granulite rock series of the Zlote Mountains in the Sudetes, SW Poland. *Archiwum Mineralogy*, **48**, 3-25.

Beane, R.J. & Connelly, J.N., 1998. $^{40}Ar/^{39}Ar$, U-Pb, and Sm-Nd constraints on the timing of metamorphic events in the Maksyutov Complex, southern Ural Mountains. *Journal of Geological Society*, London, **157**, 811-22.

Beane, R.J., Liou, J.G., Coleman, R.G. & Leech, M.L., 1995. Petrology and retrograde P-T path for eclogites of the Maksyutov complex, southern Ural Mountains, Russia. *The Island Arc*, **4**, 254-66.

Bowtell, S.A., Cliff, R.A. & Barnicoat, A.C., 1994. Sm-Nd isotopic evidence on the age of eclogitization in the Zermatt-Saas ophiolite. *Journal of Metamorphic Geology*, **12**, 187-96.

Brueckner, H.K. & Medaris, L.G. Jr., 1998. A tale of two orogens: the contrasting T-P-t history and geochemical evolution of mantle in high- and ultrahigh-pressure metamorphic terranes of the Norwegian Caledonides and the Czech Variscides. *Schweizerische Mineralogische und Petrographische Mitteilungen*, **78**, 293-307.

Caby, R., 1994. Precambrian coesite from N Mali: first record and implications for plate tectonics in the trans-Saharan segment of the Pan-African belt. *European Journal of Mineralogy*, **6**, 235-44.

Carswall, D.A., Cuthbert, S.J. & E.J. Krogh Ravna, 1999. Ultrahigh-pressure metamorphism in the western Gneiss Region of the Norwegian caledonides. *International Geology Review*, **41**, 955-66.

Carswell, D.A., Brueckner, H.K., Cuthbert, S.J., Mehta, K. & O'Brien, P.J., 2003. The timing of stabilisation and the exhumation rate for ultra-high pressure rocks in the Western *Journal of Metamorphic Geology*.

Campos Neto, M. da C. & Caby, R., 1999. Tectonic constraint on Neoproterozoic high-pressure metamorphism and nappe system south of the San Francisco craton, southeast Brazil. *Precambrian Research*, **97**, 3-26.

Chamberlain, C.P., Zeitler, P.K. & Erickson, E., 1991. Constraints on the tectonic evolution of the northwestern Himalaya from geochronologic and petrologic studies of Babusar Pass, Pakistan. *Journal of Geology*, **99**, 829-49.

Chaudary, M.N. & Ghazanfar, M., 1987. Geology, structure and geomorphology of upper Kaghan valley, Northwestern Himalaya, Pakistan. *Geological Bulletin University of Punjab*, **22**, 13-57.

Chesnokov, B.V. & Popov, V.A., 1965. Increasing volume of quartz grains in eclogites of the south Urals. *Doklady Akademii Nauk*, **162**, 176-78 (in Russian).

Chopin, C., 1984. Coesite and pure pyrope in high-grade blueschists of the Western Alps: a first record and some consequences. *Contributions to Mineralogy and Petrology*, **86**, 107-18.

Coleman, R.G. & Wang, X., 1995. Ultrahigh Pressure Metamorphism. *Cambridge University Press.*

Costa, S., Maluski, A. & Lardeaux, J.M., 1993. $^{40-39}Ar$ chronology of Variscan tectono-metamorphic events in an exhumed crustal nappe: the Monts du Lyonnais complex (Massif Central France). *Chemical Geology*, **105**, 339-59.

Cuthbert, S.J., Carswell, D.A., Krogh-Ravna, E.J. & Wain, A., 2000. Eclogites and eclogites in the Western Gneiss Region, Norwegian Caledonides. *Lithos*, **52**, 165-95.

de Sigoyer, J., Chavagnac, V., Blichert-Toft, J., Villa, I.M., Luais, B., Guillot, S., Cosca, M. & Mascle, G., 2000. Dating the Indian continental subduction and collisional thickening in the northwest Himalaya: Multichronology of the Tso Morari eclogites. *Geology*, **28**, 487-90.

de Sigoyer, J., Guillot, S., Laudeaux, J.M. & Mascle, G., 1997. Glaucophane-bearing eclogites in the Tso Morari dome, eastern Ladakh, NW Himalaya. *European Journal of Mineralogy*, **9**, 1073-83.

Di Vincenzo, G. & Palmeri, R., 2001. An ^{40}Ar-^{39}Ar investigation of high-pressure metamorphism and the retrogressive history of mafic eclogites from the Lanterman Range, Antarctica: evidence against a simple temperature control on argon transport in amphibole. *Contributions to Mineralogy and Petrology*, **141**, 15-35.

Dobretsov, N.L. & Dobretsova, L.V., 1988. New mineralogic data on the Maksyutov eclogite-glaucophane schist complex, southern Urals. *Doklady Akademii Nauk*, **300**, 111-16.

Dobretsov, N.L., Shatsky, V.S., Coleman, R.G., Lennykh, V.I., Valizer, P.M., Liou, J.G., Zhang, R.Y. & Beane, R.J., 1996. Tectonic setting and petrology of ultrahigh-pressure metamorphic rocks in the Maksyutov Complex, Ural Mountains, Russia. *International Geology Review*, **38**, 136-60.

Dobrzhinetskaya, L.F., Eide, E.A., Larsen, R.B., Sturt, B.A., Tronnes, R.G., Smith, D.C., Taylor, W.R. & Posukhova, T.V., 1995. Microdiamond in high-grade metamorphic rocks of the Western Gneiss region, Norway. *Geology*, **23**, 597-600.

Dobrzhinetskaya, L.F., Green, H.W., Mitchell, T.E. & Dickerson, R.M., 2001. Metamorphic diamonds: Mechanism of growth and inclusion of oxides. *Geology*, **29**, 263-66.

England, P.C. & Houseman, G.A., 1986. Finite strain calculations of continental deformation. II. Application to the India-Asia plate collision. *Journal of Geophysical Research*, **91**, 3664-76.

Ernst, W.G. & Liou, J.G., 2000. Ultrahigh-Pressure metamorphism and Geodynamics in Collision-type Orogenic Belts. *International Book Series, Geological Society of America*, **4**, 293.

Gardien, V., Tegyey, M., Lardeaux, J.M., Misseri, M. & Dufour, E., 1990. Crustal-mantle relationship in French Variscan chain: the example of the southern Monts du Lyonnais unit (eastern French Massif Central). *Journal of Metamorphic Geology*, **8**, 477-92.

Gebauer, D., Schertl, H.P., Brix, M. & Schreyer, W., 1997. 35 Ma old ultrahigh-pressure metamorphism and evidence for very rapid exhumation in the Dora Maira Massif, western Alps. *Lithos*, **41**, 5-24.

Ghiribelli, B., Frezzotti, M.-L. & Palmeri, R., 2002. Coesite in eclogites of the Lanterman range, Antarctica: Evidence from textural and Raman studies. *European Journal of Mineralogy*, **14**, 355-60.

Gilotti, J.A. & Ravna, E.J.K., 2002. First evidence for ultrahigh-pressure metamorphism in the Northeast Greenland Caledonides. *Geology*, **30**, 551-54.

Gilotti, J.A., 1993. Discovery of a medium-temperature eclogite province in the Caledonides of North-East Greenland. *Geology*, **21**, 523-26.

Guillot, S., de Sigoyer, J., Lardeaux, J.M. & Mascle, G., 1997. Eclogitic metasediments from the Tso Morari area, Ladakh, Himalaya. Evidence for continental subduction during India-Asia convergence. *Contributions to Mineralogy and Petorology*, **128**, 197-212.

Hacker, B.R., Ratschbacher, L., Webb, L., McWilliams, M.O., Ireland, T., Calvert, A., Shuwen, D., Wenk, H.-R. & Chateigner, D., 2000. Exhumation of ultrahigh-pressure continental crust in East-central China: Late Triassic-Early Jurassic tectonic unroofing. *Journal of Geophysical Research*, **105**, 13339-64.

Hodges, K.V., 2000. Tectonics of the Himalaya and southern Tibet from two perspectives. *Geological Society of America, Bulletin*, **112**, 324-50.

Hu, N., Zhao, D., Xu, B. & Wang, T., 1995. Petrography and metamorphism study on high-ultrahigh pressure eclogite from Guanpo area, northern Qinling Mountain. *Journal of Mineralogy and Petrology*, **15**, 1-9 (in Chinese).

Hwang, S.L., Shen, P., Chu, H.T., Yui, T.F. & Lin, C.C., 2001. Genesis of microdiamonds from melt and associated multiphase inclusions in garnet of ultrahigh-pressure gneiss from Erzegebirge, Germany. *Earth and Planetary Science Letters*, **188**, 9-15.

Jahn, B.M., Caby, R. & Monie, P., 2001. The oldest UHP eclogites of the World: age of UHP metamorphism, nature of protoliths and tectonic implications. *Chemical Geology*, **178**, 143-58.

Kaneko, Y., Maruyama, S., Terabayashi, M., Yamamoto, H., Ishikawa, M., Anma, R., Parkinson, C.D., Ota, T., Nakajima, Y., Katayama, I., Yamamoto, J. & Yamauchi, K., 2000. Geology of the Kokchetav UHP-HP metamorphic belt, Northern Kazakhstan. *The Island Arc*, **9**, 264-83.

Kaneko, Y., Katayama, I., Yamamoto, H., Misawa, K., Ishikawa, M., Rehman, H.U., Kausar, A.B. & Shiraishi, K., 2003. Timing of Himalayan ultrahigh-pressure metamorphism: sinking rate and subduction angle of the Indian continental crust beneath Asia. *Journal of Metamorphic Geology*, **21**, in press.

Kardarusman, A. & Parkinson, C.D., 2000. Petrology and P-T evolution of garnet peridotites from central Silawesi, Indonesia. *Journal of Metamorphic Geology*, **18**, 193-209.

Katayama, I., Maruyama, S., Parkinson, C.D., Terada, K. & Sano, Y., 2001. Ion micro-probe U-Pb zircon geochronology of peak and retrograde stages of ultrahigh-pressure metamorphic rocks from the Kokchetav massif, northern. *Earth and Planetary Science Letter*, **188**, 185-98.

Krogh, E.J. & Carswell, D.A., 1995. HP and UHP eclogites and garnet peridotites in the Scandinavian Caledonides. *In Ultrahigh Pressure Metamorphism*, ed. R.G. Coleman and X. Wang. Cambridge University Press, Cambridge, 244-98.

Klootwijk, C., Gee, J., Peirce, J., Smith, G. & McFadden, P., 1992. An early India-Asia contact: Paleomagnetic constraints from the Ninetyeast Ridge, ODP Leg 121, *Geology*, **20**, 395-98.

Lakhanpal, R.N., Sah, S.C.D., Kewal, K., Sharma, K.K. & Guleria, J.S., 1983. Occurrence of Livistona in the Hemis conglomerate horizon of Ladakh. In Thakur, V.C. and Sharma, K.K., eds., *Geology of Indus Suture zone of Ladakh:* Hindustan Book Publ. Corp., 179-85.

Lardeaux, J.M., Ledru, P., Daniel, I. & Duchene, S., 2001. The Variscan French Massif Central—a new addition to the ultrahigh pressure metamorphic 'club': exhumation processes and geodynamic consequences. *Tectonophysics*, **332**, 143-67.

Lee, J., Hacker, B.R., Dinklage, W.S., Gans, P.B., Calvert, A., Wang, Y., Wan, J. & Chen, W., 2001. Evolution of the Kangmar Dome, southern Tibet: Structural, petrologic, and thermochronologic constraints. *Tectonics*, **19**, 872-95.

Leech, M.L. & Ernst, W.G., 1998. Graphite pseudomorphs after diamond? A carbon isotope and spectroscopic study of graphite cuboids from the Maksyutov complex, south Ural Mountains, Russia. *Geochimica et Cosmochimica Acta*, **62**, 2143-54.

Li, S., Chen, Y., Cong, B., Zhang, Z., Zhang, R., Liou, D., Hart, S.R. & Ge, N., 1993. Collision of the North China and Yangtze Blocks and formation of coesite-bearing eclogites: Timing and processes. *Chemical Geology*, **109**, 70-89.

Liou, J.G., Zhang, R.Y. Katayama, I., Maruyama, S. & Ernst, W.G., 2002. Petrotectonic characterization of the Kokchetav Massif and the Dabie-Sulu terranes – Ultrahigh-P metamorphism in the so-called P-T Forbidden-Zone. *Western Pacific Earth Sciences*, **2**, 119-48.

Liou, J.G., 1999. Petrotectonic summary of less-intensively studied UHP regions. *International Geology Review*, **41**, 571-86.

Liou, J.G. & Zhang, R.Y., 2002. Ultrahigh-pressure metamorphic rocks. *Encyclopedia of Physical Sciences and Technology*, Third Edition, **17**, 227-44, Academia Press.

Liu, F., Xu, Z., Katayama, I., Yang, J.S., Maruyama, S. & Liou, J.G., 2001. Mineral inclusions in zircons of para- and orthogneiss from pre-pilot drillhole CCSD-PP1, Chinese Continental Scientific Drilling Project. *Lithos*, **59**, 199-215.

Lombardo, B. & Rolfo, F., 2000. Two contrasting eclogite types in the Himalaya: implications for the Himalayan orogeny. *Journal of Geodynamics*, **30**, 37-60.

Maluski, H., Matte, P. & Brunel, M., 1988. Argon 39-argon 40 dating of metamorphic and plutonic events in the North and High Himalaya belts (southern Tibet-China). *Tectonics*, **7**, 299-326.

Maruyama, S., Liou, J.G. & Terabayashi, M., 1996. Blueschists and eclogites of the world, and their exhumation. *International Geology Review*, **38**, 485-594.

Masago, H., Rumble, D., Ernst, W.G., Parkinson, C. & Maruyama, S., 2003. O^{18} depletion in eclogites from the Kokchetav massif, northern Kazakhstan. *Journal of Metamorphic Geology*, **21**, in press.

Massonne, H.J., 1999. A new occurrence of microdiamonds in quartzofeldspathic rocks of the Saxonian Erzgebirge, Germany, and their metamorphic evolution. *Proceedings of 7th International Kimberlite Conference*, Capetown, **2**, 533-39.

Massonne, H.J., 2001. First find of coesite in the ultrahigh-pressure metamorphic region of the Central Erzgebirge, Germany. *European Journal of Mineralogy*, **13**, 565-70.

Matte, O., 1986. Tectonics and plate tectonics model for the Variscan belt of Europe. *Tectonophysics*, **126**, 329-74.

Matte, P., 1998. Continental subduction and exhumation of HP rocks in Paleozoic orogenic belts: Uralides and Variscides. GFF, **20**, 209-22.

McClelland, W.C. & Gilotti, J.A., 2003. Late-stage extensional exhumation of high-pressure granulites in the Greenland Caledonides. *Geology*, **31**, 259-62.

Mposkos, E.D. & Kostopoulos, D.K., 2001. Diamond, former coesite, and supersilicic garnet in metasedimentary rocks from the Greek Rhodope: a new ultrahigh-pressure metamorphic province established. *Earth and Planetary Science Letters*, **192**, 497-506.

O'Brien, P.J., Zotov, N., Law, R., Khan, M.A. & Jan, M.Q., 2001. Coesite in Himalayan eclogite and implications for models of India-Asia collision. *Geology*, **29**, 435-38.

Ogasawara, Y., Fukasawa, K. & Maruyama, S., 2002. Coesite exsolution from supersilicic titanite in UHP marble from the Kokchetav Massif, northern Kazakhstan. *American Mineralogist*, **87**, 454-61.

Ogasawara, Y., Ohta, M., Fukasawa, K., Katayama, I. & Maruyama, S., 2000. Diamond-bearing and diamond-free metacarbonate rocks from Kumdy-Kol in the Kokchetav massif, northern Kazakhstan. *The Island Arc*, **9**, 400-516.

Okamoto, K. & Maruyama, S., 1999. The high-pressure synthesis of lawsonite in the MORB + H_2O system. *American Mineralogist*, **84**, 362-73.

Paquette, J.L., Monchoux, P. & Couturier, M., 1995. Geochemical and isotopic study of anorite-eclogite transition in the European Variscan Belt. Implications for U-Pb zircon systematics in metabasic rocks. *Geochimica Cosmochimica Acta*, **59**, 1611-22.

Parkinson, C.D., 2003. Coesite-bearing quartzites from Sulawesi, Indonesia: a first record of UHP metamorphism in SE Asia. *Tectonophysics*, in press.

Parkinson, C.D., Katayama, I., Liou, J.G. & Maruyama, S. eds., 2002. *The diamond-bearing Kokchetav Massif, Kazakhstan: Petrochemistry and tectonic evolution of an unique ultrahigh-pressure metamorphic terrane.* Universal Academy Press, Inc. Tokyo, Japan, 527.

Parkinson, C.D., Motoki, A., Onishi, C.T. & Maruyama, S., 2001. Ultrahigh-pressure pyrope-kyanite granulites and associated eclogites in Neoproterozoic Nappes of Southeast Brazil. *UHPM Workshop 2001*, Waseda University, 87-90.

Pegler, G. & Das, S., 1998. An enhanced image of the Pamir-Hindu Kush seismic zone from relocate dearth-quake hypocentres. *Geophysical Journal International*, **134**, 573-95.

Pognante, U. & Spear, D.A., 1991. First record of eclogites from the High Himalayan belt, Kaghan Valley (northern Pakistan). *European Journal of Mineralogy*, **3**, 613-18.

Reinecke, T., 1991. Very-high-pressure metamorphism and uplift of coesite bearing metasediments from the Zermatt-Saas zone, Western Alps. *European Journal of Mineralogy*, **3**, 7-17.

Rowley, D.Y., 1996. Age of initiation of collision between India and Asia: A review of stratigraphic data. *Earth and Planetary Science Letters*, **145**, 1-13.

Sachan, H., Mucherjee, B.K., Ishida, H. Muko, A., Yoshioka, N., Ogasawara, Y. & Maruyama, S., 2001. New discovery of coesite from the Indian Himalaya. In: *Fluid/slab/mantle interactions and ultrahigh-P minerals: UHPM Workshop*, 124-28, Waseda, Japan.

Searle, M.P., Cooper, D.J.W. & Rex, A.J., 1988. Collision tectonics of the Ladakh-Zanskar Himalaya. In Shackleton, R.M., Dewey, J.F., and Windley, B.F.,eds., *Tectonic evolution of the Himalayas and Tibet.* London, The Royal Society, 117-49.

Searle, M., Hacker, B.R. & Bilham, R., 2001. The Hindu Kush seismic zone as a paradigm for the creation of ultrahigh-pressure diamond and coesite-bearing rocks. *Journal of Geology*, **109**, 143-54.

Schmadicke, E., Mezger, K., Cosca, M.A. & Okrusch, M., 1995. Variscan Sm-Nd and Ar-Ar ages of eclogite facies rocks from the Erzgebirge, Bohemian Massif. *Journal Metamorphic Geology*, **13**, 537-52.

Schmadicke, E., 1991. Quartz pseudomorphs after coesite in eclogites from the Saxonian Erzgebirge. *European Journal of Mineralogy*, **3**, 231-38.

Schmidt, M.W. & Poli, S., 1998. Experimentally based water budgets for dehydrating slabs and consequences for arc magma generation. *Earth and Planetary Science Letter*, **163**, 361-79.

Shatsky, V.S., Sobolev, N.V. & Vavilov, M.A., 1995. Diamond-bearing metamorphic rocks of the Kokchetav massif, N Kazakhstan. In: *Ultrahigh Pressure Metamorphism*, ed. R.G. Coleman and X. Wang. Cambridge University Press, Cambridge, 427-55.

Smith, D.C., 1984. Coesite in clinopyroxene in the Caledonides and its implications for geodynamics. *Nature*, **310**, 641-44.

Sobolev, N.V. & Shatsky, V.S., 1990. Diamond inclusions in garnets from metamorphic rocks: a new environment for diamond formation. *Nature*, **343**, 742-46.

Song, S.G., Yang, J.S., Liou, J.G., Wu, C.L., Shi, R.D. & Xu, Z.Q., 2003. Petrology, geochemistry and isotopic ages of eclogites from the Dulan UHPM terrane, the North Qaidam, NW China. *Lithos*, in press.

Spencer, D.A. & Gebauer, D., 1996. SHRIMP evidence for a Permian age and a 44 Ma metamorphic age for the Himalayan eclogites, Upper Kaghan, Pakistan. Implications for the subduction of Tethys and the subdivision terminology of the NW Himalaya. *Himalaya-Karakoram-Tibet Workshop 11th, Flagstaff, Arizona, USA, Abstract*, 147-50.

Spencer, D.A., 1993. Tectonics of the Higher- and Tethyan Himalaya, upper Kaghan Valley, NW Himalaya, Pakistan: Implications of an early, high-pressure, eclogite-facies. metamorphism to the Himalayan belt. *PhD dissertation, Zurich*, 1050.

Stockhert, B., Duyster, J., Trepmann, C. & Massonne, H.J., 2001. Microdiamond daughter crystals precipitated from supercritical COH + silicate fluids included in garnet, Erzegebirge, Germany. *Geology*, **29**, 391-94.

Stutz, E., 1988. Geologie de la chaine du Nyimaling aux confines du Ladakh et du Rupshu, NW-Himalaya, Indes. – evolution paleogeographique et tectonique d'un segment de la marge nord-indienne. *Memorie de Geologie, Lausanne*, **3**, 149.

Sun, W.D., Williams, I.S. & Li, S.G., 2002. Carboniferous and Triassic eclogites in the western Dabie Mountains, east-central China. Evidence for protracted convergence of the North and South China Blocks. *Journal of Metamorphic Geology*, **20**, 873-86.

Tagiri, M. & Bakirov, A., 1992. Quartz pseudomorph after coesite in garnet from a garnet from a garnet-chloritoid-talc schist, northern Tien-Shan, Kirghiz, USSR. *Proceedings of the Japan Academy*, **66**, 135-39.

Tagiri, M., Yano, T., Bakirov, A., Nakajima, T. & Uchiumi, S., 1995. Mineral parageneses and metamorphic P-T paths of ultrahigh-pressure eclogites from Kyrghyzstan Tien-Shan. *The Island Arc*, **4**, 280-92.

Tonarini, S., Villa, I.M., Oberli, F., Meier, M., Spencer, D.A., Pognante, U. & Ramsay, J.R., 1993. Eocene age of eclogite metamorphism in Pakistan Himalaya: Implications for India-Eurasia collision. *Terra Nova*, **5**, 13-20.

Treloar, P.J., 1995. Pressure-temperature-time paths and the relationship between collision, deformation and metamorphism in the north-west Himalaya. *Geological Journal*, **30**, 333-48.

van Roermund, H.L.M., Drury, M.R., Barnhoom, A. & Ronde, A., 2000. Super-silicic garnet microstructures from an orogenic garnet peridotite, evidence for an ultra-deep, > 6 GPa. origin. *Journal of Metamorphic Geology*, **18**, 135-48.

van Roermund, H.L.M. & Drury, M.R., 1998. Ultra-high pressure (P > 6 Gpa) garnet peridotites in western Norway: exhumation of mantle rocks from > 185 km depth. *Terra Nova* **10**, 295-301.

van Roermund, H.L.M., Carswell, D.A., Drury, M.R. & Heijboer, T.C., 2002. Microdiamonds in a megacrystic garnet websterite pod from Bardane on the island of fjortoft, western Norway: Evidence for diamond formation in mantle rocks during deep continental subduction. *Geology*, **30**, 959-62.

Wain, A., 1997. New evidence for coesite in eclogite and gneisses: Defining an ultrahigh-pressure province in the Western Gneiss Region of Norway. *Geology*, **25**, 927-30.

Wain, A., Waters, D., Jephcoat, A. & Olijynk, H., 2000. The high-pressure to ultrahigh-pressure eclogite transition in the Western Gneiss Region, Norway. *European Journal of Mineralogy*, **12**, 667-87.

Walker, J., Martin, M.W., Bowring, S.A., Searle, M.P., Waters, D.J. & Hodges, K.V., 1999. Metamorphism, melting, and extension: Age constraints from the High Himalayan slab of southeast Zanskar and northwest Lahaul. *Journal of Geology*, **107**, 473-95.

Wang, X., Liou, J.G. & Mao, H.K., 1989. Coesite-bearing eclogite from the Dabie Mountains in central China. *Geology*, **17**, 1085-88.

Wei, C.J., Powell, R. & Zhang, L.F., 2003. Eclogites from the south Tian Shan, NW China: petrological characteristic and calculated mineral equilibria in the Na_2O-CaO-FeO-MgO-Al_2O_3-SiO_2-H_2O system. *Journal of Metamorphic Geology*, **21**, 163-80.

Weinberg, R.F. & Dunlap, W., 2000. Growth and deformation of the Ladakh Batholith, NW Himalayas: implications for timing of continental collision and origin of calc-akaline batholiths. *Journal of Geology*, **108**, 303-20.

Yang, J.J., Godard, G., Kienast, J.R., Lu, Y. & Sun, J., 1993. Ultrahigh-pressure, 60 kbar magnesite-bearing garnet peridotite from Northeastern Jiangsu, China. *Journal of Geology*, **101**, 541-54.

Yang, J.S., Xu, Z., Song, S., Zhang, J., Shi, R., Li, H. & Brunel, M., 2001. Discovery of coesite in the North Qaidam Early Paleozoic ultrahigh pressure, UHP. metamorphic belt, NW China. *Comptes Rendus de l'Academie des Sciences, Paris, Sciences de la Terre et des Planets*, **333**, 719-24.

Yang, J., Xu, Z., Dobrzhnestskaya, L.F., Green, H.W., Shi, R., Pei, X., Wu, C., Wooden, J.L., Zhang, J. & Li, H., 2002. Discovery of diamond in North Qiling: evidence for a giant UHPM belt across central China and Recognition of Paleozoic and Mesozoic dual deep subduction. *EOS, Transaction, American Geophysical Union*, **83**, F1271.

Ye, K., Cong, B. & Ye, D., 2000. The possible subduction of continental material to depths greater than 200 km. *Nature*, **407**, 734-36.

Yui, T.F., Rumble, D. & Lo, C.H., 1995. Unusually low $d^{18}O$ ultrahigh-pressure metamorphic rocks from Su-Lu terrane China. *Geochimica et Cosmochimica Acta*, **59**, 2859-64.

Zhang, J.X., Yang, J.S., Xu, Z.Q., Meng, F.C., Li, H.B. & Shi, R., 2002. Evidence for UHP metamorphism of eclogite from the Altun Mountains. *Chinese Science Bulletin*, **47**, 751-55.

Zhang, L., Ellis, D.J. & Jiang, W., 2002. Ultrahigh-pressure metamorphism in western Tianshan, China: Part I. Evidence from inclusions of coesite pseudomorphs in garnet and from quartz exsolution lamellae in omphacite in eclogites; Part II. Evidence from magnesite in eclogite. *American Mineralogist*, **87**, 853-66.

Zhang, R.Y., Liou, J.G. & Yang, J.S., 2000. Petrochemical constraints for dual origin of garnet peridotites of the Dabie-Sulu UHP terrane, China. *Journal of Metamorphic Geology*, **18**, 149-66.

Zhang, R.Y., Liou, J.G., Yang, J. & Ye, K., 2003. Ultrahigh-pressure metamorphism in the forbidden zone: the Xugou garnet peridotite, Sulu terrane, eastern China. *Journal of Metamorphic Geology*, **21**, in press.

Zhang, J.X., Zhang, Z.M., Xu, Z.Q., Yang, J.S. & Cui, J.W., 2001. Petrology and geochronology of eclogites from the western segment of the Altyn Tagh, Northwestern China. *Lithos*, **56**, 187-206.

Zhao, W., Nelson, K.D. & Team, P.I., 1993. Deep seismic reflection evidence for continental underthrusting beneath southern Tibet. *Nature*, **366**, 557-59.

Zhu, Y.F. & Ogasawara, Y., 2002a. Carbon recycled into deep Earth: evidence from dolomite dissociation in subduction-zone rocks. *Geology*, **30**, 947-50.

Zhu, Y.F. & Ogasawara, Y., 2002b. Phlogopite and coesite exsolution from super-silicic clinopyroxene. *International Geology Review*, **44**, 831-36.

13

UHT METAMORPHISM AND CONTINENTAL OROGENIC BELTS

A. MOHAN*, I.N. SHARMA* AND P.K. SINGH*

ABSTRACT

The P-T constraints across the regionally metamorphosed terranes have led to the recognition of UHT metamorphism. Regional metamorphism of deep crustal rocks that relate to UHT metamorphism have certain critical mineral associations stable at temperatures between 900–1100°C at 7–13 kbar. Among the important UHT mineral associations are sapphirine-quartz, spinel-quartz, high alumina orthopyroxene-sillimanite-quartz and osumilite etc. While more than 20 UHT terranes across the globe have been recognized in the current literature, we document only well-known UHT terranes [Anapolis – Itaucu Complex, Brazil; Eastern Ghats Belt, India; Southern Granulite Belt, India; Archaean Napier Complex, East Antarctica; Raur Group, East Antarctica; In Ouzzal Complex, Algeria; Mollendo – Camana Block, Peru and Lewisian Complex, NW Scotland] that invoke significant petrological implications in terms of tectonic setting, P-T path, heat source and their eventual exhumation. On one hand, the knowledge on these UHT terranes have resolved many current problems, on the other they have also posed many new ones that have far-reaching consequences in understanding a comprehensive geodynamic model of an UHT terrane. The fragmentation and assembly of the supercontinents is one of them.

KEY WORDS

Ultrahigh temperature metamorphism, Gondwana supercontinent, Granulites, P-T path, Petrogenetic grid.

INTRODUCTION

Ultrahigh-temperature (UHT) metamorphism refers to the regional metamorphism of deep crustal rocks which have gained temperature high enough to stabilize the critical associations sapphirine-quartz, spinel-quartz, high aluminous orthopyroxene-sillimanite-quartz and osumilite at T > 900°C at P > 9 kbar (Harley, 1998a). The pressure-temperature range for UHT metamorphism lies in between 900°C - 1100°C at moderate pressures 7-13 kbar. The Fig. 1 illustrates the domain of P-T conditions of UHT metamorphism along with the conventional metamorphic facies. UHT metamorphic rocks have important implications for the rheological and chemical behaviour of the continental crust during orogeny, mantle derived thermal pulses and the tectonic situation in which such extreme thermal conditions can arise.

* Department of Geology, Banaras Hindu University, Varanasi - 221 005, India.

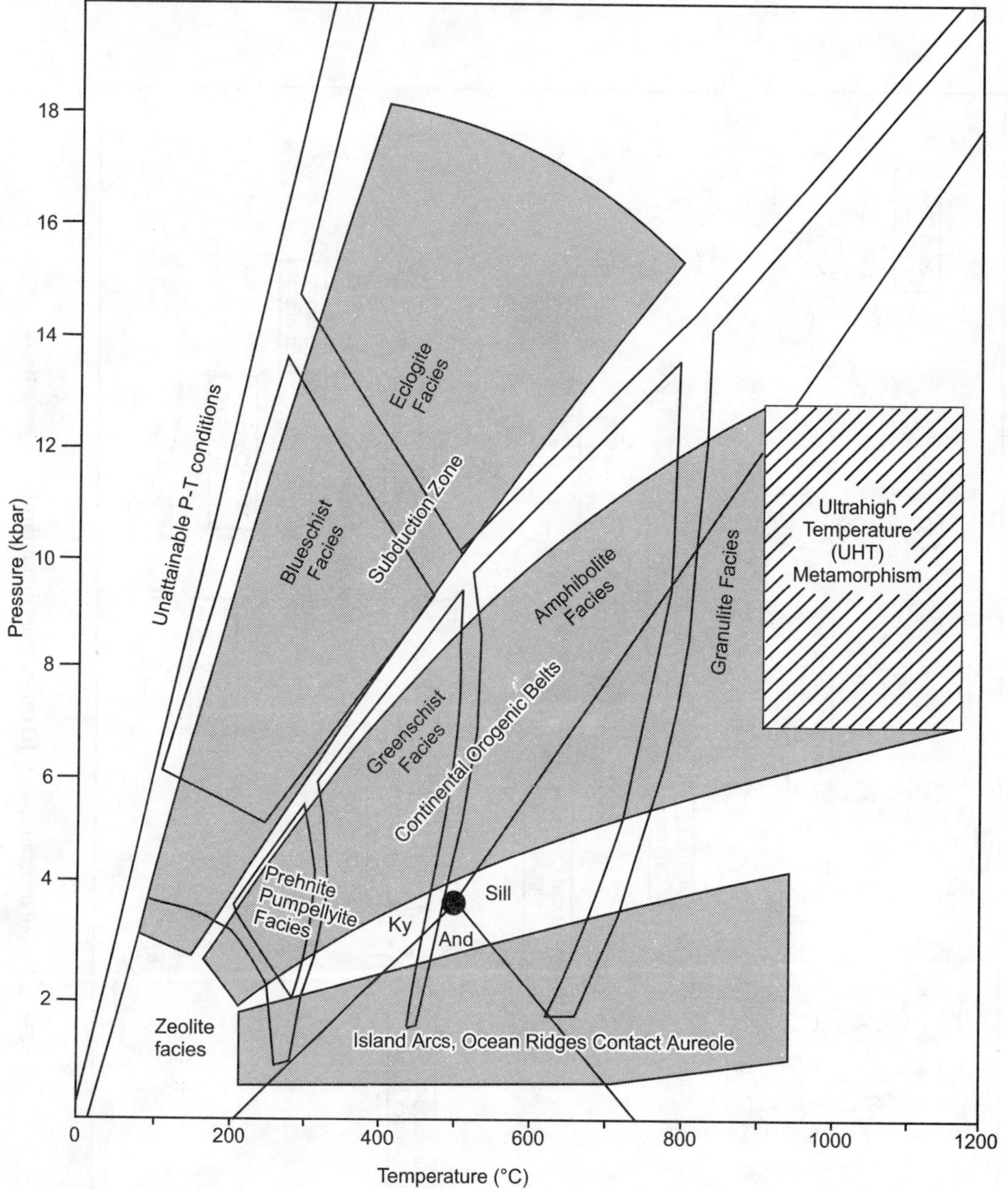

Fig. 1. Schematic P-T diagram of crustal metamorphism depicting the conventional metaporphic facies, along with P-T regime for UHT metamorphism (modified after Spear, 1993).

The aim of the present article is to present an overview of the UHT metamorphism with the illustration of petrochemical researches of various UHT terranes. The distribution along with mineralogical and petrological characteristics of the classic UHT terranes of the world is also described. Fig. 2 shows location of some well-documented classical examples of UHT metamorphic terranes.

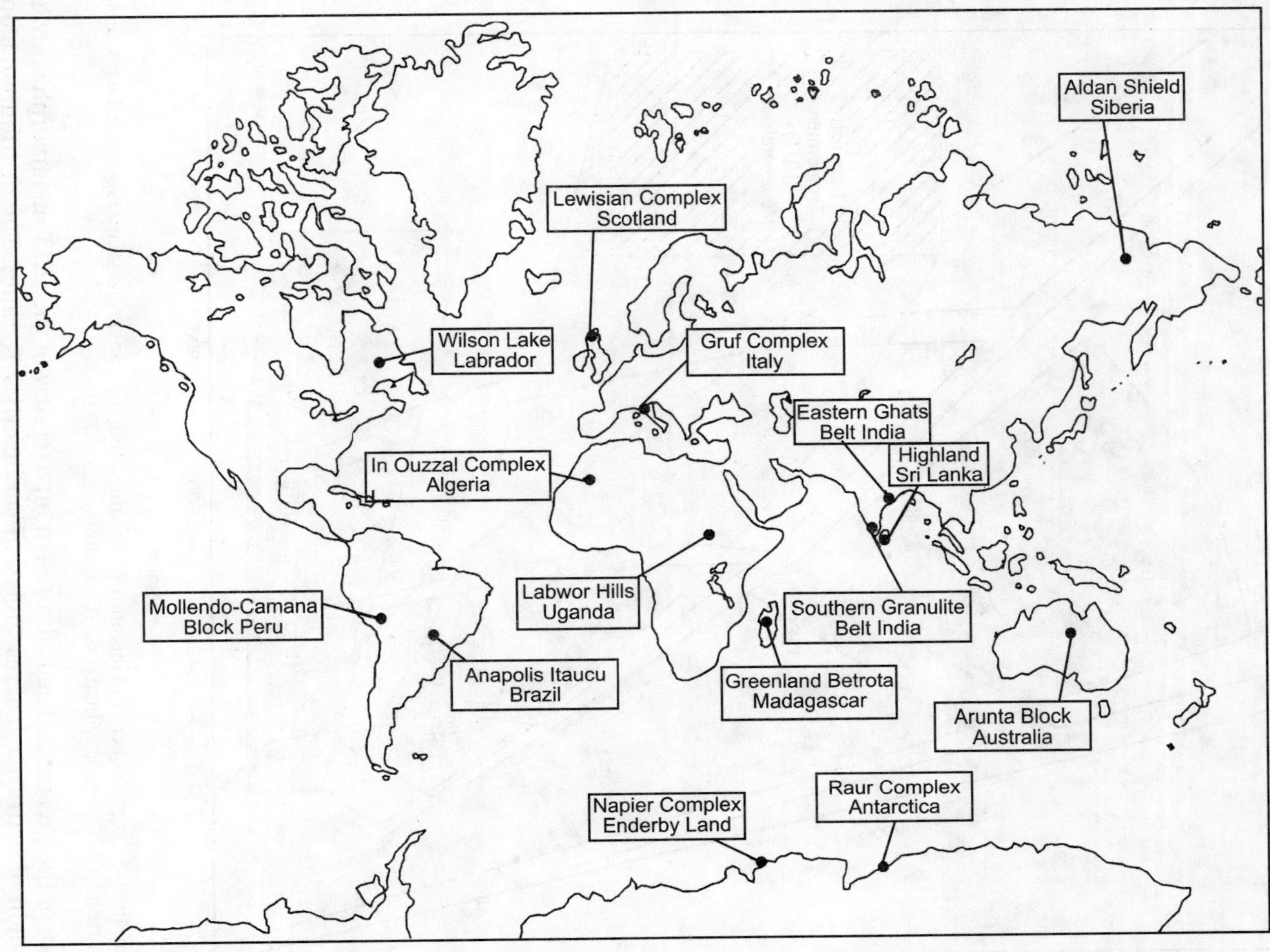

Fig. 2. Global distribution of UHT regionally metamorphosed terranes.

The key indicators of UHT metamorphism are the mineral assemblages and reaction textures in sapphirine, spinel and orthopyroxene-bearing granulites. The qualitative analysis of the changes in P-T conditions implied by frozen-in textures in the form of reaction corona and symplectites have been inferred from the petrogenetic grids of FMAS (FeO-MgO-Al_2O_3-SiO_2) and KFMASH (K_2O-FeO-MgO-Al_2O_3-SiO_2-H_2O) systems (Hensen & Green, 1973; Grew, 1980; Hensen, 1977, 1987; Hensen & Harley, 1990; Bertrand *et al.*, 1991; Audibert *et al.*, 1995; Carrington & Harley, 1995a, b; Holland *et al.*, 1996). Recent experimental data and theoretical considerations support the general applicability of Hensen & Green (1973) grid and its derivatives which identified a sapphirine-quartz stability field at temperature more than 1050ºC and defined orthopyroxene-sillimanite-quartz to occur at pressure beyond maximum cordierite stability condition of 10-11 kbar. Recent KFMASH experimental studies of biotite dehydration melting (Carrington & Harley, 1995a) and determination of osumilite stability relation in KMAS (Motoyoshi *et al.*, 1993) and KFMASH (Audibert *et al.*, 1995; Carrington & Harley, 1995b) constraint that osumilite is restricted to more than 900°C for pressure greater than 6 kbar and has an ultimate stability limit of 9 kbar in KFMAS system. The grid presented in Fig. 3 illustrates the stability field of critical associations of UHT metamorphism.

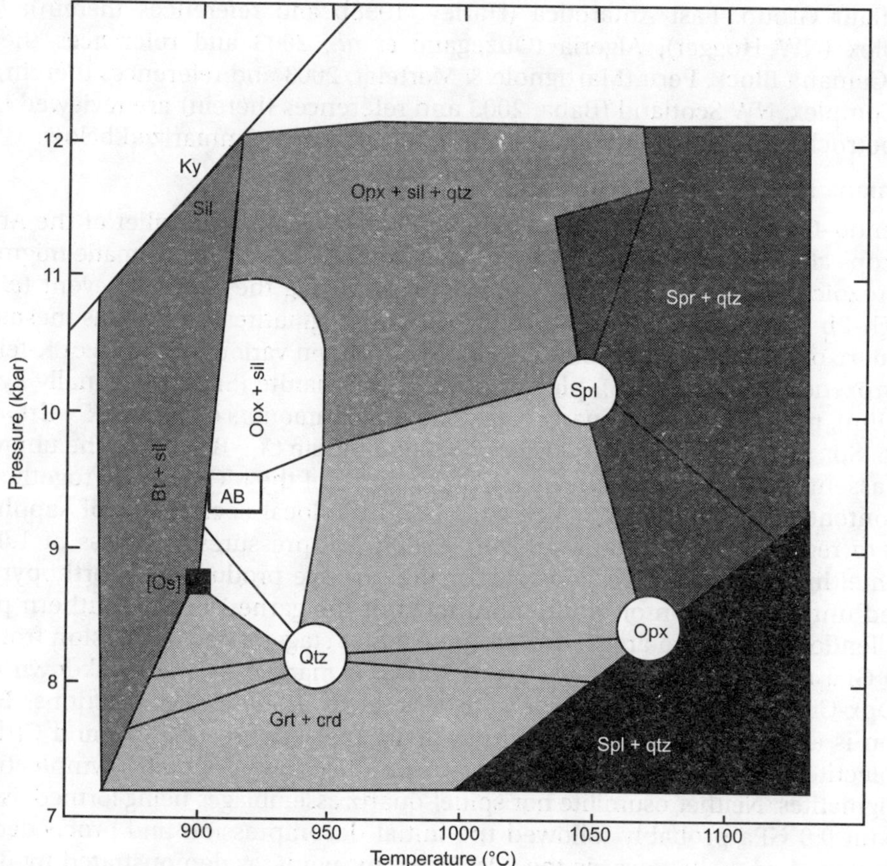

Fig. 3. Schematic P-T grid illustrating the stability field of critical associations of UHT metamorphism (modified after Harley *et al.*, 1998 b).

DISTRIBUTION OF UHT METAMORPHIC TERRANES

In the recent literature various granulite terrains have been increasingly recognised as UHT terranes and extensively reviewed by petrologists (Lal *et al.*, 1987; Dasgupta *et al.*, 1995; Mohan *et al.*, 1997; Raith *et al.*, 1997; Harley, 1998a, b; Nandakumar & Harley, 2000; Harley & Motoyoshi, 2000; Hokada, 2001; Rickers *et al.*, 2001b; Tsunogae *et al.*, 2002; Bhattacharya & Kar, 2002; Bhattacharya *et al.*, 2003; Sarkar *et al.*, 2003; Ouzegane *et al.*, 2003; Martignole & Martelat, 2003; Baba, 2003). Noted UHT terranes distributed all along the world in Eurasia, Africa, Antarctica, North and South America and Australia are shown in Fig. 2. These UHT terranes lie within major Archaean Continental orogenic belts and Proterozoic shield areas.

CLASSICAL UHT TERRANES

The global distribution of several classical UHT terranes including — (1) Anápolis-Itauçu Complex, Brazil (Moraes *et al.*, 2002); (2) Eastern Ghats Belt, India (Lal *et al.*, 1987; Dasgupta *et al.*, 1995; Mohan *et al.*, 1997; Rickers *et al.*, 2001a; Bhattacharya & Kar, 2002; Bhattacharya *et al.*, 2003; Sarkar *et al.*, 2003); (3) Southern Granulite Belt, India (Raith *et al.*, 1997; Nandakumar & Harley, 2000 and references therein); (4) Archaean Napier Complex, East Antarctica (Harley & Motoyoshi, 2000; Hokada, 2001; Tsunogae *et al.*, 2002 and references therein); (5) Raur Group, East Antarctica (Harley, 1998b and references therein); (6) In Ouzzal Complex (NW Hogger), Algeria (Ouzegane *et al.*, 2003 and references therein); (7) Mollendo-Camana Block, Peru (Martignole & Martelat, 2003 and references therein); and (8) Lewisian Complex, NW Scotland (Baba, 2003 and references therein) are reviewed in this section. The petrochemical characteristics of these terranes are summarized below:

Mollendo–Camana Block (MCB), Peru

The Mollendo–Camana Block (MCB) is a 50 × 150 km Precambrian inlier of the Andean belt that outcrops along the Pacific coast of southern Peru. It consists of stromatic migmatites of Paleoproterozoic heritage intensely metamorphosed during the Grenville event (*c.* 1 Ga; U-Pb and U-Th-Pb ages on zircon and monazite). In the migmatites, aluminous mesosomes (FMAS) and quartzofeldspathic leucosomes (KFMASH), contain various amounts of K-feldspar (Kfs), orthopyroxene (X_{Mg} Opx = 0.86), plagioclase (Pl), sillimanite (Sil; exceptionally kyanite, Ky), ilmenite (Ilm), magnetite (Mag), quartz (Qtz), and minor amounts of garnet (X_{Mg} Grt = 0.60), sapphirine (X_{Mg} Spr = 0.87), cordierite (X_{Mg} Crd = 0.92) and biotite (X_{Mg} Bt = 0.83). The ubiquitous peak mineral assemblage is Opx-Sil-Kfs-Qtz-(± Grt) in most of the MCB, which, together with the high Al content of orthopyroxene (10% Al_2O_3) and the local coexistence of sapphirine-quartz, attest to regional UHT metamorphism (> 900°C) at pressures in excess of 1.0 GPa. Fluid-absent melting of biotite is responsible for the massive production of orthopyroxene that proceeded until exhaustion of biotite (and most of the garnet) in the southern part of the MCB (Mollendo-Cocachacra areas). In this area, a first stage of decompression from 1.1–1.2 to 0.8–0.9 GPa at temperatures in excess of 950°C, is marked by the breakdown of Sil-Opx to Spr-Opx-Crd assemblages according to several bivariant FMAS reactions. High-*T* decompression is also shown by Mg-rich garnet being replaced by Crd-Spr- and Crd-Opx-bearing symplectites, and reacting with quartz to produce low-Al-Opx-Sil symplectites in quartz-rich migmatites. Neither osumilite nor spinel-quartz assemblages being formed, isobaric cooling at about 0.9 GPa probably followed the initial decompression and proceeded with massive precipitation of melts towards the (Os) invariant point, as demonstrated by Bt-Qtz-(± pl) symplectites in quartz-rich migmatites (melt + Opx + Sil = Bt + Grt + Kfs + Qtz). Finally,

Opx rims around secondary biotite attest to late fluid-absent melting, compatible with a second stage of decompression below 900°C. The two stages of decompression are interpreted as due to rapid tectonic denudation whereas the regional extent of UHT metamorphism in the area, probably results from large-scale penetration of hot asthenospheric mantle at the base of an over-thickened crust (Martignole and Martelat, 2003 and references therein).

Anápolis-Itauçu Complex, Brazil

The Anápolis-Itauçu Complex (AIC) lies within the internal zone of the Neoproterozoic Brasilia Fold Belt, which borders the western margin of Sáo Francisco Craton. The AIC forms a NW-SE elongated area of outcrop composed of mainly granulite-facies rocks, separated by a mylonite zone from Araxá Group. The AIC composed of with several lithological groups such as orthogranulites, supracrustal granulites, mafic-ultramafic rocks, granites and volcano-sedimentary sequences. Recently determined T_{DM} modal for supracrustal granulites for AIC (Pimentel *et al.*, 1999) and for the metasedimentary rocks of Araxá Group (Pimental *et al.*, 1999) are between 1.3 and 1.6 Ga in both cases. Although no geochronological data are available for UHT rocks but the age of metamorphism for surrounding 'common' granulites was determined as ~650 Ma using Sm-Nd method on whole rock and garnet along with monazite, biotite and amphibole (Fischel *et al.*, 1999) and similar age were obtained from SHRIMP method (Tassinari *et al.*, 1999). The sapphirine-quartz bearing granulites recording extreme P-T conditions: a minimum of 1030 – 1050°C at ~10 kbar. In one area, the post–peak evolution is constrained to begin at P >10 kbar by absence of cordierite and lower temperature stability of garnet – orthopyroxene – sillimanite-quartz while in second area post-peak evolution occur at P<10 kbar inferred from microstructural relations between garnet, orthopyroxene, sillimanite and coronae or symplectite involving corundum, sapphirine, spinel, cordierite, plagioclase, sillimanite, orthopyroxene, biotite and ilmenite. The retrograde segment of the P-T path was further constrained by biotite-producing reactions. A composite P-T path was constructed from samples that exhibit different amounts of retrograde reaction, reflecting different amount of melt retention. Using back-calculated composition for garnet and orthopyroxene, thermobarometry yields 1012-960°C and 9.7-8.1 kbar. In samples from both area, orthopyroxene porphyroblasts have high Al_2O_3 (12.9-9.7wt%) in cores, which suggests maximum T > 1150°C. After decompression at the metamorphic peak, the P-T path followed a near-isobaric cooling stage to < 900°C but the two paths are separated by ~ 3 kbar. The ultrahigh-temperature metamorphism of AIC occurred in a collisional tectonic setting in which the extreme thermal perturbation probably was a result of asthenosphere replacement of mantle lithosphere (Moraes *et al.*, 2002).

In Ouzzal Complex (NW Hogger), Algeria

The UHT metamorphic rocks and associated granulites of In Ouzzal complex are the rare one because these rocks preserve the prograde sequence of mineral reactions (Ozuegane *et al.*, 2003 and the references therein). The In Ouzzal series belongs to the western Hogger, which is situated between the western and eastern branches of Pan-African belt. Both the western and western shear zones represent sutures as evidenced by occurrence of eclogites in the Tin Zebane region, and of granulitic metagabbros with relict glaucophane SE of Ahnet (Caby, 1996). The inner part of this terrane composed of granulites, which preserved the unusual remnants exhibiting a great diversity of mineral reactions (Ouzegane *et al.*, 2003). Ultrahigh-temperature metamorphism occurred at 2000 Ma and was associated with two main deformation phases (Bertrand *et al.*, 1992; Ouzegane & Boumaza, 1996). The first phase of deformation was responsible for the development of a regional foliation, which was refolded during the second phase (Haddoum *et al.*, 1994). The phase relationships of UHT and associated granulites from this area are analyzed in the light of recent advancement and found that the peak temperature ranging from 850°C to more than 1100°C at about 10 kbar which is followed by decompression and cooling to about 5 kbar and 750°C. Pseudosections derived from FMAS, KFMASH, MAS and FMASH system account well to observed textures. The rise in temperature at In Ouzzal followed by near-isothermal decompression path is therefore interpreted as regional metamorphism in crust thickened by collision followed by extension with extra and localized heat sources brought by anorthosites, carbonatites, and related hot fluid (Ouzegane *et al.*, 2003).

Lewisian Complex in South Harris, NW Scotland

The Lewisian Complex is a TTG (Tonalite – Tronjhemite – Granodiorite) gneiss terrane with numerous enclaves of mafic to ultramafic and rafts of supracrustal material. This complex exposed in the central part of the Outer Hebrides and composed of high-pressure granulites facies Leverburgh belt, amphibolite facies Langavat belt and a proterozoic plutonic igneous complex. Sm–Nd dates revealed that the protoliths of these metasediment were originally post–Archaean sediments deposited at 2500 Ma – 2000 Ma (Cliff *et al.*, 1998). Recently Friend and Kinney (2001) reported age of granulite facies metamorphism around 1880 Ma. In UHT rocks of this complex sapphirine occur in two different mode of occurrence such as inclusions within porphyroblast and symplectitic grains together with secondary cordierite, plagioclase and orthopyroxene. Peak metamorphic temperature of 950±30°C and 930±30°C have been estimated using orthopyroxene isopleths and petrogenetic grids which are consistent with the stability of sapphirine–bearing association (Baba, 2003). The sapphirine inclusions indicate that they are stable during early metamorphism and broken down to produce garnet in response to pressure increase. The sapphirine breakdown textures are found only in rocks with high bulk X_{Mg} (> 0.70). Sapphirine formation and their breakdown in this complex is thought to have been dependent on (a) protolith's bulk composition and (b) restite formation via partial melting during a continuous anticlockwise P-T paths (Baba, 2003 and the references therein).

Napier Complex, East Antarctica

It is the Napier Complex (NC) where initially sapphirine-quartz association from spot height 945 (Dallwitz, 1968) had been discovered and attracted the attention of petrologist as the Earth's best preserved example of regional scale UHT crustal metamorphism (Ellis, 1980; Grew, 1980; Harley & Hensen, 1990; Harley & Motoyoshi, 2000). NC consists of Mt. Riiser-larsen, Tula Mountains and Tonagh Island. The Archean rocks of these area are formed by multiple thermal events, including Late–Archaean UHT metamorphism (Harley & Black, 1987) and was subsequently intruded by unmetamorphosed diabase dykes at 1200 Ma. The high-grade rocks of this complex include sapphirine bearing quartzite, orthopyroxene bearing felsic gneiss, garnet bearing felsic gneiss and two pyroxene bearing mafic granulites etc. These high–grade rocks contain the critical associations osumilite–garnet, (Ellis, 1980; Hensen & Motoyoshi, 1992), Mg-bearing pigeonite (Harley, 1987) sapphirine–quartz, orthopyroxene–sillimanite-quartz and thermobarometrical estimates from composition of re-integrated perthitic, mesoperthitic and antiperthitic feldspar through ternary feldspar solvus model have been used to infer peak temperature of 980°C to >1120°C (Harley & Mototyoshi 2000; Hokada, 2001; Tsunogae *et al.*, 2002 and the references therein) and pressure ranging from 7–11 kbar (Harley & Hensen, 1990; Harley & Motoyoshi, 2000). Studies of reaction texture involving the formation of orthopyroxene-sillimanite–quartz on early sapphirine bearing assemblages (Ellis *et al.*, 1980; Sandiford, 1985) and garnet coronas on orthopyroxene in a variety of peraluminous and metaluminious (Harley, 1985; Harley & Motoyoshi, 2000), mafic rock types (Ellis & Green, 1985) and the study of fluid inclusions trapped in sapphirine, garnet and quartz during UHT metamorphism (Tsunogae *et al.*, 2002) demonstrate that the NC underwent a phase of near isobaric cooling following its latest thermal peak >1120°C, implying that the 10-15 Km. thick mid-to deep-crustal section underwent and *en masse* cooling whilst remaining deeply buried at depths equivalent to 8-11 kbar at the end of Archaean (Harley, 1985; Sandiford, 1985; Harley & Black, 1987; Harley & Hensen, 1990; Harley & Motoyoshi, 2000).

Rauer Group, East Antarctica

The Rauer Group (RG) is an orthogneiss-dominated granulite terrane and which form a part of the 1100 – 1000 Ma metamorphic complex widespread in East Antarctica (Black *et al.*, 1987; Clarke, 1988; Hensen & Harley, 1990). RG separates two contrasting high-grade terrane to its north and south and having complex geological history. The P-T conditions and evolution associated with the post 1000 Ma events in the RG are constrained from garnet-orthopyroxene-felsic granulites, garnet–sillimanite metapelites and wollastonite–scopolite calc–silicates (Harley, 1988; Harley & Fitzsimons, 1991; Harley & Buick, 1992). Highly magnesium and aluminous migmatite gneisses from Mather Peninsula in the RG preserve the UHT metamorphic assemblages which include orthopyroxene – sillimanite ± quartz, garnet – sillimanite ± quartz and garnet-orthopyroxene ± sillimanite. Peak P-T conditions retrieved from Fe-Mg-Al thermobarometry involving garnet and orthopyroxene is found to lie at 1050°C and 12 kbar. Sapphirine, orthopyroxene and cordierite forms extensive symplectite and coronas on early stage whereas the specific reaction textures and assemblage involving secondary phases correlated with initial garnet, with apparent higher-pressure reaction products occurring on the more Mg garnet and are interpreted to result from an initial phase of UHT near-isothermal decompression from 12 to 8 kbar at temperature in excess of 950°C. Later texture of biotite formation, partial breakdown, along with garnet relicts, to symplectite of orthopyroxene-cordierite or cordierite-spinel may reflect hydration through back-reaction with crystallizing melts on cooling below 900-850°C, followed by isothermal decompression from 7-8 kbar at temperatures of 750-850°C. The tectonic interpretation of these textures and P-T histories is ambiguous as the RG records the effect of Archaean tectonothermal event as well as high-grade events at 1000 – 530 Ma. Late stage biotite formation and subsequent isothermal decompression can be correlated with P-T history preserved in the Proterozoic components of RG and hence with either 1000 Ma. or 530 Ma collisional orogenesis (Harley, 1998b and the references therein).

Eastern Ghats Belts, India

The Eastern Ghats Belts (EGB) lies towards the east of the Eastern Dharwar and the Bastar Cratons and south of the Singhbhum-Orissa Craton. The contact with the EGB and these cratons is marked by shear zone and is sometime referred as 'front'. The EGB is broadly divided into four longitudinal zones based on prominent lithological associations, namely (1) Western Charnockite zone (WCZ), (2) Western Khondalite zone (WKZ), (3) Central Migmatite zone (CMZ), and (4) Eastern Khondalite zone (EKZ). The region between the Cratons and the WCZ is a transition zone which consists of incipient charnockites, basic/mafic granulites and high-grade schists enclaves in pervasive orthogneisses. The WCZ probably represents the deeper part of the adjacent Cratons or reworked Cratonic gneisses (Rickers *et al.*, 2001a).

The radiometric dates recorded from the North of Godavari Graben in EGB are: (a) 3000, 2800 and 2600 Ma mostly from WCZ, which correspond to those of the adjacent Cratons. (b) ~1000 Ma pervasive granulite facies event and (c) 550 – 500 Ma late Pan-African amphibolite facies overprint (Grew & Manton, 1986; Aftalion *et al.*, 1988; Paul *et al.*, 1990; Shaw *et al.*, 1997; Mezger & Cosca, 1999; Rickers *et al.*, 2001a). In the South of Godavari Graben the Mineral dates of 1600 and 1300 Ma from granulites of the region have been reported by Mezger and Cosca (1999) and Rickers *et al.* (2001a) without any later thermal imprint of ~ 1000 and 550 – 500 Ma ages.

UHT metamorphism in EGB is revealed by the presence of sapphirine-quartz, spinel-quartz, osumilite (inferred from its breakdown reaction texture) and orthopyroxene (upto 10 wt % Al_2O_3) - sillimanite - quartz (Paderu, Lal *et al.*, 1987, Lal, 1997; and G. Madugula, Mohan *et al.*, 1997). Two isothermal decompression (ITD) paths have been inferred; high-temperature ITD from thermal peak of 950°C/8.5 kbar to 900°C/6.8 kbar from UHT granulites from Paderu (Lal, 1997) and another ITD path followed by a decompression to 740°C/5.4 kbar during which cordierite formed from spinel + quartz, moat of cordierite between orthopyroxene/garnet and sillimanite, cordierite-orthopyroxene rims on garnet and spinel rim on sapphirine followed by cordierite formation from G. Madugula (Mohan *et al.*, 1997). Bhattacharya *et al.* (2003) proposed that high temperature melting arm is related to clockwise P-T-t path with high temperature decompression during which earlier orthopyroxene + sillimanite formed cordierite at 1000°C/10 – 8 kbar and ends with a prolong cooling from above 900°C to c. 600°C at 8 kbar from metapelitic granulites from area around Paderu.

High-pressure IBC of 800°C/7-8 kbar followed by ITD upto 5 kbar from Anantgiri (Sengupta *et al.*, 1990) and Rajamundry (Dasgupta *et al.*, 1995) are reported in literature. The ITD path in quartz-absent Mg-Al granulites during which garnet breaks down to form sapphirine–sillimanite-orthopyroxene symplectite has been reported from Anakapalle by Dasgupta *et al.* (1994) and their inferred ITD passes through 900°C/8.9 kbar to 815°C/6.1 kbar, with UHT thermal peak at c. 1000°C/10 kbar (see Rickers *et al.*, 2001b). Low-pressure IBC path is also well documented in UHT granulites from all the localities mentioned above, with the formation of biotite-quartz ± sillimanite forming through orthopyroxene, garnet, cordierite and K-feldspar via 'back reactions'. In Chilka Lake area, Sen *et al.* (1995) proposed multi-stage P-T-t path in UHT granulites with three ITD ~ 1100°C/11-9 kbar, ~ 850°C/9-7 kbar and ~ 700 – 750°C/6-4 kbar separated by two IBC paths.

Towards the south of the Godavari Graben, in Kondapalle area, a suite of meta-pelitic granulites, intruded by layered gabbronorite-pyroxenite-anothosite, display UHT isobaric heating-cooling P-T-t path with thermal peak at > 1000°C/> 8 kbar (Sengupta *et al.*, 1999). The sequence of reactions inferred from the reaction textures along the prograde heating path are (i) dehydration melting of biotite and (ii) biotite-absent reactions. During the cooling path, the biotite-absent reactions are recrossed, Dasgupta *et al.* (1997) described meta-pelites from north of Chimakurthy which were thermally metamorphosed at the contact of a mafic-ultramafic intrusive complex at c. 6 ± 1 kbar along isobaric heating-cooling path within temperature interval between 750 and 1000°C. The prograde reactions inferred from the textures and phase relations are dehydration melting reaction resulting in total elimination of biotite and quartz. Similar isobaric heating-cooling path has also been reported from the UHT Mg-Al granulites of Sunkarametta area, NW of Visakhapatnam, towards the north of the Godavari Graben (Bose *et al.*, 2000). Thermal peak has been estimated at 950°C/9 kbar followed by cooling to c. 700 – 750°C. The reactions inferred from reaction textures during the prograde heating path include (i) dehydration melting of biotite (ii) symplectites of sapphirine or spinel + cordierite formed from orthopyroxene. Along the retrograde path, coronas or symplectites of orthopyroxene-sillimanite were formed by reactions involving sapphirine, spinel, MgTs, cordierite and quartz; corona of sapphirine on spinel and sillimanite; successive coronas of sillimanite and garnet on spinel and late formation of biotite and garnet-quartz symplectite through 'back reaction'. They have proposed that the source of heat for the UHT metamorphism along the isobaric heating-cooling path was basic/mafic magma in view of the close association of the Mg-Al granulites with basic/mafic granulites. This P-T-t trajectory inferred from the Mg-Al granulites towards the north of Godavari Graben is in contrast to the UHT high-pressure IBC along anticlockwise path, characterized by a 3–4 kbar increase in pressure subsequent to heating during the prograde metamorphism (Dasgupta *et al.*, 1995; Sengupta *et al.*, 1990, 1999). The UHT metamorphism with high-pressure IBC and IBC heating-cooling paths has been attributed to an earlier event ~1600 Ma and ITD path to later ~1000 Ma event (Dasgupta *et al.*, 1994; Mezger & Cosca, 1999; Sengupta *et al.*, 1999; Rickers *et al.*, 2001a).

Southern Granulite Belt, India

The Southern Granulite Belt (SGB) has been sub-divided into two parts: (a) Northern block of the SGB (NB-SGB) and (b) Southern block of SGB (SB-SGB), which are separated from each other by the Palghat-Cauvery shear zone NB-SGB lies between Moyar-Bhavani and Palghat-Cauvery shear zones. This includes the granulites of Madras (Chennai) yielding 2600, 2550 and 2450 Ma age (Bernard-Griffiths *et al.*, 1987; Unnikrishnan-Warrier *et al.*, 1995), charnockites of Nilgiri Hills with age of 2600 and 2500 Ma (Peucat *et al.*, 1989; Raith *et al.*, 1999). Ages of charnockites from Madras (Chennai) and Nilgiri Hills correspond to 2500 Ma. Towards east of the Nilgiri Hills, older dates of 2900 – 2800 Ma of meta-layered igneous complex from Sitampundi, Bhavani Sagar and Kabilmalai gneiss (Bhaskar Rao *et al.*, 1996) suggest local presence of partially reset dates of the older crustal components corresponding to 3000 Ma. Mineral dates of 730 – 770 Ma (Bhaskar Rao *et al.*, 1996) and 620 – 500 Ma (Meißner *et al.*, 2002) from these areas may be later Neoproterozoic to Pan-African tectonothermal imprint presumably related to Moyar-Bhavani and Palghat-Cauvery shears. The SB-SGB is composed of massive charnockites, khondalites and quartz-absent sapphirine-spinel granulites. Few localities of incipient charnockitization in the biotite-hornblende gneisses have been reported by Ravindra Kumar and Raghavan (1992). Preliminary radiometric dates of charnockites are 2400 – 2100 and 550 – 500 Ma (Bartlett *et al.*, 1995; Jayananda *et al.*, 1995). These dates to be the age of initial crust formation and Pan-African granulite facies metamorphism respectively have interpreted these dates. Raith *et al.* (1997), however, suggested a major granulite facies event at 2500 Ma (M II), similar to the NB-SGB, superposed by 500 Ma Pan-African thermal events. The P-T-t paths are well documented by reaction textures in Mg-Al and other granulites from the South Block of the Southern Granulite Belt (SB-SGB) in Palni Hills from localities near or south of the Palghat-Cauvery shear zone where they are characterized by CW P-T-t path from four localities, e.g. Kiranur (Lal *et al.*, 1984), Ganguvarpatti (Mohan *et al.*, 1986; Mohan and Windley, 1993), Perumalmalai (Raith *et al.*, 1997) and Palghat Gap, NW of Kodaikanal (Ravindra Kumar and Chacko 1994).

Mg-Al granulites from Kiranur (Lal *et al.*, 1984) document the evidence of (a) prograde path during which kyanite is pseudomorphed by an aggregate of sillimanite, and kyanite is also present along with staurolite in quartz-free domain within an aggregate of plagioclase. (b) ITD path: In quartz-bearing assemblage gedrite is separated from sillimanite by cordierite + plagioclase, with plagioclase forming moat on gedrite. In quartz-free rocks gedrite is isolated from sillimanite by symplectites of cordierite + sapphirine with or without spinel and Cornrupine. In some samples similar cordierite + sapphirine symplectites are formed between biotite and sillimanite. Based on their P-T estimates from associated rocks from pertinent geothermobarometers, these reaction textures suggest a CW P-T-t path with prograde loop passing through kyanite = sillimanite equilibrium at 700°C followed by nearly isobaric heating upto 750-800°C at 8 kbar, and ITD to 700°C at 5 kbar.

The ITD P-T-t path passing through the univariant reactions emerging from [Qtz] invariant point are supported by spectacular coronas and symplectites displayed from mostly quartz-absent Mg-Al granulites of Ganguvarpatti (Mohan and Windley, 1993) and Perumalmalai (Raith *et al.*, 1997). The granulites from Ganguvarpatti show textural evidence of breakdown of garnet to symplectites of orthopyroxene + sapphirine + cordierite and orthopyroxene + spinel + cordierite. Besides, the meta-pelites devoid of sapphirine document formation of cordierite + spinel, earlier garnet + sillimanite and cordierite forming moats on spinel separating the latter from quartz. An ITD P-T path from 800°C/8 kbar to 700°C/4.5 kbar has been estimated from geothermobarometry.

In Perumalmalai, the clockwise P-T path has been well documented from the reaction textures from Mg-Al granulites, with four distinct stages of evolution (Raith *et al.*, 1997). (a) The prograde path inferred by the presence of sillimanite pseudomorph after kyanite, similar to those reported by Lal *et al.* (1984) and biotite dehydration melting reactions. The UHT assemblage orthopyroxene – cordierite – sillimanite – garnet – K-feldspar – quartz – sapphirine (sapphirine in quartz-absent domain only) and biotite coexisted with melt in equilibrium at the thermal peak. (b) Later ITD have been inferred from variety of spectacular symplectites reactions involving sapphirine, cordierite, sillimanite, orthopyroxene, garnet spinel and quartz. (c) Late biotite crystallized through 'back reactions' during the IBC. (d) A second period of ITD is inferred from the partial breakdown of late biotite through dehydration reaction. The estimates of P-T are based on experimental petrogenetic grid in FMAS system of Hensen and Green (1973) and Bertrand *et al.* (1991), and Biotite dehydration melting reactions in KFMASH system and in meta-pelites of Carrington and Harley (1995a) and Vielzeuf and Montel (1994) respectively. The P-T estimates of the different stages are: (a) 800 – 900°C/12 kbar, with thermal peak at 950 – 1000°C/10 – 11 kbar, (b) ITD with decompression upto 7–8 kbar at similar temperature as given for the thermal peak, (c) 875°C at 7-8 kbar, and (d) 850°C/6-5 kbar. They proposed that the prograde limb, ITD and IBC (stages a – c) are coeval with granulite facies event at 2600/2500 (?) Ma as in the NB-SGB, while the late ITD (stage d) may be related to the Pan-African event at 500 Ma. Garnet-bearing basic/mafic granulites from Palghat Gap, NW of Kodaikanal, P-T at the thermal peak calculated from core compositions of garnet-orthopyroxene-plagioclase-quartz assemblage is 900°C/10 kbar. This was followed by breakdown of garnet to form symplectites of orthopyroxene + plagioclase during ITD at 700°C/7 kbar (Ravindra Kumar and Chacko, 1994).

It may be mentioned here that the Kerala Khondalite Belt (KKB) forms a significant portion of Southern Indian Shield area, occurring on the southwest margin of the Indian subcontinent. The KKB is essentially a Pan-African (ca. 600–500 Ma) regional granulite terrane which composed of charnockites, khondalites and migmatic leptinitic gneiss. The peak assemblages in KKB are characterized by orthopyroxene + garnet + K-feldspar + biotite + plagioclase + quartz in felsic to intermediate rocks, and by orthopyroxene + cordierite + plagioclase + quartz or garnet + cordierite + sillimanite + feldspar + quartz in semipelites and pelites (Nandakumar & Harley, 2000). The post-peak reaction textures in these assemblages often involve the production of cordierite and are consistent with a predominantly decompressional P-T history at temperature in excess of 700°C. P-T estimates based on garnet-orthopyroxene thermobarometry with application of retrieval calculation are found in range of 6.5 – 7.5 kbar and 925°C ± 25°C from the central KKB (Nandakumar & Harley, 2000). Such near–ultrahigh–temperature conditions are also supported by the recent study (Pattison *et al.*, 2003). The highest temperature attained may reflect the thermal input of dry magmas now exposed as massive charnockites within and to the extreme north of Achankovil Shear Zone, if these are synmetamorphic and similar in age to the main metamorphic event (Nandakumar & Harley, 2000 and references therein).

P–T–TIME PATHS AND THEIR TECTONIC SIGNIFICANCE

Metamorphic evolution of granulites in terms of pressure-temperature (P–T) related to time (t) exhibits orogenic P–T–t paths. These P–T–t paths are required to address the relationship between the different styles of metamorphism and tectonic settings in terms of isobarically cooled or isothermally decompressed segments along clockwise or counter-clockwise trajectories (Bohlen, 1987; Thompson and Ridley, 1987; Harley, 1989; Brown, 1993; Spear, 1993).

A P-T-time path is a loop or curve connecting all P and T points to which a rock has been subjected from the beginning of its metamorphic history until the time of its exhumation (exposure) at the surface (Fig. 4A).

Each rock or hand sample has a characteristic P-T–time path that sheds some light on the tectonic history of the area. Technically, a P-T–time path can be constructed only if the ages (or at least some age constraints) are known. If such age constraints are lacking, it is better to term the path a "P-T path", even though this path implies a continuum in time.

Clockwise P-T paths are characteristic of collisional belts, areas of thrusting and subduction zones. Clockwise P–T–t path involves attainment of P_{max} prior to T_{max} followed by an initial ITD path with large dP/dT (~ 2–4 kbar/100°C) and finally a small dP/dT (~0.2–0.5 kbar/100°C) IBC path (Fig. 4B). Clockwise path results from crustal thickening commonly attributed to collision tectonics/overthrusting/underthrusting (Thompson and England, 1984; Hodges, 1991; Huerta *et al.*, 1999) and continental arc setting (Brown, 1998). Examples of well established clockwise paths are known from southern granulite belt (Kiranur: Lal *et al.* 1984; Ganguvarpatti: Mohan and Windley, 1993; Perumalmalai: Mohan *et al.*, 1996, Raith *et al.*, 1997), Gruf complex, Italian central Alps (Droop and Bucher-Nurminen, 1984), In Hihaou, In Quzzal, Algeria (Ouzegane and Boumaza, 1996), Rundvagshetta, Lutzow-Holm Bay, East Antarctica (Motoyoshi and Ishikawa, 1998).

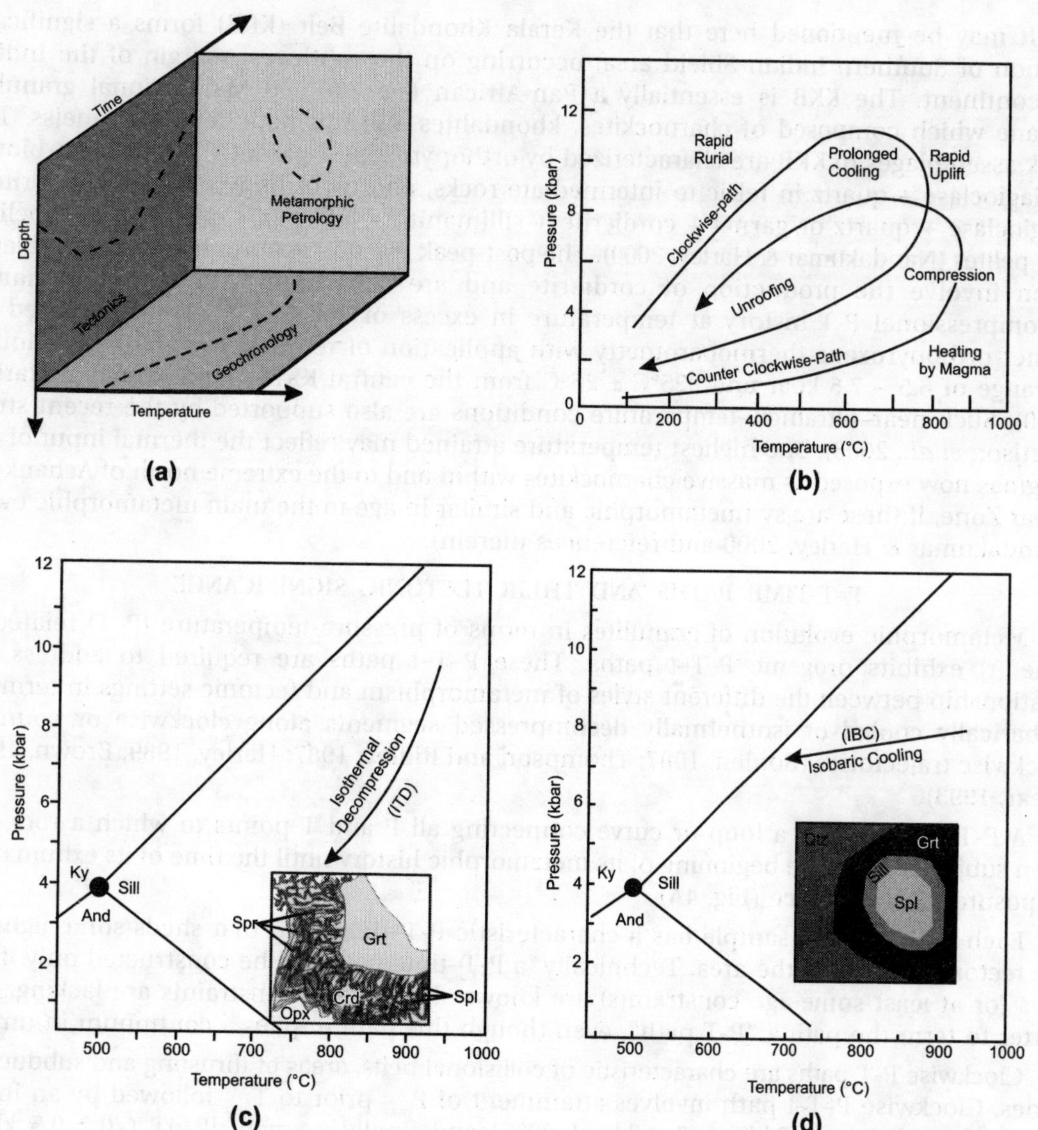

Fig. 4. (a) Projection of the P-T-t path onto the bounding surfaces produces the P-T path, thermal history and tectonic history studied by the petrologist, geochronologist and tectonicist. (b) Schematic clockwise and counterclockwise P-T paths showing stages from rapid burial to eventual unroofing via rapid uplift or compression. (c) Isothermal decompression path (ITD) reinforced by the textural evidence in the form of garnet breaking down to fine finger-like symplectites of spinel-cordierite and sapphirine-cordierite. Note orthopyroxene prisms at the extreme left corner. (d) Isobaric cooling path (IBC) sustained by the textural evidence showing a necklace of sillimanite around spinel within a garnet blast.

In contrast, counterclockwise P-T paths are characteristic of areas of burial of continental crust that is already hot (e.g. continental crust thinned by extension and igneous intrusion). Counter-clockwise P–T–t paths characterize low P/high T metamorphism, where heating of crust occurs before and during the tectonic loading such that the thermal maxima is attained prior to peak pressure conditions (Fig. 4B). These counter-clockwise paths are consequence of extensional or compressional environment where thinned normal crust is accompanied by voluminous magmatic underplating, which produce elevated geotherm (Bohlen and Mezger, 1989). Some crustal shortening accompanies late stages of prograde counter-clockwise path. Examples of well-documented counter clockwise paths are reported from Napier complex, Antarctica (Motoyoshi and Hensen, 1989), West Central New Hampshire (Spear *et al.*, 1995) and Cooma complex, Australia (Johnson and Vernon, 1995).

With the advent of new analytical tools and our improved ability to extract quantitative information from rocks, increasing complexities in the geometry of P–T–t paths have been noted. Both post-peak isobaric cooling and isothermal decompression are documented in different UHT granulites, but in the absence of well-constrained P–T histories, ambiguity remains about clockwise versus counter-clockwise P-T-t paths. Partly these ambiguities reflect difference of perception and approach. Unless these gaps are filled, the debate will rage as long as the evidence is in any way equivocal. Obliteration of prograde growth zoning and relics of prograde mineral reactions in polymetamorphic high-grade terrane makes the choice of clockwise versus counter-clockwise P–T–t path disposition a difficult one. However, distinguishing between ITD and IBC paths in relation to clockwise or counter-clockwise loops is crucial as they imply different tectonic scenarios for the origin of granulites.

The main process of return of lower crustal rocks to Earth's surface are erosion/denudation and tectonic uplift. Comparative input of each of these factors, the rates at which these processes operate, and the depth domain over which they exert influence are still unknown. It is unlikely that the granulites residing in the lower crust may be exhumed in a single orogenic cycle. When the metamorphism occurs as part of two discrete superimposed events, the eventual exhumation of granulites may occur through late tectonic event unrelated to their formation (Bohlen, 1991). In such cases, differential tectonic uplift is an important mechanism to expose lower crustal rocks at the earth's surface. Multistage P–T records that are composite decompression-cooling paths documenting stepped exhumation separated by a period of residence are furnished in Chilka lake, (Sen *et al.*, 1995) and Rayagada (Shaw and Arima, 1998) of Eastern Ghats Granulite Belt; Perumalmalai (Raith *et al.*, 1997; Mohan and Jayananda, 1999) of Southern Granulite Belt; Forefinger point, Antarctica (Harley *et al.*, 1990; Harley and Fitzsimons, 1991). We also recognize P-T trajectory of heating-cooling for lower crustal rocks (Sunkarametta of Eastern Ghats, Bose *et al.*, 2000).

HEAT SOURCE

Although it has long been established that what drives metamorphism and metamorphic processes is heat but the source of heat remains enigmatic (Jamieson *et al.*, 1998). The distribution and transport along a conveyer belt of heat through the crustal rocks have not been ascertained in as much detail (Treloar and O'Brien, 1998). Getting adequate heat to achieve the peak temperatures that is required by rocks in UHT metamorphism has been the real concern for metamorphic petrologists. It is now generally accepted that regional granulite facies metamorphism requires an input of extra heat over and above that is available from a thickened crust (Ashwal *et al.*, 1991). P–T–t paths recovered from high-grade granulites are controlled by (a) thermal perturbation, (b) relaxation of thermal perturbation towards the steady stage. In order to produce temperatures in excess of 900°C, syn-post accretion of mantle-derived magma is necessary to enhance the total heat budget. Model of magmatic underplating in the crust of normal thickness having dominance of

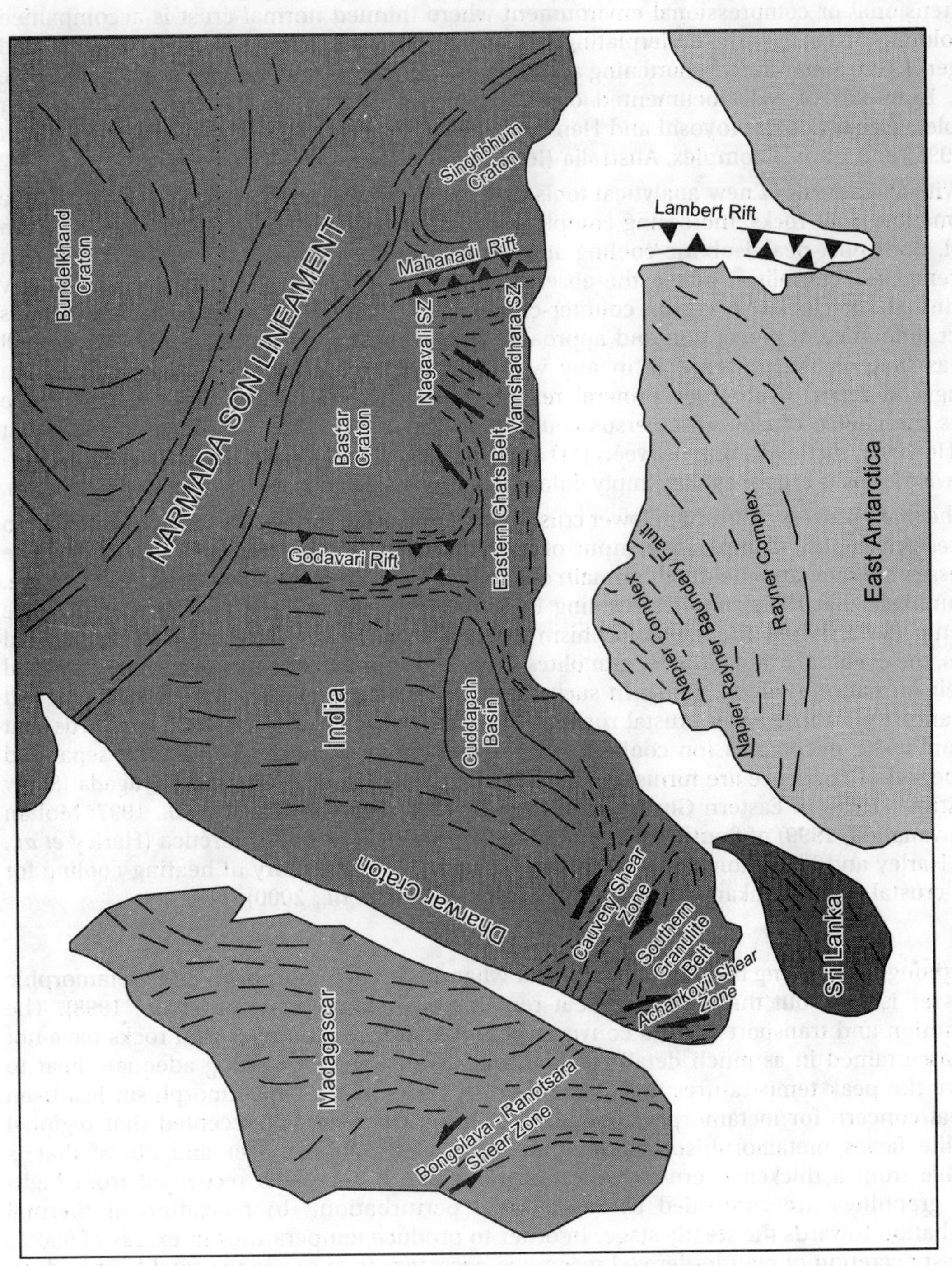

Fig. 5. Reconstruction of East Gondwana supercontinent and the continuous chain of shear zone network (modified after Chetty *et al.*, 2003)

mafic magmatic intrusives in the lowermost crust would also produce elevated geotherm. Besides, dehydration melting experiments on hydrous silicates supplement for an extra heat input and the role of dehydration melting is also a vital cause for granulite formation whose P–T evolution lies above the wet solidus for crustal rocks (Vielzuef *et al.*, 1990; Dooley and Patino Douce, 1996).

CONFIGURATION OF SUPERCONTINENTS

The assembly and break up of supercontinents within the Gondwanaland during the Earth's history has stirred a revolution for conceptual thinking on Precambrian mobile belts. The study of dispersed mobile belts provides the record of the changing map patterns of the Earth through time. The Neoproterozoic tectonothermal imprints and comparable petrological evidences in southern Granulite terrain, Eastern Ghats, Sri Lanka, Madagascar and East Antarctica granulite terrains has made a resounding impact on the assembly of Gondwana Supercontinent. Grew and Manton (1986), Katz (1989), Yoshida *et al.* (1992), have presented a model for the juxtaposition of India, Sri Lanka and Antarctica for Gondwana ensemble during Proterozoic to Mesozoic times. The ages of production, accretion and reworking in the southern Indian shield plays a very important part in the reconstruction of the ancient supercontinents (Santosh *et al.*, 2003). The Madurai Block seems to be a part of the worldwide network of orogenic belts that is centred along 1.8 Ga, hence accretion of Columbia occurred at the same time as accretion of the Madurai Block to Archaean India (Rogers and Santosh, 2002). Recent data confirm the earlier evidences that all of southern India underwent resetting of isotopic systems during the final accretion of Gondwana around Pan–African orogeny (~ 0.5 Ga). The Eastern Ghats Mobile Belt is also equally vital in East Gondwana reconstruction. The continuity of major shear zone network in the dispersed fragments (Fig. 5) further strengthen the idea of fusing them as supercontinent (Chetty *et al.*, 2003). The idea of considering Godavari rift as continuation of Napier-Rayner boundary in EGMB and new mineral ages reported by Mezger and Cosca (1999), Rickers *et al.* (2001) have opened new vistas for addressing the promising implication for the Indo-Antarctic correlation in respect of East Gondwana configuration (Sengupta *et al.*, 1999).

CONCLUSIONS

The knowledge on recently discovered Ultrahigh-temperature metamorphism and its contribution in understanding the evolution of continental orogenic belts has stolen the limelight during the recent times. Accomplishments in metamorphic petrology concern the future as well with a goal to investigate the effect of temperature and pressure on as wide range of materials, as possible to learn more about the nature of the deep earth. Today, the UHT metamorphism is in an excited stage. Geochemists, geochronologists, experimental petrologists and geophysicists have all embraced the UHT field, reaching within deep crust and beyond to mantle as well. UHT metamorphism of granulites stimulates curiosity over the possible existence of supercontinents and presents a piece of evidence that are hospitable to reconstruction of supercontinents.

ACKNOWLEDGMENTS

It is apt to describe accelerating pace in metamorphic petrology which is spurred by old-fashioned human curiosity and new-fashioned technology in honour of Professor R.S. Sharma whose wisdom and science as a teacher made an everlasting impression on our minds. We thank Dr. Harel Thomas for the invitation to write this review paper and for his subsequent patience for our delayed contribution. This work was supported by DST, New Delhi (ESS/16/077/96). The help received from Prof. R.K. Lal in the preparation of this manuscript is gratefully acknowledged.

REFERENCES

Aftalion, M., Bowes, D.R., Dash, B. & Dempster, T.J., 1988. Late Proterozoic charnockites in Orissa, India: a U-Pb and Rb-Sr isotopic study. *Jour. Geol.*, **v. 96**, 663-76.

Ashwal, L.D., Morgan, P. & Hoisch, T.D., 1991. Tectonics and heat sources for granulite metamorphism of supracrustal-bearing terranes. *Precamb. Res.*, v. **55**, 525-38.

Audibert, N., Hensen, B.J. & Bertrand, P., 1995. Experimental study of phase relations involving osumilite in the system K_2O-FeO-MgO-Al_2O_3-SiO_2-H_2O at high pressure and temperature. *Jour. Met. Geol.*, **v. 13**, 331-44.

Baba, S., 2003. Two Stages of sapphirine formation during prograde and retrograde metamorphism in the palaeoproterozoic Lewisian Complex in South Harris, NW Scotland. *Jour. Petrol.*, **v. 44**, 329-54.

Bartlett, J.M., Harris, N.B.W., Hawkesworth, C.J. & Santosh, M., 1995. New isotopic constraints on the crustal evolution of South India and Pan-African granulitemetamorphism. *Mem. Geol. Soc. India*, **34**, 391-97.

Bernard-Griffiths, J., John, B.M. & Sen, S.K., 1987. Sm-Nd isotopes and REE geochemistry of Madras granulites, India: an introductory statement. *Precamb. Res.*, **37**, 343-55.

Bertrand, P., Ellis, D.J. & Green, D.H., 1991. The stability of sapphirine-quartz and hypersthene – sillimanite – quartz assemblages: an experimental investigation in the system FeO–MgO-Al_2O_3–SiO_2 under H_2O and CO_2 conditions. *Contrib. Mineral. Petrol.*, v. **108**, 55-71.

Bertrand, P., Ouzegane, K. & Kienast, J.R., 1992. P-T-X relationships in the Precambrian Al-Mg-rich granulites from In Ouzzal, Hoggar, Algeria. *Jour. Met. Geol.*, **v. 10**, 17-31.

Bhaskar Rao, Y.J., Chetty, T.R.K., Janardhan, A.S. & Gopalan, K., 1996. Sm-Nd and Rb-Sr ages and P-T history of the Archean Sittampundi and Bhavani layered meta-anorthosite complex in Cauvery shear zone, South India: evidence for Neoproterozoic reworking of Archean crust. *Contrib. Mineral. Petrol.*, **v. 125**, 237-50.

Bhattacharya, S. & Kar, R., 2002. High–temperature dehydration melting and decompressive P-T path in a granulite complex from the Eastern Ghats, India. *Contrib. Mineral. Petrol.*, **v. 143**, 175-91.

Bhattacharya, S., Kar, R., Teixeira, W. & Basei, M., 2003. High–temperature crustal anatexis in a clockwise P-T-t path: evidence from a granulites-granitoid suite in Eastern Ghats Belt, India. *Jour. Geol. Soc. London*, **v. 160**, 39-46.

Black, L.P., Harley, S.L., Sun, S.S. & McCulloch, M.T., 1987. The Rayner Complex of East Antarctica: complex isotopic systematics within a Proterozoic mobile belt. *Jour. Met. Geol.*, **v. 5**, 1-26.

Bohlen, S.R., 1987. Pressure-temperature time path and a tectonic model for the evolution of granulites. *Jour. Geol.*, **v. 95**, 617-32.

Bohlen, S.R., 1991. On the formation of granulites. *Jour. Met. Geol.*, **v. 9**, 223-24.

Bohlen, S.R. & Mezger, K., 1989. Origin of granulite terranes and the formation of lowermost continent crust, *Science*, **v. 244**, 326-29.

Bose, S., Fukuoka, M., Sengupta, P. & Dasgupta, S., 2000. Evolution of high Mg-Al granulites from Sunkarametta, Eastern Ghats, India: evidence for a lower crustal heating-cooling trajectory. *Jour. Met. Geol.*, **v. 18**, 223-40.

Brown, M., 1993. P–T-t evolution of orogenic belts and the causes of regional metamorphism. *Jour. Geol. Soc.*, London, **v. 150**, 227-41.

Brown, M., 1998. Unpairing Metamorphic Belts: P-T Paths and A Tectonic Model for the Royoke Belt, Southwest Japan. *Jour. Met. Geol.*, **v. 16**, 3-22.

Caby, R., 1996. A review of the In Ouzzal granulitic terrane (Tuareg shield, Algeria): its significance within the Pan-African Trans-Saharan Belt. *Jour. Met. Geol.*, **v. 14**, 659-66.

Carrington, D.P. & Harley, S.L., 1995a. Partial melting and phase relations in high-grade metapelites: an experimental petrogenetic grid in the KFMASH system. *Contrib. Mineral. Petrol.*, **v. 120**, 270-91.

Carrington, D.P. & Harley, S.L., 1995b. The stability of osumilite in metapelitic granulites. *Jour. Met. Geol.*, **v. 13**, 613-25.

Clarke, G.L., 1988. Structural constraints on the Proterozoic reworking of Archaean crust in the Rayner Complex, MacRobertson and Kemp Land coast, East Antarctica. *Precamb. Res.*, **v. 40/41**, 137-56.

Cliff, R.A., Rex, D.C. & Guise, P.G., 1998. Geochronological studies of Proterozoic crustal evolution in northern Outer Hebrides. *Precamb. Res.*, **v. 91**, 91-98.

Dallwitz, W.B., 1968. Coexisting sapphirine and quartz in granulites from Eanderby Land. Antarctica. *Nature*, **v. 219**, 476-77.

Dasgupta, S., Sanyal, S., Sengupta, P. & Fukuoka, M., 1994. Petrology of granulites from Anakapalle — Evidence for Proterozoic decompression in the Eastern Ghats, India. *Jour. Petrol.*, **v. 35**, 433-59.

Dasgupta, S., Sengupta, P., Ehl, J. Raith, M. & Bardhan, S., 1995. Reaction textures in a suite of spinel granulites from the Eastern Ghats Belt, India: evidence for polymetamorphism, a partial petrogenetic grid in the system KFMASH and the roles of ZnO and Fe_2O_3. *Jour. Petrol.*, **v. 36**, 435-61.

Dasgupta, S., Ehl, J., Raith, M.M., Sengupta, P. & Sengupta Pr., 1997. Mid crustal contact metamorphism around the Chimakurthy mafic-ultramafic complex, Eastern Ghats Belt, India. *Contrib. Mineral. Petrol.*, **v. 129**, 182-97.

Dooley, D.F. & Patino Douce, A.E., 1996. Fluid–absent melting of F-rich phlogopite + rutile + quartz. *Am. Min.*, **v. 81**, 202-12.

Droop, G.T.R. & Bucher-Nurminen, K., 1984. Reaction textures and metamorphic evolution of sapphirine bearing granulites from the Gruf Complex, Italian Central Alps. *Jour. Petrol.*, **v. 25**, 766-803.

Ellis, D.J., 1980. Osumilite–sapphirine-quartz granulites from Enderby Land, Antarctica: P-T conditions of metamorphism, implication for garnet–cordierite equilibria and the evolution of the deep crust. *Contrib. Mineral. Petrol.*, **v. 74**, 201-10.

Ellis, D.J. & Green, 1985. Garnet–forming reactions in mafic granulites from Enderby Land, Antarctica–implications for geothermometry and geobarometry. *Jour. Petrol.*, **v. 26**, 633-62.

Ellis, D.J., Sheraton, J.W., England, R.N. & Dallwitz, W.B., 1980. Osumilite-sapphirine-quartz granulites from Enderby Land, Antarctica–mineral assemblages and reactions. *Contrib. Mineral. Petrol.*, **v. 72**, 123-43.

Fischel, D.P., Fuck, R.A. & Pimentel, M.M., 1999. Sm-Nd mineral isochrones and model ages of Anapolis-Itaucu Complex and associated granites, goias: new insights into the geological evolution of the Brasilia belt. *Boletim de Resumos Expandidos do 7º Simposio de Geologia do Centro Oeste e 10º Simposio de Geologia de Minas Gerais*, Brasilia, SBG, 88.

Friend, C.R.L. & Kinny, P.D., 2001. A reappraisal of the Lewisian Gneiss Complex: geological evidence for its tectonic assembly from disparate terranes in the Proterozoic. *Contrib. Mineral. Petrol.*, **v. 142**, 198-218.

Grew, E.S., 1980. Sapphirine + Quartz association from Archaean rocks in Enderby Land, Antarctica. *Am. Mineral.*, **v. 65**, 821-36.

Grew, E.S. & Manton, W.I., 1986. A new correlation of sapphirine granulites in Indo-Antarctic metamorphic terrain: Late Proterozoic dates from the Eastern Ghats. *Precamb. Res.*, **v. 33**, 129-39.

Haddoum, H., Choukroune, P. & Peucat, J.J., 1994. Structural evolution of the Precambrian In Ouzzal Massif (Central Sahara, Algeria). *Precamb. Res.*, **v. 65**, 155-66.

Harley, S.L., 1985. Garnet–orthopyroxene-bearing granulites from Enderby Land, Antarctica: metamorphic pressure–temperature–time evolution of the Archaean Napier Complex. *Jour. Petrol.*, **v. 26**, 819-56.

Harley, S.L., 1987. Precambrian geological relationships in high-grade gneisses of Rauer Island, East Antarctica. *Aust. Jour. Earth Sci.*, **v. 34**, 175-207.

Harley, S.L., 1988. Proterozoic granulites from Rauer Group, East Antarctica. I Decompressional pressure-temperature paths deduced from mafic and felsic gneisses. *Jour. Petrol.*, **v. 29**, 1059-95.

Harley, S.L., 1989. The origins of granulites: a metamorphic perspective. *Geol. Mag.*, **v. 126**, 215-47.

Harley, S.L. & Black, L.P., 1987. The Archaean geological evolution of Enderby Land Antarctica. In: Park, R.G., Tarney, J. (eds.) Evolution of the Lewisian and comparable high-grade terranes. *Geol, Soc. Lond. Spl. Pub.*, **v. 138**, 285-96.

Harley, S.L. & Buick, I.S., 1992. Wollastonite–scapolite assemblages as indicators of granulite pressure-temperature-fluid histories: the Rauer Group, East Antarctica. *Jour. Met. Geol.*, **v. 33**, 693-728.

Harley, S.L., 1998a. On the occurrence and characterization of ultrahigh-temperature crustal metamorphism. In: What drives metamorphism and metamorphic reactions? (eds. P.J. Treloar & P.J. O'Brien), *Geol. Soc., London*, **v. 138**, 81-107.

Harley, S.L., 1998b. Ultrahigh temperature granulite metamorphism (1050°C, 12 kbar) and decompression in garnet (Mg 70)–orthopyroxene–sillimanite gneisses from the Rauer Group, East Antarctica. *Jour. Met. Geol.*, **v. 16**, 541-62.

Harley, S.L., Hensen, B.J. & Sheraton, J.W., 1990. Two-stage decompression in orthopyroxene-sillimanite granulites from Forefinger Point Enderby Land Antarctica: implication for the evolution of the Archean Napier Complex. *Jour. Met. Geol.*, **v. 8**, 591-613.

Harley, S.L. & Hensen, B.J., 1990. Archaean and Proterozoic high-grade terranes of East Antarctica (40-80°E): a case study of diversity in granulite facies metamorphism. In: High–temperature Metamorphism and crustal anatexis (eds. Ashwarth, J.R. & Brown, M.) *Unwin Hyman*, London, 320-70.

Harley, S.L. & Fitzsimons, I.C.W., 1991. Pressure-temperature evolution of metamorphic granulites in a polymetamorphic terrane: the Rauer Group, East Antarctica. *Jour. Met. Geol.*, **v. 9**, 231-43.

Harley, S.L. & Motoyoshi, Y., 2000. Al zoning in orthopyroxene in a sapphirine quartzite: evidence for > 1120°C UHT metamorphism in Napier Complex, Antarctica, and implications for the entropy of sapphirine. *Contrib. Mineral. Petrol.*, **v. 138**, 293-307.

Hensen, B.J., 1977. The stability of osumilite in high-grade metamorphic rocks. *Contrib. Mineral. Petrol.*, **v. 64**, 197-204.

Hensen, B.J., 1987. P-T grid for the silica-undersaturated granulites in the system MAS (n+4) and FMAS (n+3) – tools for the derivation of P-T paths of metamorphism. *Jour. Met. Geol.*, **v. 5**, 255-71.

Hensen, B.J. & Green, D.H., 1973. Experimental study of the stability of cordierite and garnet in pelitic compositions at high pressures and temperatures. III Synthesis of experimental data and geological applications. *Contrib. Mineral. Petrol.*, **v. 38**, 151-66.

Hensen, B.J. & Harley, S.L., 1990. Graphical analysis of P-T-X relations in granulite facies metapelites. In: Ashworth, J.R. and Brown, M. (eds.) High Temperature Metamorphism and Crustal Anatexis. *Unwin Hyman*, London, 19-56.

Hensen, B.J. & Motoyoshi, Y., 1992. Osumilite–producing reactions in high-temperature granulites from Napier Complex, East Antarctica: tectonic implications. In: Yoshida, Y., Kaminuma, K. & Shiraishi, K. (eds.) Recent progress in Antarctic earth science. *Terrapub. Tokyo*, 87-92.

Hodges, K.V., 1991. Pressure-Temperature-time paths. *Annual Rev. Earth Planet. Sci.*, **v. 19**, 207-36.

Hokada, T., 2001. Feldspar thermometry in ultrahigh-temperature metamorphic rocks: Evidence of crustal metamorphism attaining ~1100°C in the Archean Napier Complex, East Antarctica. *Am. Mineral.*, **v. 86**, 932-38.

Holland, T.J.B., Babu, E.V.S.S.K. & Waters, D.J., 1996. Phase relations of osumilite and dehydration melting in pelitic rocks: a simple thermodynamic model for the KFMASH system. *Contrib. Mineral. Petrol.*, **v. 124**, 383-94.

Huerta, A.D., Royden, L.H. & Hodges, K.V., 1999. The effects of accretion, erosion and radiogenic heat on the metamorphic evolution of collisional orogens. *Jour. Met. Geol.*, **v. 17**, 349-66.

Jamieson, R.A., Beaumont, C., Fullsack, P. & Lee, B., 1998. Barrovian regional metamorphism: Where's the heat? In: What drives metamorphism and metamorphic reactions? (eds. P.J. Treloar & P.J. O'Brien), *Geol. Soc., London*, **v. 138**, 23-51.

Jayananda, M., Janardhan, A.S., Sivasubramanian, P. & Peucat, J.J., 1995. Geochronologic and isotopic constraints on granulite formation in the Kodaikanal area, Southern India. *Mem. Geol. Soc. India*, **no. 34**, 373-90.

Johnson, S.E. & Vernon, R.H., 1995. Stepping-stones and pitfalls in the determination of an anticlockwise P-T-t-deformation path: the low-P, high-T Cooma Complex, Australia. *Jour. Met. Geol.*, **v. 13**, 165-83.

Katz, M.B., 1989. Sri Lankan-Indian Eastern Ghats-Antarctica and the Australian Albany-Fraser Mobile Belt: Gross Geometry, Age Relationship and Tectonics in Precambrian Gondwanaland. *Jour. Geol.*, **v. 97**, 646-48.

Lal, R.K., 1997. Internally consistent calibrations for geothermobarometry of high-grade Mg-Al rich rocks in the MAS system and their application to sapphirine-spinel granulites of Eastern Ghats, India and Enderby Land, Antarctica. *Proc. Indian Acad. Sci. (Earth & Planet. Sci.)*, **v. 106**, 91-113.

Lal, R.K., Ackermand, D., Raith, M., Raase, P. & Seifert, F., 1984. Sapphirine-bearing assemblages from Kiranur, Southern India: A study of chemographic relationships in the Na_2O-FeO-MgO-Al_2O_3-SiO_2-H_2O system. *N. Jb. Miner. Abh.*, **v. 150**, 121-52.

Lal, R.K., Ackermand, D. & Upadhyay, H., 1987. P-T-X relationships deduced from corona textures in sapphirine-spinel-quartz assemblages from Paderu, Southern India. *Jour. Petrol.*, **v. 28**, 1139-68.

Martignole, J. & Martelat, J.E., 2003. Regional–scale Grenvillian–age UHT metamorphism in the Mollendo-Camana block (Basement of the Peruvian Andes). *Jour. Met. Geol.*, **v. 21**, 99-120.

Meißner, B., Deters, P., Srikantappa, C. & Köhler, H., 2002. Geochronological evolution of the Moyar, Bhavani and Palghat shear zones of southern India: Implications for east Gondwana correlations. *Precamb. Res.*, **v. 114**, 149-75.

Mezger, K. & Cosca, M.A., 1999. The thermal history of the Eastern Ghats Belt (India), as revealed by U-Pb and ^{40}Ar - ^{39}Ar dating of metamorphic and magmatic minerals: implications for the SWEAT correlation. *Precamb. Res.*, **v. 94**, 251-71.

Mohan, A., Ackermand, D. & Lal, R.K., 1986. Reaction textures and P-T-X trajectory in the sapphirine–spinel-bearing granulites from Ganguvarpatti, Southern India. *Nues. Jb. Miner. Abh.*, **v. 154**, 1-19.

Mohan, A. & Windley, B.F., 1993. Crustal trajectory of sapphirine-bearing granulites from Ganguvarpatti, South India: evidence for an isothermal decompression path. *Jour. Met. Geol.*, **v. 11**, 867-78.

Mohan, A., Prakash, D. & Motoyoshi, Y., 1996. Decompressional P-T history in sapphirine-bearing granulites from Kodaikanal, southern India. *Jour. Asian Earth Sci.*, **v. 14**, 231-43.

Mohan, A., Tripathi, P. & Motoyoshi, Y., 1997. Reaction history of sapphirine granulites and a decompressional P-T path in a granulite complex from the Eastern Ghats. *Proc. Indian Acad. Sci. (Earth & Planet. Sci.)*, **v. 106**, 115-29.

Mohan, A. & Jayananda, M., 1999. Metamorphism and isotopic evolution of granulites of southern India: reference to Neoproterozoic crustal evolution. *Gondwana Res.*, **v. 2**, 251-62.

Motoyoshi, Y. & Hensen, B.J., 1989. Sapphirine–quartz-orthopyroxene symplectites after cordierite in the Archaean Napier Complex, Antarctica: evidence for a counterclockwise P-T path? *Eur. Jour. Mineral*, **v. 1**, 467-71.

Motoyoshi, Y., Hensen, B.J. & Arima, M., 1993. Experimental study of high pressure stability limit of osumilite in the system $K_2O - MgO - Al_2O_3 - SiO_2$: implications for high temperature granulites. *Eur. Jour. Mineral*, **v. 5**, 439-45.

Motoyoshi, Y. & Ishikawa, M., 1998. Metamorphic and structural evolution of granulites from Rundvagshetta, Lutzow-Holm Bay, East Antarctica. In: *The Antarctic Region: Geological Evolution and Processes* (ed. C.A. Ricci), Terra Antarctica Publications, 100-10.

Moraes, R., Brown, M., Fuck, R.A., Camargo, M.A. & Lima, T.M., 2002. Characterization and P-T evolution of melt-bearing ultrahigh-temperature granulites: an example from the Anápolis–Itauçu complex of the Brasilia fold belt, Brazil. *Jour. Petrol.*, **v. 43**, 1673-705.

Nandakumar, V. & Harley, S.L., 2000. A reappraisal of the Pressure-Temperature Path of granulites from the Kerala Khondalite Belt, Southern India. *Jour. Geol.*, **v. 108**, 687-903.

Ouzegane, K. & Boumaza, S., 1996. An example of ultrahigh-temperature metamorphism: orthopyroxene–sillimanite-garnet, sapphirine-quartz and spinel-quartz parageneses in Al-Mg granulites from In Hihaou, In Ouzzal, Hoggar. *Jour. Met. Geol.*, **v. 14**, 693-708.

Ouzegane, K., Guiraud, M. & Kienast, J.R., 2003. Prograde and retrograde evolution in high–temperature corundum granulites (FMAS and FMASH systems) from In Ouzzal Terrane (NW Hoggar, Algeria). *Jour. Petrol.*, **v. 44**, 517-45.

Pattison, D.R.M., Chacko, T., Farquhar, J. & McFarlane, C.R.M., 2003. Temperatures of Granulite-facies Metamorphism: Constraints from Experimental Phase Equilibria and Thermobarometry Corrected for Retrograde Exchange. *Jour. Petrol.*, **v. 44**, 867-900.

Paul, D.K., Ray, Barman, T., McNaughton, M.J., Fletcher, I.R., Potts, P.J., Ramakrishnan, M. & Augustine, P.F., 1990. Archean-Proterozoic evolution of Indian charnockites; Sm-Nd, Pb-Pb, U-Pb and geochemical evidence from granulites of Eastern Ghats belt. *Jour. Geol.*, **v. 98**, 253-63.

Peucat, J.J., Vidal, P., Bernard-Griffiths, J. & Condie, K.C., 1989. Sr, Nd and Pb isotope systematics in the Archean low- to high-grade transition zone of southern India: syn-accretion vs post-accretion granulites. *Jour. Geol.*, **v. 97**, 537-50.

Pimentel, M.M., Fuck, R.A. & Fischel, D.P., 1999. Estudo isotopico Sm-Nd regional da porcao central da Faixa Brasilia: implicacoes para idade e origem dos granulitos do Complexo Anapolis Itaucu e rochas metassedimentares do Grupo Araxa. *Revista Brasoleira de Geociencias*, **v. 29**, 271-76.

Raith, M., Karmarkar, S. & Brown, M., 1997. Ultra-high-temperature metamorphism and multistage decompressional evolution of sapphirine granulites from the Palni Hill ranges, southern India. *Jour. Met. Geol.*, **v. 15**, 179-99.

Raith, M.M., Srikantappa, C., Buhl, D. & Koehler, H., 1999. The Nilgiri enderbites, South India: nature and age constraints on protolith formation, high-grade metamorphism and cooling history. *Precamb. Res.*, **v. 98**, 129-50.

Ravindra Kumar, G.R. & Chacko, T., 1994. Geothermobarometry of mafic granulites and metapelite from Palghat Gap, South India: petrological evidence for isothermal uplift and rapid cooling. *Jour. Meta. Geol.*, **v. 12**, 479-92.

Ravindra Kumar, G.R. & Raghavan, 1992. The incipient charnockites of transition zone, granulite zone and khondalite zone of South India: Contrasting mechanism and controlling factors. *Jour. Geol. Soc. India*, **v. 39**, 293-302.

Rickers, K., Mezger, M. & Raith, M., 2001a. Evolution of the continental crust in the Proterozoic Eastern Ghats Belt, India and new constraints for Rodinia reconstruction: implications from Sm-Nd, Rb-Sr and Pb-Pb isotopes. *Precamb. Res.*, **v. 112**, 183-10.

Rickers, K., Raith, M. & Dasgupta, S., 2001b. Multistage reaction textures in xenolithic high–MgAl granulites at Anakapalle, Eastern Ghats Belt, India: an example of contact polymetamorphism and infiltration-driven metasomatism. *Jour. Met. Geol.*, **v. 19**, 561-80.

Rogers, J.J.W. & Santosh, M., 2002. Configuration of Columbia, a Mesoproterozoic supercontinent. *Gond. Res.*, **v. 5**, 5-22.

Sandiford, M., 1985. The Metamorphic evolution of granulites at fyfe Hills: implications for Archaean crustal thickness in Enderby Land, Antarctica. *Jour. Met. Geol.*, **v. 3**, 155-78.

Santosh, M., Yokoyama, K., Biju-Sekhar, S. & Rogers, J.J.W., 2003. Multiple Tectonothermal Events in the Granulite Blocks of Southern India Revealed from EPMA Dating: Implications on the History of Supercontinents. *Gond. Res.*, **v. 6**, 29-63.

Sarkar, S., Santosh, M., Dasgupta, S. & Fukuoka, 2003. Very high density CO_2 associated with ultrahigh-temperature metamorphism in the Eastern Ghats granulite belt, India. *Geol. Soc. America.*, **v. 31**, 51-54.

Sen, S.K., Bhattacharya, S. & Acharyya, A., 1995. A multistage record in the Chilka Lake granulites: the epitome of the metamorphic evolution of Eastern Ghats, India? *Jour. Met. Geol.*, **v. 13**, 287-98.

Sengupta, P., Dasgupta, S., Bhattacharya, P.K., Fukuoka, M., Chakraborti, S. & Bhowmick, S., 1990. Petro-tectonic imprints in the sapphirine granulites from Anantagiri, Eastern Ghats mobile belt, India. *Jour. Petrol.*, **v. 31**, 971-96.

Sengupta, P., Sen, J., Dasgupta, S., Raith, M., Bhui, U.K. & Ehl, J., 1999. Ultra-high temperature metamorphism of metapelitic granulites from Kondapalle, Eastern Ghats belt: implications for the Indo-Antarctic correlation. *Jour. Petrol.*, **v. 40**, 1065-87.

Shaw, R. & Arima, M., 1998. A Corundum-Quartz Assemblage from the Eastern Ghats Granulite Belt, India: Evidence for High P-T Metamorphism? *Jour. Met. Geol.*, **v. 16**, 189-96.

Shaw, R.K., Arima, M., Kagami, H., Fanning, C.M., Shiraishi, K. & Motoyoshi, Y., 1997. Proterozoic events in the Eastern Ghats granulite belt, India: Evidence from Rb-Sr, Sm-Nd systematics and SHRIMP dating. *Jour. Geol.*, **v. 105**, 645-56.

Spear, F.S., 1993. Metamorphic phase equilibria and pressure-temperature-time-paths. *Mineral. Soc. Am.*, Michigan, U.S.A., 799.

Spear, F.S., Kohn, M.J. & Paetzold, S., 1995. Petrology of regional sillimanite zone, west-central New Hampshire, USA, with implications for the development of inverted isograd. *Am. Min.*, **v. 80**, 361-76.

Tassinari, C.G., Cordani, U.G., Correia, C.T., Nutman, A., Kinny, P. & Dias Neto, C.M., 1999. Dating of granulites by SHRIMP U-Pb systematics in Brazil: constraints for the age of the metamorphism of Proterozoic orogenics. Proceedings of 2nd South American Symposium on Isotope Geology, Cordoba. SEGEMAR, 371-73.

Thompson, A.B. & England, P.C., 1984. Pressure-temperature time paths of regional metamorphism II: their inference and interpretation using mineral assemblages in metamorphic rocks. *Jour. Petrol.*, **v. 25**, 929-54.

Thompson, A.B. & Ridley, J.R., 1987. Pressure-temperature-time (P-T-t) histories of orogenic belts. *Phil. Trans. Royal Soc. London*, A, **v. 321**, 27-45.

Treloar, P.J. & O'Brien, P.J., 1998. Introduction. In: What Drives Metamorphism and Metamorphic Reactions? P.J. Treloar and P.J. O'Brien (Eds.), *Mem. Geol. Soc. London*, **v. 138**, 1-5.

Tsunogae, T., Santosh, M., Osanai, Y., Owada, M., Toyoshima, T. & Hokada, T., 2002. Very high density carbonic fluid inclusions in sapphirine–bearing granulites from Tonagh Island in Archean Napier Complex, East Antarctica: Implication for CO_2 infiltration during ultrahigh–temperature (T > 1100°C) metamorphism. *Contrib. Mineral. Petrol.*, **v. 143**, 279-99.

Unnikrishnan-Warrier, C., Yoshida, M., Kagami, H. & Santosh, M., 1995. Sm-Nd and Rb-Sr mineral isochron ages of Madras charnockite, and evaluation of garnet Isochrometry in granulite. *Mem. Geol. Soc. India*, **No. 34**, 25-45.

Vielzeuf, D. & Montel, J.M., 1994. Partial melting of metagreywackes. Part I. Fluid-absent experiments and phase relationships. *Contrib. Mineral. Petrol.*, **v. 117**, 375-93.

Vielzeuf, D., Clemens, J.D., Pin, C. & Moinet, E., 1990. Granites, granulites and crustal differentiation. In: *Granulites and crustal evolution* (eds. Vielzeuf, D. & Vidal, Ph., Kluwer Academic Publishers, 59-85.

Vernon, R.H., 1996. Problems with inferring P-T-paths in low-P granulite facies rocks. *Jour. Met. Geol.*, **v. 14**, 143-53.

Yoshida, M., Funaki, M. & Vitanage, P.W., 1992. Proterozoic to Mesozoic East Gondwana: the juxtaposition of India, Sri Lanka and Antarctica. *Tectonics*, **v. 11**, 381-91.

SECTION – V

GEOCHRONOLOGY & GEOCHEMISTRY

14

SINGLE ZIRCON DATING OF HYPERSTHENE-BEARING GRANITOID FROM BALARAM-ABU ROAD AREA, SOUTHERN PART OF THE ARAVALLI MOUNTAINS, NW INDIA — IMPLICATIONS FOR MALANI MAGMATISM RELATED THERMAL EVENT

A.B. ROY[1], ALFRED KRÖNER[2], VIVEK LAUL[3] AND RITESH PUROHIT[4]

ABSTRACT

Single zircon dating was undertaken on a hypersthene-bearing granitoid, which was emplaced together with a suite of gabbro-norite bodies in the Balaram-Abu Road area in the southern part of the Aravalli Mountains, northwest India. A mean age of 757.8 ± 0.9 Ma for the hypersthene-bearing granitoid reflects a thermal (igneous) event, which post-dated the closing of the Mesoproterozoic Delhi basin as well as the emplacement of 850 Ma synorogenic Erinpura granites in the region. We consider this young thermal event in the southern part of the Aravalli Mountains to be related to Malani magmatism recorded in southwestern Rajasthan.

KEY WORDS

Hypersthene-bearing granitoid, Single Zircon dating, Malani Magmatism and thermal event.

INTRODUCTION

An association of mafic plutonic rocks along with hypersthene-bearing granitoid and metasedimentary/metavolcanic rocks occurs in the southernmost part of the Aravalli Mountains and constitutes a triangular, wedge-shaped outcrop east and southeast of Abu Road in the southern Rajasthan and northern Gujarat area of the northwestern India (Fig. 1 A-C). The association comprises gabbro-norites, mafic granulites (meta-norite) along with hypersthene-bearing granitoid including charnockite, their hydrated varieties, pelitic and calc-silicate granulites. The plutonic association (loosely described as granulites in this contribution) occurs in tectonic contact with the lower-grade metasedimentary and metavolcanic rocks of the Delhi Supergroup in the east. The western contact of these granulites with the granitoids

1. INSA Senior Scientist, JNV University, Jodhpur. *Mailing Address*: 9 Mayurvan, Panerionki Madri, Udaipur 313 002. e-mail: abroy_g@yahoo.com. *Present Address*: Niloy Apartment, Flat 2G, 46A, Ramanath Das Road, Kolkata-700 031.
2. Institut für Geowissenshaften, Universität, Mainz, 55099 Mainz, Germany.
3. House No. 805, Sector 10, Malviya Nagar, Jaipur-302 017.
4. Department of Geology, Sirohi Govt. College, Sirohi, Rajasthan.

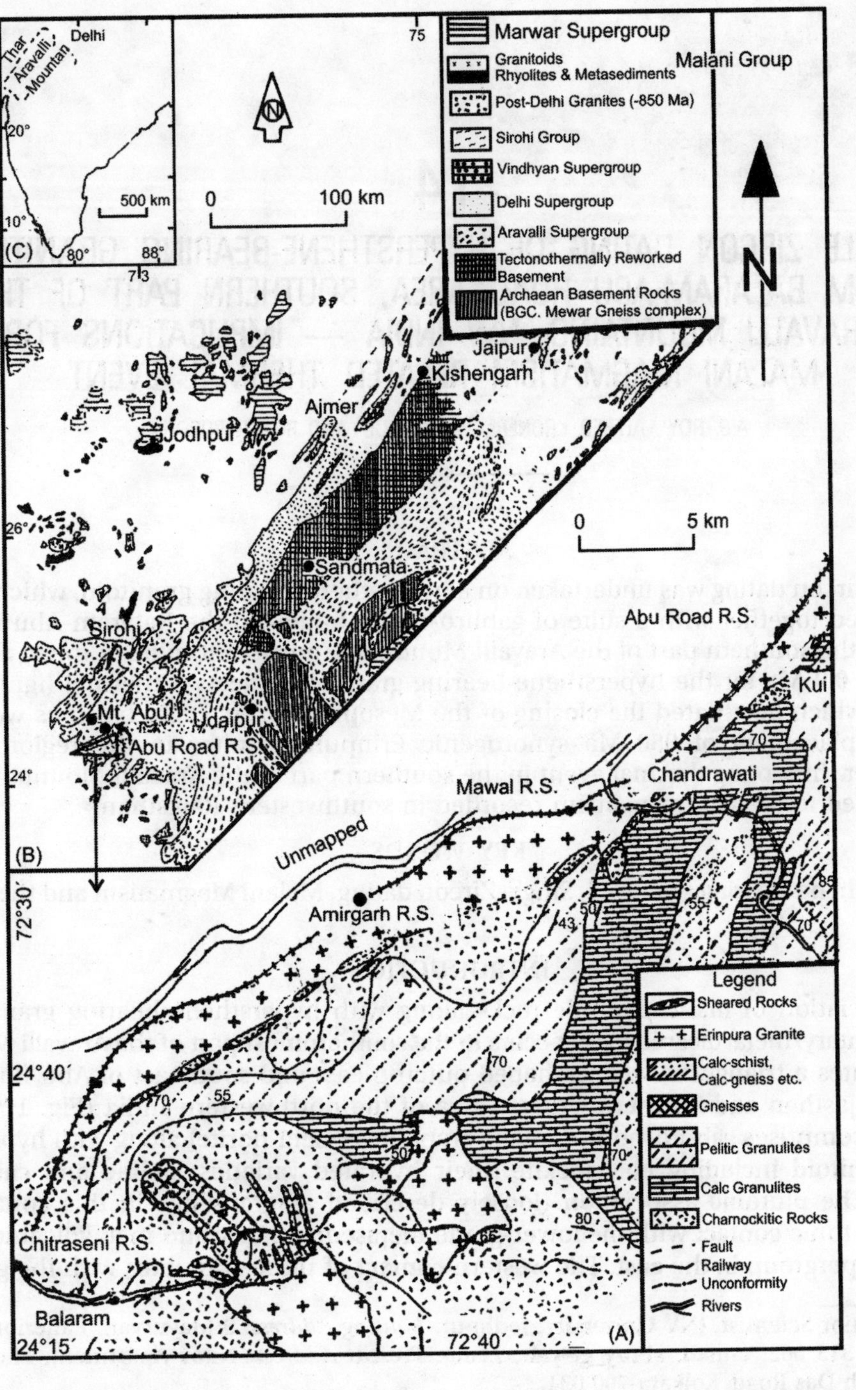

Fig. I. A. Geological map of the Balaram-Abu Road area, in southern Rajasthan and northern Gujarat. Modified after Desai *et al.* (1978). B. Geological map of the Aravalli Mountains showing the location of the map area in Fig. 1A. After Roy and Jakhar (2002). C. Map of India showing location of the Aravalli Mountains.

of the Mount Abu 'massif' in the west is marked by a major fault, known as the Kui-Chitraseni Fault (Desai *et al.*, 1978). The southern margin of the triangular outcrop is against the Jaisalmer-Barwani lineament, which marks the abrupt truncation of the Aravalli Mountains in the south.

The lithostratigraphic relationship of the plutonic rocks with the Mesoproterozoic Delhi Supergroup (Roy and Jakhar, 2002) remains highly controversial. Amongst the earlier workers, Coulson (1933), Sharma and Nandy (1936), and Heron and Ghosh (1938) considered the present group of rocks as a part of the 'Delhi system' (re-designated by the later workers as the Delhi Supergroup; Gupta *et al.*, 1980; Sinha-Roy *et al.*, 1998). According to Crawford (1970), the presence of granulites near the southern termination of the Aravalli Mountains is related to the increasing metamorphism from amphibolite-facies near Ajmer in the north to the granulite-facies rocks in the south. However, other workers, mainly on the consideration of metamorphic grade, correlated the granulites with the Palaeoproterozoic to Archaean pre-Delhi basement (Desai and Patel, 1992; Desai *et al.*, 1978; Limaye and Desai, 1995, 1999; Patel *et al.*, 1985; Srikarni *et al.*, 1999; 2000). Biswal (1998), on the other hand, suggested that the granulites are the products of continental magmatism during subduction-related event associated with the closure of the basin into which the Delhi Supergroup rocks were deposited.

We undertook the single zircon dating of a hypersthene-bearing granitoid of the Balaram-Abu Road area in order to resolve the temporal relationships with the Delhi Supergroup and the possible tectonic implications of the granulite wedge in southern Aravalli Mountains.

GEOLOGICAL SETTING

The geological map of the Balaram-Abu Road area (Fig. 1 A) is based on the pioneering works of Desai *et al.* (1978), and shows a dominance of 'charnockitic rocks' in association with the pelitic and calc-silicate granulites and with distinct bodies of granites and granitoid gneiss of various dimensions. Re-examination of these rocks in the field and under microscope revealed that the charnockites mapped by Desai *et al.* (1978) actually represent a suite of gabbro-norite and mafic granulites (meta-norite) surrounded by hypersthene-bearing felsic rocks including charnockite. A narrow wollastonite-bearing thermal aureole was observed where the mafic bodies are capped by the calcareous metasediments. At several places, the foliation planes in the adjacent metasedimentary rocks wrap around the igneous bodies (Biswal, 1988). In the western most part of the granulite wedge, these mafic bodies occur as a series of isolated 'domes' that abut against the Kui-Chitraseni Fault. Veins of hypersthene-bearing felsic rocks, which also include charnockite and their retrograde variants, occur within the gabbro-norite-mafic granulite bodies (Biswal *et al.*, 1998).

A complex deformation pattern involving five phases of folding and ductile shearing has been recorded in the pelitic and calc-silicate rocks of the area (Biswal, 1988), whereas most of the gabbro-norite bodies are massive and display no macroscopic evidence of deformation. The plutonic bodies also do not show evidence of metamorphic recrystallisation. The complexity observed in the structural geometry of the supracrustal host rocks, according to Roy and Jhakhar (2002), is the consequence of multiple plutonic intrusions of the gabbro-norite suite in the region post-dating the deformation and metamorphism associated with the Delhi orogenic cycle. Granulite facies metamorphism in the area, has been interpreted to be spatially and temporally related to the emplacement of the gabbro-norite (mafic granulite) intrusions (Biswal *et al.*, 1998).

ANALYTICAL PROCEDURES

Separation of zircon grains followed standard methods of crushing followed by gravity and magnetic separation and final hand picking of the grains. Most grains were studied under the binocular microscope for their shapes and crystal characteristics.

Isotopic measurements were carried out on a Finnigan-MAT 261 mass spectrometer at the Max-Planck-Institut für Chemie at Mainz. For isotopic measurements of single zircons, the method developed by Kober (1986, 1987) was used, which involves repeated evaporation and deposition of Pb from chemically untreated single zircons in a double-filament arrangement. The precise laboratory procedures as well as comparisons with conventional and ion-microprobe zircon ages are summarized in Kröner and Todt (1988), and Kröner and Hegner (1998).

The calculated ages and uncertainties are based on the means of all ratios evaluated and their 2σ (mean) errors. The mean ages and errors for several zircons from the same sample are presented as weighted means of the entire population. The analytical data are presented in Table 1, whereas the $^{207}Pb/^{206}Pb$ age spectra are shown in the histogram that permits visual assessment of the data-distribution from which the ages are derived (Fig. 2). The evaporation technique provides only Pb isotope ratios, and there is no *a priori* way to determine whether the measured $^{207}Pb/^{206}Pb$ ratio reflects a concordant age. Thus, the $^{207}Pb/^{206}Pb$ age determined by this method is necessarily a minimum age.

SAMPLE COMPOSITION, RESULTS AND DISCUSSION

A sample of hypersthene-bearing granitoid (BAL 99/2, Table 1) was collected from a quarry face north of the Balaram Temple. The dominant rock in the quarry is a noritic gabbro showing igneous textures under the microscope. The sample was collected from one of the several large and small dykes of hypersthene-bearing granitoid rocks, which intruded into the gabbro-norite body. The mineral composition of sample BAL 99/2 is as follows: quartz-plagioclase ± hypersthene-hornblende-biotite. Stringers and patches of hypersthene

Table 1: *Isotopic data from single grain zircon evaporation of sample BAL 99/2, Rajasthan, India.*

Sample Number	Zircon colour and morphology	Grain #	Mass scans[1]	Evaporation temp. in °C	Mean $^{207}Pb/^{206}Pb$ ratio[2] and 2-σm error	$^{207}Pb/^{206}Pb$ age and 2-σm error
BAL 99/2	clear, transparent,	1	88	1598	0.064471±44	757.1±1.5
	long-prismatic,	2	132	1599	0.064480±23	757.4±0.8
	idiomorphic	3	165	1597	0.064494±17	757.9±0.5
		4	165	1598	0.064521±19	758.8±0.6
		5	176	1600	0.064485±16	757.6±0.5
mean of 5 grains		1-5	726		0.064493±10	*757.8±0.9

1 Number of $^{207}Pb/^{206}Pb$ ratios evaluated for age assessment.

2 Observed mean ratio corrected for non-radiogenic Pb where necessary. Errors based on uncertainties in counting statistics.

* Error based on reproducibility of internal standard.

occur as relics surrounded by hornblende and biotite. The granitoid rock passes laterally into massive charnockite.

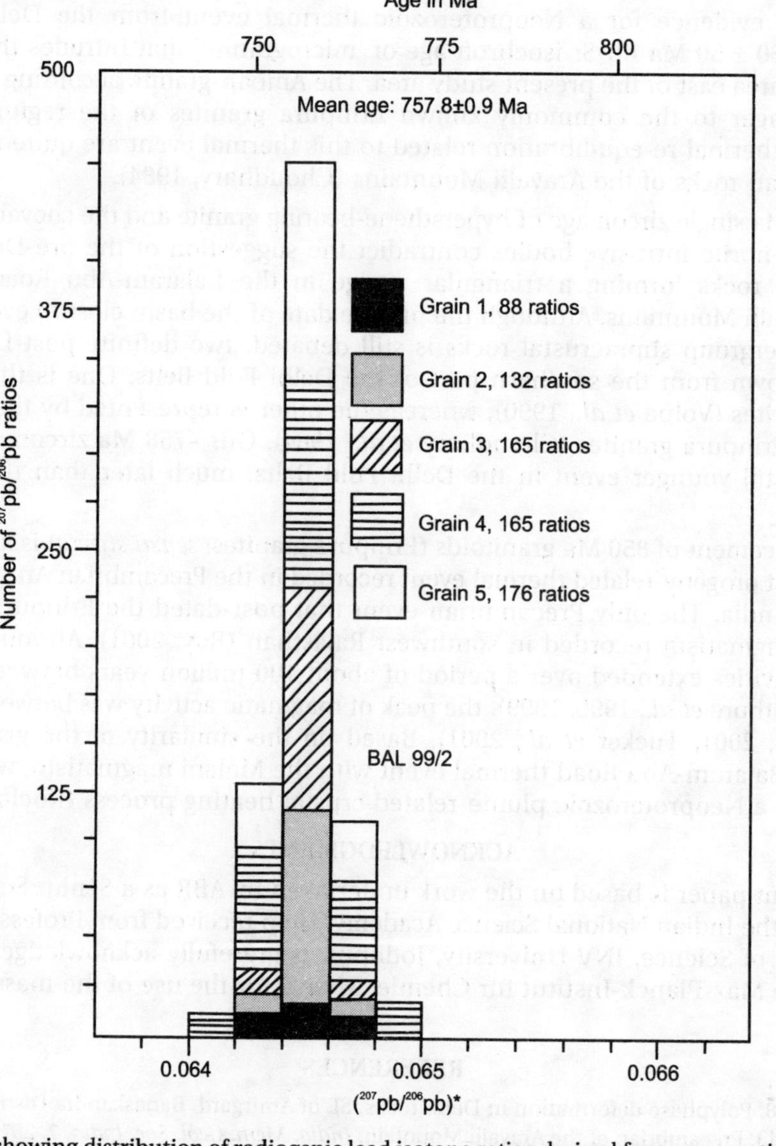

Fig. 2. Histogram showing distribution of radiogenic lead isotope ratios derived from evaporation of five single zircons from the hypersthene-bearing granitoid sample BAL 99/2, Balaram Temple, south of Mt. Abu, integrated from 269 ratios. Mean age is given with 2 sigma (mean) error.

Zircons separated from sample BAL 99/2 are light brown to light honey-yellow in colour and are perfectly euhedral with well-developed pyramidal faces. Five grains were evaporated individually and yielded identical isotopic ratios that combine to a mean $^{207}Pb/^{206}Pb$ age of 757.8 ± 0.9 Ma (Table 1, Fig. 2). In view of the igneous nature of the zircons, we interpret this age as reflecting the time of emplacement of the granitoid. Correlation of the zircon age with emplacement of the gabbro-norite (mafic granulite) bodies is based on the suggestion that the intrusions were coeval (cf. Biswal *et al.*, 1998).

Additional evidence for a Neoproterozoic thermal event from the Delhi Fold Belts comes from 760 ± 50 Ma Rb-Sr isochron age of 'microgranite' that intrudes the Delhi rocks in the Ambaji area east of the present study area. The Ambaji-granite according to Choudhary (1984) is younger to the commonly known Erinpura granites of the region. Records of mineral scale thermal re-equilibration related to this thermal event are quite widespread in the Precambrian rocks of the Aravalli Mountains (Choudhary, 1984).

The ~758 Ma single zircon age of hypersthene-bearing granite and the coeval emplacement of the gabbro-norite intrusive bodies contradict the suggestion of the pre-Delhi basement status for the rocks forming a triangular wedge in the Balaram-Abu Road area of the southern Aravalli Mountains. Although the precise date of the basin closure event deforming the Delhi Supergroup supracrustal rocks is still debated, two definite post-Delhi intrusive events are known from the southern part of the Delhi Fold Belts. One is the intrusion of ~1000 Ma diorites (Volpe *et al.*, 1990), whereas the other is represented by the intrusions of the ~850 Ma Erinpura granites (Choudhary *et al.*, 1984). Our ~758 Ma zircon age, therefore, represents a still younger event in the Delhi Fold Belts, much later than the Delhi basin closing event.

The emplacement of 850 Ma granitoids (Erinpura granites, *sensu stricto*) is also considered as the youngest orogeny-related thermal event recorded in the Precambrian Aravalli Mountains of northwest India. The only Precambrian event that post-dated the Erinpura intrusions is the Malani magmatism recorded in southwest Rajasthan (Roy, 2001). Although the Malani magmatic activities extended over a period of about 100 million years between ca. 780 and ca. 680 Ma (Rathore *et al.*, 1996; 1999), the peak of magmatic activity was between 750-755 Ma (Torsvik *et al.*, 2001; Tucker *et al.*, 2001). Based on the similarity of the granite ages, we correlate the Balaram-Abu Road thermal event with the Malani magmatism, which has been interpreted as a Neoproterozoic plume-related crustal heating process (Kochar, 1984).

ACKNOWLEDGMENTS

The present paper is based on the work undertaken by ABR as a Senior Scientist scheme sponsored by the Indian National Science Academy. Help received from Professor B.S. Paliwal, Dean, Faculty of Science, JNV University, Jodhpur, is gratefully acknowledged. We are also grateful to the Max-Planck-Institut für Chemie, Mainz, for the use of the mass spectrometer facilities.

REFERENCES

Biswal, T.K., 1988. Polyphase deformation in Delhi rocks, SE of Amirgard, Banaskantha District of Gujarat. In: A.B. Roy (ed.), Precambrian of the Aravalli Mountain, *India. Mem. Geol. Soc. India*, **7**, 267-78.

Biswal, T.K, Gyani, K.C., Parthasarathy, R. & Pant, D.R., 1998. Implications of the geochemistry of the pelitic Granulites of the Delhi Supergroup, Aravalli Mountain belt. Northwestern India. *Precam. Res.*, **87**, 75-85.

Choudhary, A.K., 1984. Precambrian geochronology of Rajasthan, Western India. Unpublished Ph.D. thesis, Gujarat University, Ahmedabad, 182.

Choudhary, A.K., Gopalan, K. & Sastry, C.A., 1984. Present status of the geochronology of the Precambrian rocks of Rajasthan. *Tectonophysics*, **105**, 131-40.

Coulson, A.L., 1933. The geology of Sirohi State, Rajputana. *Mem. Geol. Surv. India*, **63**, Pt-1, 166.

Crawford, A.R., 1970. The Precambrian geochronology of Rajasthan and Bundelkhand, Northern India. *Canadian J. Earth Sci.*, **7**, 91-110.

Desai, S.J. & Patel, M.P., 1992. Granulite rocks of Banaskantha Dist., N. Gujarat, India. In: A Barbo-Kyriadidis (ed), *High-grade metamorphics*. Theophrastus Pub., Athens, Greece, 103-21.

Desai, S.J., Patel, M.P. & Merh, S.S., 1978. Polymetamorphites of Balaram-Abu Road area, north Gujarat and southwest Rajasthan. *J. Geol. Soc. India,* **19**, 383-94.

Gupta, S.N., Arora, Y.K., Mathur, R.K., Iqballuddin, Prasad, B., Sahai, T.N. & Sharma, S.B., 1980. Lithostratigraphic map of Aravalli range. *Geol. Soc. India*. Hyderabad.

Heron, A.M. & Ghosh, P.K., 1938. The geology of Palanpur, Danta and part of Idar State. *Rec. Geol. Survey India,* **79**, 367-412.

Kober, B., 1986. Whole-grain evaporation for $^{207}Pb/^{206}Pb$ age investigation on single zircons using a double-filament thermal ion source. *Contrib. Miner. Petrol.*, **93**, 482-90.

Kober, B., 1987. Single-zircon evaporation combined with Pb^+ emitter-bedding for $^{207}Pb/^{206}Pb$ age investigations using thermal ion mass spectrometry, and implications to zirconology. *Contrib. Miner. Petrol.*, **96**, 63-71.

Kochar, N., 1984. Malani Igneous Suite: Hot-spot magmatism and cratonization of the northern part of the Indian Shield, India, *J. Geol. Soc. India,* **42**, 449-76.

Kröner, A. & Todt, W., 1988. Single zircon dating constraining the maximum age of Berberton Greenstone belt, southern Africa. *J. Geophysic. Res.*, **93**, 5329-37.

Kröner, A. & Hegner, E., 1998. Geochemistry, single zircon ages and Sm-Nd systematics of granitoid rocks from Gory Sowie (Owl) Mts. Polish West Suetes: evidence for early Palaeozoic arc-related plutonism. *J. Geol. Soc. London,* **155**, 711-24.

Limaye, M.A. & Desai, S.J., 1995. Petrological studies of metamorphites of Danata area, north Gujarat. In: N. Desai and S. Ganapati (Eds.), *Proc. Nat. Sem on Recent Researches in Geology of Western India.* MS University, Vadodara, 87-94.

Limaye, M.A. & Desai, S.J., 1999. Metamorphic history of Precambrian rocks of Danta area, Banaskantha District, North Gujarat. In: N.G.K. Murthy and V. Rama Mohan (eds.), Charnockite and Granulite facies Rocks. *Geologists' Association of Tamil Nadu, Chennai, India,* 127-34.

Patel, M.P., Desai, S.J. & Merh, S.S., 1985. Structural and metamorphic studies of the Precambrian rocks of Balaram-Swaroopgunj area of North Gujarat and SW Rajasthan. *Proc. Indian Geol. Cong.* (5th Session, I.I.T., Mumbai) 65-75.

Rathore, S.S., Venkatesh, T.R. & Srivastava, R.K., 1996. Rb-Sr and Ar-Ar systematics of Malani volcanic rocks of southwest Rajasthan: evidence for a younger post-crystaline thermal event. *Proc. Indian Acad. Sci. (Earth Planet. Sci.),* **105**, 131-41.

Rathore, S.S., Venkatesh, T.R. & Srivastava, R.K., 1999. Rb-Sr isotope dating of Neoproterozoic (Malani Group) magmatism from southwest Rajasthan, India: evidence of younger Pan-African thermal event by Ar^{40}-Ar^{39} studies. *Gond. Res.*, **2 (2)**, 271-81.

Roy, A.B. & Jakhar, S.R., 2001. Geology of Rajasthan-Northwest India: Precambrian to Recent. Scientific Publishers (India), Jodhpur, 421.

Sharma, N.L. & Nandy, N.V., 1936. A note on the petrological classification of the basic intrusives of the Danta state. N. Gujarat. *Proc. Ind. Acad. Sci.*, **3(B)**, 366-76.

Sinha-Roy, S., Malhotra, G. & Mohanty, M., 1998. Geology of Rajasthan. *Geol. Soc. India,* Bangalore, 278.

Srikarni, C., Desai, S.J. & Patel, M.P., 1999. Tectono-stratigraphic significance of granulites of North Gujarat, W. India, In: N.G.K. Murthy and V. Ram Mohan (eds.), Charnockite and Granulite facies Rocks. Geologists' Association of Tamil Nadu, Chennai, India, 175-86.

Srikarni, C., Desai, S.J. & Limaye, M.A., 2000. Plumose symplectites—A case of osumilite break-down in Pelitic granulites of North Gujarat, W. India. In: K.C. Gyani and P. Kataria (eds.), Tectonomagmatism, Geochemistry and Metamorphism of Precambrian Terrains. Department of Geology, ML Sukhadia University, Udaipur, 265-74.

Torsvik, T.H., Carter, L.M., Ashwal, L.D., Bhushan, S.K., Pandit, M.K. & Jamtveit, B., 2001. Rodinia redefined or obscured: palaeomagnetism of the Malani igneous suite (NW India). *Precambrian Res.*, **108**, 319-33.

Tucker, R.D., Ashwal, L.D. & Torsvik, T.H., 2001. U-Pb geochronology of Seychelles granitoid: a Neoproterozoic continental arc fragment. *Earth and Planetary Science Letter*, **187**, 27-38.

Volpe, A.M. & Macdaugall, J.D., 1990. Geochemistry and isotopic characteristic of mafic (Phulad Ophiolite) and related rocks in the Delhi Supergroup, Rajasthan, India: implications for rifting in the Proterozoic. *Precambrian Res.*, **48**, 167-91.

15

GEOCHEMISTRY AND PETROGENESIS OF THE HIGH GRADE GRANULITES FROM KODAIKANAL, SOUTH INDIA

D. PRAKASH[1] AND H. THOMAS[2]

ABSTRACT

The Precambrian shield of southern India represents a classic terrane of Archean continental crust where different crustal levels are exposed due to differential uplift, erosion and tectonic fragmentation. The Kodaikanal is a part of the granulite facies terrain of the south Indian shield which contains an ensemble of charnockites, mafic granulites, garnet-cordierite gneisses, leptynites, migmatites, sapphirine-bearing granulites, calc-silicates and quartzites of which the mafic granulites are found to be metamorphosed tholeiitic basalts that show REE fractionation and Fe-enrichment trend in the FeO-MgO-(Na_2O+K_2O) diagram.

The different views on the charnockite genesis have been initially evaluated in light of textural relations, geochemical data, phase-relations, fluid inclusion studies and experimental constraints on the mineral equilibria and estimation of metamorphic conditions. Evidence favours that the charnockite formation was by the reaction:

$$\text{phl} + \text{qtz} = \text{en} + \text{kfs} \pm \text{grt} + \text{v}$$

under the reduced activity of water. The proposition is consistent with the experimental data of Hoffer and Grant (1980), Wendlandt (1981) and Peterson and Newton (1989, 1990).

The bulk composition of sapphirine-granulites from Kodaikanal is unusual. These rocks are rich in Al_2O_3, (MgO+FeO), but poor in CaO. The SiO_2-content is rather high compared to most silica-deficient granulites and the (Na_2O+K_2O) content is also raised on account of the presence of phlogopite, K-feldspar and plagioclase in Kodaikanal sapphirine granulites. The different views proposed so far have been critically examined on the basis of phase petrology and bulk chemistry. Waters' calculation (1988) show that a rock with initial 10 vol % biotite would produce only small 10-20% of melt at elevated temperature ($\geq 850^{o}$C), so that melt extraction is probably not effective in granulites facies desiccation. On the similar lines, it is possible to write a dehydration melt reaction for Kodaikanal sapphirine granulites

$$\text{bt} + \text{sil} = \text{spr} + \text{crd} + \text{kfs} + \text{L}$$

1. Department of Geology, Banaras Hindu University, Varanasi-221 005, U.P.
2. Department of Geology, Mizoram University, Aizawl, Mizoram-796 012.

Such a mechanism suitably explains the bulk chemistry of sapphirine rocks of Kodaikanal but further experiments are required to substantiate this hypothesis.

KEY WORDS

Geochemistry; Petrogenesis; mafic granulite, Kodaikanal.

INTRODUCTION

The study area situated in the Kodaikanal ranges of the Madurai block lies between the Palghat-Cauvery and the Achankovil shear zones. Mafic granulites form an important lithounit in the area around Kodaikanal hills. They occur as smaller enclaves or pods extending for 4-6 meters having a width of 2-3 meters within charnockitic country rocks. Both garnet-free and garnet-bearing mafic granulites are recognized from the Kodaikanal. The charnockite forms an integral part not only of the southern India but also of other Precambrian high-grade granulite terranes in the world. The opinions on the genesis of charnockites vary widely. It is expected that geochemical attributes of this rock, supported by experimental data would provide independent information on the geologic processes operative in the deep continental crust. Despite their importance, geochemistry and petrogenesis have rarely been attempted for these rocks. The occurrence of sapphirine-bearing granulites has been recorded from Perumalmali, Ganguvarpatti, Kambam, Panrimalai and Usilampatti. Knowledge on the petrotectonic evolution of these continental fragments have vastly upgraded over recent years, but further advancement in understanding the geochemistry and petrogenesis of these high-grade rocks have been restrained. Attempt is here made to identify the geochemical signature of the precursors of the granulites through the bulk and REE analyses. Petrogenesis of mafic granulites, charnockites and sapphirine granulites has been discussed here.

Analytical procedure

Twenty representative samples were selected for major oxide determinations and limited ten samples were chosen for trace element and rare earth geochemistry. The major elements were analysed using XRF (X-ray fluorescence) technique at WIHG, India. USGS standards viz. G-2, BHVO-1 were used as reference material and at least duplicate results were taken for each oxide determination. Rare earth elements concentration (ppm) were determined by inductively coupled plasma-mass spectrometry (ICP-MS) at NGRI, India. Dissolution of samples were accomplished by the acid decomposition method (Balram, 1991 and 1995). The bulk analyses and REE contents are given in Tables 1 and 2.

The geochemistry of the representative rock suits from Kodaikanal area is described in this study under the following heads:

Mafic granulites

The mafic granulites of the study area occur as concordant lenses within the charnockites. Mineralogically, the mafic granulites predominantly contain orthopyroxene, clinopyroxene, plagioclase feldspar (An_{31-33}) and little amount of quartz.

As early as 1900, Holland proposed that the rocks of the charnockite series, including basic/mafic and ultramafic types, represent a magmatic differentiated suite (see Holland, 1900). Howie (1955) also advocated similar origin for the charnockite rock series on the basis of observed chemical variation. However, other workers differ from this view on the

Table 1: *Chemical analyses of representative rock types*

Rock Type	Sapphirine-bearing granulites					Charnockites		
Sample No.	1111	1100	1101	1109	1103	341	327	334
Ref. No.	10	1	26	9	6	17	15	16
SiO_2	63.57	60.04	60.47	58.04	59.15	73.24	72.92	69.69
TiO_2	0.58	0.60	0.76	0.49	0.55	0.34	0.24	0.4
Al_2O_3	17.02	18.52	16.95	19.4	18.7	13.74	13.7	14.63
FeO	6.5	6.03	7.49	8.64	6.94	3.94	3.39	4.31
MnO	0.12	0.08	0.09	0.15	0.08	0.06	0.03	0.08
MgO	6.02	5.23	6.67	6.12	6.24	0.32	0.68	0.84
CaO	1.21	1.82	1.39	1.50	1.51	3.17	1.94	3.46
Na_2O	3.48	4.10	3.48	3.96	4.00	2.73	3.71	2.74
K_2O	0.67	2.29	1.82	0.82	1.93	1.12	1.73	2.54
P_2O_5	0.45	0.45	0.52	0.75	0.62	0.10	0.10	0.23
Total	99.62	99.28	99.64	99.87	99.72	98.76	98.44	98.92

Rock Type	Charnockites				Grt-Charnockites		Migmatitic gneiss	
Sample No.	370	1214	1261	1262	1113	1131	1125	1126
Ref. No.	19	27	28	30	11	12	13	14
SiO_2	73.46	67.39	64.05	67.77	61.70	57.68	64.81	63.00
TiO_2	0.46	0.53	0.72	0.7	0.06	0.76	0.57	0.74
Al_2O_3	14.36	14.72	14.73	14.44	12.64	15.39	15.47	16.96
FeO	4.02	7.45	8.93	8.91	14.26	14.97	8.89	7.91
MnO	0.04	0.17	0.39	0.16	1.06	0.35	0.26	0.23
MgO	0.62	2.02	2.85	2.19	3.65	4.55	4.29	3.27
CaO	2.66	3.39	4.51	1.41	3.66	1.78	3.28	4.72
Na_2O	2.39	1.53	1.9	1.98	0.47	0.48	1.78	1.93
K_2O	2.37	1.87	1.18	1.96	0.53	1.92	0.21	0.23
P_2O_5	0.14	0.31	0.49	0.26	0.40	0.72	0.47	0.57
Total	99.52	99.38	99.75	99.78	98.43	98.6	100.03	99.56

Rock Type	Grt-crd-sil gneisses		Mafic granulites	
Sample No.	1116	1256	353	1213
Ref. No.	20	29	18	22
SiO_2	52.41	59.88	52.41	46.36
TiO_2	0.90	1.30	0.60	0.97
Al_2O_3	21.84	15.75	15.28	13.32
FeO	12.05	10.80	12.45	12.73
MnO	0.17	0.17	0.14	0.18
MgO	7.95	3.59	6.53	9.61
CaO	1.00	1.34	8.02	11.53
Na_2O	0.46	3.26	2.44	2.83
K_2O	2.45	2.04	1.27	0.83
P_2O_5	0.37	0.65	0.61	0.33
Total	99.60	98.78	99.48	98.69

Table 2: *Geochemical data on various rock types*

Rock Type	Charno-ckite	Mafic granulite		Grt-charno-ckite	Migmatitic gneisses		Grt-crd-sil gneisses		Spr-bearing granulites	
Sample No.	341	353	1213	1131	1125	1126	1116	1256	1100	1101
Ref. No.	17	18	22	12	13	14	20	29	1	26
Sc	3.40	26.29	44.87	27.62	33.27	30.29	21.00	19.50	10.94	12.55
V	94.21	202.98	271.86	228.53	152.25	169.97	143.24	161.72	123.81	154.11
Cr	31.75	326.79	259.66	307.30	127.29	121.80	398.35	191.58	61.87	219.42
Co	9.02	35.09	50.18	35.77	20.75	19.70	32.50	28.26	13.18	17.31
Ni	28.11	74.60	66.24	58.91	25.50	31.13	107.14	41.62	32.08	51.09
Cu	285.80	467.16	152.37	115.54	78.00	152.12	176.51	55.40	46.34	131.05
Zn	192.06	352.87	150.97	154.02	74.56	128.12	142.89	90.92	66.33	128.07
Ga	19.08	17.82	15.27	16.04	16.47	18.14	20.98	21.42	19.66	24.15
Rb	15.97	17.55	13.38	71.54	16.92	2.42	90.08	171.09	87.31	107.75
Sr	433.80	479.07	151.26	258.73	109.25	130.39	116.47	170.34	75.96	76.48
Y	4.23	23.17	23.86	79.17	44.75	63.88	68.85	24.45	8.86	9.26
Zr	77.75	32.22	59.21	45.70	30.54	43.50	30.97	44.36	47.48	61.46
Nb	2.70	4.47	3.49	11.78	1.62	3.50	15.71	11.35	4.89	5.80
Ba	748.39	496.39	183.43	1515.00	109.15	33.68	289.10	170.00	194.37	184.11
La	44.80	25.78	5.66	62.82	21.19	33.75	12.40	56.83	39.84	28.07
Ce	64.69	48.14	9.86	81.99	34.79	58.39	17.09	90.27	59.99	41.25
Pr	6.70	6.19	1.40	8.31	4.06	7.55	1.80	9.70	6.36	4.85
Nd	22.13	28.03	7.01	29.67	13.97	29.62	6.70	36.50	22.70	15.86
Sm	3.26	5.77	1.60	6.02	3.62	7.29	3.11	6.67	3.59	2.90
Eu	1.23	1.31	0.71	1.54	0.75	1.61	0.76	1.93	0.75	0.56
Gd	2.18	4.73	2.73	7.47	3.98	8.13	5.29	6.65	2.61	2.04
Dy	1.00	4.01	3.96	10.15	6.24	9.19	8.13	4.84	1.96	1.34
Er	0.46	2.50	1.65	7.15	4.12	5.19	5.34	1.98	0.88	0.66
Yb	0.36	1.92	2.05	6.40	4.64	5.04	5.05	1.32	0.75	0.58
Lu	0.09	0.35	0.35	1.06	0.67	0.81	0.80	0.22	0.13	0.13
Hf	2.06	1.07	1.35	1.09	0.94	1.16	0.73	1.08	1.32	1.50
Ta	0.17	0.64	0.31	0.92	0.32	0.56	1.54	0.46	0.43	0.45
Pb	9.53	12.37	7.05	10.25	5.03	5.24	6.50	13.01	5.92	7.75
Th	6.68	14.41	0.69	17.08	12.70	19.83	1.21	15.66	27.38	21.54
U	0.42	0.18	0.28	0.48	0.44	0.50	0.32	0.65	0.65	0.62
Σ REE	146.90	128.73	36.98	222.58	98.03	166.57	66.47	216.91	139.56	98.24
La_N/Yb_N	84.02	9.06	1.86	6.63	3.08	4.52	1.66	29.08	35.88	32.69
La_N/Sm_N	8.65	2.81	2.23	6.57	3.68	2.91	2.51	5.36	6.99	6.09
Gd_N/Yb_N	4.89	1.99	1.07	0.94	0.69	1.30	0.85	4.07	2.81	2.84
Eu_N/Eu^*	1.41	0.77	1.04	0.70	0.60	0.64	0.57	0.89	0.75	0.70

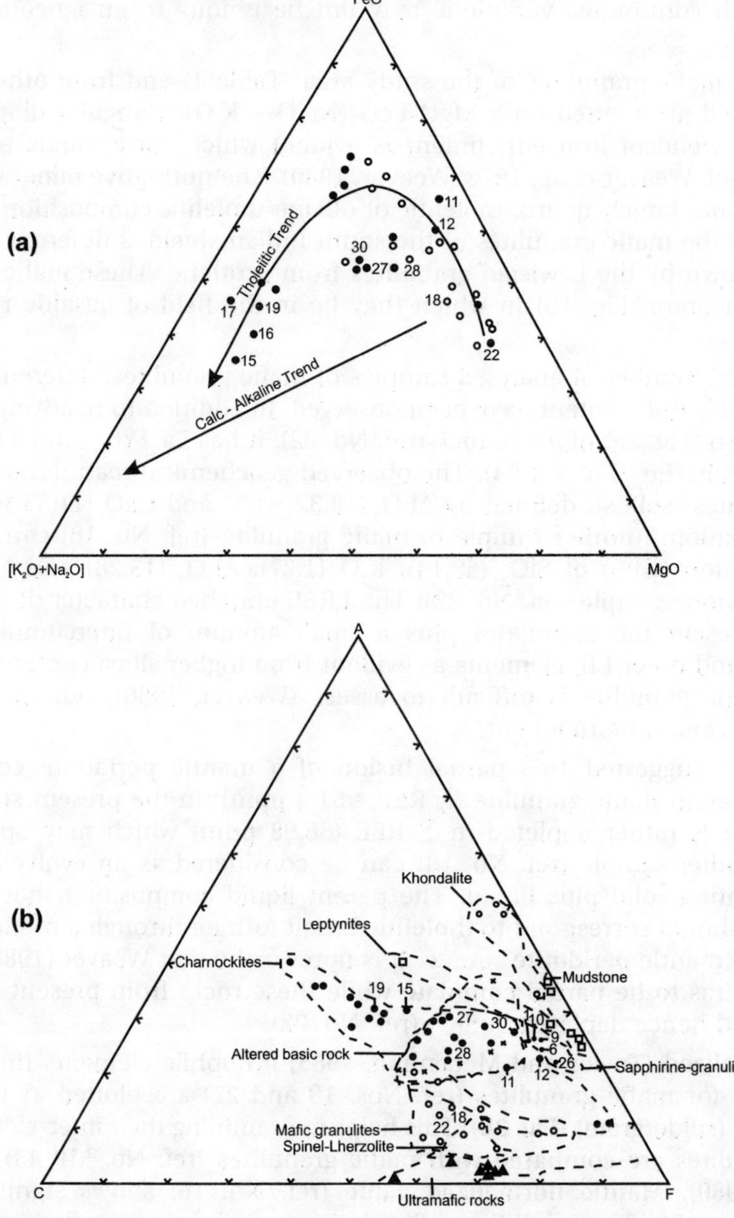

Fig. 1. (a) Triangular variation diagram FeO-MgO-(Na_2O+K_2O) showing the plots of mafic granulites (open circle) and charnockites (solid circle). Numbered samples are from Kodaikanal area. Other plots are based on analyses given by Mohan (1983). (b) ACF diagram showing the plots of sapphirine-bearing granulites (solid circle), charnockites (big solid circle), mafic granulites (open circle), Khondalite (encircled plus), leptynites (encircled cross) from Mohan (1983). Mudstone associated with evaporates (plus) from Schreyer (1977), altered mafic rock (cross) from Valance (1967), Spinel-Lherzolite (grey with circle) from Varne (1977), Ultramafic rocks (solid triangle) from Jan and Howie (1981). Number samples are from Kodaikanal area. Unnumbered analysed plots of charnockites, leptynites, khondalites and mafic granulites are based on analysis given by Howie and Subramanian (1957), Weaver *et al.* (1978), Weaver (1980), Janardhan *et al.* (1982).

grounds that such continuous variations may not be unique to an igneous suite (Heier, 1960).

The analysed mafic granulites of the study area (Table 1) and from other areas of the south Indian shield are plotted on a MgO-FeO-(Na_2O + K_2O) triangular diagram (Fig. 1a). Here the general trend of iron enrichment is evident which corresponds to the trend of tholeiitic basalts (cf. Weaver *et al.*, 1978; Weaver, 1980). The normative minerals of the mafic granulites also favour largely quartz-tholeiitic or olivine-tholeiitic composition. The observed tholeiitic trend of the mafic granulites of the south Indian shield is different from the calc-alkaline trend shown by the Lewisian granulites from Scotland. These mafic granulites are plotted in ACF diagram (Fig. 1b) in which they lie in the field of basaltic rocks (Winkler, 1987).

Despite limited number of analysed samples of mafic granulites, different levels of SiO_2 concentration and Σ REE content have been observed. In addition to relatively flat chondrite normalized pattern (Fig. 2a) of mafic rock (ref. No. 22), it has La_N/Yb_N ratio 1.86 and slightly positive Eu-anomaly (Eu_N/Eu^* = 1.04). The observed geochemical feature noted above may represent cumulates (solids), defined by Al_2O_3 (13.32 wt%) and CaO (11.53 wt%) favouring feldspar accumulation. Another sample of mafic granulite (ref. No. 18) contains relatively higher concentration (wt%) of SiO_2 (52.14), K_2O (1.27), Al_2O_3 (15.28) and Σ REE (128.73) compared to previous sample (ref. No. 22). The LREE enriched character of mafic granulite may largely represent the cumulates plus a small amount of intercumulus felsic melt enriched in K_2O and other LIL elements as evident from higher silica content. Metasomatic effect in the mafic granulite is difficult to assess (Weaver, 1980), but the possibility of K-metasomatism cannot be ruled out.

Weaver (1980) suggested 10% partial fusion of a mantle peridotite composition for generation of a parent mafic granulite (Σ REE – 51.4 ppm). In the present study, the mafic rock (ref. No. 22) is rather depleted in Σ REE (36.98 ppm) which may approximate the cumulates. The other sample (ref. No. 18) can be considered as an evolved composition, possibly representing solid plus liquid. The parent liquid composition that evolved these mafic granulites should correspond to tholeiitic basalt formed through a moderate degree of partial melting of mantle peridotite source. It is noteworthy that Weaver (1980) found mafic granulites of Madras to be partial cumulate while these rocks from present area represent true cumulate and hence depleted in SiO_2 (ref. No. 22).

Mantle normalized (Taylor and McLennan, 1985) lithophile elements (in order of their relative mobility) for mafic granulites (ref. Nos. 18 and 22) are plotted as multi-elements variation diagram (spidergram) (Fig. 2b). This helps in examining their inter-element variation. Kodaikanal granulites are compared with mafic granulites (ref. No. MP 43) of Pallavaram area (Weaver, 1980). Mantle normalized value (ref. No. 18) shows significantly higher concentration of Ba, Rb, Th, K, LREE and Sr relative to Pallavaram mafic granulite (MP 43). The high field strength elements such as Zr, Hf, Ti, Y and Yb are enriched in one sample (Ref. No. 22) and also in Madras granulite (MP 43). Further it has been noted that U, Th and Ga are high relative to mantle concentration which indicate that the concentration of these elements possibly represents the original protolith composition. Transition metals (Ni, Cr and Co) are depleted relative to the mantle value, indicating an early fractionation or removal of olivine, chrome diopside and chrome spinel prior to the emplacement of basic magma. Likewise V and Ti show a slight enrichment by a factor ranging between 1.5 and 6,

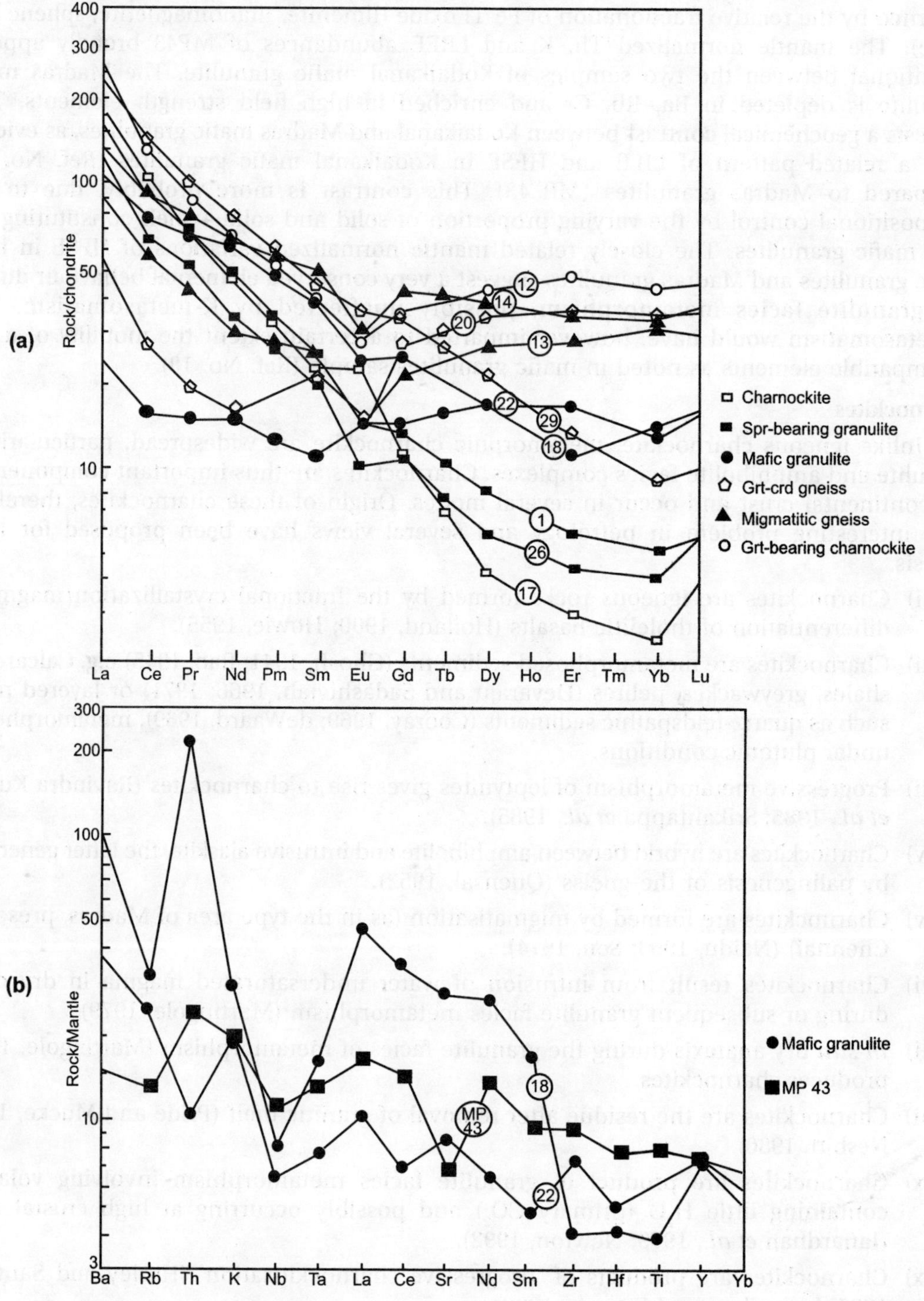

Fig. 2. (a) Chondrite normalized REE pattern for different rock types of the study area. (b) Normalized trace element plots for mafic granulites of the study area.

governed by the relative fractionation of Fe-Ti oxide (Ilmenite, titanomagnetite, sphene and rutile). The mantle normalized Th, K and LREE abundances of MP43 broadly appears transitional between the two samples of Kodaikanal mafic granulite. The Madras mafic granulite is depleted in Ba, Rb, Cr and enriched in high field strength elements. This suggests a geochemical contrast between Kodaikanal and Madras mafic granulites, as evident from a related pattern of LILE and HFSE in Kodaikanal mafic granulites (Ref. No. 22) compared to Madras granulites (MP 43). This contrast is more probably, due to the compositional control by the varying proportion of solid and solid + melt constituting the bulk mafic granulites. The closely related mantle normalized variations of HFSE in both mafic granulites and Madras granulites suggest a very conserved elemental behaviour during the granulite facies metamorphism, possibly unaffected by K-metasomatism. The K-metasomatism would have, however imparted to a certain extent the mobility of a few incompatible elements as noted in mafic granulites sample (Ref. No. 18).

Charnockites

Unlike igneous charnockite, metamorphic charnockite are widespread, particularly in granulite and amphibolite facies complexes. Charnockites are thus important component of the continental crust and occur in several modes. Origin of these charnockites, therefore, pose interesting problem in petrology and several views have been proposed for their genesis.

(i) Charnockites are igneous rocks formed by the fractional crystallization/magmatic differentiation of tholeiitic basalts (Holland, 1900; Howie, 1955).

(ii) Charnockites are metamorphosed sediments (Ghosh, 1941; Rao, 1945) e.g. Calcareous shales, greywackes, pelites (Devaraju and Sadashiviah, 1966; 1971) or layered rocks such as quartz-feldspathic sediments (Cooray, 1969; deWaard, 1969), metamorphosed under plutonic conditions.

(iii) Progressive metamorphism of leptynites gives rise to charnockites (Ravindra Kumar *et al.*, 1985; Srikantappa *et al.*, 1985).

(iv) Charnockites are hybrid between amphibolite and intrusive alaskite, the latter generated by palingenesis of the gneiss (Quensal, 1952).

(v) Charnockites are formed by migmatisation (as in the type area of Madras, presently Chennai) (Naidu, 1963; Sen, 1974).

(vi) Charnockites result from intrusion of water undersaturated magma in dry crust during or subsequent granulite facies metamorphism (Martingole, 1979).

(vii) *In situ* dry anatexis during the granulite facies of metamorphism (Martingole, 1979) produces charnockites.

(viii) Charnockites are the residue after removal of granitic melt (Pride and Mucke, 1981; Nesbitt, 1980).

(ix) Charnockites are product of granulite facies metamorphism involving volatiles containing little H_2O (primary CO_2) and possibly occurring at high crustal level (Janardhan *et al.*, 1979; Newton, 1992).

(x) Charnockites are products of 'progressive charnockitisation' (Harley and Santosh, 1995; Janardhan and Francis, 1996).

(xi) 'Arrested growth' mechanism, specially for the patchy charnockites (Dobmeier and Raith, 2000).

(xii) Internally controlled dehydration process of arrested charnockitization (Raith and Srikantappa, 1993) which is in contrast to the model of externally derived fluids proposed by a number of other workers (cf. Newton, 1992).

It may be mentioned here that both fluid-deficient thermal processes (dehydration melting) and fluid-present (CO_2-rich) thermal processes can account in principle for the reduced conditions of H_2O needed to produce charnockite from precursor crustal rocks (Newton, 1992). The ideas on the origin of arrested charnockites in southern India and Sri Lanka, and their genetic linkage to regional scale granulites, have been considered by the influence of 'CO_2-wave' scenario of Newton *et al.* (1980). Arrested charnockites are seen to represent the initial stage of dehydration of amphibolite-grade crustal lithologies in the front of a 'CO_2-wave' ascending from deep-seated (mantle) sources (e.g. Friend, 1981; Janardhan *et al.*, 1982; Hansen *et al.*, 1984, 1987; Santosh *et al.*, 1990, 1991).

Carbonic fluid inclusions

An important factor in south India and Sri Lanka charnockite occurrences is the abundance of CO_2-rich fluid inclusions (Srikantappa *et al.*, 1985; Hansen *et al.*, 1987; Stähle *et al.*, 1987; Santosh and Tsunogae, 2003). The compositions determined by microthermometry are almost pure CO_2, with little evidence for immiscible liquid H_2O. Fluid inclusion studies in charnockites from Kodaikanal indicate the presence of primary monophase high density CO_2-rich (0.968-1.014 g/cm^3) inclusions in quartz (Mohan *et al.*, 1996). Therefore, the role of 'CO_2-wave' scenario of Newton *et al.* (1980) deserves special mention, especially when charnockite occurs on a large tract.

Experimental constraints

The process of charnockite formation can be discussed on the basis of phase equilibrium relation in the model system K_2O-MgO-Al_2O_3-SiO_2-H_2O (KMASH) with and without CO_2. The experimental work has been carried out in the systems KMASH (Luth, 1967; Bohlen *et al.*, 1983; Peterson and Newton, 1989), KMASH-CO_2 (Hoffer and Grant, 1980). The transition from amphibolite facies to granulite facies can be modeled by subsolidus reaction in KMASH system:

phlogopite + quartz = enstatite + K-feldspar + vapour ...(1)

Bohlen *et al.* (1983) placed this reaction near 855°C at 5 kbar and Peterson and Newton (1989) at 805°C at the same pressure. The earlier study suggested K-feldspar as a product rather than a reactant, in contrast to the finding of the later study.

The equilibrium is stable with a pure H_2O vapour only at low pressures. The experimental work in the KMASH-CO_2 system by Wendlandt (1981) determined the univariant melting reaction:

phlogopite + K-feldspar + quartz + ($CO_2 - H_2O$) vapour = enstatite + liquid ...(2)

A remarkable feature of Wendlandt's experiment is that the melt can occur under CO_2-rich vapours (80 mol % CO_2, 20 mol % H_2O at 3-6 kbar) at temperatures just above the pure H_2O solidus. Recent experiments by Peterson and Newton (1990) have confirmed Wendlandt's work. Wendlandt (1981) also studied the reactions:

K-feldspar + enstatite + quartz = liquid ...(3)

K-feldspar + enstatite + quartz + CO_2 = liquid ...(4)

Both these reactions occur at ultrahigh temperatures (≥ 1000°C). Hoffer and Grant (1980) and Grant (1986) experimentally studied the dehydration reaction in Fe-Mg bearing system,

biotite + quartz = K-feldspar + hypersthene + H_2O ...(5)

The dehydration reaction (5) valid for the condition $P_{H_2O} = P_{total}$ would be displaced to lower temperatures under reduced water activity. Using the internally consistent thermodynamic data sets of Berman (1988, 1991) and his updated computer program (1992), precise pressure, temperature and water activity were obtained for the charnockite rock (Fig. 3). The relative displacement of the reaction (5) at reduced activity of water (<0.2) is shown in Fig. 3. Most likely these charnockites were formed through the dehydration reaction (5) under the conditions of reduced activity of water caused by the influx of CO_2 with or without the presence of Cl, F, N_2 etc. However Charnockites from Eastern Ghats in India have been interpreted as the product of dehydration melting (Kar, 2001).

Different possible CO_2 sources have been described in the literature. They include residual pore fluids, decarbonation reactions in the deep crust, CO_2 release from crystallizing magmas or from deep-crustal granulite fluid inclusions or even mantle degassing.

Geochemically, the highest absolute ∑ REE content (222 ppm) has been noted in garnet charnockite sample (ref. No. 12) from Kodaikanal (Table 2). The charnockites bear moderate positive Eu anomaly (Eu_N/Eu^* = 1.41). A steeply inclined REE pattern is noted in garnet charnockite sample (ref. No. 12) with a low degree of positive Eu-anomaly (Eu_N/Eu^* = 0.70) due to the limited accumulation of plagioclase (16 wt % modal) relative to another charnockite (ref. No. 17) that exhibits moderate negative Eu-anomaly and a nearly flat HREE

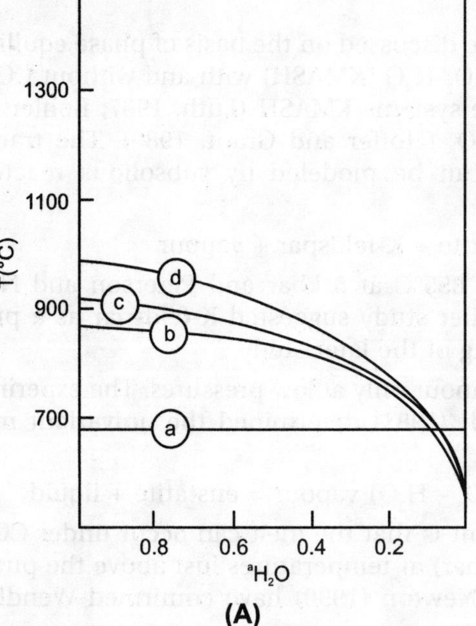

(A)

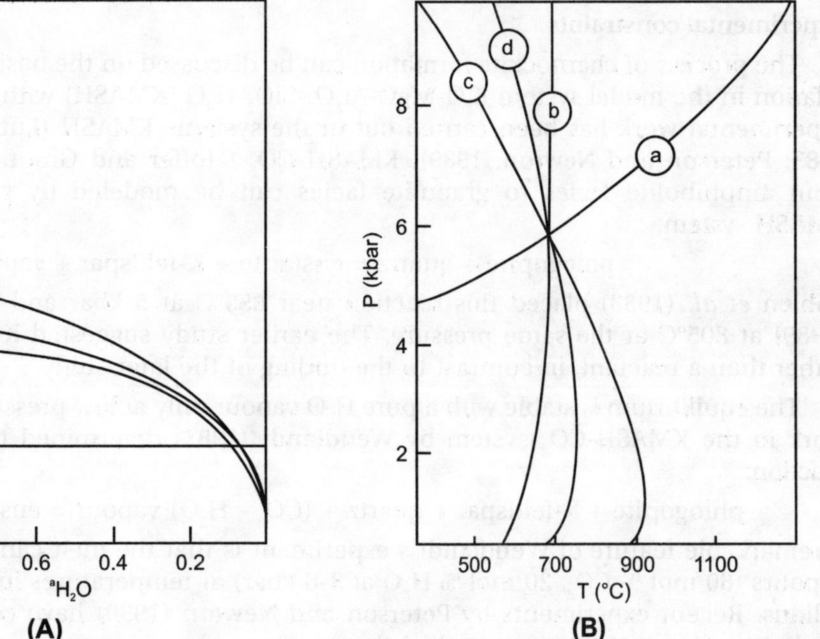

(B)

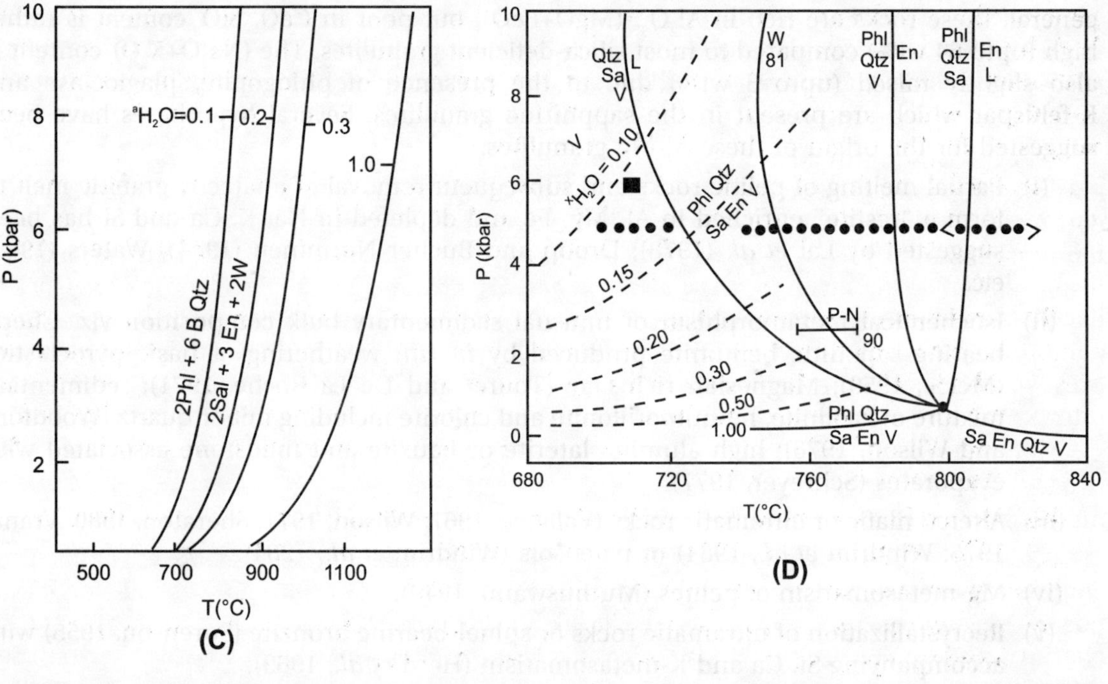

Fig. 3. (A) Equilibria plotted in Figure are: (a) 3En + 3An = Grs + 2Prp + 3βQtz (ΔS = 75.7 JK^{-1}, ΔV = 6.61 J bar^{-1}) (b) 2Sa + 3En + 2W = 2Phl + 6βQtz (ΔS = 332.0 JK^{-1}, ΔV = –0.39 J bar^{-1}) (c) Grs + 2Sa + 2Prp + 2W = 3 βQtz + 2Phl + 3An (ΔS = 256.2 JK^{-1}, ΔV = –7.00 J bar^{-1}) (d) 2 Grs + 2Sa + 4Prp + 2W = 2Phl + 3En + 6An (ΔS = 180.5 JK^{-1}, ΔV = –13.62 J bar^{-1}). Phases used in calculation are: Anorthite, Orthoenstatite, Grossular, Phlogopite, K-feldspar (high), Pyrope, β-Quartz. Assumed P = 5.8 Kbar, Temperature = 677°C, Activity of H_2O = 0.08.

(B) Equilibria plotted in figure are: (a) Grs + 2Prp + 3βQtz = 3En + 3An (ΔS = –77 JK^{-1}, ΔV = –6.74 J bar^{-1}) (b) 2Phl + 6 βQtz = 2Sa + 3En + 2W (ΔS = -78.60 JK^{-1}, ΔV = –6.74 J bar^{-1}) (c) 2 Grs + 2Sa + 4Prp + 2W = 2Phl + 3En + 6An (ΔS = 233 JK^{-1}, ΔV = –17.1 J bar^{-1}) (d) Grs + 2Sa + 2Prp + 2W = 3 βQtz + 2Phl + 3An (ΔS = 156 JK^{-1}, ΔV = –10.33 J bar^{-1}). Phases used in calculation: Anorthite, Orthoenstatite, Grossular, Phlogopite, K-feldspar (high), Pyrope, β Quartz. TWEEQU results based on intersection of all equilibria are P = 5.8 kbar, T = 678°C.

(C) Note the shift in the univariant curves calculated at reduced water activity. Phases used in calculation: Orthoenstatite, Phlogopite, K-feldspar (High), β Quartz.

(D) Experimental phase equilibrium relations in the model granite system K_2O-MgO-Al_2O_3-SiO_2-H_2O-CO_2 (KMASH-CO_2), from Wendlandt (1981) (W) and Peterson and Newton (1989, 1990) (P-N). Lighter lines are univariant melting equilibria in the pure H_2O (KMASH) system. The bold cuvre is a univariant melting curve deduced by Wendlandt (1981): Phlogopite (Phl) + Quartz (Qtz) + K-feldspar (Sa) + CO_2-H_2O vapour (V) = orthopyroxene (En) + liquid (L). The horizontal bracket is an experimental reversal of this equilibrium by Peterson and Newton (1990). The dashed lines are calculated contours of $X^V_{H_2O}$.

pattern (Fig. 2A). The variation in HREE pattern among charnockites shows a notable range. This range of REE abundance in charnockites, implies an extremely heterogeneous source, either chemical, mineralogical or physical (P, T and a_{H_2O}). Such a view is also being supported by Weaver *et al.* (1978) and Weaver (1980).

Sapphirine granulites

The bulk composition of sapphirine granulites from Kodaikanal (Table 1) is unusual in the sense that they do not correspond to any normal igneous or sedimentary rocks. In general, these rocks are rich in Al_2O_3, (MgO+FeO), but poor in CaO. SiO_2 content is rather high (upto 64 wt%) compared to most silica-deficient granulites. The (Na_2O+K_2O)-content is also slightly raised (upto 3 wt%) due to the presence of phlogopite, plagioclase and K-feldspar which are present in the sapphirine granulites. Several hypotheses have been suggested for the origin of these Al-Mg granulites:

(i) Partial melting of pelitic rocks and subsequent removal of anatectic granitic melt to form a "restite" enriched in Al, Mg, Fe and depleted in Na, K, Ca and Si has been suggested by Lal *et al.* (1978); Droop and Bucher Nurminen (1984), Waters (1989) etc.

(ii) Isochemical metamorphism of unusual sedimentary bulk composition viz. Chert-bearing saponitic bentonite produced by *in situ* weathering of basic pyroclastics (Mckie, 1959); Magnesian rich clay (Touret and De La Roche, 1971); sedimentary mixture of kaolinite, montimorillonite and chlorite including minor quartz (Woodford and Wilson, 1976); high alumina laterite or bauxite and mudstone associated with evaporates (Schreyer, 1977).

(iii) Altered mafic or ultramafic rocks (Valance, 1967; Wilson, 1971; Sheraton, 1980; Vrana, 1975; Windrim *et al.*, 1984) or paleosols (Windrim *et al.*, 1984).

(iv) Mg-metasomatism of pelites (Muthuswami, 1949).

(v) Recrystallization of ultramafic rocks or spinel-bearing bronzite (Sorenson, 1955) with accompanying Si, Ca and K-metasomatism (Herd *et al.*, 1969).

Although, isochemical metamorphism of pelitic rocks of unusual composition is feasible, this view ignores possible chemical changes in rocks during high-grade metamorphism and anatexis.

The bulk compositions of sapphirine granulites from Kodaikanal area (Table 1) along with other lithotypes are plotted in ACF diagram (Fig. 1b). The plots of sapphirine granulites mark a distinct separate region and thus these are not comparable to any of the association.

Extensive Mg-metasomatism in the area does not suitably explain the higher Al_2O_3 and relatively lower MgO and FeO-contents in sapphirine rocks. There is also a problem of source for Mg-rich solutions.

Restite origin is being advocated by several workers for sapphirine-bearing silica-deficient granulites. The process involves partial melting of pelitic rocks and subsequent removal of anatectic melt to form 'restite' that is enriched in Al, Fe and Mg but impoverished in K, Na, Ca and Si. Application of 'restite origin' has been demonstrated by Lal *et al.* (1978); Droop and Bucher-Nurminen (1984) and Water (1989) and others. This hypothesis holds good for the sapphirine-bearing silica-deficient granulites from Ganguvarpatti (Mohan, 1983). However, the higher concentration of SiO_2 and (Na_2O+K_2O) prevents the sapphirine granulites of Kodaikanal to be of typically restitic origin. The geochemical features of sapphirine granulites from Kodaikanal area suggest close compositional resemblance to the average crust (Taylor and McLennan, 1985), as evident atleast from the high field strength elements such as Nb, Ga, Ta, Y and LREE etc. (Table 2). Other elements may have been mobilized during the metamorphism of the protolith, since those elements are depleted when compared to the crustal composition. Hence these sapphirine granulites cannot be truly restitic in origin as they largely represent peraluminous composition (molar A/C+N+K>1.7) that would require melt retention with residue. In the absence of major extraction of silicate melt, Waters

(1988) proposed an alternative mechanism of vapour absent dehydration melting reactions as plausible method for development of granulite facies assemblages. In such dehydration melt reactions, water from the hydrous phase enters the melt, leaving an anhydrous solid assemblage. Water (1988) showed by mass balance calculation that the relative amount of melt generated by melting reaction near the solidus is slightly greater than the volumetric amount of biotite originally present, a conclusion confirmed by experimental work on KMASH system (Peterson and Newton, 1989). Waters' (1988) calculations for Namaqualand Migmatitic gneiss initially having 10 vol % biotite show that only small 10-20% of melt has been generated at temperatures $\geq$ 850°C. This amount is probably too small to segregate and rise efficiently (Wickham, 1987), so that melt extraction was probably not effective in granulite facies desiccation in Namaqualand (Waters, 1988).

It is possible to write a similar dehydration melt reaction for the Kodaikanal sapphirine granulites.

biotite + sillimanite = cordierite + sapphirine + K-feldspar + Melt ...(6)

Presence of low biotite modal amount and lower temperature conditions attained by Kodaikanal sapphirine rocks produced relatively small proportion of melt that was probably insufficient to migrate, suggesting only a limited melt extraction. Such a mechanism suitably explains the rather high SiO_2 concentration and crystallization of K-feldspar due to segregation of melt. So far no experimental work exists, for the above reaction but experiments will help in resolving such complex reactions.

ACKNOWLEDGMENTS

We wish to express our sincere thanks to Prof. M. Arima and Dr. M. Ishikawa, for their kind support during field work. Financial assistance to DP from JSPS in form of, as Post-doctoral fellow (Award No. P 01198) is gratefully acknowledged. DP would like to specially thank Dr. Anand Mohan for introducing him to the fascinating world of granulites.

REFERENCES

Balram, V., 1991. Determination of rare earth elements in geological samples by Inductively Coupled Plasma-Mass Spectrometry. *Journal of Indian Chemical Society*, **68**, 600-03.

Balram, V., 1995. Developments and trends in Inductively Coupled Plasma Mass spectrometry and its influence on the recent advances in trace element analysis. *Current Science*, **69**, 640-48.

Berman, R.G., 1988. Internally consistent thermodynamic data for stoichiometric minerals in the system Na_2O-K_2O-CaO-FeO-Al_2O_3-SiO_2-TiO_2-H_2O-CO_2. *Journal of Petrology*, **29**, 445-22.

Berman, R.G., 1991. Thermobarometry using multi-equilibrium calculations. A new technique with petrological applications. *Canadian Mineralogist*, **29**, 833-55.

Bohlen, S.R., Boettcher, A.L., Wall, V.J. & Clemens, J.D., 1983. Stability of phlogopite-quartz and sanidine-quartz: A model for melting in the lower crust. *Contributions to Mineralogy and Petrology*, **83**, 270-77.

Cooray, P.G., 1969. Charnockites as metamorphic rocks. *American Journal of Science*, **267**, 969-82.

Dobmeier, C. & Raith, M., 2000. On the origin of 'arrested' charnockitization in the Chilka lake area, Eastern Ghats belt, India: a reappraisal; *Geological Magzine*, **137**, 27-37.

Devaraju, T.C. & Sadashivaiah, M.S., 1966. Pyroxene-quartz-magnetite rocks from Satnur-Halaguru area, Mysore State. *Journal of the Geological Society of India*, **7**, 70-85.

Devaraju, T.C. & Sadashivaiah, M.S., 1971. Some orthopyroxene bearing rocks constituting an integral part of high grade metapelites of Satnur-Halaguru area, Mysore State. *Journal of the Geological Society of India*, **12**, 1-13.

De Waard, D., 1969. The occurrence of charnockite in the Adirondacks: a note on the origin and definition of charnockite. *American Journal of Science*, **267**, 983-87.

Droop, G.T.R. & Bucher Nurminen, K., 1984. Reaction textures and Metamorphic evolution of sapphirine-bearing granulites from the Gruf complex, Italian central Alps. *Journal of Petrology*, **25**, 766-803.

Friend, C.R.L., 1981. The timing of charnockite and granite formation in relation to influx of CO_2 at Kabbaldurga, Karnataka, South India. Nature, **294**, 550-52.

Ghosh, P.K., 1941. The charnockite series of Bastar State and western Jeypore. *Records of Geological Survey of India*, 75.

Grant, J.A., 1986. Quartz-phlogopite-liquid equilibria and origins of Charnockites. *American Mineralogist*, **71**, 1071-75.

Hansen, E.C., Newton, R.C. & Janardhan, A.S., 1984. Fluid inclusion in rocks from the amphibolite-facies gneiss to charnockite progression in Southern Karnataka, India; direct evidence of the fluids of granulite facies metamorphism. *Journal of Metamorphic Geology*, **2**, 249-64.

Hansen, E.C., Janardhan, A.S., Newton, R.C., Prame, W.K.B.N. & Ravindra Kumar, G.R., 1987. Arrested charnockite formation in southern India and Sri Lanka. *Contributions to Mineralogy and Petrology*, **96**, 225-44.

Harley, S.L. & Santosh, M., 1995. Wollastonite at Nuliyam, Kerala, South India: a reassessment of CO_2 infiltration and charnockite formation at a classic locality, *Contributions to Mineralogy and Petrology*, **120**, 83-94.

Heier, K.S., 1960. Petrology and geochemistry of high-grade metamorphic and igneous rocks on Langy, northern Norway. *Norges Geologiske Undersokelse*, **207**, 7-246.

Herd, R.K., Windley, B.F. & Ghisler, M., 1969. The mode of occurrence and petrogenesis of sapphirine-bearing and associated rocks of west Greenland. *Geologiske Undersokelse*, **24**, 1-44.

Hoffer, E. & Grant, J.A., 1980. Experimental investigation of the formation of cordierite-orthopyroxene parageneses in pelitic rocks. *Contributions to Mineralogy and Petrology*, **73**, 15-22.

Holland, T.H., 1900. The Charnockite series, a group of Archaean hypersthenic rocks in Peninsular India. *Geological Survey Indian Memoir*, **28**, Pt. 2.

Howie, R.A., 1955. The geochemistry of the charnockite series of Madras, India. *Transactions of the Royal Society of Edinburgh*, **62**, 725-68.

Howie, R.A. & Subramanium, A.P., 1957. The paragenesis of garnet in charnockite, enderbite and related granulites. *Mineral. Mag.*, **31**, 565-86.

Janardhan, A.S., Newton, R.C. & Smith, J.V., 1979. Ancient crustal metamorphism at low P_{H2O}: Charnockite fomation at Kabbaldurga, South India. *Nature*, **278**, 511-14.

Janardhan, A.S., Newton, R.C. & Hansen, E.C., 1982. The transformation of amphibolite facies gneiss to charnockite in Southern Karnataka and Northern Tamil Nadu, India. *Contributions to Mineralogy and Petrology*, **79**, 130-49.

Janardhan, A.S. & Francis Anto, K., 1996. Transition zone charnockites-Harbingers of granulite metamorphism? Examples from South India, *Gondwana Research Group Memoir-3*: The Archaean and Proterozoic terrain in South India within East Gondwana. (eds) M. Santosh and M. Yoshida, 145-55.

Kar, R., 2001. Patchy charnockite from Jenapore, Eastern Ghats granulite belt, India: Structural and petrochemical evidences attesting to their relict nature. *Proceedings of Indian Academy of Sciences*, **110**, 337-50.

Lal, R.K., Ackermand, D., Seifert, F. & Haldar, S.K., 1978. Chemographic relationship in sapphirine-bearing rocks from Sonapahar, Assam, India. *Contributions to Mineralogy and Petrology*, **67**, 169-87.

Luth, W.C., 1967. Studies in the system $AlSiO_4$-SiO_2-H_2O. Inferred phase relations and petrologic applications. *Journal of Petrology*, **8**, 372-16.

Martingole, J., 1979. Charnockite gneiss and Proterozoic crust. *Precambrian Research*, **9**, 303-10.

Mckie, D., 1959. Yoderite, a new hydrous magnesium iron aluminosilicate from Mautia Hill. Tanganyika. *Mineralogical Magazine*, **32**, 282-307.

Mohan, A., 1983. Mineralogy and Petrology of the area around Ganguvarpatti, Madurai District, Tamil Nadu, Unpublished Ph.D. thesis, B.H.U., Varanasi.

Mohan, A., Prakash, D. & Sachan, H.K., 1996. Fluid inclusions in charnockites from Kodaikanal massif (south India): P-T record and implications for crustal uplift history. *Mineralogy and Petrology*, **57**, 167-84.

Muthuswami, T.N., 1949. Sapphirine, Madurai, *Proceeding of Indian Academy of Science*, **30**, 295-301.

Naidu, P.R.J., 1963. A layered complex in Sittampundi, Madras State, India, *Min. Soc. Am. Spec. Pap.*, **1**, 116-23.

Nesbitt, H.W., 1980. Genesis of the New Quebec and Adirondack granulites: evidence for their production by partial melting. *Contributions to Mineralogy and Petrology*, **72**, 303-10.

Newton, R.C., Smith, J.V. & Windley, B.F., 1980. Carbonic metamorphism, granulites and crustal growth. *Nature*, **288**, 45-50.

Newton, R.C., 1992. Charnockite alteration: evidence for CO_2 infiltration in granulite facies metamorphism. *Journal of Metamorphic Geology*, **10**, 383-400.

Peterson, J.W. & Newton, R.C., 1990. Experimental biotite-quartz melting in the KFMASH-CO_2 system and the role of CO_2 in the petrogenesis of granite and related rocks. *American Mineralogist*, **75**, 1029-42.

Peterson, R.C. & Newton, R.C., 1989. Reversed experiments on biotite-quartz-feldspar melting in the system KMASH: implication to crustal anatexix. *Journal of Geology*, **97**, 465-85.

Pride, C. & Mucke, G.K., 1981. Rare earth element distributions among coexisting granulite facies minerals, Scourian complex, NW Scotland. *Contributions to Mineralogy and Petrology*, **76**, 463-71.

Quensal, P., 1952. The charnockite series of the Varberg on the southwest coast of Sweden. Arkiv för Mineralogie und Petrographie, **1**, 229-32.

Raith, M. & Srikantappa, C., 1993. Arrested charnockite formation at Kottavattam, southern India. *Journal of Metamorphic Geology*, **11**, 815-33.

Rao, R.B., 1945. The charnockitic rocks of Mysore (Southern India). *Mysore Geol. Dept. Bull.*, 18.

Ravindra Kumar, G.R., Srikantappa, C. & Hansen, E., 1985. Charnockite formation of Ponundi. *Nature*, **313**, 207-09.

Santosh, M. & Tsunogae, T., 2003. Extremely high density pure CO_2 fluid inclusions in a garnet granulite from Southern India. *Journal of Geology*, **111**, 1-16.

Santosh, M., Harris, N.B.W., Jackson, D.H. & Mattey, D.P., 1990. Dehydration and incipient charnockite formation: a phase equilibria and fluid inclusion study from south India. *Journal of Geology*, **98**, 915-26.

Santosh, M., Yoshida, M. & Nanda-Kumar, V., 1991. Fluid characteristics across a gneiss-charnockite reaction front in Sri Lanka: implications for granulite formation in the Gondwanian deep crust. *Journal of Mineralogy, Petrology and Economic Geology*, **86**, 27-44.

Schreyer, W., 1977. White schists: Their compositions and pressure-temperature regimes based on experimental, field and petrographic evidence. *Tectonophysics*, **43**, 127-44.

Sen, S.K., 1974. A review of some geochemical characters of the type area (Pallavaram, India) Charnockites. *Journal of the Geological Society of India*, **15**, 413-20.

Sheraton, J.W., 1980. Geochemistry of Precambrian metapelites from east Antarctica: secular and metamorphic variations. DRM *J. Austr. Geol. and Geoph.*, **5**, 279-88.

Serensen, H., 1955. On sapphirine from West Greenland. *Meddr. Greenland*, **137**, 1-32.

Srikantappa, C., Raith, M. & Spiering, B., 1985. Progressive charnockitization of a leptynite-khondalite suite in southern Kerala, India. Evidence for formation of charnockites through decrease in fluid pressure. *Journal of the Geological Society of India*, **26**, 849-72.

Stähle, H.J., Raith, M., Hoerness, S. & Delfs, A., 1987. Element mobility during incipient granulite facies formation at Kabbaldurga, Southern India. *Journal of Petrology*, **28**, 803-34.

Taylor, S.R. & McLennan, S.M., 1985. The continental crust. Its composition and Evolution. Blackwell, Oxford, 312.

Touret, J.L.R. & De La Roche, H., 1971. Sapphirine a Snaresund, pres de Trede strand (Norvege meridionate). *Norsk Geologisk Tidsskrift*, **51**, 169-75.

Valance, T.G., 1967. Mafic rock alteration and isochemical development of some cordierite-anthophyllite rocks. *Journal of Petrology*, **8**, 84-96.

Vrana, S., 1975. Magnesian-aluminous rocks, the associated ore mineralization and the problem of magnesium-iron metasomatism. *Krystalinikum*, **11**, 101-14.

Water, D.J., 1988. Partial melting and the formation of granulite facies assemblages in Namaqualand, South Africa. *Journal of Metamorphic Geology*, **6**, 387-404.

Water, D.J., 1989. Metamorphic evidences for the heating and cooling oath of Namaqualand granulites. In: Daly, J.S. Cliff, R.A., Yardely, B.W.D. (eds.), Evolution of metamorphic belts. *Geol. Soc. Lon., Special publication*, **43**, 357-63.

Weaver, B.L., Tarney, J., Windley, B.F., Slugavaram, E.B. & Venkata, Rao, 1978. Madras granulites geochemistry and P-T conditions of crystallization. In Windley, B.F. and Naqvi, S.M. (eds.), Archaean geochemistry developments in Precambrian geology, Elsevier, Amsterdam, 177-204.

Weaver, B.L., 1980. Rare earth elements geochemistry of Madras granulites. *Contributions to Mineralogy and Petrology*, **71**, 271-79.

Wendlandt, R.F., 1981. influence of CO_2 on melting of model granulite facies assemblages: A model for the genesis of Charnockite. *American Mineralogist*, **66**, 1164-74.

Wickham, S.M., 1987. The segregation and implacement of granitic magmas. *Journal of Geological Society of India*, **144**, 281-97.

Wilson, A.F., 1971. Some geochemical aspects of the sapphirine bearing pyroxenites and related highly metamorphosed rocks from the Archaean ultramafic belt of South Quairading, Western Australia. *Geol. Soc. Aust. Spec. publ.*, **3**, 401-11.

Windrim, D.P., McCulloch, M.T., Chappell, B.W. & Cameron, W.E., 1984. Nd isotopic systematics and geochemistry of central Australian sapphirine granulites: an example of the rare earth element mobility. *Earth Planet. Sci. Lett.*, **70**, 27-39.

Winkler, H.G.F., 1987. Petrogenesis of metamorphic rocks. Springer-Verlag, New York.

Woodford, P.J. & Wilson, A.F., 1976. Sapphirine-hogbomite, Kornerupine, Surinamite from aluminous granulite, north-eastern, Strangway range, central Australia. *Neues Jahrbuch für Mineralogie*, 15-35.

16

THE LOWER CRUST OF THE INDIAN SHIELD, ITS CHARACTERISTICS AND EVOLUTION

T.M. MAHADEVAN*

ABSTRACT

The Indian shield had a polychronous evolution with a core region of Archaean-Early Proterozoic rocks, encircled by high-grade domains (HGDs) evolved largely in the Proterozoic. Exhumation from deep crustal levels of 10-12 kbars to 4-5 kbs and then uplift to the present surface aided by erosion are perhaps the most characteristic feature of these terrains. They however are differentially exhumed and uplifted in both time and space, most of the exhumation having been completed by the Mid- to late Proterozoic. The south Indian HGD and the Eastern Ghat HGD represent the most extensively exhumed LC but exhumed respectively in the early and late Proterozoic. In the NW, central and NE parts of the Shield, the LC still remains relatively less exhumed. The exhumation may be traced to the lowering of the density of the LC by metamorphic transformation under the influence of episodic and extensive acid and alkaline magmatism and fluid activity and/or a lighter lithospheric mantle and the consequent buoyancy aided by isothermal decompression. These processes may be traced to the interaction between the evolving crust and the thermal regimes of the mantle.

Several recently acquired geophysical characteristics such as velocity structure, reflection fabrics and bimodal gravity signatures and depth extent of the magnetic crust need to be viewed in the context of the above evolution and merely relating them to a generalized collision regime may be simplistic. On the other hand they seem to be well explained by recognition of the transformations effected in the LC by several Proterozoic episodes of magmatism that can impart buoyancy to the LC. Proterozoic exhumation may have brought up the Archaean LC to shallow levels (? 4 kb level). Distensional tectonics that led to the formation of the Vindhyan basin in the Neoproterozoic, and the Mesozoic Gondwana basins may have played an important role in the further exhumation of the LC, already exhumed to shallow depths in the Proterozoic. The other episodes that may have further influenced exhumation/uplift of the LC include (i) the pericontinental rifting, magmatism and accompanying magmatic underplating and crustal tilting during the Cretaceous followed by erosion and (ii) the continuing uplift of the Western Ghats, the Satpura Range and the Chhotanagpur Plateau through epeirogenic processes.

* Sree Bagh, Ammankoil Road, Kochi-682 035. E-Mail: enk_mahadev@sancharnet.in

KEY WORDS

Lower Crust; Indian Shield; Exhumation and Evolution.

INTRODUCTION

The standard continental crust is generally defined as comprising an "Upper Crust" (UC) ~ 18 km thick with a relatively more siliceous composition that permits seismic P-wave velocities generally lower than 6.6 km/sec, and a Lower Crust (LC) with a thickness of ~ 20 km and a debated "basaltic" composition, characterised by seismic velocities of P > 6.6 km/sec. The CONRAD disconformity separates the UC from the LC, though it is not very persistent and universal. Such a definition becomes simplistic when large segments of the continent are involved in tectonics that lead to the exhumation of the LC that eventually modifies the stratification implied in the above definition. Where the exhumation raises the LC significantly, erosion may remove the overlying crust including the CONRAD, as is the case in the southern Indian High-Grade Domain (HGD) and the Eastern Ghat HGD. In areas of lesser exhumation, erosion may remove much of the UC that then may become thinner, as seems to be the case in a major part of the Son-Narmada belt of Central India. Another significant factor relevant to seismic modeling is that the velocity structure that distinguishes the UC from the LC becomes less diagnostic of the LC, when the LC is transformed or modified by Proterozoic tectonics, magmatism and metallogeny. Often these processes are contemporaneous with the processes of exhumation. The structure and composition of the LC gets then transformed on a massive scale. Lower crustal rocks are expected to have a granulite facies metamorphic impress regionally and when the exhumation is due, especially due to isothermal distension, the rocks are subjected to high temperatures but lower pressures inducing re-equilibration of mineral assemblages. Under the influence of fluids, the lower crustal rocks get more hydrated and subjected to mainly alkaline metasomatism and imbibe an amphibolite or even lower metamorphic facies. The resulting transformed crust, being less dense, may be expected to have lower seismic wave velocities than the unaffected rocks of the lower crust.

More than one temporally and geologically distinct phase may have operated in bringing the Lower Crustal rocks to the present surface. Exhumation that operated in the Proterozoic may have been itself episodic and brought the LC from depths of over say 10-12 kbars to shallow depths of say 3 to 5 kbars. Subsequent uplift to the current levels may be possibly the result of Phanerozoic rifting, underplating of the crust leading to uplift and often tilting and finally, epeirogenic uplift and erosion. Clear distinctions have not yet been made regarding the operation of these processes in any of the regions and the assumptions of depths of exhumation and subsequent uplift and their timing are now arbitrary (see England and Molnar, 1990 for definitions of exhumation and uplift). However it is necessary to recognise these possibilities and the possible distinctions in their scales to understand the present LC-UC configurations in continents.

The Indian shield provides an excellent example of differential exhumation and uplift of the lower crust and its polychronous transformation. The Proterozoic orogeny considerably modifies the UC-LC stratification compositionally and a lack of appreciation of these processes often renders geophysical modeling of the crustal structure unreal. For example modeling based purely on the velocity structure can be misleading, as the seismic velocities cannot distinguish between an UC and an exhumed transformed LC. Gravity fields will be influenced by the total history of the terrains and cannot be viewed only on the basis of the Precambrian history. On the other hand, the available geophysical features that have been gathered on large tracts of exhumed and exposed crust can provide insights into what may be anticipated below other areas of exhumed or partially exhumed crust. In such tracts the P-wave velocities for the exhumed crust fall in the range of ~ 6.2 to 6.8 km/sec. In regions with a exhumation history this range of velocity may be taken to represent the exhumed crust, even though it may be tentative.

The most exhaustively studied Lower Crustal rocks of the Indian Shield, both geologically and geophysically, are the large exhumed exposures of charnockitic and associated rocks of the south Indian shield. It is proposed to discuss in this paper the recent insights gained into the characteristics of the large exhumed crust of south India and compare them with the known features of the LC rocks in other parts of the shield to arrive at a comprehensive picture.

UC-LC ARCHITECTURE

The Indian peninsular shield is characterised by a central core region (CCR) comprising an Archaean–Early Proterozoic greenstone-granite domain (GGD) circumscribed rather transitionally by high-grade domains (HGD) that are possibly differentially exhumed from mid- to lower crustal levels and transformed during the Proterozoic (Mahadevan, 2000) (Fig. 1). Seismic velocity structure of the CCR is of an idealised continental crust with a well-developed UC underlain by the LC (Fig. 2 A to D). The LC rocks, however, are partially exhumed and outcrop within the CCR along the fringes of the rifted Mesozoic Gondwana basins of the Godavari and Mahanadi valleys, consistent with an episode of crustal thinning during the process of rifting. These exposures and the general velocity structure away from the rift zones are consistent with the view that the high-grade charnockitic rocks underlie the GGD in the CCR and, along the rifts, they are exposed due to crustal thinning.

The HGDs around the CCR may be classified into (i) the southern HGD; (ii) the Eastern Ghat Mobile Belt on the east and (iii) the Chhotanagpur HGD on the NE. In the Shillong Plateau, further NE, charnockitic rocks, possibly representing exhumed LC are known along the fringes of the Shillong series of rocks; but are poorly exposed. In Central India, exposures of rocks of a granulite facies metamorphic impress (charnockitic rocks) are restricted to a belt some 100 km long, fringing the high-grade Sausar Formations, that have a lower to upper amphibolite metamorphic facies and bound in the south by the low-grade Sakoli Formations with a metamorphic impress of greenschist to amphibolite facies. To the NW of the CCR high-grade granulites and charnockitic rocks are exposed in the Balaram-Abu area, SW of the Aravalli Range, in the Sandmata Complex Thana-Gyangarh and also along the eastern and western side of the Delhi fold belt. The HGDs transit from below the greenstone-granite domains of the CCR, through a transition of amphibolite facies rocks. The deeper facies of the HGDs comprise invariably of charnockitic rocks with their plutonic features and khondalite group of metasediments and are often associated with plutonites of ultrabasic rocks, layered as well as massive anorthosite bodies and alkaline syenites and granites of more than one generation.

SOUTH INDIAN HGD: GEOLOGICAL SETTING

The south Indian HGD provides an oblique cross-section across the Indian continental crust (Fermor in Pascoe, 1954; Fountain and Salisbury, 1981) and is of considerable interest, being one of the most extensive of exhumed lower crustal rocks in the world. The extensive

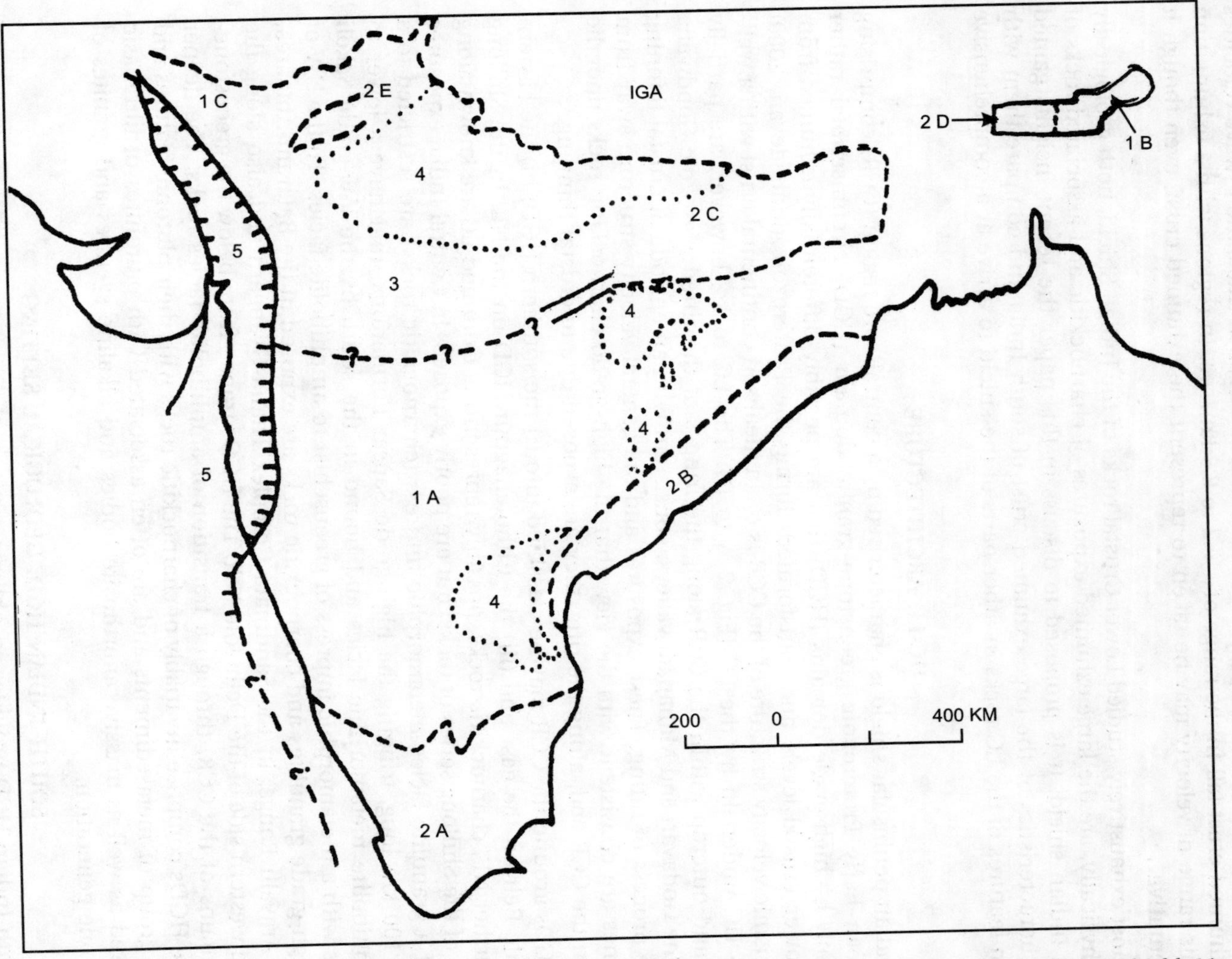

Fig. 1. Spatial relationship between the Archaean craton and the Proterozoic High Grade Domains (HGD) in the Indian shield. 1A. Archaean central cratonic region (CCR); 1B. Archaean Shillong outlier; 1C. Archaean Aravalli outlier; 2A. HGDs of South India; 2B. HGD of Eastern Ghats; 2C. HGD of Chhotanagpur Plateau; 2D. HGD of Shillong Plateau; 2E. HGD of Rajasthan; 3. Lowly exhumed HGD of Central India; 4. Middle to late Proterozoic Supracrustals (Puranas); 5. Western pericratonic belt of Mesozoic active rifts (Modified after Mahadevan, 1996).

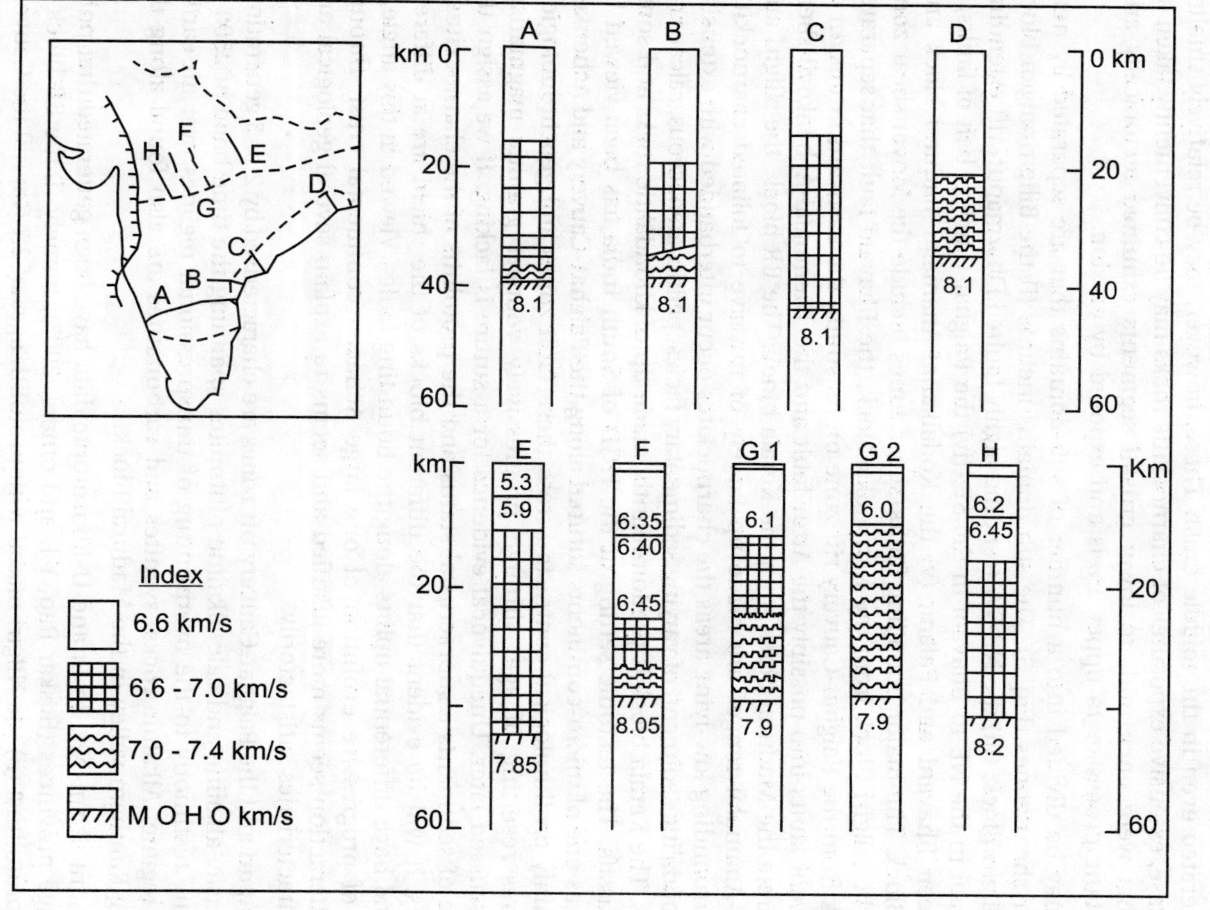

Fig. 2. A representative seismic velocity–depth function derived from DSS profiles (see inset). A-Udipi-Kavali; B. Alampur-Konika; C. Narsapur-Paloncha (Godavari graben); D. Beliamba-Jagannathpur (Mahanadi Gondwana graben); E. Hirapur-Mandla; F. Kajuriakalan-Phulgaon; G1. Ujjain – Dorwa; G2. Dorwa-Mohan; H. Tuadara – Sindad Profiles. (Compiled from Kaila and Krishna, 1992 and papers cited therein). The velocity bands in the index are chosen as generally accepted boundaries for the UC, LC and its high velocity layer (Modified after Mahadevan, 1996).

regional exposures of charnockitic rocks preserve the metamorphic imprint of the LC and enclaves with more ancient protolithic ages than the host rocks that give a clue to their ancestry. More recently, more than one generation of charnockitic rocks have been identified (Santosh, 1989), among which neocharnockites developing from high partial pressures of CO_2 can be generated even in the middle crust. These, however, may be relatively smaller bodies. In any case, extensive exposures of charnockitic rocks may be conveniently taken as representing what were once mid- to lower crustal segments exhumed tectonically and uplifted by isostatic processes to upper levels and exposed by erosion.

The HGD may be divided into a number of sub-domains that are separated by near vertical faults or shear zones (Fig. 3). The sub-domains include (i) the Biligirirangan block (BRB); (ii) the Nilgiri Block; (iii) the NE-SW trending highly faulted Dharmapuri rift extending from Vaniyambadi in the NE to Bhavani in the SE; (iv) the Palghat-Cauvery belt of faults, a rift zone, between Bhavani and Palani; (v) the Kodaikanal-Anamalai-Madurai block and (vi) the Kerala Block. The major bounding faults/shear zones include the Moyar shear zone that separates the Nilgiri Block from the Karnataka block; the Bhavani fault that separates the Nilgiri Block from the Palghat-Cauvery rift zone on its south; the eastward extension of the Moyar-Bhavani faults into possibly the Attur fault and the Thenmalai-Achenkovil shear zones that separate the Madurai block from the Kerala block. The BR block, the Nilgiri and the Kodaikanal-Anamalai ranges are made up largely of massive to foliated charnockitic rocks. In the surrounding low-lying areas the charnockites occur interbanded with extensive exposures of khondalitic sediment, of varying sedimentary facies, pelitic, arenaceous, calcareous and ferruginous. The Kerala block is predominantly made up of khondalitic rocks and some charnockitic massifs. The tectonic setting of the HGD of South India has been viewed by several workers as one of micro-continents sutured along the Palghat-Cauvery and Achenkoil shear zones mainly on lithological contrasts across these fault zones and geochronological data that ignores resetting of ages through progressively younger granitic magmatism southwards, discussed later. Unequivocal evidence for a suture is lacking. If we assign the khondalite suite of sediments a supracrustal status and the protoliths of the charnockites a basement status, it will be evident that the different blocks of the shield are at different crustal levels and have differential uplifts along the bounding faults. Viewed in this angle, a unitary model of progressive evolution of one large Achaean continental crust, through crust-mantle interaction seems more justified and seems to explain several geological and geophysical characteristics satisfactorily.

The Dharmapuri and the Palghat-Cauvery rift zones are characterised by 2 or 3 generations of ultrabasic, basic, alkaline and calc-alkaline plutonites spanning the time band of 2500 to 550 Ma, standing testimony to the overprinting of transpressional regimes over an earlier compressional regime. Alkaline rocks, syenites and carbonatites are also found along the NE-SW trending Kambam valley in the Madurai block.

A large amount of Sm-Nd, Rb-Sr and U-Pb isotopic data have been generated that point to 4 different age provinces (Baskar Rao, Y.J. and others, 2003), namely (i) the BR block of mid-Archaean protoliths; (ii) the Nilgiri massif with protolith ages of 2.7 Ga; (iii) the Cauveri-Noyil-shear zone of 3.2 to 2.7 Ga ages and (iv) the belt of Kodaikanal-Oddanchatram block of 2.7 – 1.7 – 0.5 Ga ages. A synthesis of the available Sm-Nd, Rb-Sr and U-Pb isotope systematics point to (i) protolith ages of ~3.0 to 2.8 Ga overprinted by the younger ages of 2.5 ± 0.4 Ga dominantly in the Dharwar craton and the Dharmapuri belt and (ii) ages of

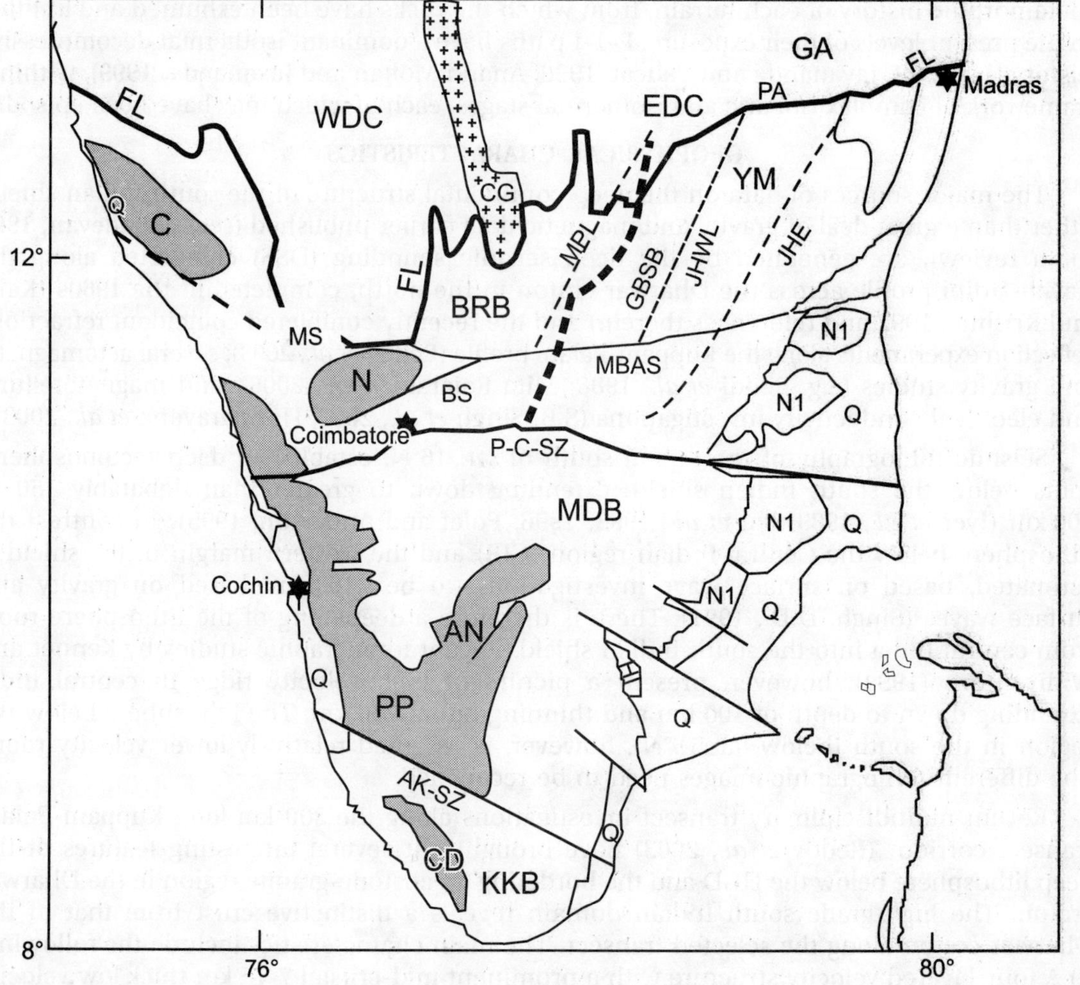

Fig. 3. Major shear zones and faults in the south Indian high-grade domain. (Modified after Gopalakrishnan, 1996)

~ 750 - 550 Ma in the Dharmapuri and Palghat-Cauvery belts. The region further south, not covered by the transect in the Madurai and Kerala blocks has a dominant overprinting by ~ 550 Ma ages. Older protolith ages of 2.7 Ga are noted in the Madurai block but no ages older than the Pan-African event (750–550 Ma) have so far been recorded from the Kerala block, possibly due to incomplete sampling and a large overprinting by the last orogeny.

Geothermometric and geobarometric studies on several mineral pairs point to palaeopressures of 4 kbar (= ~12 km palaeodepth) in the GGD in the Karnataka Craton; 9 kbar (= 27-30 km palaeodepth) in the Nilgiri Mountains and in the Biligirirangan, 8 kbar (24 to 27 km palaeodepth) in the Madurai Block in an early phase followed by a 4.5 kbar (14 to 16 km palaeodepth) and 700°C and 4.5 to 6 kbar in the temperature range of 700-750°C in the KKB. The palaeodepths indicate the minimum depths of evolution registered in the metamorphic history of each terrain, from which the rocks have been exhumed and uplifted to the present levels of their exposure. P-T-t paths have a dominant isothermal decompressive history (ITD), (cf. Jayananda and Peucat, 1996; Anand Mohan and Jayananda, 1999), within a framework of complex isobaric and isothermal stages, each of which may have been episodic.

GEOPHYSICAL CHARACTERISTICS

The major sources of data on the deep continental structure of the south Indian shield, other than a great deal of gravity and magnetic data earlier published (see Mahadevan, 1994 for a review), are generated by the deep seismic sounding (DSS) completed along the Kavali-Udipi profile across the Dharwar craton in the north, completed in the 1960s (Kaila and Krishna, 1992 and references therein) and the recently completed coincident refraction-reflection experiments along the Kuppam-Palani profile (Reddy *et al.*, 2003); several aeromagnetic and gravity studies (e.g. Reddi *et al.*, 1988; Mita Rajaram *et al.*, 2003); and magnetotelluric and electrical conductivity investigations (S.B. Singh *et al.*, 2003; Harinarayana *et al.*, 2003).

Seismic tomography of the region south of lat. 16°N, established deep tectonospheric roots below the south Indian shield, extending down to greater than debatably 150 to 400 km (Iyer *et al.*, 1989; Rai *et al.*, 1993; 1996; Polet and Anderson, 1995). In contrast the lithosphere below the Central Indian region (CIR) and the western margin of the shield is estimated, based on surface wave investigations, to be ~100 km based on gravity and surface waves (Singh, D.D., 1991). There is therefore a deepening of the lithosphere roots from central India into the south Indian shield. Recent tomographic studies by Kennet and Widiyantoro (1999), however, present a picture of high velocity ridge in central India extending down to depth of 100 km and thinning out at 250 km. The lithosphere below the region in the south (below lat. 15°N), however, is assigned relatively lower velocity roots. The different tomographic images need to be reconciled.

Recent multidisciplinary transect investigations along the 300 km long Kuppam-Palani transect corridor (Reddy *et al.*, 2003) have brought out several interesting features of the deep lithosphere below the HGD and the bordering greenstone-granite region in the Dharwar craton. The high-grade south Indian domain reveals a distinctive crust from that of the Dharwar craton along the selected transect. The main characteristics include the following. (i) A four-layered velocity structure with a prominent mid-crustal 7-15 km thick low velocity zone (LVZ) (at a depth of ~ 20 km), confirmed by gravity modeling. A general correspondence is observed, though with some exceptions, in the order of electrical conductivity with the LVZ of high electrical conductivity. (ii) Resistivity at depth is strongly influenced by the deep faults in the region that result in crustal fracturing and fluid migration. The distribution of resistivity pattern is complex and resistivity contrasts have been noticed on either side of the Palghat-Cauvery Shear Zone. (iii) The crust is highly reflective below ~10 km from the surface and the reflective layers show some reversals in dip giving rise to regional convergence, apart from being domical on a smaller scale. (iv) The depth to the MOHO ranges from 41 to 45 km with local upwarps and therefore points to a thickened crust. (v) The crust has distinctive magnetic properties in that the crust below 22 km does not contribute significantly to the magnetic anomalies, in contrast to the crust below the Dharwar craton, which is magnetic over its entire thickness. This may be either due to compositional changes or Curie isotherm destroying the magnetic field. Interestingly the crustal layering generally coincides with the LVL established in seismic profiles and therefore may be due to the compositional changes in the crust with depth. (vi) Heat flow measurements specially prepared boreholes drilled in the block north of the Palghat-Cauvery rift zone works out to 36 ± 4 (SD) mWm^{-2} and for the block south of it 44 ± 6 (SD) mWm^{-2}. Based on measured values of heat generation, it is inferred that the heat-flow from the mantle would be 26-29 mWm^{-2}. In contrast, the heat-flow from the mantle below the Dharwar craton is estimated at 12-19 mWm^{-2}. These results are consistent with an enriched mantle underlying the SGT, possibly generated by post-Archaean (early Proterozoic to Pan-African) mantle metasomatism (Roy *et al.*, 2003).

BEARING ON CRUSTAL EVOLUTION

Continental evolution is the history of crust-mantle interaction in response to the thermal evolution of the mantle interior. The Dharwar craton had a dominant evolution in the Late Archaean-Early Proterozoic period and the mantle below the Archaean Dharwar craton was possibly depleted of the large ion lithophile (LIL) elements and High Charge elements due to their transfer through generation of large volumes of partial melts/magmas of komatiitic and tholeiitic composition, that form an integral part of the Archaean Dharwar crust. On the other hand, the HGD seems to have had a dominant Proterozoic evolution, with a possibly Archaean ancestry, evidenced by several isotopic protoages. The Proterozoic mantle below the HGD shows evidences of having been enriched in these elements during the whole of the Proterozoic, thereby suggesting an active and thermally resurgent mantle.

The major unique features, recently deciphered, namely, a four-layered velocity-density structure above the MOHO; a mid- to lower crustal low velocity layer (LVL) with distinctive electrical conductivity, testifying the changes that may have been brought about in the lower crust by the action of fluids and a change in the character of magnetic properties. A high reflectivity, both domical and dipping concordantly with the MOHO and distinctive in both its pattern and intensity compared to many known standard reflectivity patterns known over continental regions characterises these rocks. Such reflectivity patterns possibly arise from the neofabric the rocks have imbibed through recrystallisation during their differential exhumation from greater depths under isothermal decompression. A distinctive thermal structure arising from both crustal and mantle heat flow/heat generation can also influence crustal rheology and consequently the dynamics. The fact that the magnetic crust is thin has to be viewed in the context of a coinciding LVL and high conductivity layer. It points to possibly a compositional change resulting from transformations that have rendered the deeper levels non-magnetic during the last phases of the exhumation and under the influence of the alkaline and granitic material introduced or generated in the crust. If the causative magnetic bodies are the large masses of charnockites, this implies that they may be more or less rootless or may have transformed roots. Modeling of crustal structure using assumptions of standard gravity or velocity structure for the whole crust would indeed be inappropriate if the assumptions of homogeneity are challenged by such transformation of the deeper crust.

The divergent reflection fabric of the HGD along the seismic profile has been related to a collision tectonics (Reddy *et al.*, 2003). Bi-modal gravity and magnetic anomalies across steep dipping shear zones are interpreted as evidence of subduction zones. These interpretations are empirical and ignore the extensive and marked isothermal and episodic exhumation of the HGD from great depths, a process that has large potential to erase several older geophysical signatures and tectonic implications of its own. It would be more rational to relate the reflectivity and velocity of the deep crust along especially the rift zones to the

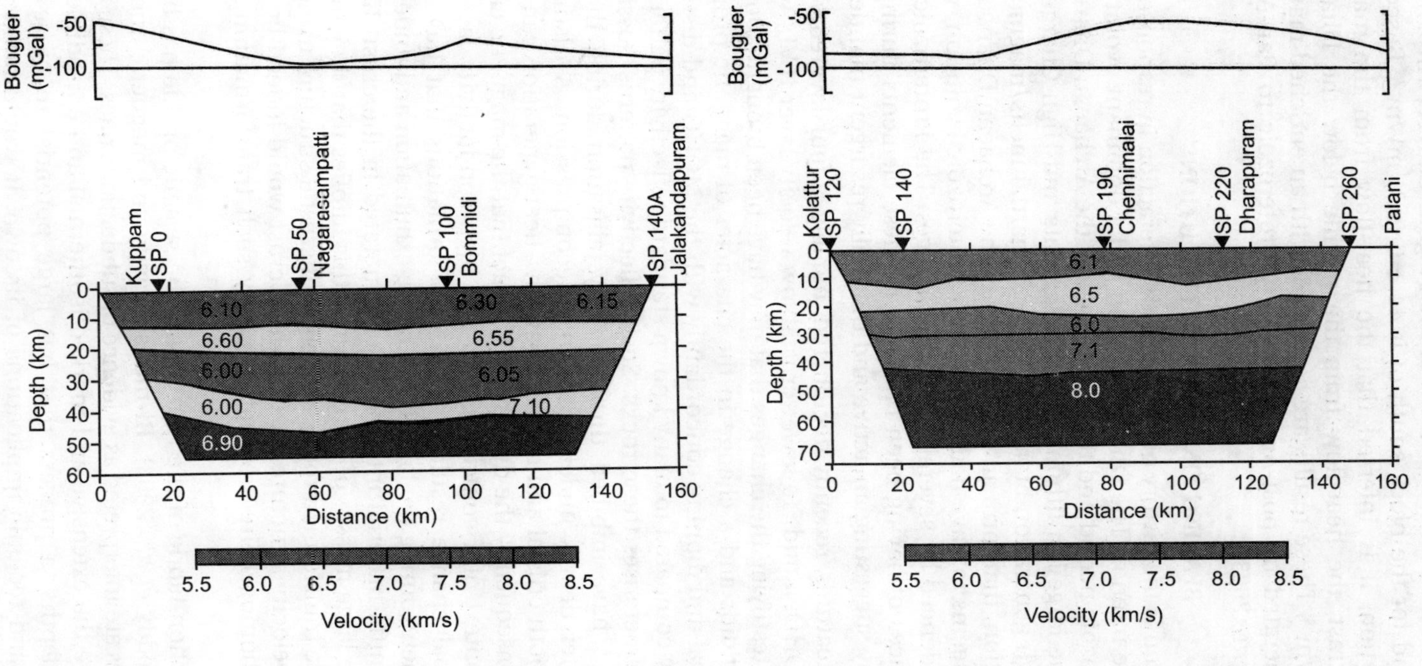

Fig. 4. Four-layered velocity structure below (a) the Kuppam-Kumarapalaiyam segment; (b) Kolattur Palani segment (Redrawn after Reddy, P.R. *et al.*, 2003).

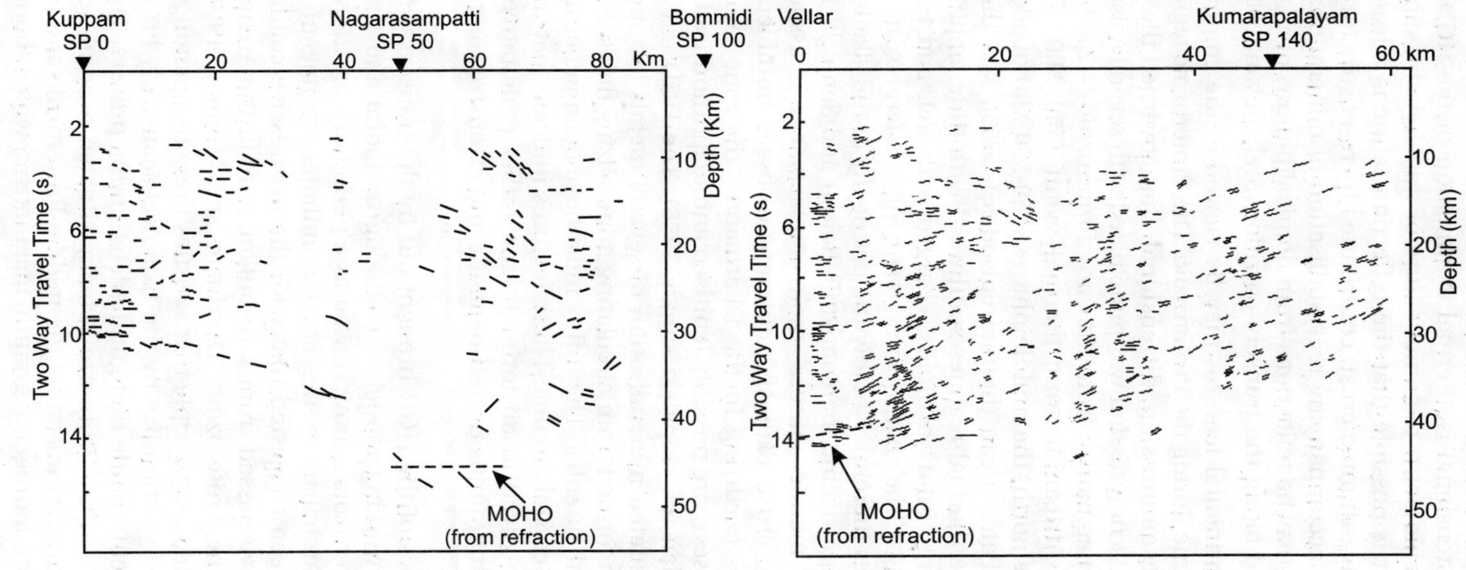

Fig. 5. Line drawings of stack section along (A) Kuppam-Bommidi segment and (B) Vellar-Kumarapalaiyam segment. The depth scale is approximate and is based on travel time to depth conversion with a constant velocity of 6.0 km/s throughout (Redrawn after Reddy, P.R. *et al.*, 2003).

events of transtension and differential exhumation of the LC. An alternative to this model of collision arises from the distensional isothermal exhumation of the HGD that is, perhaps, the most pervasive synorogenic petro-tectonic feature of this domain, superimposed on all the earlier tectonic fabric. It is possible that the reflection fabric is a feature intrinsic to a differentially and episodically exhumed crust, contributed by reset structures during upward isothermal exhumation and accompanying recrystallisation of mineral phases. Some of the smaller wavelength features may be reflections from domical plutons of alkaline and granitic rocks so profusely developed along the transect corridor (Mahadevan, 2003).

The distinctive deep continental features may be viewed in the context of an intensive crust-mantle interaction regime during the Proterozoic. The Proterozoic regime was expressed in an episodic isothermal decompression, distensional, that governed the exhumation path of the high-grade rocks. The decompression was punctuated with several episodes of ultrabasic, basic, alkaline and granitic magmatism from the early Proterozoic (~ 2500 Ma), culminating in the last phases of Pan-African tectono-thermal event (550 Ma). The early phases (< 2500 Ma) dominate in the north, the middle phases (750-550 Ma) of alkaline influxes the central parts and the final Pan-African thermal episodes (~> 550 Ma) the southern parts. This implies a shifting of the thermal regimes southward with time, implying cratonisation progressing southwards. Large influx of fluids, charged with CO_2 and volatiles promoted both the making and breaking of the charnockites, testifying to a large-scale mantle degassing (Santosh, 1989 and references therein). Notably the petrogeny of the alkaline rocks has been related to the process of "taphrogenic magmatism" (Rajesh and Santosh, 1996). Even if the alkaline rocks have deeper mantle sources, their emplacement is associated with large mobilization of fluids in the crust. Extensive outcrops of fissile hornblendic gneisses and epidote-hornblende gneisses occurring in the Dharmapuri rift zone (along the Kuppam-Bommidi section of the transect) are possibly fenitised and diaphthoritised earlier basement rocks of the charnockites clan (cf. Gopalakrishnan, 1996). The HGD has also experienced rather extensive but disseminated mineralisation of several metals like molybdenum, rare-earths, uranium and thorium formed from circulating mineralizing fluids that have generated pegmatitic and hydrothermal aureoles. The influx must have been penetrative as evidenced by reset ages of even single crystal zircons. These features indicate that the HGD was, as a whole, an "open system" involved in an intimate and rather penetrative fluid influx and magmatism, with the transfer of fluids from the mantle and extensive chemical mobilization of crustal elements.

The several characteristics of the HGD brought out by the transect programme may be viewed in the context of a working model of changing temporal and spatial lithosphere-mantle interaction. The Proterozoic is now viewed as a period of transition from the mantle dominated regime of the Archaean to the crust dominated regime of the Phanerozoic (Dickin, 1995, 100). It has been suggested that with the progressive cooling of the mantle, the convection cells may have passed from a turbulent state in the Archaean to organised stable convection cells in the Proterozoic (Durrheim and Mooney, 1994). The Southern Indian Shield possibly provides an example that satisfies these two paradigms. The Dharwar craton was, perhaps, the first to cratonise by the Late Archaean-Early Proterozoic and may have evolved above a turbulent mantle that got depleted in the process. The HGD remained mobile throughout the whole of the Proterozoic, being situated over an upwelling enriched mantle (Fig. 6). It was possibly cratonised by the middle Proterozoic and was later involved in the Pan-African episodes of anorogenic granitic magmatism, which then carried the torch of deep crustal transformations, leaving several vestiges of the Archaean protoliths. It, therefore, seems that the Dharwar craton and the HGD were part of a continental segment, which responded to a differentially cooling mantle. The tectonics and magmatism and perhaps metallogeny of the HGD is one of a thickened lithosphere closely interacting with a hot organized actively convective enriched mantle.

The deep-rooted Dharwar-HGD continental tectosphere seems to have remained quite stable till late in the Mesozoic. The impact of the Rajmahal Volcanism of early Cretaceous and the accompanying rifting along the eastern margins of the Indian Shield is less evident on the features of the Southern Indian Shield. The rifting and accompanying Deccan continental basalt magmatism had a profound impact on the shield and led possibly to magmatic under plating resulting in crustal uplift on the west and a tilt eastwards. The thermal impacts of the mantle on a major part of the western half of the shield seem to be an ongoing process resulting in the continuing uplift of the Nilgiri and Anamalai Plateaus. These processes have an obvious impact on the geomorphology of the region and on many geophysical characteristics, especially the potential fields that call for a careful modeling.

COMPARISON WITH OTHER SHIELD SEGMENTS

Rajasthan

The Precambrian geological evolution of the Rajasthan region is different from that of the southern Indian shield as is evident from the absence of the spatial polarity between Archaean greenstone–granite domain and the HGD domain characterising the south Indian shield. At least three major orogenic cycles have been penetrative through the Precambrian crust of Rajasthan, all of which must have contributed to the exhumation and modification of the LC. These are represented by the Archaean Banded Gneissic Complex (BGC), that forms the basement for the Paleoproterozoic Aravalli belt and Mesoproterozoic Delhi belt. The three belts cradle between the unmetamorphosed platformal Meso/Neoproterozoic Vindhyan Basin (VB) and the Neoproterozoic Marwar Basin (MB) in the east.

Belts of high-grade rocks of the granulite facies are restricted to the large exposures of charnockitic rocks in the Balaram-Abu Road area in the SW part of the Aravalli-Delhi fold belt; to the Sandmata Complex and the Thana Gyangarh basement complex on the eastern flanks of the Delhi fold belt and to the flanks of the Delhi fold belt (Raja Rao, 1971; Desai *et al.*, 1978; Ram Sharma, 1988, 1999; and Harel Thomas, 1995). Whether they are restricted to small belts or are part of an underlying larger charnockitic mass is yet to be established but the available evidences point to their exhumation from great depths. The P-T-t paths of the Sandmata Complex has a mineral assemblage that stabilized at 650-850°C and 8-11 kb pressure, suggestive of a minimum burial of, say, 20 km from which it was exhumed to shallow crustal levels. Sharma (1988) suggests an early anticlockwise prograde path followed by near-isothermal exhumation to 5 kb level and 600°C temperature. Sharma suggests a non-collision setting compatible with some magmatic underplating. The subsequent uplift of the Complex may be due to distension events in the late Precambrian. The Sandmata Complex of granulite is dated 1600 Ma and considering the fact this is the age of their transformed state, the age may fix a stage in the exhumation of the Sandmata Complex. It may be relevant to point out that the basement of the Vindhyans in the SONATA belt of Central India could have possibly been the exhumed charnockitic rocks (as discussed later here) and if the Vindhyans are older than 1100 Ma, as fixed by the kimberlitie intrusives into the Semri group of the Vindhyans, then the exhumation of what are the basement may

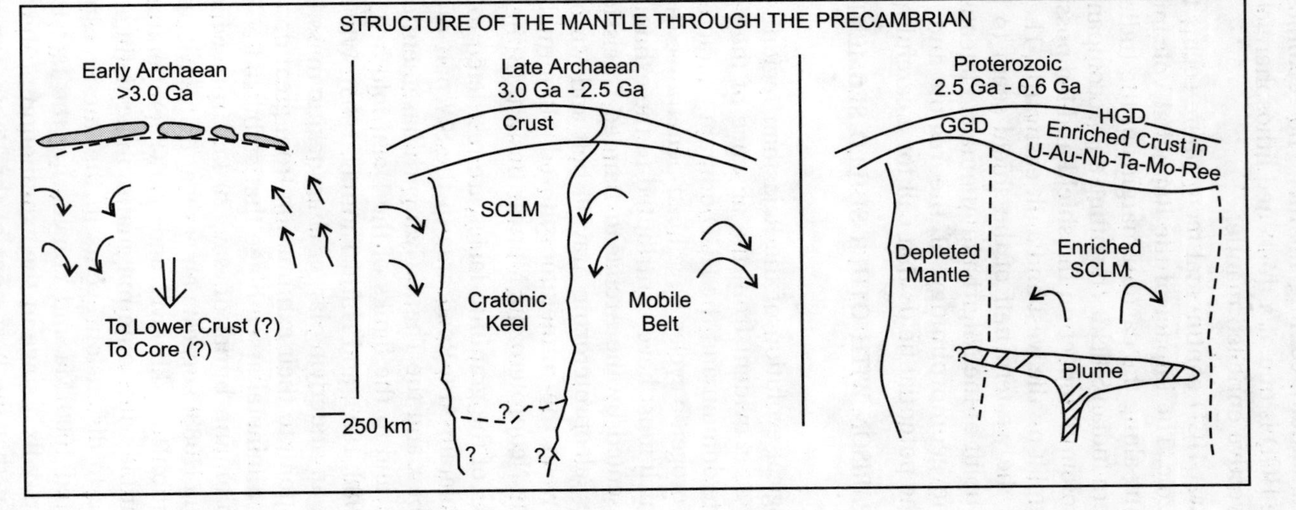

Fig. 6. A working model of the changing temporal and spatial lithosphere-mantle interaction below the South Indian Peninsular shield from the Archaean through the Proterozoic. (See text for explanation)

be older (*i.e.* >1100 Ma). The phase of uplift from the depths of 5 kb to the surface may have been aided by the distension associated with rifting and formation of the Vindhyan basin. The time span of 1600 to 1100 Ma fixed by the above dates then implies that the exhumation of the LC in Rajasthan was an event contemporaneous with the Delhi Orogeny, opening up thereby the questions of cause and effect. The formation of the Delhi basin may itself be related to the history of the exhumation of the lower crust. The charnockites in other parts of Rajasthan are not dated; a common presumption is that they could have been part of the Archaean basement.

The orogenic evolution of the Rajasthan region is punctuated by repeated acid magmatism manifested in the Berach granites (2.6-2.5 Ga), Darwal and Amet granites (2.0-1.9 Ga); Ajmer, Alwar granites (1.7-1.5 Ga); Kishangarh syenite (1.5 Ga); Erinpura granite (850 Ma) and the Malani Igneous Suite (750 Ma). The western fringes of the Delhi Fold belt also experienced the impact of the 550 Ma Pan-African thermal event and the mafic and alkaline magmatic activity associated with the Deccan magmatism of the late Mesozoic. If the features presented by the exhumed LC in southern Indian shield are a guide, one may expect extensive transformations in the LC due to these magmatic episodes and the tectonics associated with their emplacement.

The results of the recent seismic, gravity, magnetic and magnetotelluric studies along the Nagaur-Jhalawar transect of the Proterozoic Aravalli Delhi Fold Belt region in the NW Indian shield have significantly enhanced our understanding of the deep continental crust. There is, however, a need for a more reliable velocity structure to be established along the profile that confirms the gravity layering now proposed, on limited wide angle reflection. Despite this lacuna in the data, some interesting features of relevance to understanding the LC of the Rajasthan region emerge from these studies. Crustal seismic reflection fabric shows a number of dipping reflections from either side at several crustal depths including the MOHO. Some of these reflections with opposite dip constitute a divergent reflection fabric, as in the case of the south Indian transect already discussed. Two such reflection fabrics observed in the region may be related to Paleoproterozoic Aravalli and Mesoproterozoic Delhi Fold Belts. A thickened 48-km thick crust with a high velocity (7.3 km/s) lower crust is suggested under the Delhi Fold Belt and Sandmata Complex. A mid crustal domal shaped reflector demarcates the upper boundary of the high velocity layer. A 70 mGal gravity high with conspicuous lows on either side correlates well with the observed seismic structure of these parts. The boundaries of various shear zones of the region are well identified in the reflectivity patterns. A deep penetrating crustal scale thrust fault extending upto Moho depths and a high conductive dipping interface, identified from the magnetotelluric studies, correlate well with each other (Reddy *et al.*, 1995; Tewari *et al.*, 1998; Rajendra Prasad *et al.*, 1998; Vijaya Rao *et al.*, 2000). Tewari and Vijaya Rao relate the seismic reflection fabric and the density structure modeled to the plate tectonic processes that dominated the orogenic cycles in the Rajasthan region. Though such an explanation is in keeping with several earlier models of plate tectonics applied to this region (Sinha-Roy, S., 1988; Sugden *et al.*, 1990; Sychanthavong and Desai, 1977), it is very generalized and ignores the reflective fabric that the LC can imbibe through the several changes that accompany the tortuous evolution of the total crustal structure through three orogenies and a Mesozoic distensional episode associated with Deccan magmatism. As in the south Indian profile the reflection fabric should be related to the distensional exhumation of the LC during which the rocks would have developed a mineral fabric in response to the changing P-T-t paths of their exhumation. The shear zones in the reflection profile may have resulted during the differential solid state ductile exhumation of these rocks and may not be specific to any tectonic feature. There is thus a parallel with the South Indian HGD as already discussed.

CENTRAL INDIA

Central Indian Region (CIR)

The major elements in the Precambrian crust of the CIR may be grouped into two blocks, viewed by some as distinct protocontinental regimes (Jain *et al.*, 1991 and Radhakrishna and Naqvi, 1986). The northern Bundelkhand block/Protocontinent comprises the Bundelkhand Granitic Complex, the Tirodi Gneisses, the high-grade (amphibolite facies) Sausar metasediments and the charnockitic rocks outcropping along a NE-SW-trending 100 km long belt along the southern fringes of the Sausar Group of rocks. Granulite facies cordierite metapelites and mafic granulites also occur along the northern margin of the Sausar basin in the Ramkona-Katangi area (Bhowmik *et al.*, 1998). These are classified as pre-Sausar group of younger granulites by Roy *et al.* (2000). The southern Deccan Protocontinent comprises the low-grade (greenschist facies) Sakoli and Dongargarh Groups and the associated Amgaon gneisses, the Dongargarh granites and the Chhattisgarh granites. The two blocks are separated by a major shear zone, the Central Indian Shear (CIS) that has been traced from south of Nagpur to the north of Chilpi and beyond. The interrelation between various formations and between the two blocks or proposed protocontinents is rendered uncertain by the extensive younger cover. The cover includes the arcuate Neoproterozoic Vindhyan basin fringing the Bundelkhand Granite Complex, the Gondwana sediments and the Deccan Trap cover.

Though exposures of granulite facies rocks are very restricted, rocks with lower- mid-crustal seismic velocities but that they occur at shallow depths in the several seismic profiles in this region. Two large crustal segments have been investigated in detail by geological and geophysical methods. These comprise (i) the Precambrian formations of the Bundelkhand–Sausar-Bandara-Chilpi Ghat-Nagpur region that have a complex Precambrian succession and (ii) the Narmada-Son-Tapti lineaments (the SONATA belt) which is noted for a Precambrian ancestry overprinted by Mesozoic sedimentation (Gondwana), Deccan Volcanism and, perhaps, continuing tectonics as evidenced by the ongoing Satpura uplift and current seismicity.

A deep seismic reflection/refraction, wide angle reflection survey along 155 km long Mungwani-Seoni-Katangi-Kalimati (NW-SE) profile brings out several interesting features (Reddy *et al.*, 2000; Reddy *et al.*, 2003). The crustal velocity model along the profile portrays a five-layer model, in which the crust-mantle boundary is at depths varying between 42 and 44 km, suggestive of a thickened crust as many other HGDs. The depth to the Moho boundary obtained from the refraction data agrees well with the depth of the reflection Moho. Two adjacent reflection domains across the CIS dip towards each other and the reflection MOHO is off-set by the suture. The pattern of reflections is same from upper crust to MOHO boundary, although the dips of the reflectors decrease with depth. The reflection segments below Kalimati, in the upper mantle appear to belong to the subducted oceanic block situated between these Protocontinents before the convergence started. Reddy *et al.* (2000) relate the reflection fabric to a continental collision-suture developed during Proterozoic.

Five deep seismic sounding (DSS) profiles each about 250 km long have been shot by the National Geophysical Research Institute across the SONATA belt (Kaila and Krishna, 1992 and references therein). The broad crustal velocity structure obtained from these studies is presented in Fig. 2 E-H (compiled from Kaila and Krishna, 1992; Kaila and Sain, 1997; Reddy *et al.*, 1997a and Reddy *et al.*, 1999) and compared with the velocity structure of the CCR as obtained in other seismic transects. It may be seen that the upper crust in the SONATA belt is relatively thinner compared to the CCR. The seismic sections reveal a highly reflective crust, generally characteristic of the LC, with large number of reflectors at various depths down to the MOHO. The velocity structure points to an exhumed lower crust. In the profile between Damoh and Katangi, a high velocity between 6.5 to 6.7 km/s is observed immediately below the sediments, and may correspond to the exhumed LC that forms the basement for the overlying Vindhyan basin (Mahadevan, 1995b). Presence of high velocity of 6.7 km/s at shallow depths of 2 to 8 km in the Katangi–Jabalpur block and between Batiagarh and Damoh also indicate an exhumed shallow LC. If the P-wave velocity of the exhumed LC can be as lower as 6.2 km/sec, as is seen in seismic profile across the HGD of South India, then the thickness of the LC will be even more.

There is a dominance of 7.0 to 7.4-km/sec layers, which may be a denser part of the lower crust. Such a layer is not pronounced in the SIS or the CCR and indicates that the Lower crust in the CIR is denser (cf. Verma and Banerjee, 1992). This may explain why the LC did not get exhumed to higher levels as in the south Indian HGD. There is a notable absence of a 7.8 km/sec layer that acts as a rift cushion in the Cutch region further west and this distinguishes the SONATA from the Cutch basin. The MOHO depths are of the order of 38–45 km, implying a thickened crust, not consistent with the rifted nature of the SONATA belt. It indicates therefore a more complex evolution of the SONATA belt than by mere rifting, wherein a thinned crust during early rifting gets thickened later. One possibility is the underplating of the crust during the Deccan magmatism. However the velocity structure does not give a clear evidence of such underplating, except through a thin 7.0 –7.4 km/sec velocity layer.

In contrast, in the Cambay-Billimora region, as in the latter, the crustal section reveals a clear horst and graben structure and the MOHO shallows considerably reaching a depth of about 18 km near Billimora, south of the Narmada. In the crustal column, high velocities like 7.0 to 7.26 km/s are encountered at shallow depths of about 10 to 12 km down to the MOHO, indicating a possible rift cushion of basaltic rocks. It may therefore be inappropriate to club the seismic profiles in the Cambay-Billimora region with that of the SONATA east of 74°E longitude. However the pericontinental western profiles also indicate the scope for exhumation of the LC through the Mesozoic distension events.

Two deep-rooted conductive bodies are delineated along the Damoh-Jabalpur-Mandla profile in the depth range of about 5 km, upto almost about 50 km and beyond and have resistivity of about 10 – 5000 m. The seismic reflection patterns observed at depth range of 10 and 20 km seem to envelope the southern and more prominent of these two conductors. The high conductive features are attributed to the intrusion of mantle material and subsequent partial melting or the release of saline fluids into the shallow crust (Gokarn *et al.*, 2001).

In the absence of data on the P-T-t paths and geochronological data on the LC rocks of Central India, no discussion of their exhumation history is possible. However the distensional episodes related to the Vindhyan and Gondwana basin formation may be expected to have influenced the uplift of an already exhumed Proterozoic LC crust, which was also modified, as in south Indian HGD, by the impact of granitic magmatism associated with the Dongargarh (~2300 Ma) and Malanjhkhand granites.

EASTERN GHAT MOBILE BELT

The Eastern Ghat mobile belt (EGMB) (so named commonly because of its marked Proterozoic mobility) is cradled to the east of the CCR. It is in fact a continuation of the eastern segment of the south Indian HGD, interrupted by the presence of the Cuddapah basin. The continuity is evidenced by the presence of the charnockite and khondalite outcrops from north of Madras (present Chennai) along the eastern margins of the Cuddapah (east of Ongole) extending northwards into the Eastern Ghats Ranges; the extension of the alkaline and anorthositic rocks from these ranges into the Vellore-Palani region of South India (falling in the same age group of 1200-550 Ma); a continuity of structural elements from the EGMB into the eastern segment of the South Indian HGD and a similarity in high temperature-low pressure granulite facies of metamorphism. The EGMB therefore seems to be the exhumed Lower crust from below the CCR very much as the south Indian HGD. The two regions also have been placed in a single gravity province. NE trending steep shear zones that may be mantle faults mark the boundary between the EGMB and the CCR. Recent suggestions that the EGMB is a distinct microplate from the south India HGD seem to ignore these aspects.

The P-T-t paths of evolution of the EGMB granulites, that can throw light on the exhumation trends, have been extensively studied. An anti-clockwise prograde path to a P-T_{max} of ≅ 10 kb and 960-1100°C is suggested for the Gokavaram and, perhaps, the Ananthagiri areas of SE EGMB. In the northern part the retrograde path has a characteristic (anticlockwise) overturned Y-shaped configuration (Mukherjee, 1989). In the Chilka area a combination of 3 ITD and two IBC paths from 8-10 kb and 1000°C down to 4.5 kb and 650°C has been suggested (Sen *et al.*, 1995). The charnockitic rocks of EGMB seem to have remained under isobaric cooling conditions for a long period from 2500 Ma to 1400-1000 Ma when it underwent rapid decompression to ~ 5kb depths. The final exhumation may be due to later events (Sengupta *et al.*, 1990; Das Gupta *et al.*, 1995). The P-T-t paths of the HGD on EGMB seems to be greatly influenced by the thermal influence of the anorthosite and alkaline rocks. There is also a suggestion of plume-induced tectonics and petrogenesis from 1500 Ma onwards (Mukherjee, 1996). Recently reported protolith ages of > 3.0 Ga for the EGMB granulites bestow on them an Archaean ancestry, similar to those of the South Indian HGD. The exhumation of the HGD in the EGMB also may be rather episodic.

No seismic profiles are available for the EGMB, except the eastward extensions of the Cuddapah, Godavari and Mahanadi basin DSS profiles (Kaila and Krishna, 1992). Gravity estimates place the thickness of the crust below the EGMB at ~ 40 km (Qureshy, 1970; Mahadevan, 1994). The DSS profile extensions point to crustal depths of 43-45 km below Nizampatnam and also below the Godavari graben (Fig. 2C). However the crustal thickness is 31-34 km in the Mahanadi graben (Fig. 2D), which may be due to Phanerozoic thinning, not evident in the Godavari graben. There is, in general, a similarity in crustal thickness below the south Indian HGD and the EGMB, except for the Mahanadi region.

The uplift of the EGMB may have been influenced by the late Mesozoic rifting of the eastern continental margin of India and the recent discoveries of equivalents of the Rajmahal Basalts in the Mahanadi and Godavari offshore strengthen this possibility.

CHHOTANAGPUR PLATAEU

The Chhotanagpur Plateau in Eastern India comprises the Singhbhum Greenstone-granite domain (SGD) of green-schist-lower amphibolite facies rocks in the south and the Chhotanagpur Granitic Gneissic Complex (CGGC) that comprises rocks of the amphibolite and granulite facies of metamorphism (Fig. 7) (Mahadevan, 1992, 2003). Though the boundary of the former is marked by the Khatra-Tamor fault, a deep mantle fault, there is a metamorphic gradation from the SGD to the CGC through lower amphibolite facies. The charnockitic and anorthositic rocks, which are considered as typical of the lower-middle crust, in regions to the north and south of the Damodar Valley Gondwana belts, and also in the Palamau district in the west, extending westwards into the Sarguja District of Madhya Pradesh (Fig. 7). The more prominent outcrops studied are of the Adra belt (Purulia and Bankura districts) in the south and the Dumka-Mayurakshi Reservoir region (Monghyr district) in the north, west of the Rajmahal Volcanic belt. The charnockitic rocks of the Adra region are dated ~1200 Ma and the Mayurakshi region ~1500 Ma (Ray Burman *et al.*, 1994). These rocks have been intruded by later hornblende granites and are in stages of retrogressive metamorphism and, therefore, the dates may be that of the later thermal events and also of the stages in the exhumations (see discussion in Mahadevan, 2002). The extensive Gondwana cover masks the nature of the basement, which is of charnockitic rocks could imply a larger extent of these rocks. The large extent of the amphibolite facies rocks prograding into the charnockitic belts implies exhumation of the lower-middle crust to shallower levels than the deeper crust exhumed in south Indian Shield or the Eastern Ghats. The exhumation is, however, from greater depths than the Central Indian region. The distribution of the metamorphic prograding northwards is suggestive of a southerly tilt and the concentration of deeper facies rocks (charnockites) in the east may imply a westerly tilt too, possibly during the rifting related to Rajmahal volcanism.

No seismic profiles are available for the region that can throw light on the deep continental structure. In the absence of data on P-T-t paths, the styles of late Proterozoic exhumation is not understood. The CGGC has witnessed three orogenies, the Chhotanagpur orogeny (1500-1600 Ma); the Satpura Orogeny (900-850 Ma) and the milder Munger Orogeny (350-420 Ma) and these may have influenced the exhumation of the LC. Subsequent uplift of the CGGC may have been influenced by the Gondwana distension events culminating in the rifting of the eastern pericontinental region and the Rajmahal Volcanism. Considering the large extent of the CGGC and the metamorphic gradation from Singhbhum greenstone-granite domain in the south, there is a possibility of a tectonic tilt southwards that could bring up the HGD from deeper levels.

DISCUSSION

The LC is viewed in this paper as the deeper of a binary division of the crust (lithosphere above the MOHO) that is petrologically defined by the granulite facies of metamorphism. It is seismologically defined by virgin P-wave velocities exceeding 6.6 km/sec, which, however, get changed to lower values of upto, say, 6.2 km/sec by thermal, magmatic and fluid-charged diaphthoretic interactions affecting the LC during the Proterozoic evolution and consequent exhumation. The LC is preserved below the CCR of the Indian shield as evidenced from velocity structure and exposures of charnockites along the rift collars of the Gondwana basins, brought about by crustal thinning. The exhumed lower crust outcrops around the CCR and is differentially exhumed, constituting the Proterozoic Mobile Belts. The most

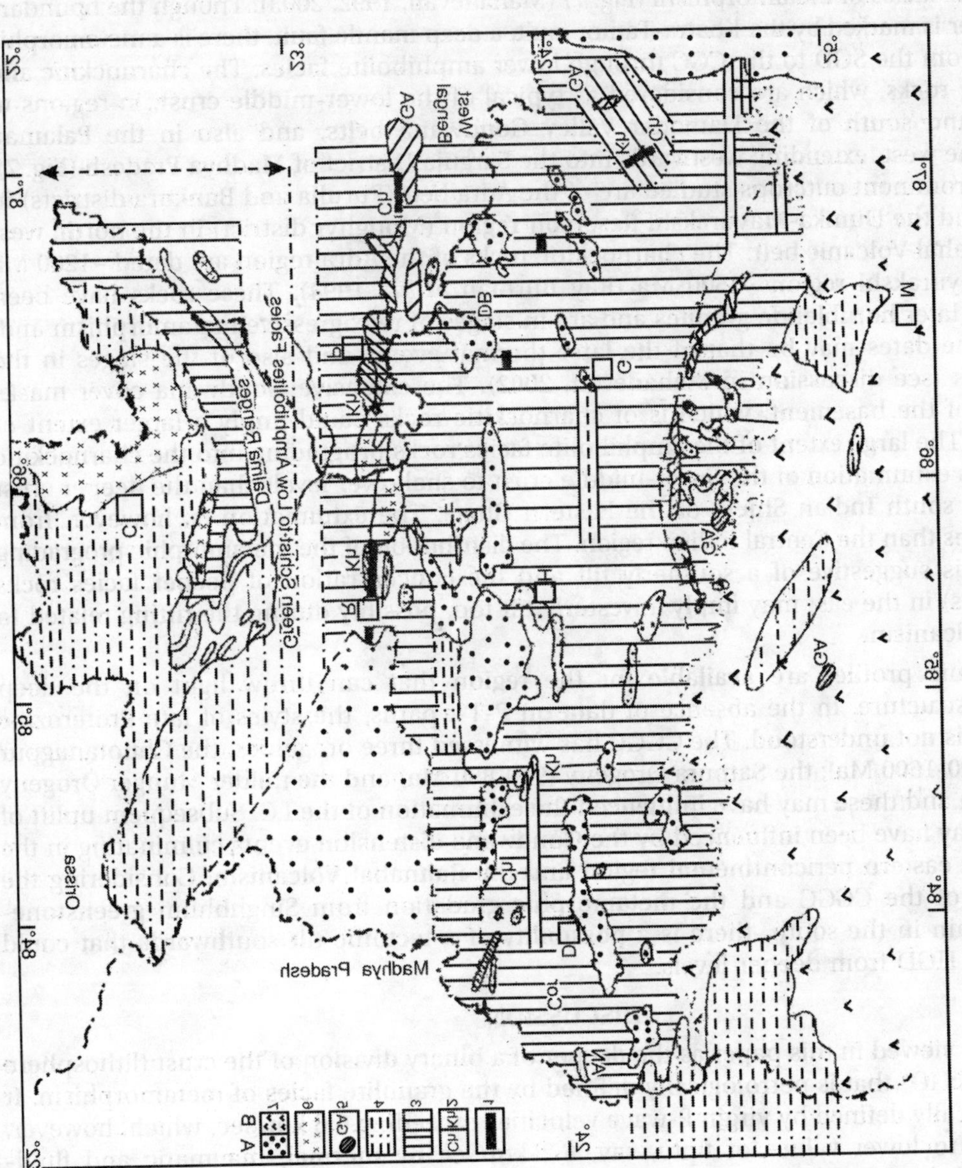

Fig. 7. Metamorphic facies distribution in Jharkhand and Bihar. Index: 1 Granulite facies; 2 Ch-Charnockite exposures; Kh-Khondalitic schists and gneisses (garnet-sillimanite rocks with graphite); 3. Amphibolite facies; 4. Greenschist facies; 5. Gabbro-anorthosite bodies; 6. Major granitic bodies; 7A. Phanerozoic over (Gondwana and Rajmahal Fms. and soils); 7B. Ganga Alluvium; Cor-Cordierite; AN-Andalusite. LOCALITIES: C – Chaibasa; D – Daltonganj; DB – Dhanbad; G – Giridih; H Hazaribagh; M – Monghyr. (After Mahadevan, 2003).

highly exhumed LC segments are those of South India and Eastern Ghats. To the NE, N and NW of the CCR, the exhumation of the LC is of a lower magnitude, amphibolite facies rocks of the middle crust dominating over the granulite facies rocks. The least exhumed of the three is the central region, where low grade supracrustal sequences predominate, covered to a large extent by Gondwana sediments and Deccan trap flows.

The present exposures of HGDs are due to both exhumation from deeper crustal levels in the mid- to late Proterozoic and the subsequent uplift due to several distensional episodes in largely Phanerozoic.

Exhumation can be achieved in several ways. These include (i) repeated phases of compression, shortening, thickening of the crust associated with each of the orogenies followed by denudation; (ii) extensional stretching, crustal thinning to different degrees and denudation; (iii) through an elevator mechanism by underplating by basic igneous rocks and denudation; (iv) transpressional uplift from a vertical component of motion along large transcurrent faults and (v) hypervelocity meteoritic impact causing rebound of deep crustal materials along major fault networks (Dewey, 1986 and Percival *et al.*, 1990). A synorogenic rheomorphic model has been proposed by Weber (1986) wherein decollemation along a number of crustal boundaries accompanied by crustal shortening may lead to thrusting up of intensely folded granulite facies rocks.

Available structural and petrological data indicate that the HGDs comprising the massive/foliated charnockitic and the khondalitic sedimentary components must have formed from early crust of largely basic igneous rocks of tholeiitic and komatiitic composition and sediments of siliceous cherts, clays and carbonates, some ferruginous. They must have undergone burial to lower crustal depths, folding, metamorphism and exhumation. The exhumed LC rocks have followed different P-T-t trajectory paths and exhumation has been differential both in time and space. However generally isothermal decompression processes seem to have dominated in a generally complex milieu at least in the later stages. The middle and late phases of the evolution of the HGDs is marked everywhere by the emplacement of several generations of large volumes of granitic rocks and alkaline rocks and anorthosites and gabbros. While there is differences in the sequence and scale of emplacement, these events point to the impact of a thermal regime of the mantle both on the crust and the lithospheric mantle that had the effect of generating partial melts emplaced into the crust in response to essentially the dictates of gravity. The diaphthoretic changes that are seen in the exposed south Indian HGD are very extensive and portray the manner in which the LC has reacted to the episodes of fluid metasomatic activity under conditions of isothermal decompression. The several isotopic age groups in the South Indian HGD may be viewed as representing the different stages in the transformation of the LC and should be viewed as geochronological memories left in the episodic evolution, the last phase being the Pan-African events. The reduction in the density of the crust by the alkaline reactions and diaphthoresis would bring about significant isostatic imbalances and could lead to the exhumation of the transformed LC. Processes of delamination of the LC can de-insulate the LC from the thermal regime of the mantle and could have played a role in the initiation of isothermal decompression. Greater insights into the sub-crustal lithosphere, now not available, may throw light on such possibilities. Seismic tomography reveals that the HGD is lighter but thicker than the Greenstone-Granite terrain (southern part of CCR) that is thinner and denser (Rai *et al.*, 1993). More recent results of Kennet and Widiyantoro (1999) point to a lighter lithosphere below the southern part of the shield, though, as pointed out earlier, these results are not consistent with the earlier tomographic results. These results are consistent with the diaphthoretic changes in the charnockitic rocks and the consequent lowering of density. This may imply that HGD in the south may have buoyancy that then could support exhumation. The thin magnetic crust (~ 20km) (Mita Rajaram *et al.*, 2003) below the South Indian HGD coinciding with the low velocity layer of high electrical conductivity, indicative of high fluid content in the rocks is consistent with a transformed LC underlying even the massive charnockitic bodies outcropping in the region. This implies that the charnockitic bodies that are the source for the magnetic anomalies may be rootless or have roots transformed to poorly magnetic amphibolite facies rocks. More detailed geophysical experiments are necessary to confirm this possibility.

Intrinsic to the scenario of exhumation of the LC are the numerous shear zones, (often called lineaments) that dissect the HGDs. These include the Achenkoil, Thenmalai shears, the Palghat-Cauvery and Kabbini-Bhavani and Moyar shears, the Malkanagiri shear zone of Orissa separating the EGMB from the CCR in the west, the Khatra-Tamar boundary fault between the Singhbhum greenstone-granite and CGC regimes of Jharkhand and the Central Indian shear, to cite the more important among them. These have been often assigned the status of a suture that welds two micro-plates, but lack the petrotectonic assemblages that one expects along a suture, are steep and with no marked formational distinctions. Models of suturing do not explain why the formations bounding these shears have a lower crustal petrological character and structure. The shears do separate blocks with distinct overprinted geochronology and isotope signatures. These could as well develop when Proterozoic magmatism migrates from one segment of the HGD to the other over the Proterozoic time band, as explained in the discussions on the south Indian HGD and also from an enriched mantle source. The shears are therefore better explained as ductile shears developed in a setting of differential exhumation now left exposed in levels of brittle deformation.

A major factor in the exhumation and uplift of the HGDs of the Southern Indian shield is a possible northerly tilt of the continent as suggested by Fermor, who also considered the possibility of an alternative to tectonic exhumation (*In* Pascoe, 1954). Perhaps of equal significance is the easterly tilt that the shield underwent due to underplating along the western borders during the Deccan magmatism. The tilt may explain in part the uplift and exposure by erosion of the large massive charnockites along the whole of the Western Ghats.

A third mechanism which seems to be relevant is the continuing uplift seen in the Western Ghats as a whole and the Nilgiri and Anamalai hills in particular, evidenced by a remarkably youthful topography that stands testimony to continuing uplift. Such uplift is the result of epeirogenic processes that are poorly understood but which may be the result of continuing lithosphere-mantle interaction deep down. To sum up the exhumation of the HGD of the south may be attributed to isothermal decompression, lowering of crustal density by metasomatic changes that confer a buoyancy to the LC and also due to ongoing epeirogenic process that involves a mantle-lithosphere interaction.

Recent interpretations of dipping reflections in seismic profiles as supportive of subduction and collision tectonics (Reddy *et al.*, 2003) ignore the fact that the LC within the Proterozoic mobile belts encircling the CCR has experienced extensive mobilization. Isothermal decompressional exhumation of the LC generates reorientation of the mineral fabric by recrystallisation of the earlier minerals under changed physico-chemical conditions. Where the exhumation is differential, several dipping surfaces can be generated depending on the ductile flow dynamics of exhumation. Emplacement of younger granites and alkaline complexes can generate domical structures that, therefore, dip away from the center of the domes. Dipping reflectors, therefore, need not be unique to collision zones. Similarly bimodal gravity signatures across thrusts are being relied upon to infer a subduction scenario. But many of the thrusts are steep ductile shears or mantle faults that could have resulted during the differential exhumation of the LC. Exhumation being a near vertical emplacement, bimodal gravity signatures can result across differentially exhumed segments. Bimodal gravity signatures need not, therefore, be unique to plate subduction boundaries, especially when the faults are near vertical.

The episodes of magmatism in the evolution of the EGMB have been quite matching those in the South Indian HGD. EGMB has been also exhumed to almost the same extent as the latter. The main difference is that the large massive charnockitic bodies are restricted to a few exposures with the Eastern Ghats and the Malkanagiri Ranges on the western border and the Eastern Ghat mobile belt had witnessed a relatively younger orogeny (mid-late Proterozoic).

The highly reflective crust and the velocity structure brought out by the several DSS profiles and reflection seismic profiles in Rajasthan and Central India and the SONATA belt point to a thinner UC and differentially exhumed LC. The low levels of exhumation of the LC in the regions to the NE, N and NW of the CCR cannot be well explained from present knowledge of the deep continental structure. A fundamental difference in the nature of the lithosphere in these regions results from the two large magmatic episodes in the Mesozoic, namely the Rajmahal magmatism in the east and the Deccan magmatism in the west. Perhaps consistent with these distinctive episodes in the geological history of the terrain, the lithosphere in these regions presents distinctive seismic wave velocity structure, though we still do not have a unified picture of the exact nature of the velocity distributions (Iyer *et al.*, 1989; Polet and Anderson, 1995; Kenneth and Widiyantoro, 1999). The relatively poor exhumation of the LC in the northern regions may be due to different causes, such as for example, a heavier lithosphere/crust underplated by denser ultrabasics/basics or insulation of the crust to the thermal regime of the mantle due to a underplated layer. More critical geophysical data is needed to answer the question of low degree of exhumation. What however is evident from the Late Proterozoic and Phanerozoic history of these terrains is that the crust underwent considerable distension during the basin formation associated with the Vindhyan and Gondwana sedimentation. A distensional milieu is associated also with the Rajmahal and Deccan Volcanism. These distensional episodes could have thinned the lithosphere and raised the LC from deeper levels. The presence of rocks of LC velocity below the Vindhyan basin forming the basement supports such a view and even helps to date some part of the exhumation event as pre-Vindhyan. The large presence of charnockitic rocks on either flanks of the Damodar Valley Gondwana basins is also consistent with exhumation of the LC due to the crustal distension during the Gondwana sedimentation. The major rifting along the pericontinental margin of the Indian shield during the Rajmahal Volcanic episode may have also played a role in the exhumation of the lower crust. Recent discoveries of Rajmahal basalts in the offshore of Orissa and Andhra Coast indicate that EGMB have been close enough to the rift collar as to have the LC exposed. The Phanerozoic distensional events may not have been of a magnitude to raise an LC from depths of

> 20 kms but may have augmented the uplift of an already exhumed LC in the early or mid-Proterozoic.

ACKNOWLEDGMENT

I deem it a privilege to be invited to contribute this paper to a volume in honour of Prof. R.S. Sharma, an outstanding petrologist and a great teacher. I have known him since some three decades and we have been associated in some joint tasks for promoting research in Earth Sciences in India. I am grateful to Prof. Harel Thomas for inviting me to make this presentation. Prof. Thomas has shown great patience in tolerating a delayed submission and I thank him for this courtesy. The comments of an anonymous referee helped to improve the text of this paper and he is thanked.

REFERENCES

Anil Kumar, Padmakumari, V.M., Dayal, A.M., Murthy, D.S.N. & Gopalan, K., 1993. Rb-Sr ages of Proterozoic kimberlites of India. Evidence for contemporaneous emplacement. *Precamb. Res.*, **vol. 62**, 227-37.

Bhaskar Rao, Y.J., Janardhan, A.S., Vijaya Kumar, T., Narayana, B.L., Dayal, A.M., Taylor, P.N. & Chetty, T.R.K., 2003. Sm-Nd model ages and Rb-Sr isotopic systematics of charnockites gneisses across the Cauvery shear zone, south India: implication for Archaean, Neoproterozoic terrane boundary in the granulite terrain. *Mem. Geol. Soc. India*, **vol. 50**, 297-317.

Bhowmik, A.K., Pal, T. & Roy, A., 1998. Reworking of older granulites during Sausar orogeny: an evidence from northern margin of Sausar mobile belt in Central India. *Ind. Mineral*, **vol. 32**, 19-20.

Das Gupta, S., Sen Gupta, P., Juergen, E.H.L., Raith, M. & Bardhan, S., 1995. Reaction textures in a suite of spinel granulites from the Eastern Ghats belt, India: Evidence for polymetamorphism, a partial petrogenetic grid in the system KFMASH and the roles of ZnO and Fe_2O_3. *Jour. Petrology*, **vol. 36**, 435-61.

Desai, S.J., Patel, M.P. and Merh, S.S. 1978. Polymetamorphites of Balaram-Abu Road area, north Gujarat and southwest Rajasthan. *Jour. Geol. Soc. India*. **vol. 19**, 383-94.

Dewey, J.F., 1986. Diversity in the lower continental crust. *Geol. Soc. London*. Spl. Pub., 71-78.

Dickin, A.P., 1995. Radiogenic isotope geology. Cambridge University Press; 490.

Durheim, R.J. & Mooney, W.D., 1994. Evolution of the Precambrian Lithosphere: geological and geochemical constraints. *J. Geophys. Res.*, **vol. 99**, 15359-74.

England, P. & Molnar, P., 1990. Surface uplift, uplift of rocks and exhumation of rocks. *Geology*, **vol. 18**, 1173-77.

Fountain, D.M. & Salisbury, M.H., 1981. Exposed cross-sections through the continental crust: implications for crustal structure, petrology and evolution. *Earth Planet. Sci. Lett.*, **vol. 56**, 263-77.

Gokaran, S.G., Rao, C.K. & Singh, B.P., 1995. Crustal structure in southeast Rajasthan using Magnetotelluric techniques. In: S. Sinha-Roy and K.R. Gupta (Eds.), Continental Crust of NW and Central India. *Geol. Soc. India, Mem.*, **vol. 31**, 373- 81.

Gokaran, S.G., Rao, C.K., Gautam Gupta, Singh, B.P. & Yamashita, M., 2001. "Deep crustal structure in the Damoh-Jabalpur-Mandla region using magnetotelluric studies. *Geophys. J. International*, **vol. 144**, 685-94.

Gopalakrishnan, K., 1996. An overview of southern granulite terrain India – Constraints in reconstruction of Precambrian assembly of Gondwanaland. Ninth International Gondwana Symposium, Hyderabad; January 1994. Gondwana Nine, **vol. 2**, 1003-26.

Harinarayana, T., Naganjaneyalu, K., Manoj, C., Patro, B.P.K., Kareemunnisa Begum, S., Murthy, D.N., Madhusudhana Rao, Kumaraswamy, V.T.C. & Virupakshi, G., 2003. Magnetotelluric investigations along Kuppam-Palani geotransect, south India—2-D modeling results. *Mem. Geol. Soc. India*, **vol. 50**, 107-24.

Iyer, H.M., Gaur, V.K., Rai, S.S., Ramesh, D.S., Rao, C.V.R. *et al.*, 1989. High velocity anomaly beneath the Deccan Volcanic Province: evidence from seismic tomography. *Proc. Ind. Acad. Sci.*, **vol. 98**, 31-60.

Jain, S.C., Yedekar, D.B. & Nair, K.K.K., 1991. Central Indian Shear zone: A major Precambrian crustal boundary, *Jour. Geol. Soc. Ind.*, **vol. 37**, 521-31.

Jayananda, M. & Peucat, J-J., 1996. Geochronological framework of southern India. Gondwana. *Res. Mem.*, **No. 3**, 53-75.

Kaila, K.L. & Krishna, V.G., 1992. Deep seismic sounding studies in India and major discoveries, *Curr. Sci.*, **vol. 62**, 117-54.

Kaila, K.L. & Sain, K., 1997. Variation of crustal velocity structure in India as determined from DSS studies and their implications on regional tectonics, *J. Geol. Soc. Ind.*, **vol. 49**, 395-407.

Kennett, B.L.N. & Widiyantoro, S., 1999. A low seismic wave speed anomaly beneath northwestern India: a seismic signature of the Deccan plume? *Earth Planet. Sci. Letters*, **vol. 165**, 145-55.

Mahadevan, T.M., 1992. Geological evolution of the Chhotanagpur gneissic complex in a part of Purulia District, West Bengal. *Ind. Jour. Geol.*, **vol. 64**, 1-22.

Mahadevan, T.M., 1994. Deep continental structure of India: *A review. Mem. Geol. Soc. India*, **vol. 28**, 569.

Mahadevan, T.M., 1995a. Deep continental structure and Precambrian evolution of the Indian Peninsular Shield. In Ed. Santosh and Yoshida. Gondwana Research Group Mem. 3; The Archaean Proterozoic terrains in southern India within East Gondwana, 39-52.

Mahadevan, T.M., 1996. Deep continental structure of northwestern and Central Indian peninsular Shield - *A review, Mem. Geol. Soc. India*, **No. 31**, 1-35.

Mahadevan, T.M., 2003. Geology of Bihar and Jharkhand. Text Book Series. *Geological Society of India*, Bangalore. 563.

Mahadevan, T.M., 2002. Geological evolution of south Indian shield-constraints on modeling. *Mem. Geol. Soc. India*, **vol. 50**, 25-46.

Mita Rajaram, Harikumar, P. & Balakrishnan, T.S., 2003. Thin magnetic crust in the southern granulite terrain. *Mem. Geol. Soc. India*, **vol. 50**, 165-76.

Mohan, A. & Jayananda, M., 1999. Metamorphism and isotopic evolution of granulites of southern India: Reference to Neoproterozoic crustal evolution. *Gondwana Res.*, **vol. 2, No. 2**, 151-62.

Mukherjee, A., 1989. P-T time history and thermal modelling of an anorthosite granulite interphase, Eastern Ghats metamorphic belt, India. *Geol. Soc. London. Special Publ.*, **No. 43**, 265-74.

Mukherjee, A., 1996. Thermo-tectonic evolution of the Proterozoic anorthosites towards a unified theory. *Ind. Jour. Geol.*, **vol. 68**, 133-56.

Mooney, W.D. & Meissner, R., 1992. Multi-genetic origin of crustal reflectivity: A review of seismic reflection profiling of the continental lower crust and moho. In: D.M. Fountain, R. Arculus and R.W. Kay (Eds.), *Continental Lower Crust*, Elsevier, 45-79.

Pascoe, E.H., 1954. The manual of the geology of India and Burma. Manager of Publications, Govt. of India, Delhi.

Percival, J.A., Fountain, D.M. & Salisbury, M.H., 1990. Exposed crustal cross-sections as windows on lower crust. In: Eds. D.M. Fountain, R.J. Arkulus and R.W. Kay. *Continental Lower Crust*. Elsevier, Amsterdam, 317-62.

Polet, J. & Anderson, D.L., 1995. Depth extent of cratons as inferred from tomographic studies. *Geology*, **vol. 23(3)**, 205-08.

Qureshy, M.N., 1970. Relation of gravity to elevation, geology and tectonics in India. Proc. II Symposium on Upper Mantle Project. *Nat. Geophys. Res. Inst.*, 1-20.

Radhakrishna, B.P. & Naqvi, S.M., 1986. Precambrian continental crust of India and its evolution, *Jour. Geol.*, **vol. 94**, 145-66.

Rai, S.S., Sree Nagesh, D. & Gaur, V.K., 1993. Granulite evolution in South India – A seismic tomographic perspective. *Mem. Geol. Soc. India*, **vol. 25**, 235-63.

Rai, S.S., Rajagopala Sarma, P.V.S.S., Prakasam, K.S. & Rao, V.K., 1996. Seismic evidence for a thick and underplated late Archaean crust of Eastern Dharwar craton. *Proc. Ind. Acad. Sci.* (EPS), **vol. 105**, 431-39.

Raja Rao, C.S., Poddar, M.C., Basu, K.K. & Dutta, A.K., 1971. Precambrian stratigraphy of Rajathan – *A review. Rec. Geol. Surv. India*, **vol. 101 (20)**, 52-79.

Rajesh, H.M. & Santosh, M., 1996. Alkaline magmatism in Peninsular India. Chapter 7; *Gondwana Res. Group Memoir*, **3**, 91-115.

Ray Burman, T., Bishnui, P.K., Mukhopdhyaya, K. & Ray, J.N., 1994. Rb-Sr geochronology of the high grade rocks from Purulia, West Bengal and Jamua-Dumka sector, *Bihar. Ind. Min.*, **vol. 48 (1&2)**, 45-60.

Reddi, A.G.B. & Ramakrishna, T.S., 1988. Subsurface structure of the Shield area of Rajasthan-Gujarat as inferred from gravity. In: A.B.Roy (Ed.), Precambrian of the Aravalli Mountain, Rajasthan, India. *Geol. Soc. India, Mem.*, **vol. 7**, 279-84.

Reddy, P.R., Murthy, P.R.K., Rao, I.B.P., Prakash Khare, Mall, D.M., Koteswara Rao, P., Kesava Rao, G., Raju, S., Sridhar, V. & Reddy, M.S., 1996. Coincident deep seismic reflection/refraction survey across Central Indian Suture (CIS) along Mungwani-Seoni-Katangi-Kalimati profile-Central India. Technical Report: Category B-NGRI-96-Lithos-188.

Reddy, P.R., Mall, D.M. & Prasad, A.S.S.S.R.S., 1997a. Subhorizontal layering in the lower crust and its tectonic significance in the Narmada-Son region, India. *Pure and Applied Geophysics*, **vol. 149**, 525-40.

Reddy, P.R., Koteswara Rao, P. & Kalachand Sain, 1999. Crustal configuration of the Narmada-Son lineament in Central India from Deep Seismic Sounding studies. *Mem. Geol. Soc. India*, **vol. 43**, 353-65.

Reddy, P.R., Rajendra Prasad, B., Vijaya Rao, V., Khare, P., Rao, G.K., Murty, A.S.N., Sarkar, D., Raju, S., Rao, G.S.P. & Sridhar, V., 1995. Deep seismic reflection profiling along Nandsi-Kunjer section of Nagaur-Jhalawar transect, preliminary results. In: S. Sinha-Roy and K.R. Gupta (Eds.), Continental Crust of NW and Central India. *Geol. Soc. India, Mem.*, **vol. 31**, 353-72.

Reddy, P.R., Rajendra Prasad, B., Vijaya Rao, V., Kalachand Sain, Prasada Rao, P., Prakash Khare and Reddy, M.S., 2003. Deep seismic reflection and refraction/wide angle reflection studies along Kuppam-Palani transect in Southern granulite terrain in India. *Mem. Geol. Soc. India*, **vol. 50**, 79-106.

Roy, A., Ramachandra, H.M. & Bandopadhyay, B.K., 2000. Supracrustal belts and their significance in the crustal evolution of central India. *Spl. Publ.* **no. 55**; *Geol. Surv. India*, 361-80.

Roy, S., Ray, L., Senthil Kumar, P., Reddy, G.K. & Srinivasan, R., 2003. Heat flow and heat production in the Precambrian Gneiss- Granulite Province of southern India. *Mem. Geol. Soc. India*, **vol. 50**, 177-92.

Santosh, M., 1989. Alkaline plutons, decompression granulites and late Proterozoic CO_2 influx in Kerala, South India. *Mem. Geol. Soc. India*, **No. 15**, 177-88.

Sen, S.K., Bhattacharya, S. & Acharya, A., 1995. A multi-stage pressure-temperature record in the Chilka Lake granulites: the epitome of the metamorphic evolution of Eastern Ghats, India? *J. Metamorphic Petrol*, **vol. 13**, 287-98.

Sengupta, P., Das Gupta, S., Bhattacharya, P.K., Fukuoka, M., Chakraborti, S. & Bhowmick, S., 1990. Petro-tectonic imprints in the sapphirine granulites from Anantagiri, Eastern Ghat Mobile belt, India. *J. Petrology*, **vol. 31**, 971-96.

Sharma, R.S., 1988. Patterns of metamorphism in the Precambrian rocks of the Aravalli Mountain belt. In: A.B. Roy (Ed.), Precambrian of the Aravalli Mountain, Rajasthan, India. *Geol. Soc. India, Mem.*, **vol. 7**, 33-75.

Sharma, R.S., 1999. Crustal development in Rajasthan Craton. *Indian J. Geology*, **vol. 71**, 65-80.

Singh, A.P., Misra, D.C., Vijay Kumar, V. & Vyaghreswara Rao, M.B.S., 2003. Gravity-magnetic signatures and crustal architecture along Kuppam-Palani geotransect, south India. *Mem. Geol. Soc. India*, **vol. 50**, 139-63.

Singh, S.B., Stephen, J., Singh, V.K., Srinivas, Yasok Babu, G., Singh, K.P. & Jana Reddy, E., 2003. Electrical signatures in the high grade metamorphic terrain of south India using deep resistivity sounding studies. *Mem. Geol. Soc. India*, **vol. 50**, 125-38.

Singh, D.D., 1991. An elasticity of the crust and upper mantle beneath north and central India from the inversion of observed Love and Rayleigh wave attenuation data. *Pageoph.* **vol. 135**, 545-58.

Sinha-Roy, S., 1988. Proterozoic Wilson cycles in Rajasthan. In: A.B. Roy (Ed.), Precambrian of the Aravalli Mountain, Rajasthan, India. *Geol. Soc. India, Mem.*, **vol. 7**, 95-108.

Sugden, T.J., Deb, M. & Windely, B.F., 1990. The tectonic setting of mineralization in Proterozoic Aravalli Delhi orogenic belt, NW India. In: S.M. Naqvi (Ed.), Precambrian Continental Crust and its Economic Resources, Developments in Precambrian Geology, **8**, *Elsevier Pub.*, 367-90.

Sychanthavong, S.P.H. & Desai, S.D., 1977. Protoplate tectonics controlling the Precambrian deformations and metallogenic epochs of NW Peninsular India. *Mineral Sci. Engg.*, **vol. 9**, 218-37.

Tewari, H.C., Divakar Rao. V., Narayana, B.L., Dixit, M.M., Madhava Rao, N., Murthy, A.S.N., Rajendra Prasad, B., Reddy, P.R., Venkateswarlu, N., Vijaya Rao, V., Mishra, D.C. & Gupta, S.B., 1998. Nagaur-Jhalawar geotransect across the Delhi/Aravalli fold belt in NW India. *J. Geol. Soc. India*, **vol. 52**, 153-61.

Thomas, H., 1995. Pressure Temperature considerations for granulite form Thana Gyangarh, District-Bhilwara, Rajasthan: Implications for Crustal Evolution. In: R.K. Srivastava & R. Chandra (ed.) Magmatism in relation to Diverse Tectonic setting. A.A. Balkema Pub. & Oxford & IBH Publishing Com Pvt. Ltd., 439-56.

Verma, R.K. & Banerjee, P., 1992. Nature of continental crust along the Narmada-Son lineament inferred from gravity and deep seismic sounding data, Tectonophysics, **vol. 202**, 375-97.

Vijaya Rao, V., Rajendra Prasad, B., Reddy, P.R. & Tewari, H.C., 2000. Evolution of Proterozoic Aravalli Delhi fold belt in the northwestern Indian shield from seismic studies. *Tectonophysics*, **vol. 327**, 109-30.

Weber, K., 1986. Metamorphism and crustal rheology–implications for the structural development of the continental crust during prograde metamorphism. *Geol. Soc. of London, Sp. Publ.*, **No. 24**, 95-106.